THE SCIENTIFIC
NATURE OF
GEOMORPHOLOGY

THE BINGHAMTON SYMPOSIA IN GEOMORPHOLOGY

List of Proceedings Volumes

THE SCIENTIFIC NATURE OF GEOMORPHOLOGY

Proceedings of the 27th Binghamton Symposium in Geomorphology held 27–29 September 1996

Edited by

BRUCE L. RHOADS

and

COLIN E. THORN

University of Illinois at Urbana-Champaign, USA

JOHN WILEY & SONS

Chichester • New York • Brisbane • Toronto • Singapore

Other Wiley Editorial Offices

John Wiley & Sons, Inc., 605 Third Avenue,
New York, NY 10158-0012, USA

Jacaranda Wiley Ltd, 33 Park Road, Milton,
Queensland 4064, Australia

John Wiley & Sons (Canada) Ltd, 22 Worcester Road,
Rexdale, Ontario M9W 1L1, Canada

John Wiley & Sons (Asia) Pte Ltd, 2 Clementi Loop #02-01,
Jin Xing Distripark, Singapore 129809

Library of Congress Cataloging-in-Publication Data

Binghamton Symposium in Geomorphology (27th : 1996)
    The scientific nature of geomorphology : proceedings of the 27th
    Binghamton Symposium in Geomorphology, held 27–29 September, 1996 /
    edited by Bruce L. Rhoads & Colin E. Thorn.
        p. cm.
    Includes bibliographical references and index.
    ISBN 0-471-96811-0
    1. Geomorphology—Congresses.   I. Rhoads, Bruce L.   II. Thorn,
Colin E.   III. Title.
GB400.2.B57   1996
551.4'1—dc20   96-14602
                                                  CIP

British Library Cataloguing in Publication Data

A catalogue record for this book is available from the British Library

ISBN 0-471-96811-0

Typeset in 10/12pt Times by Techset Composition Ltd, Salisbury, Wiltshire
Printed and bound in Great Britain by Bookcraft (Bath) Ltd, Midsomer Norton, Somerset
This book is printed on acid-free paper responsibly manufactured from sustainable forestation,
for which at least two trees are planted for each one used for paper production.

Contents

Contributors

Philip J. Ashworth, School of Geography, University of Leeds, Leeds, West Yorkshire LS6 9JT, UNITED KINGDOM.

Victor R. Baker, Department of Geosciences, Building #77, Gould-Simpson Building, The University of Arizona, Tucson AZ 85721, USA.

Bernard O. Bauer, Department of Geography, University of Southern California, University Park, Los Angeles, CA 90089-0255, USA.

James L. Best, Department of Earth Sciences, University of Leeds, Leeds, West Yorkshire LS6 9JT, UNITED KINGDOM.

Keith Beven, Centre for Research on Environmental Systems and Statistics, Institute of Environmental and Biological Sciences, Lancaster University, Lancaster LA1 4YQ, UNITED KINGDOM.

Harold I. Brown, Department of Philosophy, Northern Illinois University, DeKalb, IL 60115, USA.

Michael Church, Department of Geography, #127–1984 West Mall, The University of British Columbia, Vancouver, British Columbia V6T 1Z2, CANADA.

Ronald I. Dorn, Department of Geography, Arizona State University, PO Box 870104, Tempe, AZ 85287-0104, USA.

William L. Graf, Department of Geography, Arizona State University, PO Box 870104, Tempe, AZ 85287-0104, USA.

Peter K. Haff, Department of Geology, Center for Hydrologic Science, PO Box 90227, Duke University, Durham, NC 27708-0230, USA.

Cliff R. Hupp, US Geological Survey, Reston, VA, USA.

Michael J. Kirkby, School of Geography, University of Leeds, Leeds, West Yorkshire LS6 9JT, UNITED KINGDOM.

Deborah S.L. Lawrence, Postgraduate Research Institute for Sedimentology, The University of Reading, PO Box 227, Whiteknights, Reading RG6 6AB, UNITED KINGDOM.

Waite R. Osterkamp, US Geological Survey, 1675 W. Anklam Road, Tucson, AZ 85745, USA.

Jeffrey Peakall, Department of Earth Sciences and School of Geography, University of Leeds, Leeds, West Yorkshire LS6 9JT, UNITED KINGDOM.

Jonathan D. Phillips, Department of Geography, East Carolina University, Greenville, NC 27858, USA.

Bruce L. Rhoads, Department of Geography, 220 Davenport Hall, University of Illinois at Urbana-Champaign, 607 South Mathews Avenue, Urbana, IL 61801, USA.

Keith Richards, Department of Geography, University of Cambridge, Downing Place, Cambridge CB2 3EN, UNITED KINGDOM.

Douglas J. Sherman, Department of Geography, University of Southern California, University Park, Los Angeles, CA 90089-0255, USA.

Colin E. Thorn, Department of Geography, 220 Davenport Hall, University of Illinois at Urbana-Champaign, 607 South Mathews Avenue, Urbana, IL 61801, USA.

C. Rowland Twidale, Department of Geology and Geophysics, University of Adelaide, Adelaide, South Australia 5005, AUSTRALIA.

Preface

Geomorphology is concerned primarily with generating knowledge about the terrestrial surface of the earth, excursions into planetary 'geomorphology' and submarine geomorphology notwithstanding. Whether or not all contributions to knowledge derive from science is a philosophical issue, but it is reasonable to assert that most geomorphic knowledge is derived scientifically. Numerous articles have appeared over the last few years expressing concern about the future of geomorphology or voicing opinions about the character of geomorphic enquiry (e.g. Richards 1990, 1994; Baker and Twidale 1991; Yatsu 1992; Baker 1993; Rhoads and Thorn 1993, 1994; Bassett 1994; Rhoads 1994), but these articles, while divergent in many respects, agree on one point – that geomorphology is a science and, therefore, a scientific discipline. What seems to be at issue is what this status as a science implies for geomorphologic theory and methodology. The current debate appears to reflect tension between a research tradition rooted firmly in geology/physical geography and an emerging approach grounded more directly in the scientific principles and methods of physics and chemistry. This tension has initiated a period of introspection as geomorphologists search for a way to reconcile traditional and emerging perspectives.

The nature of the current debate cannot be completely appreciated without placing it in the appropriate historical context. Much of modern (American) geomorphology emerged during the turn of the century, a period when territorial expansion and exploration in western North America were rife, and was accompanied by the scientific work of such renowned geomorphologists as C. E. Dutton, G. K. Gilbert, and J. W. Powell. The result in the USA was a scientific discipline pervaded by a preoccupation with fieldwork. However, the most famous geomorphologist of this era (arguably of any era) William Morris Davis, is renowned for his unifying conceptual framework. Indeed, the other great geomorphologist of the same period, G. K. Gilbert, is also revered for his conceptual approach, although appreciation of his remarkable insights took much longer to develop.

Conceptually, turn-of-the-century geomorphology was dominated by the biologically inspired, all-embracing 'Geographical Cycle' of William Morris Davis (1899). By virtue of widespread support and limited opposition, this grand vision of landform development dominated geomorphological research between about 1900 and 1945, and geomorphological teaching for many more years thereafter. The demise of the 'Geographical Cycle' as the dominant overarching conceptualization of landscape development did not result in the emergence of another, similarly dominant, model. It is true that John Hack's (1960) resurrection of G. K. Gilbert's concept of dynamic equilibrium became very influential in the 1960s and remained so, as did Strahler's (1952) emphasis on process. However, whereas Davis's ideas were a strictly geomorphological theory, those of Hack and Strahler were more generalized principles designed to guide research.

Today, it is reasonable to characterize geomorphology as a burgeoning scientific discipline of increasing societal significance which, while embracing the very latest concepts in chemistry, physics, mathematics, and computer modeling, lacks not only a unified body of theory, but, more importantly, a clear and unifying sense of disciplinary identity. In particular, there is a growing schism between those focusing upon reconstruction of the development of individual landscapes (often Quaternarists) and those seeking general principles governing landscape dynamics (most frequently numerical modelers, or process-oriented geomorphologists). This conflict is not an inherently valid one intellectually, but is a perceived one that commonly rests on the different scales at which the two groups work and the varying techniques they employ. Nevertheless, whether valid or perceived, the schism, if allowed to grow through inadvertence, could promote fragmentation of geomorphology.

The driving force behind the production of this volume is the belief that by overtly focusing on the discipline's methodological and philosophical underpinnings, geomorphology can thwart the tendency toward fragmentation. Indeed, it could be argued that the failure to engage such issues in the past has in part led to the current situation. Geomorphology already is over 100 years old, yet this volume is the first concerted attempt to bring together a diverse group of practitioners to systematically explore the methodology and philosophy of geomorphology. If geomorphologists are to develop a profound and robust sense of disciplinary identity, they must attend to foundational issues more fully than they have in the past.

The purpose of this volume is to initiate a broad examination of contemporary perspectives on the scientific nature of geomorphology. This initial exploration of methodological and philosophical diversity within geomorphology is viewed as a necessary first step in the search for common ground among the diverse group of scientists who consider themselves geomorphologists. The volume aims to simultaneously advance, enhance, and strengthen geomorphology as it enters the twenty-first century by clarifying the bases for internal debate, by showing how geomorphology fits into the realm of science at large, by examining the relationship of the discipline to other areas of knowledge, by providing an improved understanding of methodological diversity within the discipline, and by identifying potential bases for disciplinary unity.

To achieve these goals, contributors were asked to address specific topics in a manner that illuminates contemporary conceptual issues or problems and that casts light upon desirable or potential developments. It was not the editors' expectation or aspiration that the contributors would speak with a single voice; indeed, they have addressed issues which can variously be described as methodology, theory, philosophy, or amalgamations of all of these components and of less clearly identifiable ingredients as well. Given this situation, the organization of the chapters of this book into distinct groups is based not on some essential set of criteria, but instead reflects the editors' perception of prominent commonalities among individual contributions. The first group, which is loosely labeled 'philosophical', commences by addressing fundamental issues in the philosophy of science and moves on to consider philosophical issues in geomorphology. The second group of 'methodological' chapters addresses in one form or another scale issues, encompassing not only how we attempt to arrange our ideas of scale, but how they influence our scientific methodology, as well as the limitations that our dating metrics impose upon our scientific understanding. The third group focuses on geomorphological modeling. As models form the primary link between what we are able to measure and study directly, namely the world as it is, and how we think about the world, their role in the discipline is both pivotal and critical. The fourth, and final, group of chapters addresses geomorphology's position in the web of academic disciplines, as well as illustrating an important future role – geomorphology in the service of society. Each group of chapters is prefaced by a brief introduction.

One final note – this volume should be seen as inceptive, rather than definitive. There are, no doubt, many important aspects of geomorphology that receive no attention herein. No apology is offered for these omissions. Rather, the hope is that others will feel compelled to champion their importance elsewhere.

REFERENCES

Baker, V.R. 1993. Extraterrestrial fluvial geomorphology: science and philosophy of Earthlike planetary landscapes, *Geomorphology*, **7**, 9–36.

Baker, V.R. and Twidale, C.R. 1991. The reenchantment of geomorphology, *Geomorphology*, **4**, 73–100.

Bassett, K. 1994. Comments on Richards: the problems of 'real' geomorphology, *Earth Surface Processes and Landforms*, **19**, 273–276.

Davis, W.M. 1899. The geographical cycle, *Geographical Journal*, **14**, 481–504.

Hack, J.T. 1960. Interpretation of erosional topography in humid temperate regions. *American Journal of Science*, **258-A**, 80–97.

Rhoads, B.L. 1994. On being a 'real' geomorphologist, *Earth Surface Processes and Landforms*, **19**, 269–272.

Rhoads, B.L. and Thorn, C.E. 1993. Geomorphology as science: the role of theory, *Geomorphology*, **6**, 287–307.

Rhoads, B.L. and Thorn, C.E. 1994. Contemporary philosophical perspectives on physical geography with emphasis on geomorphology, *Geographical Review*, **84**, 90–101.

Richards, K.S. 1990. 'Real' geomorphology, *Earth Surface Processes and Landforms*, **15**, 195–197.

Richards, K.S. 1994. 'Real' geomorphology revisited. *Earth Surface Processes and Landforms*, **19**, 277–281.

Strahler, A.N. 1952. Dynamic basis of geomorphology, *Geological Society of America Bulletin*, **63**, 923–938.

Yatsu, E. 1992. To make geomorphology more scientific, *Transactions, Japanese Geomorphological Union*, **13**, 87–124.

To *Kathy* and *Carole*
for patience and support

To *Jamie* and *Steven*
for being a joyful distraction from the rigors of academe

To *Stephen* and *Jeffrey*
from a proud father

Acknowledgements

We wish to thank the Binghamton Geomorphology Symposium Steering Committee for providing us with the opportunity to create this symposium under their rubric. We have received financial support from the College of Liberal Arts and Sciences at the University of Illinois at Urbana-Champaign, as well as from the Department of Geography at UIUC. The National Science Foundation was the primary contributor to the speakers' travel costs. We thank all of these organizations for their generosity.

Each paper has been reviewed by at least two reviewers, in the overwhelming number of cases these consisted of one other symposium participant and one reviewer external to the symposium; in some instances three reviewers were used, and in others both reviewers were external to the symposium. Reviewers put forth a tremendous amount of effort in their reviews and we thank them all very heartily – they measurably improved both individual papers and the entire symposium.

We also thank Carole Thorn for performing various editing and typing chores, and Barbara Bonnell for some typing.

PHILOSOPHICAL ISSUES

Philosophy embraces the issues of what exists (ontology) and how we can know what exists (epistemology). The philosophy of science attempts to resolve these issues in the sphere of science at large as well as for specific scientific disciplines. Geomorphologists have not readily embraced philosophical discussion, an attitude reflected in Schumm's (1991) understated remark that 'most earth scientists do not find philosophical discussion of their field very interesting'. Traditionally, most geomorphologists have had, at best, only limited formal exposure to philosophy, and, what fleeting exposure they have had has been limited mainly to the tenets of logical positivism or critical rationalism. Unfortunately, the normative qualities of these philosophical doctrines have tended to irritate practicing scientists, resulting in a generation of geomorphologists that has shunned philosophy of science. In recent years, many traditional philosophical doctrines have been challenged as philosophy of science has shifted from a highly normative posture to a more naturalized one. Today, philosophy of science can in many instances be characterized as a 'science of science.' If geomorphologists are to develop a better understanding of their science, they must subject it to critical scrutiny. Here contemporary philosophical analysis can play an important role. This session was conceived as contributing to this task.

As an active participant in the creation of a naturalized philosophy of science, Harold Brown brings to the symposium a fountain of knowledge and experience of the present situation in philosophy of science. In providing a survey of his discipline Brown is able to show that scientific theories are an integral part of scientific methodology, and, consequently, that methodological and theoretical development in science progress hand in glove.

The great tradition of fieldwork in geomorphology has placed observation in a revered position in geomorphological inquiry. Bruce Rhoads and Colin Thorn examine recent ideas on observation in the philosophy of science and use these ideas as a filter or lens through which to view observation in geomorphology. They conclude that despite an undercurrent of radical empiricism in the discipline, observation in geomorphology is inherently theory-dependent. They also show how objectivity of geomorphologic inquiry can be preserved in the face of theory-dependent observations.

The logical positivist school believed that logic played no role in the discovery of new scientific ideas, but only in the justification of scientific knowledge. Victor Baker reviews the philosophical ideas of the American philosopher Charles S. Peirce, who held views different from those expressed by the logical positivists, but similar to those espoused by contemporary proponents of naturalized philosophy of science. Peirce paid great heed to abductive reasoning, which he believed was fundamental to the conception of hypotheses. Therefore, his philosophy, unlike logical positivism, permits a real role for the philosophy of science in scientific discovery.

Because it is a human enterprise, science is at all times and places conducted in a social context. Consequently like philosophy, the sociology of science plays an important role in the manner in which knowledge is created and science is structured. Douglas Sherman takes up this issue directly for the discipline of geomorphology. He shows that a case can be made that the history of geomorphology is as much a reflection of the influence of individual 'fashion leaders' as any other factor.

The final chapter in this section, by Bruce Rhoads and Colin Thorn, was not presented at the Binghamton symposium from which this volume derives. It was conceived as an extension of the philosophy session after reading the papers in the editorial process. As such it does not attempt any definitive statements, but rather is devoted to pointing out philosophical issues or themes where geomorphologists may well garner important insights into the scientific nature of geomorphology. Its scope is broad, embracing natural kinds, laws, causality, causal explanation, theory and models, discovery, gender issues, and applied geomorphology.

REFERENCE

Schumm, S.A. 1991. *To Interpret the Earth – Ten Ways to be Wrong*, Cambridge University Press, Cambridge, 133 pp.

1 The Methodological Roles of Theory in Science

Harold I. Brown

Department of Philosophy, Northern Illinois University

ABSTRACT

Philosophy of science in this century can be roughly divided into two periods, positivist and postpositivist. Positivists maintained that scientific theories should be evaluated solely on the basis of observational data in accordance with a set of formal methodological rules. According to the positivists, both the data and the methodological rules are known independently of any scientific theories. Data are acquired by sense perception and they provide the empirical foundation for science. Methodological rules are established a priori and these rules provide a universal, permanent framework for the evaluation of scientific theories. One central postpositivist theme is that data and formal rules are not sufficient for evaluating scientific theories because they leave too many options open. Many different generalizations are compatible with any given body of data and, when an observation contradicts a theory, many responses remain possible. Postpositivists argue that, in scientific practice, these options are reduced because established theories take on a methodological role. These theories serve as guiding assumptions that, for a time, are not open to empirical challenge. Instead, these assumptions provide additional criteria for evaluating hypotheses, developing observational procedures, and interpreting the outcomes of these procedures. Still, since guiding assumptions are themselves empirical theories, they can be rejected and replaced with new guiding assumptions as science develops. Such replacements may result in a reconsideration of lower-level theories and a reinterpretation of observational data. The main result of this discussion for present purposes is that scientific theories are seen to provide an important part of the methodology of science, so that the development of methodology is an integral part of the development of science.

The development of philosophy of science in the twentieth century can be roughly divided into two major periods: the 'positivist' period, which has roots in the nineteenth century and dominated the field until the late 1950s, and the 'postpositivist' period, which

The Scientific Nature of Geomorphology: Proceedings of the 27th Binghamton Symposium in Geomorphology held 27–29 September 1996. Edited by Bruce L. Rhoads and Colin E. Thorn. © 1996 John Wiley & Sons Ltd.

began to develop in the 1950s. Positivism was not a single doctrine. There were disputes among positivists on many issues of detail and substantial changes in aspects of the positivist program over time.[1] Still, there were some central doctrines that remained constant and that provided a unified framework for the philosophical analysis of scientific knowledge. These central doctrines include the view that there is a sharp distinction between theories and observational data, and a division of labor between science and philosophy of science. Positivists were strict empiricists with regard to scientific knowledge: scientific claims stand or fall solely on the basis of their ability to explain and predict our sensory experience. Philosophy of science, however, was conceived of as an a priori discipline. Positivists considered the proper subject matter of philosophy to be very narrow, but it included the methodology of science. Methodology provides norms for evaluating empirical claims and these norms, it was argued, must be known independently of any empirical research, and must provide a permanent, universal guide for the practice of science.

Postpositivist philosophy of science has challenged every major positivist theme and there is little unity among postpositivists. One exception concerns the relation between specific scientific theories and methodology. A radical alternative to the positivist view emerged at the beginning of the attack on positivism and has remained a central postpositivist theme. This alternative holds that methodology and science are deeply intertwined and that the evaluation and development of methodology are part of the process by which science develops. In this chapter I will examine the new view of methodology and some of its consequences.

For present purposes I will use the term 'theory' to refer to any generalization that goes beyond available data. Thus theories will range from straightforward universal generalizations, to theories such as those of Newton or Maxwell that involve several interlocked generalizations which cannot be applied or tested individually, to sweeping doctrines that cut across specific sciences, such as that every event has a cause. I will begin with an account of the positivist doctrines that bear on the relation between scientific theories and methodology.

THE POSITIVIST FRAMEWORK

P1

Within science there is a fundamental distinction between data and theory: theories make claims that go beyond available data, and thus yield predictions about what will occur in cases that have not yet been examined. Familiar examples of theoretical claims include: 'All objects fall to the earth with (the same) constant acceleration (provided that we can neglect air resistance),' 'Acceleration is always proportional to force,' 'All energy transfers take place at velocities less than the velocity of light *in vacuo*,' and 'Information always travels from nucleic acid to proteins, never in the reverse direction.' These examples include some claims that were once held to be universally true but are now rejected, and I have included them to illustrate a point. A major task of scientific research is to establish such theories, but since theories make claims that go beyond available evidence, it is difficult – perhaps impossible – to establish theoretical claims once and for all. Universal claims that have been supported by all the evidence available at a given time have sometimes turned out to be either quite wrong or to provide only approximately correct results in a limited range of situations – much more limited than had been intended by the scientists who introduced the theory. This poses the problem of assessing when it is appropriate to accept a theory.

To understand the positivist approach to this problem, we should begin by noting that in modern logic universal generalizations are treated as hypothetical statements. The first of the above examples becomes: 'If an object is falling to the earth (and air resistance is insignificant), then the object falls with a constant acceleration.' Empirical support for a universal generalization consists of finding cases that meet the conditions stated in the antecedent and consequent of the generalization. Empirical refutation requires finding a case that meets the condition stated in the antecedent but does not meet the condition of the consequent. Positivists considered empirical refutation to be logically straightforward since refutation requires only a single counterinstance. Thus they focused their attention on confirmation which seemed much more problematic since we can find some supporting cases for virtually any universal generalization. The traditional problem of induction is to determine when we have accumulated enough evidence to be confident in a generalization.

From a historical perspective the confidence with which earlier scientists and philosophers accepted theories now looks somewhat naive since we have many examples of theories that are supported by an enormous body of evidence, but turned out to be wrong. In many case theories failed for reasons that no one had imagined in the period when the theory seemed beyond challenge. For example, the main problem with Newton's second law is that the proportionality constant – the object's mass – turned out not to be a constant after all, but a function of velocity.

P2

The situation I have been discussing concerns *logical relations* between a universal generalization and a body of evidence. I must say a few words about logic because I will be using this term in a narrower and more precise way than it is used in everyday conversation. Logic primarily concerns *relations* between statements, in particular relations between a body of evidence (typically referred to as 'premises') and a conclusion that is supposedly supported by those premises. Please note the following five points.

1. Modern logic is a subdiscipline of mathematics and is thus a formal subject. The point can be illustrated by the rule for finding the derivative of expressions of the form x^n: given any expression of this form, its derivative will be of the form nx^{n-1}. This rule holds independently of what x and n stand for. Whether a particular situation in the world can be correctly described by a law of the form x^n is not the mathematician's concern.

In an analogous fashion, logic is concerned with formal relations between premises and a conclusion. The gathering of evidence and the assessment of whether the description of the evidence is accurate does not concern the logician. Indeed, logicians do not even concern themselves with what the statements are about – just as mathematicians do not care if x^n concerns gravitation, magnetism, or what have you.

2. *Deductive logic* studies a particular relation between premises and conclusion: a *valid* deductive argument is one in which all true premises *guarantee* a true conclusion. Note that this definition concerns a relation that is independent of whether the premises and conclusion are in fact true. This is why logic, like the rest of mathematics, is an a priori

discipline. Logic abstracts from any actual features of the world and studies formal relations between formal expressions. Let me illustrate the idea in another context: in court, an attorney may present an argument to establish a conclusion. The opposing attorney can attack this argument in two ways: the truth of his/her opponent's premises can be questioned or the relevance of these premises to the conclusion can be questioned. Logic is concerned only with relevance relations and deductive logic is concerned with assessing the presence or absence of the strongest possible relevance relation. Few, if any, of the arguments that Sherlock Holmes and Mr Spock refer to as 'deductions' fit this account. It is not that they are misusing the word. In everyday talk 'deduce' is used as a synonym for 'infer'. Logicians and mathematicians use 'deduce' in a narrower, more precise way; our concerns here require that narrower usage.

3. I want to underline some immediate consequences of this account of deductive validity. First, the fact that an argument is invalid does not show that the conclusion is false; invalidity allows us to conclude only that we have failed to prove the conclusion. Here is an example of an obviously invalid argument: $2 + 3 = 5$, therefore smoking causes cancer. If pointing out the invalidity of the argument were sufficient to show that the conclusion is false, life and research would be much simpler than they are. Unfortunately, invalidity is not always so obvious.

By the same token, a valid argument in which one or more of the premises are false shows nothing about the conclusion. In general (i) a valid deductive argument in which (ii) all of the premises are true, guarantees a true conclusion. Given a failure of either condition, we learn nothing from the argument. It is a particularly poignant fact, with direct applications to our attempts to understand the world, that we can validly deduce true conclusions from false premises.[2]

4. An important consequence of our account of deduction is that if we know we have a valid argument with a false conclusion, we can be sure that the premises are not all true. This result provides the logical basis for testing scientific theories; I will have more to say about its significance shortly. For the moment let me emphasize a picky, but important, point: valid deduction of a false conclusion from a set of premises shows that the premises are not all true; it does not show that the premises are all not true.

5. Now consider *inductive logic*. I will use this term to cover all cases in which a body of evidence is supposed to support a conclusion on the basis of a weaker relation than that of deductive validity. In other words, we are in the inductive realm when we claim that our premises support a conclusion even though it is still possible that all of our premises are true but the conclusion is false. The classic illustration is the claim that all swans are white, a conclusion that turned out to be false even though, at one time, an enormous quantity of evidence available to Europeans uniformly supported this conclusion.

There are many kinds of inductive arguments but I will note only two of them here. I have already mentioned the most familiar type: a number of positive instances are taken to support a simple universal generalization. The second kind occurs when we *deduce* true conclusions from a theory and take those conclusions as evidence for the truth of the theory. While deduction of true conclusions does not guarantee that the theory is true, it does seem to provide support for that theory. It should be clear that scientists are mostly concerned with inductive support although, as the case just mentioned indicates, there is an intimate relation between deduction and induction. Moreover, the second kind of inductive support is fundamental when we are dealing with a theory of any degree of complexity.

P3

There is an important respect in which logic is a *normative* rather than a descriptive enterprise. Logic is concerned with the relations that *ought to hold* between evidence and a conclusion if the evidence is to support that conclusion. The point can be made by considering, once again, more familiar parts of mathematics. The rules for calculating a derivative were not arrived at by conducting an empirical study of the ways in which various people in fact carry out this calculation. The rules specify the correct way to determine a derivative, and anyone who follows different rules is just mistaken. How, you may ask, do we arrive at these norms? For present purposes I will consider only part of the positivists' answer to this question. There is a crucial tie between the claim that our knowledge of mathematics and logic is a priori and the claim that these disciplines provide norms. If results in these disciplines were established by gathering evidence about how people in fact calculate or which arguments they actually accept, further data on actual behavior would be relevant to assessing mathematical and logical results. This suggests that one necessary condition for a normative discipline is that it be an a priori discipline.

P4

Let us look at these points about logic and norms from a slightly different direction. Suppose one wishes to evaluate a scientific theory on the basis of a body of evidence. The scientists concerned have formulated the theory and gathered the evidence, but one more element is required in order to carry out the evaluation: we still need rules for determining what evaluation this body of evidence confers on the theory. It is these rules of evaluation that logic is supposed to provide and positivists considered these rules to be the core of 'the scientific method'.[3] Since this is a much more austere notion than is generally associated with the rubric 'scientific method,' I want to describe two developments that led to the adoption of this view.

The first was the extraordinary development of deductive logic that began in the middle of the nineteenth century. For more than 2000 years logic had been viewed as a finished subject to be taught in essentially a single course; professional logicians concerned themselves with polishing up a few rough edges. At present, logic is a living research subject that has provided (among other things) some rather dramatic surprises, such as Gödel's theorems. One feature of this development has been a reconceiving of the nature of logic. In particular, the view that logic studies 'laws of thought' has been abandoned and replaced by the view that logic studies certain formal structures. Questions about how people actually think is a concern of psychology, an empirical science. Questions about how people ought to think have also been left aside, but the reasons for this require that we consider the second development mentioned above.

An older view of scientific method is that it concerns techniques for making new discoveries. This is the view that we typically trace back to Bacon and Descartes, and it endures well into the nineteenth century. Major reasons for abandoning this view of method come, first of all, from developments within science – especially the unanticipated transformations in our understanding of several domains brought about by the appearance of evolutionary biology, relativity, quantum mechanics, molecular biology, and more.

Looking back on the circumstances under which these developments appeared, there does not seem to have been any single method that the innovators followed. Nor is there any reason for thinking that there is a method that would allow less gifted researchers to have made these discoveries.

This reflection must be tied back to the developments in logic noted above. One consequence of these developments is a much clearer conception of what one is seeking when looking for a method. A notion of method has emerged that is extremely powerful and well-understood. This is the notion of an algorithm – a set of rules that guarantee a solution of a problem in a finite number of steps. From this perspective, a discovery method would be a set of rigorous rules that generate new discoveries. Now there are certain limited domains in which such rules exist. Rules of arithmetic provide routines that allow us to find answers to an enormous number of calculation problems; there is a set of rules that allow us to calculate the derivative of any expression; and, up to a point, there are rules of deductive logic that allow us to determine whether a given argument is valid. Unfortunately, we quickly run out of cases of this sort. There are, for example, no rules we can follow to integrate any expression whatsoever, or even to determine if an expression has an integral. And in deductive logic, once we reach a particular degree of complexity, there are no longer any rules that will allow us to apply an automatic procedure to assess whether a given argument is valid.[4] Nice examples of this situation are provided by apparently simple mathematical conjectures that long resisted resolution, such as the four-color theorem, Fermat's last theorem, and Goldbach's conjecture. From a logical point of view, each of these is a problem about whether a particular proposition follows validly from a set of premises. Although two of these problems have now been resolved, none were resolved by applying an algorithm to test if the arguments in question are valid.

However, while there are no algorithms for finding answers in these cases, there are algorithms that allow us to answer another question. Given a proposed integral for an expression, the rules of differential calculus provide an algorithm for deciding if the proposal is correct. In a similar way, given a proposed proof of a conclusion from a set of premises, there are algorithms that allow us to evaluate whether that proof is valid.[5] Algorithms of this latter sort became the focus of positivist studies of scientific method. In order words, the idea that we might find a discovery method – an algorithm – that would solve such problems as the causes of cancer or the origin of the universe came to be seen as utterly implausible (cf. Hempel 1966, pp. 14ff). But the idea that we could find algorithms that would allow us to assess a proposed solution to a problem seemed eminently plausible. In particular, this enterprise seemed well worth pursuing in the case of inductive arguments where we have a theory and a body of evidence, and we want to assess whether, or to what degree, the evidence supports the theory. Since there are no known algorithms for carrying out this project, it seemed that there was work to be done.

For those who are familiar with the jargon, this is what the distinction between *context of discovery* and *context of justification* is all about. Positivist philosophers of science concluded that their proper domain was logic and that as logicians they had nothing to say about discovery; their work was to be confined to establishing rules for assessing logical relations of confirmation. Moreover, since adequate rules for the assessment of deductive arguments were already known, these rules provided the model on which positivists sought to construct an inductive logic of confirmation. Thus positivists undertook the development of a formal inductive logic – one that would permit assessment of nondeductive evidence relations solely on the basis of formal relations between a formalized theory and a formalized set of evidence reports.

P5

The positivist picture that I have been sketching leads directly to a division of labor between philosophers and scientists. Scientists are in charge of the entire empirical realm. It is their concern to gather evidence and formulate empirical theories. Philosophers maintain control of the a priori realm. For the philosophers I am discussing this is a very limited realm; it concerns only formal logic plus meaning relations that result from definitions – and only the first of these concerns us here. Nevertheless, this a priori realm is especially important because it alone provides the norms that justify scientific procedures. Without these norms we lack any basis for believing that science is a rational enterprise.[6]

POSTPOSITIVISM

I hope that at least the outlines of this picture are familiar and that I have succeeded in filling in some of the details because I now want to consider why, over the past 40 years or so, a number of philosophers, historians, and others who are interested in the overall development of science have argued that this picture is seriously defective.

I will begin with two points about induction. The first point has already been mentioned and I will not dwell on it. So far there is no acceptable set of rules of induction that provide clear criteria for deciding when a theory should be accepted or for determining the probability that a theory is true. To be sure, scientists do accept laws and theories, but not as a result of applying a formal inductive calculus; so far philosophers have not succeeded in providing an inductive logic that can stand alongside deductive logic.

The second point is commonly referred to as 'underdetermination of theory by evidence', although I think it more revealing to describe it as 'underdetermination of theory by evidence plus logic'. An unlimited number of alternative generalizations are equally well supported by any finite body of data. One way of seeing this is to think of data points that have been plotted on a pair of coordinate axes. On formal grounds alone, every curve that passes through these points is supported equally by the data. Observational data and formal logic provide no grounds for preferring one generalization over another – and thus no grounds for preferring one prediction for the unexamined cases over a multitude of competitors.

Reflection on the curve-fitting example will suggest an immediate response: some curves that fit the data are more plausible than others and additional criteria can be invoked for choosing among them. Kuhn (1977) has discussed five such criteria that are commonly used by scientists: a good theory should be *accurate* within its domain, *consistent* both internally and with other accepted theories, have *wide scope* (i.e. consequences that go beyond the data it was introduced to explain), be *simple*, and be *fruitful* as a basis for further research. Note, however, that this list introduces criteria for evaluating theories that go beyond the resources of observation and formal logic. This point is clear in at least three respects.

1. With one limited exception, there are no known formal rules for applying these criteria, although there have been many attempts to formalize some of them. The exception is consistency. In logic there is a precise sense of the term 'consistent': a set of statements (which may have only one member) is inconsistent if it implies both p and not-p for some statement p. Otherwise, the set is consistent. We can see the point by considering a set that consists of two self-consistent statements: these statements are mutually consistent unless one of them contradicts the other. One consequence of this definition is that two statements that have nothing to do with each other must be mutually consistent. Now the requirement that a theory be consistent with other accepted theories includes the need to avoid formal inconsistency, but it includes more than this. In particular, a new theory must 'fit' with existing views, but there is no formal analysis of what counts as an appropriate fit – there is no formal analysis of consistency in this sense of the term.

In addition, even if we limit ourselves to consistency in the formal sense, the situation is more complex than it might seem at first glance. Internal inconsistency is a defect in a theory that gives grounds for seeking a better theory. Proving that a theory is inconsistent has the same import as showing that it has false empirical consequences: inconsistency guarantees that some part of the theory is false. Unfortunately, this does not require that we throw out the entire theory. Inconsistency only requires some modification in the theory, and logic provides little guidance as to what modification is appropriate. A classic example will illustrate the point.

Special relativity has two key postulates: the laws of physics are the same for all frames of reference moving with constant relative velocities, and the velocity of light is the same in all reference frames. Other physicists besides Einstein contemplated the reconstruction of physics on the basis of these two postulates, but abandoned the idea because it seems easy to show that these postulates are mutually inconsistent. Einstein notes the problem early in his first relativity paper, 'On the electrodynamics of moving bodies', and announces that he will show that these are only apparently inconsistent. He then argues that the inconsistency actually requires a third proposition: whether two events (at a distance from each other) are simultaneous is an objective fact. Before Einstein's analysis, physicists had not considered this claim to be a hypothesis subject to reconsideration, and this is the hypothesis that Einstein rejects in order to eliminate the inconsistency.

There is an additional twist that we should note. An inconsistent theory may be extremely useful if it is deployed with sufficient care. The classic example here is Bohr's theory of the atom which postulates that certain electron orbits are stable, but does so in the context of classical electrodynamics which implies that no orbits are stable. The theory's inconsistency was recognized as a defect and this defect was eliminated by the new quantum theory of Schrödinger and Heisenberg. But inconsistency is not a defect that immediately blocks legitimate application of the theory. In general, inconsistency is a formal defect in a theory, but the discovery that a theory is inconsistent does not require any specific response from theorists.

Simplicity is another example worth considering. There have been many attempts to provide a formal analysis of simplicity and none have withstood criticism. Part of the problem is that there are different senses in which a theory may be called 'simple'. One of these is psychological, i.e. one theory is easier to understand and use than another. But psychological simplicity is not a formal property of a theory. Psychological simplicity varies among individuals and among historical periods: which theory is easiest to use can depend on features of an individual's psychology and on just how one was trained.

Another kind of simplicity requires limiting the number of independent premises in a theory. Note, however, that this may compete with psychological simplicity. In addition, it is not always clear how to count the premises. Purely logical manipulations can alter the apparent number of assumptions involved in a theory.[7]

A third approach is captured in Occam's razor: postulate as few entities as possible. But again we have a kind of simplicity that can compete with other versions, and historical experience should make us wonder if the theory that postulates the smallest number of entities is more likely to be true than one which postulates a larger set of entities. For example, fundamental particle physics has now faced several generations of proliferating basic entities followed by a move to a deeper level that reduced the number of basic entities, followed by another proliferation. The particles at the current deepest level are tied together by an elegant mathematical structure – but this may involve yet another kind of simplicity that I will not pursue in detail.

The last example suggests another interesting point about simplicity: if we apply induction to the history of science, there does not seem to be any good reason for believing that the simplest theories are the most likely to prevail. The progression in planetary theory from circles to ellipses to perturbed ellipses that do not fit any simple curve is a progression of, at least, increasing mathematical and psychological complexity. The move from Newton's gravitational theory to Einstein's involves an increase from one gravitational potential to 10; there are other respects in which the new theory is arguably more complex, and others in which it is arguably simpler, than the older theory. Of course, these changes were not made arbitrarily. At the very least they were made in the pursuit of greater accuracy – which brings me to the second point that was mentioned some while back.

2. As Kuhn points out, all five of these criteria can compete with each other. I have just cited cases in which accuracy and simplicity compete, but it would be a mistake to jump to the conclusion that accuracy always prevails. Sometimes theories that are accurate but out of harmony with established approaches will have a hard time getting a hearing.[8] There is no reason for assuming that we will eventually find formal rules for deciding how these criteria should be applied – and there is no reason for thinking that scientists have been impotent in the absence of such criteria. Assessment of which criteria should prevail in a particular case requires reflection and judgment that must be based on a detailed understanding of the situation in the field. Such scientific judgment lacks the certainty of results arrived at by means of formal rules, but this does not make judgment arbitrary or utterly unreliable.

3. Now consider why these five criteria are appropriate. Kuhn's response is that historical research shows them to be widely used in science and the position I will develop below suggests that this is the right way to go about finding the relevant criteria. For the moment I want to stress that these criteria are themselves nonformal and that the process of recognizing the appropriate criteria also requires considerations that go beyond the application of a formal calculus.

POPPER

Before we leave the topic of formal methodological rules, I want to back up historically and consider Popper's attempt to preserve the idea that evaluations of theories are solely a matter of evidence and formal logic. Popper's starting point was his conviction that there is no such thing as inductive support for a theory – a result that he maintained had already been demonstrated by Hume.[9] There is, according to Hume, a psychological process that results in our believing a generalization as we encounter confirmations, but there is no basis in logic for this process. As a result, Popper concluded, the logical basis for the evaluation of scientific theories must be found in deductive logic. Popper built his account of scientific method on a feature of deduction that we have already noted: the deduction of a false conclusion from a set of premises proves unequivocally that something is wrong with those premises. If we focus our attention on falsification, rather than on confirmation, the only logic we require is deductive logic, whose status is not in doubt. This leaves us with the central characteristic of Popper's philosophy of science: the defining character of scientific propositions is that they are empirically falsifiable – not that they are empirically confirmable – and the proper method of scientific research is to propound hypotheses and then attempt to falsify them.

I want to emphasize a point that is often missed by scientists who turn to Popper for an account of the nature of science. Popper insists that there is no such thing as induction and that there is no process by which the testing of a scientific theory yields reasons to believe that theory. It is alright to use a theory that has passed severe tests, but we ought also to be continually attempting to disprove that theory – keeping in mind that one false consequence provides sufficient grounds for rejecting the theory while no run of correct predictions provides any grounds for believing the theory to be true. The only time we learn something definitive from the scientific process is when we discover that we are wrong.

There are many problems with the Popperian approach, but I will focus on those that generate further forms of underdetermination. Recall that while the deduction of a false consequence from a theory proves that something is wrong with that theory, it does not pick out what is wrong or even how many revisions we must make. The only clear result is that at least one revision is required. Consider a realistic example of the deduction of an observable result in Newtonian mechanics: the prediction of the location of Uranus in its orbit at a specific time. This prediction requires Newton's laws of motion and gravitation plus a large number of initial conditions: the masses of Uranus and the sun, the distance between them at some point in time, Uranus' tangential velocity, and its location in its orbit at some point in time. If we want a really accurate calculation, we will need values for the masses of other planets and their distances from Uranus.[10] In addition, any instrumentation that is required to gather the observational data will depend on other scientific results that – on Popperian grounds – have not been established as correct. When we encounter an incorrect prediction, logic tells us that at least one member of our total set of assumptions is mistaken, but logic does not provide a clue as to which should be challenged.

Moreover, even if we leave aside questions raised by the theoretical basis of our instruments, Popper holds that observational data are part of the body of science and thus falsifiable. The upshot is that we should consider the complex consisting of a scientific theory, the initial and boundary conditions needed to arrive at a testable prediction, any theories required by the test procedure, and the observational outcome of the test as a single set of claims. The discovery that the observational result contradicts the prediction amounts to the discovery that this set is inconsistent. From a Popperian perspective, empirical refutation and the discovery of an internal inconsistency are variations on a single theme. In either case, once we discover an inconsistency, logic has done all it can for us, and we do not have a clue as to the source of the inconsistency. The contemporary solar neutrino problem will underline the point that this is a genuine difficulty. The anomalous result of the first solar neutrino experiment was announced almost 30 years ago and it shows that there is a genuine problem somewhere in a large body of science. There have been dozens of proposals as to the source of the problem and research aimed at narrowing down the choices continues, but methodology, as understood by positivists and Popperians, has little guidance to offer (see Bahcall 1989 for a comprehensive, although already somewhat out of date, review).

Popper was aware of this problem and attempted to solve it by introducing a number of additional methodological rules, such as that one should always reject the most general propositions in a set (see Popper 1968, especially sections 11 and 20, and Ch. 5). There is much interesting material here, and perhaps a few surprises for the reader, but the outcome is to underline again the limits of methodology when one attempts to capture this subject wholly within the confines of formal logic.

METHODOLOGY RECONSIDERED

Please note, I am not claiming that observation and logic are unimportant for scientific method, only that they are not sufficient because at every key juncture they leave too many options open. The point is nicely stated in a recent essay by Antony (1993, p. 211): 'The problem of paring down the alternatives is the defining feature of the human epistemic condition.' The number of alternatives to be explored at a given time must be limited if effective research is to take place. This is also one of the key problems that Kuhn (1970) addressed in *The Structure of Scientific Revolutions* and the answer that Antony offers is essentially the same one that Kuhn proposed: at any period in the development of knowledge there are theories that are generally accepted by the relevant community and are taken as fixed. These fixed theories guide research by providing criteria for deciding what questions are worth asking, what observations are worth making, what phenomena are problematic, and what counts as a legitimate solution to a problem. Theories that play this role function as part of the methodology of a discipline. I will call theories that function as methodology 'guiding assumptions'.[11]

Guiding assumptions have two key features that pull in opposite directions and that must be balanced against each other. On one hand, methodology is supposed to provide a stable set of rules for the pursuit of science. The idea that methodology is known a priori makes it not just stable, but permanent, by taking it completely outside the realm of empirical evaluation. Guiding assumptions share this feature of the a priori in that they are immune to being undermined by empirical results – but only temporarily. For guiding assumptions are still theories that can be challenged empirically and replaced under appropriate circumstances. Guiding assumptions are the locus of 'the essential tension' between conservatism and innovation that Kuhn maintains is central to scientific research.

The idea that science requires guiding assumptions originates with Kant's notion of synthetic a priori propositions, although Kant thought that he had identified a set of propositions that necessarily provide the framework for science and thus that can never be challenged. Kant's two most important examples – Euclidean geometry and the causal principle – illustrate how guiding assumptions that endure for substantial periods of time can eventually be challenged by empirical results. These two examples, along with Einstein's treatment of simultaneity, also illustrate the way in which scientists can rely on guiding assumptions without realizing that they are assumptions. Hopefully, work since the 1950s in the history and philosophy of science has placed us in a somewhat more reflective situation with respect to the role of guiding assumptions in science. Nevertheless, I think it will be worthwhile to examine a few examples.

Perhaps the most widely discussed example is the response to discrepancies between the calculated and observed orbits of Mercury and Uranus that emerged in the nineteenth century. I will not tell this story yet again, but I will note the nice way in which it illustrates the two key features of guiding assumptions. In the case of Uranus, Adams and Leverrier assumed that Newtonian mechanics was correct and used the resources of this theory to solve the problem; the outcome was the discovery of Neptune. In the case of Mercury, Leverrier's attempt to overcome the anomaly by introducing yet another planet failed, and the anomaly was not resolved until Einstein introduced his new theory of gravitation. I want to add three more examples that will illustrate various aspects of the role of guiding assumptions in scientific research.

During the nineteenth century a great deal of careful empirical research by highly skilled chemists was guided by Prout's thesis that the weight of every chemical element is an integral multiple of the weight of hydrogen. When a sample presented an anomaly it was assumed that the sample was insufficiently pure and that further laboratory refinement was required. After the discovery of isotopes, weights that were previously considered anomalous were no longer considered problematic and this entire line of research was no longer significant. Frederick Soddy, who won a Nobel Prize in chemistry for his work on isotopes, summed up the situation this way (1932, p. 50):

> There is something, surely, akin to if not transcending tragedy in the fate that has overtaken the life work of that distinguished galaxy of nineteenth-century chemists, rightly revered by their contemporaries as representing the crown and perfection of accurate scientific measurement. Their hard-won results, for the *moment at least* [italics added], appears as of as little interest and significance as the determination of the average weight of a collection of bottles, some of them full and some of them more or less empty.

I want to suggest, however, that it is a mistake to describe this outcome as tragic. Rather, it is a normal outcome of research given human epistemic conditions. This is the same fate that befell Ptolemaic astronomers, phlogiston chemists, Leverrier's attempt to account for the orbit of Mercury, Einstein's 30-year quest for a theory to unify gravitation and electro-magnetism, and many others. It is also a fate that may befall many contemporaries if their guiding assumptions are undermined. The key point is that we must get beyond the view that valuable scientific work consists only in the production of true results. In many cases, the most important outcome of scientific work is found in the contributions it makes to undermining the guiding assumptions on which it is based.

As a more recent example, consider the use of red-shift data in determinations of the recession velocities of celestial objects. There is much to say about the theoretical elements involved in breaking light up into a spectrum and interpreting an unfamiliar spectrum as a red-shifted version of a familiar spectrum, but I want to focus on a different point. Once it has been decided that we are dealing with a Doppler shift, we need a formula to convert that shift into a recession velocity; Newtonian physics and relativity give different formulas for this conversion. Which one should we use? Obviously, since relativity is the currently accepted theory, we use the relativistic formula – with its built-in guarantee that no recession can achieve the velocity of light.[12] One might reflect on how different current cosmology would be if we still used the Newtonian formula. There is no guarantee that future developments will never lead to another revision in this formula along with a reinterpretation of masses of data already collected. But, at the present time, the methodological role that was once played by Newtonian mechanics is being played by relativity. The theory is used to interpret data rather than being subjected to evaluation on the basis of that data.

For my final example I turn to the solar neutrino experiment. The idea of measuring the neutrino flux from the sun in order to test the standard model of stellar energy production is a good example of the way in which even the best established of theories may be subjected to empirical evaluation when a new kind of test becomes available. Davis proposed the experiment in the early 1960s, only a few years after the ability to detect neutrinos was established.[13] Still, there was considerable debate over whether the test was worth the expenditure of limited resources given the absence of any serious doubt about the standard theory. It is pretty likely that the experiment – which is not very expensive by contemporary standards – would not have been done if its cost had been significantly higher. (The relevant history is reviewed in Bahcall and Davis 1989.) Bahcall (1989, p. 5) also describes the central role of the standard model as a source of guiding assumptions

> that astronomers use everyday in their research. In interpreting astronomical observations (made by detecting photons), and in constructing astronomical theories, we use the theory of stellar evolution to determine the ages of the stars, to interpret their compositions, to infer the evolution of galaxies, and to place limits on the chemical composition of the primordial material of the universe. Each of these basic astronomical industries is called into question by at least one of the proposed solutions of the solar neutrino problem.

A large number of additional theories are implicated in the design of the experiment and the interpretation of its results. The key instrument in Davis's experiment is a large tank of cleaning fluid in a South Dakota gold mine. The tank was placed about a mile underground because contemporary physics indicates that neutrinos are the only particles of astronomical origin that could penetrate the earth and reach the tank. The material in the tank is C_2Cl_4 and the relevant reaction is reverse beta decay: occasionally a neutrino will transform a chlorine atom into argon which appears in the tank as a dissolved gas that can be periodically removed. Since the resulting isotope of argon is radioactive with a known half-life, the amount recovered can be measured by studying its decay. All of the rather complex chemistry involved in extracting the minute amounts of argon produced, plus the physics involved in measuring how much has been extracted, are included in the complex of guiding assumptions required for this experiment. From the point of view of formal logic alone, any of these theories can be viewed as having been challenged by the anomalous result. But at no point are logic and the data sufficient to settle the source of the anomaly; at each stage of evaluation, productive research requires that a substantial number of these theories are not being actively questioned. Indeed, one must accept a large set of theories (e.g. the existence of inverse beta decay, our understanding of radio-active decay and of the operation of a proportional counter) for there to be an anomaly in the first place.[14] Consider, also, the role that guiding assumptions play in turning a tank of cleaning fluid into a sensitive scientific instrument.

I want to return now to a previous topic, the relation between methodology and discovery. The positivists held that there is no connection between these at all. However new ideas appear, the role of methodology is to guide the process of sorting out which proposals are to be accepted and which rejected. Now if one identifies methodology with formal logic, and identifies a rational discovery process with the application of an algorithm, then there is indeed no rational basis for pursuing scientific discovery. But once we expand our notion of methodology to include established theories, then it becomes clear that the process of introducing hypotheses worth taking seriously receives considerable methodological guidance. No further comment should be needed to make this point with regard to the role of Newtonian theory in guiding the search for reasonable explanations of the anomalies in the orbits of Uranus and Mercury, nor with regard to a large body of physics in the development of the solar neutrino experiment. But consider the case of general relativity, which involved significant departures from previous physics. Many features of Einstein's program are continuous with developments in physics over the previous three centuries. For example, Einstein was pursuing the ideal of a single theory that would encompass as wide a range of physical phenomena as possible – an approach that had achieved several notable successes since Galileo. In addition, Einstein relied on the standard mathematics that had been so successful in physics (although he introduced the use of tensors, which were not familiar to physicists) and on familiar observational results (although special relativity had already required reinterpretation of the Michelson–Morley experiment). In other words, neither Einstein nor any of the other scientists who initiated scientific revolutions, worked in a methodological vacuum even when their results overturned key features of previously accepted guiding assumptions. But the methodology they pursued came from accepted science not from formal logic.

Still, the case of scientific revolutions does raise additional questions about the role of methodology in science. Revolutions involve challenges to established guiding assumptions. As a result, the set of alternatives that can be seriously considered expands as previously accepted criteria for limiting the range of acceptable alternatives are challenged. This raises serious questions about the existence of sufficient methodological guidance to resolve revolutionary disputes. Within the limits of this chapter I can only indicate the way that this question should be approached. Note, first, that while guiding assumptions are challenged in revolutionary situations, they are not completely eliminated; only a particular subset is challenged in a specific case. The range of alternatives is increased, but the research community is not in the position of considering any proposal whatsoever. Second, recall our earlier discussion of Kuhn's five criteria for evaluating theories. Whatever guidance these criteria give, they do not provide algorithms. The way they are to be applied, and their relative weighting, depend on the professional judgment of the scientists concerned – which is why Kuhn (1977, p. 331) says that these criteria 'function not as rules, which determine choice, but as values, which influence it'. It is this scientific judgment, operating within the framework of a scientific community, that provides the ultimate court of appeal in the development of science. This court of appeal is far from infallible, but it is also far from arbitrary.[15]

This leads me to the final topic I wish to discuss. If methodology is fallible and subject to change, what happens if scientists spend their entire careers working on the basis of guiding assumptions that turn out to be false? Does this not make their work utterly worthless? The short answer is, 'No,' for the reasons I indicated in response to Soddy's remarks. To see why, we must keep in mind that scientists, like other human beings, are fallible cognitive agents – which does not mean that they are without cognitive resources. One of the most powerful of these resources is the ability to develop and test scientific theories, and it is through this process that we discover mistakes and develop new proposals. Most scientists make their contributions through this process and none can expect to bring any important part of this process to its completion. Consider some of our examples once again. The anomalies in the orbits of Uranus and Mercury were discovered as a result of generations of scientific work. This work included the development of improved telescopes, better calculation techniques, and long-term accumulation of observations. In the case of Mercury, the anomaly was identified only after the telescopic observation of hundreds of orbits. Everyone now alive will be dead long before a single orbit of Pluto has been observed.

An even more sobering thought for those who think that valuable scientific work requires definitive solutions to problems in their lifetime is provided by the more recent discovery of deterministic chaos. We now know of the existence of previously unnoticed chaotic solutions to fairly simple equations that have been in use for substantial periods of time. It is far from clear how much this discovery will alter our view of nature and of scientific method. But the exploration of chaos – and other nonlinear phenomena – is vitally dependent on modern computing power, so that if the discovery had come much earlier (Poincaré saw at least the outlines), not much could have been done with it.

CONCLUSION

I will end this discussion by stressing one central contrast between the approach with which I began and that at which I have arrived. Positivists and Popper, along with the mainstream empiricist tradition in philosophy, held that we have no a priori insight into nature. Whatever we know about the world must be learned through experience. But they also held that the empirical study of nature must be guided by a methodology that is known a priori. The view I have defended here is a more radical form of empiricism which denies that we have a priori insight into how nature should be studied. One of the things that we must learn is how to learn.[16] We have had to learn, for example, that observation is not just a matter of using our senses. The world is full of entities and processes that we cannot sense but that we can still study by building appropriate instruments. We have also learned that items outside the range of our senses are often more important for understanding nature than are the items we can detect with our unaided senses. We have learned the power of controlled experiments and of how to achieve the virtues of controlled experiments in realms where literal control is not possible (consider, yet again, the solar neutrino experiment). We have learned the power of statistical methods and found applications for mathematics that is so abstract that it once seemed clear it would always be just an intellectual game. From the perspective I have been presenting, all of these and many other developments are improvements in the methodology, as well as in the content, of science. In other words, instead of the older picture of scientists exploring nature within a fixed methodology, we have arrived at the conclusion that improving methodology is a central part of the scientific process.

Finally, if scientists create their own methodology, where does this leave the philosopher? I hope that this chapter has provided an illustration of one major task for the philosophy of science: to attempt to understand the scientific process by looking at science from a broader logical and historical perspective than is required for work within a single scientific discipline. This, too, is an empirical endeavor that works in terms of guiding assumptions that are subject to challenge when science goes off in a new direction, as well as from internal philosophical considerations. Given the empirical nature of this study, it is important to extend its data base in another way. Positivists took physics as the paradigmatic science and – since the thesis that methodology is the same for all sciences was one of their guiding assumptions – they did not see any need to examine other sciences. In recent decades, however, we have seen philosophers paying increasing attention to other scientific fields. From the perspective of this chapter, the present interaction between philosophy of science and geomorphology is one from which both subjects should benefit.[17]

NOTES

1. See Scheffler (1963) and Brown (1979) for discussions of the development of positivism from, respectively, a positivist and a postpositivist point of view.
2. Here is a trivial example: all dogs are guppies, all guppies are mammals, therefore all dogs are mammals.
3. That theories are to be evaluated empirically is also part of scientific method, but this was not considered problematic, and it dropped into the background of actual discussions among positivists.
4. This point is reached when we have existential quantifiers trapped inside universal quantifiers. See Quine (1982, especially Parts II and III), for a systematic discussion of the limits of algorithms in assessing validity.
5. Those familiar with discussions of the recent proof of Fermat's last theorem will recognize that even here the situation is more complex than positivists imagined.
6. Coffa (1976, p. 207), writing specifically about Carnap, offers the following description: 'the linguistic world of knowledge was, for Carnap, an Upstairs–Downstairs world. Upstairs, in the object language, live the masters, the scientists, those whose task it is to find out the facts about the world. Downstairs, in the metalanguage, live the servants, the philosophers, whose task it is to keep the house of knowledge clean and to set an occasional Upstairs novice straight on matters of taste and manners.'
7. Consider another kind of problem: the introduction of vector and tensor notation allows us to capture sets of equations in a single equation. Does this reduce the number of independent premises?
8. See, for example, Polanyi's (1969) discussion of the reasons why his potential theory of adsorption was largely ignored for several decades – along with his defense of the practices that led to his own theory being ignored at first.

9. The original German-language edition of Popper (1968) appeared in 1934, well before the problem of choosing between equally well-supported theories came to prominence. However, Popper's complete rejection of any notion of inductive support makes the latter problem moot.

10. If we pursue this degree of accuracy we also move into a realm in which a rigorous deduction from a list of all of these premises is not possible. Approximations must be introduced and these approximations provide additional reasons why a prediction may have been incorrect. Newton faced this problem in calculating the orbit of moon. For one key parameter he got a result that was twice the observed value (Newton 1971, p. 147). Clairaut initially thought that the problem should be solved by adding an inverse fourth-power term to the gravitation law. 'But before proceeding far on his new theory he took the precaution to carry his calculation, using Newton's law, to a higher degree of approximation, and, behold, in 1749 he reached results agreeing with observation' (Cajori 1971, p. 650).

11. This term was introduced in Laudan et al. (1986) to cover this central theme that is shared by many postpositivist writers. See also Donovan et al. (1988). In his 'Postscript' Kuhn (1970, p. 182) notes that 'Scientists themselves would say they share a theory or a set of theories, and I shall be glad if the term can ultimately be recaptured for this use.' Kuhn's hesitation to use 'theory' was generated by the narrow use of this term that was standard in the late stages of positivism.

12. For a measured red-shift of S, the relativistic formula is $\beta = (S^2 - 1)/(S^2 + 1)$, where β is the recession velocity expressed as a fraction of the velocity of light *in vacuo*. The Newtonian formula is $\beta = S - 1$. Although it is mathematically impossible for β to exceed one in the relativistic formula, the use of this formula in an observation procedure does not automatically prevent that procedure from yielding a result that can contradict the relativistic limitation on velocities. See Brown (1993, 1994a) for discussion.

13. The first experimental detection of neutrinos was reported in 1953, but the experimenters were not fully confident of their result until 1956. See Reines and Cowan (1953), Cowan et al. (1956), Reines (1979).

14. See Bahcall (1989) for these and many other details. For discussion of the implications of this experiment for our understanding of scientific observation see Brown (1987, 1995) and Shapere (1982).

15. I have developed this view at greater length in Brown (1988) and have improved on several aspects of that discussion in Brown (1994b).

16. The idea that we must learn how to study nature has been a consistent theme of naturalistic epistemology. See, for example, Hooker (1987), Laudan (1987), and Shapere (1984).

17. For some examples from geomorphology of the view of methodology I have proposed, see Rhoads and Thorn (1996).

REFERENCES

Antony, L.M. 1993. Quine as feminist: the radical import of naturalized epistemology, in *A Mind of One's Own*, edited by L.M. Antony and C. Witt, Westview Press, San Francisco, pp. 185–225.

Bahcall, J.N. 1989. *Neutrino Astrophysics*, Cambridge University Press, Cambridge, 576 pp.

Bahcall, J. and Davis, R, 1989. An account of the development of the solar neutrino problem, in *Neutrino Astrophysics*, edited by J.N. Bahcall, Cambridge University Press, Cambridge, pp. 488–530.

Brown, H.I. 1979. *Perception, Theory and Commitment*, University of Chicago Press, Chicago, 203 pp.

Brown, H.I. 1987. *Observation and Objectivity*, Oxford University Press, New York, 255 pp.

Brown, H.I. 1988. *Rationality*, Routledge, London, 244 pp.

Brown, H.I. 1993. A theory-laden observation can test the theory, *British Journal for the Philosophy of Science*, **44**, 555–559.

Brown, H.I. 1994a. Circular justifications, in *PSA 1994* vol. 1, edited by D. Hull, M. Forbes and R.M. Burian, The Philosophy of Science Association, East Lansing, Mich., pp. 406–414.

Brown, H.I. 1994b. Judgment and reason, *Electronic Journal of Analytic Philosophy*, **2**.

Brown, H.I. 1995. Empirical testing, *Inquiry*, **38**, 353–399.

Cajori, F. 1971. An historical and explanatory appendix, in *Principia* by I. Newton, translated by A. Motte, revised by F. Cajori, University of California Press, Berkeley, pp. 626–680.

Coffa, J.A. 1976. Carnap's sprachanschauung circa 1932, *PSA 1976*, Vol. 2, edited by F. Suppe and P.D. Asquith, The Philosophy of Science Association, East Lansing, Mich., pp. 205–241.

Cowan, Jr, C.L., Reines, F., Harrison, F.B., Kruse, H.W., and McGuire, A.D. 1956. Detection of the free neutrino: a confirmation, *Science*, **124**, 103–104.

Donovan, A., Laudan, L., and Laudan, R. 1988. *Scrutinizing Science*, Kluwer, Dordrecht, 379 pp.

Hempel, C.G. 1966. *Philosophy of Natural Science*, Prentice-Hall, Englewood Cliffs, NJ, 116 pp.

Hooker, C.A. 1987. *A Realistic Theory of Science*, State University of New York Press, Albany, 479 pp.

Kuhn, T.S. 1970. *The Structure of Scientific Revolutions*, 2nd edn, University of Chicago Press, Chicago, 210 pp.

Kuhn, T.S. 1977. Objectivity, value judgment, and theory choice, in *The Essential Tension*, University of Chicago Press, Chicago, pp. 320–339.

Laudan, L., Donovan, A., Laudan, R., Barker, P., Brown, H., Leplin, J., Thagard, P., and Wykstra, S. 1986. Scientific change: philosophical models and historical research, *Synthese*, **69**, 141–223.

Laudan, L. 1987. Progress or rationality? the prospects for normative naturalism, *American Philosophical Quarterly*, **24**, 19–31.

Newton, I. 1971. *Principia*, translated by A. Motte, revised by F. Cajori, University of California Press, Berkeley, 680 pp.

Polanyi, M. 1969. The potential theory of adsorption, in *Knowing and Being*, edited by M. Grene, University of Chicago Press, Chicago, pp. 87–96.

Popper, K. 1968. *Logic of Scientific Discovery*, 2nd English edn, Harper & Row, New York, 480 pp.

Quine, W. 1982. *Methods of Logic*, 4th edn, Harvard University Press, Cambridge, 333 pp.

Reines, F. 1979. The early days of experimental neutrino physics, *Science*, **203**, 11–16.

Reines, F. and Cowan, C. 1953. Detection of the free neutrino, *Physical Review*, **92**, 830–831.

Rhoads, B. and Thorn, C. 1996. Observation in geomorphology, Ch. 2 this volume.

Scheffler, I. 1963. *The Anatomy of Inquiry*, Bobbs-Merrill, Indianapolis, 337 pp.

Shapere, D. 1982. The concept of observation in science and philosophy, *Philosophy of Science*, **49**, 485–525.

Shapere, D. 1984. *Reason and the Search for Knowledge*, Reidel, Dordrecht, 438 pp.

Soddy, F. 1932. *The Interpretation of the Atom*, John Murray, London, 355 pp.

2 Observation in Geomorphology

Bruce L. Rhoads and Colin E. Thorn

Department of Geography, University of Illinois at Urbana-Champaign

ABSTRACT

Observation traditionally has occupied a central position in geomorphologic research. The prevailing, tacit attitude of geomorphologists toward observation appears to be consistent with radical empiricism. This attitude stems from a strong historical emphasis on the value of fieldwork in geomorphology, which has cultivated an aesthetic for letting the data speak for themselves, and from cursory and exclusive exposure of many geomorphologists to empiricist philosophical doctrines, especially logical positivism. It is, by and large, also an unexamined point of view.

This chapter provides a review of contemporary philosophical perspectives on scientific observation. This discussion is then used as a filter or lens through which to view the epistemic character and role of observation in geomorphology. Analysis reveals that whereas G.K. Gilbert's theory-laden approach to observation preserved scientific objectivity, the extreme theory-ladenness of W.M. Davis's observational procedures often resulted in considerable subjectivity. Contemporary approaches to observation in geomorphology are shown to conform broadly with the model provided by Gilbert. The hallmark of objectivity in geomorphology is the assurance of data reliability through the introduction of fixed rule-based procedures for obtaining information.

INTRODUCTION

Geomorphologists traditionally have assigned great virtue to observation. The venerated status of observation can be traced to the origin of geomorphology as a field science in the late 1800s. Exploration of sparsely vegetated landscapes in the American West by New-berry, Powell, Hayden, Gilbert, and others inspired perceptive insights about landscape dynamics that provided a foundation for the discipline (Baker and Twidale 1991). Con-

The Scientific Nature of Geomorphology: Proceedings of the 27th Binghamton Symposium in Geomorphology held 27–29 September 1996. Edited by Bruce L. Rhoads and Colin E. Thorn. © 1996 John Wiley & Sons Ltd.

temporary geomorphology continues to have, as it should, a strong field component. The importance of this component and its relationship to observation is nicely summarized by Ritter (1986, p. 3): '...the real test of geomorphic validity is outdoors, where all the evidence must be pieced together into a lucid picture showing why landforms are the way we find them and why they are located where they are. A prime requisite for a geomorphologist is to be a careful observer of relevant field relationships.'

Although observation clearly is assigned privileged epistemological status in geomorphology, little, if any, formal justification of this status has been provided by geomorphologists. Instead, the privileged role of observation appears to be an implicit unquestioned presupposition. Recently, debate has arisen concerning the relative importance of theory and observation in geomorphology. Whereas Baker and Twidale (1991) emphasize the primacy of observation over theory, Rhoads and Thorn (1993) have argued that this view is misleading because virtually all scientific observation is inherently intertwined with theoretical presuppositions. The latter position generally has not been widely embraced within science at large, including geomorphology, because of the fear that objectivity will be compromised if the observations used to test a theory depend strongly on that theory (e.g. Feyerabend 1993). If such dependence was common, the power of empirical testing would be undermined because no neutral body of facts would exist against which competing theories could be compared. Under these conditions, theory choice becomes subjective and scientific knowledge relative, a view that is not widely held by scientists in general and geomorphologists in particular. On the other hand, most contemporary philosophers of science fully acknowledge that scientific observation is inherently theory-laden, yet have defended objectivity by defining a naturalistic perspective on scientific observation that accords with its function in actual scientific practice.

The purpose of this chapter is to examine the epistemic role of observation in geomorphology in the light of recent developments in the philosophy of science. The first section of this chapter provides a context for subsequent discussion by reviewing traditional philosophical perspectives on scientific observation. The next section highlights the basic tenets of the emerging, naturalistic perspective on observation and illustrates how this perspective differs from traditional philosophical views on observation. The final section of the chapter explores the relevance of philosophical perspectives for understanding the science of geomorphology by examining G.K. Gilbert's and W.M. Davis's ideas on observation and by analyzing the character and function of observation in contemporary geomorphologic research.

PHILOSOPHIC PERSPECTIVES ON SCIENTIFIC OBSERVATION

Since the origin of modern science during the Renaissance, observation has been viewed by both practicing scientists and philosophers of science alike as a, and in some cases the, fundamental building block of science. Observation occupied center stage in Sir Francis Bacon's early description of the methods of science (Haines-Young and Petch 1986). It also played a foundational role in perhaps the most influential model of modern science: the view developed between 1900 and 1960 by the logical empiricists (also known as the logical positivists) (Suppe 1977).

The Empiricist View of Observation

Logical empiricism is founded on a clear dichotomy between observation and theory. In particular, this language-based analysis of science distinguishes between observation statements and theoretical statements. According to proponents of this view, the vocabulary for observation statements is acquired ostensively and statements made in this vocabulary can be verified (or disconfirmed) empirically through direct observation. In contrast, statements made in the theoretical language are about unobservable phenomena and thus only have truth value if they are related to the observational vocabulary via explicit definitions or correspondence rules.

Logical empiricists never specified a definitive set of criteria governing the distinction between observable and unobservable phenomena, and generally assumed that the notion of something being directly observable is unproblematic (Suppe 1977, p. 46). It is also clear that they associated direct observation with sensory perception:

> To a philosopher, 'observable' has a very narrow meaning. It applies to such properties as 'blue', 'hard', 'hot'. These are properties directly perceived by the senses (Carnap 1966, pp. 225–226).

According to the logical empiricists, assertions about directly observable phenomena can be verified intersubjectively without recourse to theoretical presuppositions; thus, the observation language is theoretically neutral and the truth of statements made in this language is unproblematic (Suppe 1977, p. 48). This argument implies that observation is independent of *all* theory.

Because the truth of observation statements can be determined on the basis of theory-neutral observations, the observations used to verify these statements provide a foundation for scientific knowledge. All other claims to knowledge must rest on the foundation of this base knowledge. Science grows through the steady accumulation of facts and empirical generalizations based on direct observation, and then later introduces theoretical postulates (tied to verified observation-language assertions) to organize these facts and generalizations. This upward growth from facts to theories is similar to the inductive Baconian model of scientific progress (Suppe 1977, p. 15).

Postpositivist empiricism has moved away from a language-based analysis of science toward a view that emphasizes the important role of models in scientific practice (van Fraassen 1980, 1989). However, this form of empiricism still retains observation as an epistemological foundation for knowledge: science is justified only in assigning truth to the claims it makes about observable phenomena, not the claims it makes about unobservable phenomena. Postpositivist empiricists also have not fully articulated what it means for something to be classed as 'observable'. Van Fraassen (1980, pp. 16–17) states that 'observable' is a vague predicate, but maintains that the distinction between observable and unobservable phenomena is nevertheless useful provided it has distinct cases and countercases. He counts seeing something with the unaided eye as a clear case of an observable phenomenon and refers to subatomic particles detected in a cloud chamber as unobservable phenomena. Apparently, the distinction van Fraassen has in mind depends strongly on human perception (Suppe 1989, p. 26).

Problems with the Empiricist View of Observation

Logical empiricism has few supporters within contemporary philosophy of science. The downfall of this philosophical doctrine centered on its distinction between the observational and theoretical. Attacks on this distinction consist of three types: attempts to show that the distinction between observational and theoretical *terms* is untenable (e.g. Putnam 1962; Achinstein 1968), attempts to show that no sharp ontological distinction exists between observable and unobservable entities (e.g. Maxwell 1962), and attempts to show that observation is inherently theory-laden and thus the notion of theory-neutral observation or observation statements must be denied (e.g. Hanson 1958; Feyerabend 1958; Kuhn 1962; Popper 1968). Although not all of these arguments are convincing (Suppe 1977, pp. 80–86), it is now generally accepted within philosophy of science that the methods and language of scientific observation are theory-dependent (Boyd et al. 1991, pp. xi–xiv).

Recognition that observation is theory-dependent led initially to the emergence of other epistemologies that emphasize the relativistic nature of scientific knowledge. The best-known alternatives are those by Kuhn (1970) and Feyerabend (1993). Both of these views embody the notion that the content of observations or observation statements used to test theories is at least partly determined by the theoretical commitments which a group of scientists accepts. Differences in such commitments between groups or over time within a given group will result in differences in the meaning of observational statements and in what constitutes relevant observational evidence for each theory. Thus, disputes about competing theories cannot necessarily be resolved by appeal to a shared body of observational evidence, thereby undermining the adjudicatory power normally accorded to observation. Instead, knowledge becomes relative to the social, cultural, and historical settings of a particular group of scientists. The result is incommensurability of theories, and relativism about what can be known (epistemological relativism), about what is accepted or regarded as being known (epistemic-avowal relativism), but not about what exists (metaphysical relativism or subjective idealism) (Suppe 1989, pp. 302–328).

The relativist views of Kuhn and Feyerabend provide the foundation for what is now known as social constructivism. This epistemological stance derives from ideas about language espoused by Wittgenstein (1969, 1980), who, ironically, earlier in his life contributed to the development of logical empiricism (Wittgenstein 1922). The scope of constructivism is broad, ranging from neo-Kantian idealism, which contends that scientists actively construct theories (and thus the reality they investigate) through social practices (Knoor-Cetina 1983; Collins 1983) to epistemic relativism, which concedes that scientists interact with an objective reality, but maintains that the acceptability or unacceptability of claims to knowledge about phenomena is relative to a particular scientific community (Longino 1990; Pickering 1990; Nelson 1994). Constructivist epistemologies are aligned closely with postmodernist philosophy, which discounts the capacity of human rationality to provide a 'view from nowhere' on the world (Rorty 1979; Parvsnikova 1992).

Observation Naturalized

Over the last 20 years, efforts to formulate a naturalized epistemology of scientific knowledge (see Shapere 1987; Nickles 1987) have led first to an extended view of observation that attempts to accommodate the application of this term to unobservable objects and processes by practicing scientists. In turn, in response to perceived short-comings in the extended view of observation, more recent philosophical work has focused on the way in which scientists empirically evaluate scientific claims using *data* or *evidence*. These two views are examined in the next two subsections, followed by a discussion of their epistemological implications.

The extended view on observation

A major shortcoming both with logical empiricism and with many relativist views of science is that observation is equated with human sensory perception. In science, on the other hand, this term is used in a less restrictive sense. Many scientists claim to observe or even directly observe phenomena in situations where sensory perception plays only a minor role (see examples in Shapere 1982; Brown 1987a). Whereas the logical empiricists avoided this seeming contradiction by claiming that scientists and philosophers simply use the terms in different ways (e.g. Carnap 1966), such a stance becomes difficult to defend if philosophical usage limits or unnecessarily complicates attempts to understand science as an epistemic activity. Although debate about the theory-ladenness of human perception continues, spilling over from pure philosophy into the domains of neurophysiology, cognitive psychology, and visual psychophysics (see Werth 1980; Weckert 1985; Fodor 1984, 1988; Churchland 1988; Gilman 1992; Brewer and Lambert 1993), it is becoming increasingly evident that philosophical perspectives that equate observation with perception fail to capture the full richness of the epistemic role of observation in science (Suppe 1989, pp. 38–77).

Scientific observation involves obtaining information about an item in the world through a causal chain linking the (senses of the) observer to that item. This chain can be either short or long. Perceiving something through the use of our native senses alone is an example of a short chain, whereas exploration of the world through the use of sophisticated instrumentation is an example of a long chain. In any case, background information (i.e. accepted theories and beliefs that have proved successful and are free from specific and compelling doubt) is an integral component of observation because it is only within the context of such information that a causal chain between the observer and the observed can be established and that the outputs of such a chain can be interpreted in a scientifically meaningful (epistemically relevant) fashion (Shapere 1982; Brown 1987b). Thus, background information is an essential epistemological ingredient in our ability to observe items in the world.

The reliance on background information acknowledges that observation is at least to some extent theory-laden; however, this aspect of the observational process does not necessarily threaten scientific objectivity, an issue that is explored in detail in the next section. Furthermore, unlike traditional empiricist views, this naturalist perspective maintains that short causal chains are not inherently more reliable than long causal chains. As Brown (1987b) points out in a detailed analysis, human senses are instruments that operate according to natural laws in the same manner as artificial instruments; thus, the senses, like other instruments, are not only fallible (in contrast to logical empiricist claims to the contrary), but also are limited in their capacity to interact with items in the world. In fact, Brown (1987b) argues that the senses are fairly crude instruments compared to many artificial ones. For this reason, short causal chains involving the human senses alone often produce less reliable and/or more limited observations than the long complex causal chains associated with sophisticated artificial instrumentation.

Data and empirical evidence

The quintessence of extended observation is to eliminate the distinction between the roles of instrumentation and human perception in scientific investigation. In many ways this perspective represents an attempt to salvage a term, observation, and a distinction, observation versus theory, that have been deeply embedded and pivotal in the philosophy of science. Perceived shortcomings with this effort have spawned alternative naturalized epistemologies that differ from the extended view of observation in several important respects. A common concern of those criticizing attempts to retain observation is that this concept, even in the extended form, is epistemologically inadequate, i.e. it fails to capture the full role of empirical results in evaluating scientific claims to knowledge.

An examination of actual scientific practice suggests that the fundamental distinction is between 'data' and 'phenomena', not between observation and theory (Bogen and Woodward 1988, 1992; Woodward 1989). Whereas scientists use terms such as 'observations' and 'observed', they do so in a sense that refers to data or data collection procedures designed to detect phenomena. Data and phenomena differ in important ways.

Data are the products of procedures designed to gather information about the world. These products are accessible to the human perceptual system and thus available to public inspection. Data can be derived solely from human sensory perception; however, this situation usually occurs only in the early stages of development of a scientific discipline. More commonly, the procedures for producing data involve instruments or detection devices that have been developed artificially using background knowledge. Even in cases where data are based on sensory perception alone, such data do not consist entirely of the raw perceptual experiences of individual observers. In all cases, the character of data is tied intimately to the peculiarities of the research design, investigative methods, and instruments or apparatus used to collect the data (including human senses) (Bogen and Woodward 1988). The production of data also tends to be 'noisy' in the sense that virtually all data collection efforts will be influenced by a variety of idiosyncratic causal factors. Controlling for these confounding factors is one of the primary concerns of science.

Phenomena in contrast are objects, entities, events, and processes that exist in the world. Whereas data tend to be idiosyncratic to particular investigative contexts, phenomena are relatively uniform stable features with regular properties or characteristics in a variety of contexts. According to Bogen and Woodward (1988) phenomena cannot be observed, but nevertheless are the foci of explanatory theories. This perspective contrasts sharply with antirealist forms of empiricism, which maintain that theories are merely devices developed to explain or 'save' observable aspects of the world. Instead, it is consistent with the realist notion that science seeks the truth about observable and unobservable aspects of the world. This view does maintain a role for empiricism in the sense that data provide an evidential basis for evaluating scientific claims about unobservable phenomena. Scientists produce data in an effort to infer the existence of phenomena or to test theoretical assertions about phenomena. These inferences and tests occur within the context of background information and theoretical assumptions that govern the reliability of data and that provide connections between data and phenomena. It is the truth or approximate truth of this background knowledge that allows scientists to delve beyond the senses and explore 'hidden' aspects of reality.

Another shortcoming of the extended view of observation is that the definition of observation it puts forward requires an investigator to causally interact with something that actually exists in the world in order for an observation to occur (Brown 1995). In many cases, however, the lack of an observation (i.e. interaction with the world) will, in fact, still produce data, which can provide powerful evidence for theory evaluation. For example, suppose a specific theory under consideration predicts the occurrence of a particular phenomenon. In turn, a well-established background theory indicates that this type of phenomenon, if it exists, should causally interact with a particular instrument, thereby producing a certain type of instrument response. In this case, the failure to detect the expected response would be powerful evidence against the existence of the inferred phenomenon. This type of situation, along with the distinction between data and phenomena discussed above, shifts the focus away from observation as a key scientific concept, and instead emphasizes the important epistemic role that evidence, a concept that subsumes observation, plays in scientific practice.

The epistemological nature of data or evidence

Naturalized perspectives on data and evidence have led to widespread recognition that theory penetrates all aspects of scientific activity, including determinations of (1) what can be measured or observed, (2) which data are important, (3) how 'raw' data are processed to produce 'refined' data, (4) how data are interpreted, and (5) how interpreted data are used in empirical testing (Nickles 1987). Given that most contemporary philosophers subscribe to the notion that a symbiosis exists between theory and data, a key issue concerns the potential effect of the theory-ladenness of data on scientific objectivity. How can scientific objectivity be preserved in the light of this theory-ladenness? An examination of this question requires a definition of objectivity. According to Brown (1987a) objectivity includes two necessary (but not sufficient) conditions: relevance and independence. In other words, the evidence used to evaluate specific claims about the world must be relevant to and independent of these claims.

Relevance demands that a scientific claim about a phenomenon be evaluated using appropriate evidence, i.e. evidence produced by a procedure that is capable of interacting causally with the phenomenon of interest. In turn, determination of which procedures, instruments, analytical techniques, etc. are capable of interacting causally with a phenomenon depends on the availability of appropriate background theories. Thus, for example, geomorphic claims that the Channeled Scablands are the result of a catastrophic paleoflood requires that such claims be evaluated using information collected in the Channeled Scablands, not from landscapes in southeastern Pennsylvania. Moreover, the procedures used to generate appropriate evidence from the Channeled Scablands should rely on established hydraulic and sedimentologic principles as well as knowledge of the glacial history of the region, rather than on genetic theory or quantum mechanics. Relevance does not require that background theories be correct, only that they be appropriate. Advances in hydromechanics someday may show that what was once believed to be relevant geomorphic evidence concerning the origin of the Channeled Scablands has become irrelevant to this problem. Thus, relevance can only be judged within a particular historical setting. This requirement also implies that acceptable background theories exist at the time the decision of relevance is made. It is not appropriate to claim at present that the Channeled Scablands were produced by aliens repeatedly draining the waste disposal tanks on board their spaceships at this location on our planet because no well-established body of background knowledge exists to causally connect evidence from the Scablands to that claim. On the other hand, this explanation cannot be completely ruled out a priori, no matter how ridiculous it may seem, because the development of new background knowledge supporting this claim is not impossible. The key point is that this explanation currently has no epistemic import given that no background theory about alien visits to earth now exists. Should such aliens suddenly show up and tell us that the Scablands is where they like to drain their waste disposal tanks, we would have to reevaluate current geomorphic claims about this area.

The other major factor influencing objectivity is independence of observations from the scientific claim under consideration. The issue of independence relates to the concern raised in the introduction to this chapter: if all scientific observation inherently depends on theory, as is now widely accepted within contemporary philosophy of science, does this theory-ladenness threaten scientific objectivity? Brown (1995) has conducted a perspicacious analysis of this question and has concluded that the answer to it is contingent upon the manner in which a particular observation depends on theory. According to him, observation can depend on theory in at least six different ways (Table 2.1).

The first two forms of theory dependence (TD1, TD2) provide the basis for constructivist theories of scientific knowledge. Acceptance of TD1 and TD2 leads to radical relativism and incommensurability of competing theories. Although recent experimental studies provide some support for the influence of TD2 at the level of an individual or small group (Brewer and Chinn 1994), the major problem with accepting TD1 and TD2 at the level of an entire scientific community is that anomalies, which the history of science demonstrates play an important role in scientific discovery, become difficult to explain.

The third type of theory dependence (TD3) involves two separate elements: (a) that observational tests of a theory presuppose that theory, and (b) that use of the theory undermines the objectivity of testing the theory because it is impossible for the test to yield an outcome that will contradict a theory that has been presupposed. While Brown (1995) acknowledges that the first element is often the case in scientific testing, he argues

Table 2.1.   Six different ways in which observation can depend on theory (from Brown 1995)

TD1. The items we perceive are already infected with material from the theories we accept
TD2. Scientists ignore evidence that contradicts their favored theories
TD3. Observations that are undertaken to evaluate a comprehensive theory presuppose that very theory in a way that prevents an objective test of that theory
TD4. a. All scientifically significant observations assume theories besides the theory being evaluated; b. it is always possible to protect a favored theory by challenging these auxiliary theories
TD5. Which observations scientists undertake is determined by accepted theory
TD6. Observation reports must be expressed in the language of the theory being tested if they are to be relevant to the evaluation of that theory

persuasively that this element does not logically imply the second and thus the two elements are not necessarily connected as is often assumed. He describes an example from astronomy where evidence that presupposes the theory of relativity still plays an important role in an objective test of relativity (Brown 1993). In many scientific investigations, a hypothesis derived from a theory is assumed valid and then is used, in combination with observational data, to construct (or compute) evidence that provides a test of another hypothesis derived from the theory, a method of empirical testing known as the bootstrapping method (Glymour 1980). As long as reliance on the theory under consideration does not guarantee confirmatory evidence, the testing can be considered objective. One factor that appears to be important for producing disconfirmatory evidence is fixedness of the theoretical considerations that undermine observational procedures within a particular area of science. As long as these considerations are fixed, they will provide constraints on expectations, and it is against this backdrop of theoretical predilections that involuntary, unanticipated evidence occurs (Hudson 1994).

The fourth type of theory dependence also has two parts. Auxiliary theories play a part in virtually all observational procedures and also are required to derive theoretical predictions that can be compared with data. Because in many cases these background theories can be viewed as largely or wholly independent of the theory under consideration (see Hacking 1983; Kosso 1989), this type of theory-ladenness does not inherently bias empirical testing. Greenwood (1990) draws the distinction between *exploratory* and *explanatory* theories; the former enable scientific observation and interpretation, but in many cases are largely independent from the latter, which directly generate testable scientific claims within a specific scientific discipline. In general, exploratory theories are uncontroversial and can be considered established knowledge relative to explanatory theories. It is this characteristic of exploratory theories that provides the basis for addressing the second part of TD4. The notion that a favored theory can always be defended against disconfirmation by modifying an auxiliary hypothesis is known as the Duhem–Quine thesis. This argument represents the greatest challenge to Popper's falsification methodology (cf. Putnam 1974). The greater epistemic force of exploratory theories compared to explanatory theories, however, usually precludes pursuit of the Duhem–Quine strategy when the scientist is faced with a situation where observations clash with predictions of explanatory theories. Exploratory theories often have widespread use in science that transcends the range of explanatory theories of particular disciplines; thus, modification of an exploratory theory to 'save' a favored explanatory theory will often have consequences across a broad range of science. Changing an exploratory theory usually will undermine prior observational evidence not only for the explanatory theory under consideration, but also for explanatory theories (and perhaps other accepted exploratory theories as well) in other disciplines. An extreme example would be if geomorphologists tried to protect a favored geomorphic theory by modifying the principles of Newtonian mechanics. Such a maneuver would have ramifications not only throughout other areas of geomorphology but throughout science at large. Clearly, this type of maneuver generally is unacceptable. The interconnectedness among accepted background knowledge (i.e. exploratory theories) within science is referred to as the web of belief (Greenwood 1990). When scientists are faced with recalcitrant observations, this web of belief tends to preclude modification of auxiliary hypotheses (exploratory theories) to preserve an explanatory theory.

The distinction between exploratory and explanatory theories is consistent with Woodward's (1989) division of scientific labor between those who gather data in support of phenomena (experimentalists or field scientists) and those who attempt to explain or predict phenomena (theorists). Whereas theorists focus mainly on theory construction, either to explain an existing phenomenon or to predict the existence of some new phenomenon, data gatherers are concerned primarily with methods of generating reliable data that can be used to test theoretical predictions about phenomena or infer the existence of new phenomena. Separation of these activities is necessary in science because it helps to ensure independence of data and phenomena. When a theorist devises a theory that predicts the existence of a new phenomenon, the details of the procedures, devices, and research design required to produce reliable data to detect the existence of this phenomenon typically are not prescribed by the theory, but are left to the data gatherers to determine (Woodward 1989). The considerations involved in such determinations are related more closely to potential exploratory theories and to the issue of data reliability, than to the content of the explanatory theory. Thus, theory and data are not connected by correspondence rules or explicit definitions in the manner envisioned by the logical empiricists; instead, an explanatory theory often provides little guidance about how one should go about obtaining evidence of a phenomenon. Conversely, data collectors not only engage in theory confirmation, but also search for new phenomena in a manner largely independent of explanatory theory. Although this activity generally does not occur in a complete theoretical vacuum, it is data-driven (inductive) in the sense that it is fueled by the development of new instruments, techniques of data analysis, investigative procedures, or methods for controlling error. This type of research, which also draws on exploratory theories and theories governing data reliability, generates discoveries of phenomena that subsequently require theoretical elaboration. Discoveries often occur when puzzling data are produced. The problematic character of data is determined by a conflict with accepted background theories and beliefs (Shapere 1984, p. 283). An excellent geological example of the way in which 'bottom-up' construction of evidence using exploratory theories interacts with the 'top-down' deduction of theoretical implications from explanatory theory is discussed by Kaiser (1991).

The selectivity aspect of theory dependence (TD5) is closely related to the issue of relevance. Theory plays a key role in determining which observations, out of the infinite variety of possible observations, scientists actually make. It not only serves to focus attention on aspects of reality that are relevant to scientific inquiry, but also is used to sort out the relative importance of observations that are considered relevant. Whereas this aspect of theory dependence may *delay* challenges to theory because scientists deem certain types of information as irrelevant to a problem based on theoretical presuppositions, it does not determine the content of scientific evidence. Thus, TD5 does not inherently threaten scientific objectivity.

The final form of theory dependence (TD6) relates to the requirement of logical positivism that observation reports be described in theory-neutral language. Arguments against the existence of such a language are numerous and persuasive, and few contemporary philosophers of science defend the view that all theoretical predictions and empirical evidence can be expressed in some universal language. The issue at hand is whether the lack of a theory-neutral language prevents comparison of rival theories. Brown (1995) argues rather convincingly that in the absence of a universal language, theories can still be compared as long as proponents of competing theories agree on a body of relevant evidence and then use that evidence to compare the relative success (agreement between predictions and evidence) of the two theories. Even if each proponent does not understand anything about how their rival generated predictions, they should still be capable of discerning whether the predictions of their rival's theory conform with the evidence to a greater or lesser extent than the predictions of their own theory. In this case, neither scientist understands the language of the other. All that is required is some shared ability to communicate with one another (e.g. pointing, recognition of symbols). Obviously, comparison will be easier if one investigator understands the theoretical perspective of the other or if they share some common language independent of their theoretical commitments. Scientific examples of each of these situations indicate that this form of theory dependence also does not inherently compromise objectivity (Brown 1995).

In sum, only the first two forms of theory dependence seriously compromise scientific objectivity. Most scientists and many philosophers of science view TD1 and TD2 as inconsequential given the importance of anomalies in science. The remaining genuine forms of theory dependence do not by necessity lead to subjectivity. On the other hand, TD3, TD4, and TD6 individually and collectively undermine the epistemic and semantic foundationalism of observation reports inherent to logical empiricism. The conclusions are that inherent epistemic relativism appears untenable, foundational forms of empiricism cannot be sustained, and despite the symbiosis of theory and observation, science can still produce objective knowledge. This naturalized perspective does emphasize, however, that science is considerably more complex, internalized, and fallible than logical empiricism would lead us to believe.

OBSERVATION IN GEOMORPHOLOGY

Superficially, the strong emphasis placed on fieldwork in geomorphology accords with the foundational view of observation embodied in logical empiricism. Geomorphologists who emphasize the importance of field observations (e.g. Baker and Twidale 1991), like classical empiricists, tend to draw an alleged distinction between theoretical and non-theoretical components of the scientific process. The basic message is that the true nature of the 'observed' (i.e. perceived) landscape will be revealed to those who venture forth into the field unshackled by theoretical constraints. This view undoubtedly can be attributed in part to the long-standing positivist traditions in geology and geography, the parent disciplines of geomorphology. The value of inductive reasoning from (implicitly theory-neutral) observation to theory has long been cherished in geology (e.g. Oliver 1991). According to Kitts (1977, p. xiv) 'the geological tradition has fostered from the beginning a radically empirical and therefore, in the minds of many geologists, a notably untheoretical geology'. The accuracy of this comment is reflected in the recent debate about the increase in deductive theoretical modeling in geology, which has caused some to yearn for a return to the golden age of Hutton or Gilbert, when the emphasis was on developing generalizations from field-based observational facts (Brown 1974; Baker and Twidale 1991). In geography, many geomorphologists subscribe to the logical empiricist view of science unwittingly because they have been trained in the methods of science by human geographers (e.g. Harvey 1969), who tend to view the physical sciences in a positivist vein (Rhoads and Thorn 1994). The long tradition of inductive empiricism in physical geography is still very much alive. A recent textbook emphasizes that the process of discovery in physical geography involves a pathway from observation of geographic phenomena to theoretical explanation (Bradshaw and Weaver 1993, pp. 27–29).

The persistence of an implicit undercurrent of radical empiricism within geomorphology is somewhat ironic, given that the discipline lacks any formal philosophical statements in support of this perspective. Perhaps the strongest indicator that geomorphologists have assigned a foundational status to observation is the paucity of serious debate about this aspect of scientific inquiry. Geomorphologists appear to have reached an implicit consensus about the adjudicatory primacy of observation, and thus do not regard this element of scientific inquiry as a topic in need of further elaboration. However, taken together, these comments suggest a dilemma. If geomorphologists have not engaged in serious discussion about observation, but have implicitly deemed it central to geomorphic inquiry, on what grounds is this judgment based? As noted by Schumm (1991, p. 2) many earth scientists seem content 'to go about their scientific endeavors without giving much thought to the manner in which they proceed'. Glib comments such as 'study nature, not books' have a certain intuitive appeal among field scientists, but in the end merely discourage philosophical introspection and in the process raise suspicion that as a group geomorphologists fear what such introspection might reveal.

Because geomorphologists have been reluctant to engage in serious philosophical examination of their scientific activities, it is not possible to turn to a substantial body of previous commentary by geomorphologists-cum-philosophers. Instead, insights must be gleaned from a small body of formal commentary about the process of geomorphic inquiry, from analysis of case studies, and from general deliberation on the character of the discipline. Certainly, a comprehensive examination of observation in geomorphology would require a depth of treatment beyond the scope of a single chapter. The following discussion seeks merely to determine whether particular philosophical perspectives inform the role of observation in geomorphology *in at least some instances*. The choice of specific topics included in the discussion is admittedly subjective, but includes historical as well as contemporary examples. A brief survey of the views of G.K. Gilbert and W.M. Davis on observation examines the character and role of observation in the research of these two founding fathers. This discussion provides a basis for evaluating the role of observation within contemporary geomorphology.

Observation in the Science of G.K. Gilbert

Most of G.K. Gilbert's formal commentary about scientific methodology is contained in two papers: Gilbert (1886, 1896). His emphasis in these papers on observation as a first step in scientific inquiry has led some to associate his methodology with pure Baconian inductivism (e.g. Gilluly 1963); however, his own words reveal a more sophisticated insight into the relation between observation and theory:

> Scientific observation...endeavors to discriminate the phenomena observed from the observer's inference in regard to them, and to record the phenomena pure and simple. I say 'endeavors' for in my judgement he does not ordinarily succeed. His failure is primarily due to subjective conditions; perception and inference are so intimately associated that a body of

inferences has become incorporated in the constitution of the mind. And the record of an untainted fact is obstructed not only directly by the constitution of the mind, but indirectly through the constitution of language, the creature and imitator of the mind. But while the investigator does not succeed in his effort to obtain pure facts, his effort creates a tendency, and that tendency gives scientific observation and its record a distinctive character (Gilbert 1886, p. 285).

Gilbert clearly recognized that all observation is theory-laden to some extent, but he did not feel that this characteristic seriously compromised objectivity. His comment also suggests that he associated observation with perception; a conjunction that is not supported by his research. Gilbert (1886) went on to claim that observation is not indiscriminate, but rather is guided by theoretical presuppositions about classification. These presuppositions sharpen 'the vision for the detection of matters that are unnoticed by the ordinary observer' (Gilbert 1886, p. 285).

Gilbert (1886) introduced the concept of multiple working hypotheses as a safeguard against extreme forms of theory-ladenness that threaten objectivity. The purported advantages of multiple working hypotheses are that:

1. An investigator is less likely to be biased by excessive attachment to a 'pet' theory (thus avoiding TD1 and TD2).
2. Consideration of more than one hypothesis helps the investigator to determine which observations are most relevant for discriminatory tests (TD5) (Table 2.1).

Not only has the method of multiple working hypotheses enjoyed sustained support in geomorphology (Chamberlin 1890, 1897; Johnson 1933; Haines-Young and Petch 1983), it has been embraced within science at large under the rubric 'strong inference' (Platt 1964). Recently, the method has been the target of minor criticism within geology (Johnson 1990) and has been called into question more generally on rational and empirical grounds within the philosophy of science (McDonald 1992). In any case, geomorphologists have accepted this notion uncritically because of its intuitive appeal and have not rigorously evaluated its validity in actual practice. At a superficial level the method appears to foster healthy scientific skepticism; however, without rigorous empirical evaluation its true efficacy cannot be determined.

Background knowledge played an important role in shaping Gilbert's scientific methodology. His approach to hypothesis formulation and observation was greatly influenced by combined training in Newtonian mechanics and geology (Pyne 1980, pp. 95–103; Sack 1991). He saw the earth as a natural machine, operating according to mechanistic principles in which equilibrium played a central role. A hallmark of Gilbert's work is that he never presumed knowledge of cause, except in the sense that underlying causal mechanisms consisted of natural processes or events that obeyed the laws of physics and chemistry. Instead, he formulated various hypotheses about causal events and then proceeded to determine which evidence is relevant to, yet independent from, these hypotheses. Both of these steps relied heavily on background knowledge (Gilbert 1886, 1896). Gilbert's engineering mindset is also reflected in the value he placed on precise, reliable data, especially those produced by instruments. Although in formal discussion of scientific method Gilbert equated observation with perception (e.g. Gilbert 1886), the 'observations' he collected in actual practice are synonymous with contemporary notions of data and evidence.

Gilbert's analysis of Coon Butte (now known as Meteor Crater), which Gilbert (1896) himself presented as an archetype of his research methodology, illustrates how background knowledge played an important role in his research. In this investigation Gilbert drew upon his background knowledge in the earth sciences and physics to formulate two competing hypotheses about the possible origin of this feature, namely by volcanic explosion or by meteorite impact. Based on deductive reasoning and his familiarity with certain types of field techniques, Gilbert (1896, p. 5) then argued that 'any observation which would determine the presence or absence of a buried star might therefore serve as a crucial test'. Gilbert tested only the meteorite hypothesis explicitly. Although he did not directly test the volcanic origin hypothesis, he did present some qualitative arguments in support of it.

Gilbert noted that exploratory drilling beneath the crater floor would produce the best evidence for or against the meteorite impact hypothesis; however, because this option was too costly, he proposed two other options. His first test was drawn from the physical principle of conservation of mass; he deduced that if the crater was caused by a meteorite impact, part of the crater would be occupied by material from the meteorite, and the volume of the material surrounding the crater would be greater than the volume of the crater. Calculations based on Gilbert's topographic survey of the crater and its surrounding rim revealed that the two volumes are equivalent, which constituted disconfirmatory evidence for the hypothesis.

The second test was based on the assumption that the meteorite consisted primarily of iron, an assumption suggested by the scattered masses of iron around the crater, and on background knowledge concerning the magnetic properties of iron. The data in this case were drawn from a magnetic survey and consisted of 'observation . . . of the three magnetic elements: the horizontal component of the direction, or the compass bearing; the vertical component of direction, or the inclination of the dip needle; and the intensity of the magnetic force' (Gilbert 1896, pp. 8–9). Gilbert predicted that the presence of an iron mass would produce systematic deviations in the compass readings; however, he (1896, p. 6) found that within the limits of measurement error the magnetism was constant in direction and intensity around the crater. Gilbert did not automatically view this result as evidence against the impact hypothesis; instead, he conducted a series of experiments at the Navy Yard in Washington, DC to determine the sensitivity of the compass to the presence of iron bodies of different sizes and different depths of burial. Gilbert's effort to determine the sensitivity of the compass to an iron body shows that he was deeply concerned about the reliability of the data as evidence for or against the impact hypothesis (e.g. Woodward 1989). Based on the results of the calibration experiments Gilbert (1896, p. 9) concluded that 'the theory of a great iron meteor is negatived by the magnetic results, unless we may suppose that the meteor was quite small as compared to the diameter of the crater, or that it penetrated to a very great depth'. Clearly, here is a case supporting Brown's (1995) contention that the failure to causally interact with something in the real world (in this case the failure of the compass needle to interact causally with a large iron meteorite) can be important evidence in an empirical test.

Although Gilbert was forced to conclude that the available evidence did not support the meteorite impact hypothesis, his critical attitude did not lead him to presume that the tests were conclusive. Gilbert proposed several possible scenarios that could account for this evidence even if the crater was produced by a meteorite impact. Although he cautioned that scientific results are 'ever subject to the limitations imposed by imperfect observation'

(Gilbert 1896, p. 12), it was not his observations that were to blame; the same results would be obtained today were the tests to be repeated. Rather, it was limitations in the background knowledge available at the time that led Gilbert to a conclusion (albeit tentative) about the origin of meteor crater that today is viewed as incorrect.

The Coon Butte example shows how Gilbert used the term 'observation' to refer to data. He did rely on visual information to infer that the depression known as Coon Butte was a crater; however, the evidence used in hypothesis testing was based not on visual information but on precise data collected with instruments. Whereas the functioning of the instruments relies on theories governing their operation, knowledge of these theories is not necessary to use the instruments. Moreover, Gilbert did not use raw data produced by these instruments directly in hypothesis testing, but instead employed theories of method and analysis to transform the raw data into evidence. He used surveying theory to convert instrument readings into volumetric values that could be used to test the volumetric hypothesis, which itself depended on conservation principles. In a similar fashion, the individual compass readings had to be arranged in a measurement scheme and the absence or presence of a spatial pattern be determined before the readings constituted evidence.

The Coon Butte example also shows that Gilbert was using data or evidence in a search for a causal phenomenon. This phenomenon could not be observed directly because it occurred in the geologic past; moreover, the infrequency of the phenomenon prevented Gilbert from 'observing' directly its effects on the earth's surface. In this geohistorical investigation, Gilbert was employing a combination of abductive reasoning and deductive inference. To develop his hypotheses he inferred possible cause (initial condition) from effect (antecedent condition) and background knowledge about geological processes (meteorite impacts, volcanic eruptions). The evidence for the effect was the characteristic shape of the depression, which Gilbert's background knowledge permitted him to recognize as a crater and consequently the product of only a limited number of possible geologic processes. The hypothesis under test (i.e. the origin of the crater) was a singular causal statement, rather than a generalization or law. This pattern of inference from effect to potential cause based on background knowledge of cause–effect relations is known as abductive inference; such reasoning is common in geohistorical investigations (Kitts 1963; von Englehardt and Zimmerman 1988), including those conducted in geomorphology (Rhoads and Thorn 1993). Abductive inference is greatly strengthened when it is accompanied by deductive reasoning in which the abductively inferred initial conditions are combined with physical principles to deduce other effects associated with the inferred cause (Kitts 1963; Rhoads and Thorn 1993). Gilbert followed this strategy when he combined the meteorite impact hypothesis with laws governing magnetism and mass conservation to deduce hypotheses, the testing of which relied on evidence independent of that involved in the initial abductive inference.

Although Gilbert's methods of hypothesis formulation, data collection, and data analysis depended on background theories, the type of data that resulted from the Coon Butte investigation was not strongly influenced by the hypothesis under test. Perhaps the ultimate measure of objectivity is whether one can explicitly state the conditions under which the hypothesis could be disconfirmed by the data. Gilbert clearly stated these conditions both for the survey and the compass data. Gilbert also elaborated in detail the relevance of the data to the hypothesis; again background theories (conservation principles; magnetic theory) played a key role.

In sum, Gilbert's example shows that his investigative style generally avoided the more serious forms of theory-ladenness (TD1, TD2) (Table 2.1). Gilbert's mechanical view of the world certainly influenced the manner in which he conducted his research, but it did not predetermine what he would 'observe' about the world, nor did it cause him to ignore evidence that contradicted what was most likely his favored hypothesis prior to the investigation, i.e. the meteorite impact hypothesis (El-Baz 1980). Although he used auxiliary theories to guide hypothesis testing, data collection, and analysis (TD3, TD4a), and reported results in the language of the hypothesis being tested (TD6), neither of these factors seriously compromised the objectivity of testing. Gilbert could have protected the meteorite impact theory by challenging auxiliary theories (TD4b); and in fact identified how such theories could be challenged, but in the end chose not to do so. The fact that Gilbert could clearly state the conditions under which the data would disconfirm the meteorite impact hypothesis suggests that the data do not presuppose this hypothesis in any viciously circular way (TD3). The method of multiple working hypotheses seems to have helped prevent vicious circularity by providing valid alternatives for explaining the existence of the crater; TD3 may be more of a concern where only a single theory is considered plausible, as in the relativity example cited by Brown (1993).

Perhaps the most important lesson to be learned from Gilbert's investigation of Coon Butte is that objectivity alone does not guarantee truth. Instead, improvements in the accuracy of hypothesis testing in the earth sciences occur as objectivity is combined with development and growth of background knowledge from other areas of science. Although objective testing did not prevent Gilbert from reaching an erroneous conclusion, the fault appears to lie, not in Gilbert's investigative style, but in the inadequate, relevant background knowledge in the late nineteenth century. Had Gilbert been armed with current theories of shock metamorphism and impact energy (e.g. Chao et al. 1960; Shoemaker 1960), he probably would have reached the conclusion that Coon Butte is the result of a meteorite impact (von Engelhardt and Zimmermann 1988, pp. 278–282, 285).

Observation in the Science of W.M. Davis

Whereas Gilbert's view was imbued with the principles of mechanics, Davis's perspective on geomorphology was influenced heavily by evolutionary theory (Stoddart 1966). During the late nineteenth century, the concept of evolution had a distinct nonbiological connotation wherein antecedent forms were considered as stages in an orderly sequence leading to subsequent forms (Chorley et al. 1973). In such a view, time itself is a process, not merely a framework within which events occur. This characterization suggests that the general concept of evolution in the late 1800s was more consistent with contemporary notions of ecological succession than it was with the modern biological theory of evolution. The concept of orderly sequential change obviously is one Davis drew upon to formulate his theory of the cycle of erosion, which he originally referred to as 'the cycle of life' (Davis 1885). The cycle of erosion, with its progressive development of landscapes through a series of stages, is clearly an example of a developmental theory, or one that specifies an irreversible process 'characterized by a certain direction or tendency so that beginning and end stages as well as intermediary stages may be differentiated' (von Engelhardt and Zimmermann 1988, pp. 322–323). Developmental theories often are substituted for causal theories (i.e. those that relate processes or mechanisms specifically to effects) for systems with complex interactions between structure and process (e.g. geomorphic systems) (von Engelhardt and Zimmermann 1988, p. 177).

Previous criticisms (e.g. Strahler 1950; Judson 1960; Chorley 1962, 1965; Flemal 1971; Higgins 1975; Beckinsale 1976) of the cycle have focused on the sociological factors or conceptual shortcomings associated with its downfall. Here discussion examines the manner in which theory and observation are interrelated within the Davisian scheme of geomorphic inquiry and how this interrelationship may have contributed to its eventual demise.

Davis's formal comments about the relationship between theory and observation were marked by inconsistency, suggesting that he saw this relationship as complex. At times, he drew a sharp contrast between theory and observation, at others, he discussed these concepts as intertwined. Davis claimed that 'the scheme of the cycle . . . is by intention a scheme of the imagination and not a matter for observation, yet it should be accompanied, tested, and corrected by a collection of actual examples that match as many of its elements as possible' (Davis 1905, p. 152). Theory testing to Davis did not represent an attempt to compare independent evidence with precisely specified evidential conditions; instead it involved an effort to *match* elements of the cycle with selective pieces of morphological information from actual landscapes, which were then subsequently construed as evidence for these theoretical elements (Davis 1915).

Although Davis paid lip-service to the importance of proposing various hypotheses to explain observed facts (see Davis 1922), his dogmatic allegiance to the cycle of erosion conformed more closely with Chamberlin's (1897) notion of the method of the 'ruling theory', than with the method of multiple working hypotheses. Davis was willing to entertain various hypotheses about the development of a particular landscape only to the extent that they conformed with sacrosanct tenets of the cycle (e.g. rapid uplift followed by progressive, sequential change in landforms). Thus, his method of landscape analysis became a rigid scheme in which 'each element of the landscape is treated as the surface of a structural mass which has been carried forward from an initial form to some specified stage of development in the cycle of erosion . . . ' (Davis 1911, p. 30). In other words, the cycle served as a regulative principle for developing geohistorical theories for specific landscapes in much the same sense that mechanics served as a regulative principle for Gilbert. The important issue is that whereas by the late nineteenth century physical principles of mechanics had been extensively tested and widely applied to natural systems, such was not the case for the sequential, directed evolutionary concepts embodied in the cycle of erosion. More recent analysis by Chorley (1965) reveals that in fact the cycle can be reduced to a physical interpretation, and that this interpretation is inconsistent with contemporary knowledge of the physical mechanisms that govern the geomorphic processes responsible for landform change. Thus, not only was the background knowledge supporting the cyclic conception much weaker than that supporting Gilbert's mechanical conception, but the cyclic conception also could not readily accommodate the infusion of new background information into the discipline from other areas of science.

Bishop (1980) has argued that the cycle of erosion is inherently nonfalsifiable because of the irrefutability of the central concept of stage. This assessment is a bit misleading. Certainly, one can conceive of types of evidence, such as experimental studies of drainage basin evolution (e.g. Schumm et al. 1987), contemporary information on plate tectonics, rates of orogenic uplift, and isostatic adjustment (e.g. Summerfield 1991), and absolute dating of preserved paleolandscapes (e.g. Young and McDougall 1993), that, had they been available in the late nineteenth century, may have caused Davis and his followers to modify or even abandon completely the basic tenets of the cycle. Because the cycle is an evolutionary theory, the best type of evidence for testing it is temporal data. The problem with the cycle is not that it is inherently irrefutable, but that it was assessed with a type of evidence that in essence made it irrefutable (i.e. information on existing morphology). Moreover, important types of temporal evidence required to test it rigorously often could not be readily produced due to the unavailability of appropriate techniques for generating these data. It is also true that when applied to the interpretation of a *specific* landscape, the evidence required to refute a cycle-based explanation may for all practical purposes be impossible to obtain, no matter what information-gathering procedures are available.

Another reason the cycle of erosion was difficult to refute is that Davis lacked a healthy dose of skepticism concerning the validity of his ideas. In contrast to Gilbert, Davis never specified exactly the conditions or type of evidence required to refute the basic tenets of the cycle. Instead, when faced with observational evidence that was seemingly inconsistent with the theory, Davis introduced 'various kinds and degrees of interruption at various stages in a cycle' with the remarkable result that 'the variety of possible combinations becomes so great that there is no difficulty whatever in matching the variety of nature' (Davis 1905, p. 156). As Bishop (1980) has correctly noted, such modifications were often *ad hoc* in the sense that they made the cyclic theory less, rather than more, falsifiable. This strategy was adopted because of Davis's bold assurance in the fundamental correctness of the basic tenets of his theory. On one occasion he even proclaimed that 'explanatory concepts, deduced from general principles, are much more intimately and reasonably knowable than empirical concepts or even than facts of observation usually are, and in this quality of being intimately and reasonably knowable lies their highest value' (Davis 1912, p. 106). Davis recognized the limited power of our senses in observation, but unlike most scientists, he felt that theorizing, and not the development of sophisticated nonsensory observational techniques, was the best way to overcome this limitation:

> If we wish really to understand the natural world, surely those of its phenomena which are not immediately detectable by our limited senses must be detected in some way or other; and the way usually employed is – theorizing (Davis 1922, p. 198).

Davis clearly associated observation with visual perception, rather than with the collection of field data with instruments (Beckinsale 1976). In this particular sense, he was a consummate empiricist. A global traveler, Davis based his theories mainly on personal examination of many landscapes around the world. He also relied heavily on maps, which he used primarily to supplement visual information on landscape form, rather than as a source of quantitative data (Chorley et al. 1973, p. 232). Whether he was in the field or inspecting maps, Davis's genetic perspective strongly influenced his way of observing the landscape. It is this aspect of his methodology that is problematic.

The genetic-historical approach of Davis involved morphogenetic classification, i.e. assigning landforms to categories based on their morphology and genesis. Davis certainly was not the first to use morphogenetic terms to describe observations; however, by embedding such terms within an overarching evolutionary model, he developed this approach ('explanatory description') to an extreme. Although Davis (1913, pp. 686–687) claimed to recognize the 'explicit distinction between . . . inferred conclusions . . . and the observed facts on which the conclusions are based', in many cases this distinction became blurred when morphogenetic terms were used in conjunction with cyclic principles to interpret the landscape.

Davis's method of analysis in specific instances often was complex (Chorley et al. 1973, p. 212); however, some general aspects can be identified. Davis commonly stressed the importance of confronting theory with observed facts (e.g. Davis 1909, pp. 342–343). To Davis, 'observed facts' were uninterpreted morphologic features, such as accordant ridge tops. Yet assignation of a feature to a category, such as 'accordant ridge tops', is based in part on inference, an inference that must be supported by evidence independent of the theory used to explain the development of the landscape of which the ridge tops are a part. Davis, on the other hand, viewed the first stage of inference as occurring at a higher level when the observed evidence is used to assign the feature to a morphogenetic category, e.g. when accordant ridge tops are recognized as the remnants of a former peneplain. This inference from 'morphologic fact' to 'landform associated with a particular stage of the cycle of erosion' was an abductive one within which the cycle itself played the role of law. Davis readily acknowledged that morphogenetic terms are inherently theory-laden in that they implicitly involve inference about the genesis of the feature being observed. Nevertheless, he advocated the use of morphogenetic terms on the grounds that their theory-ladenness promoted deeper insight into the nature of landscapes than mere empirical description (Davis 1912).

The first stage of inference was followed by a second, more complex stage of reasoning, the purpose of which was to adduce the history of the landscape based on (1) the results of the morphogenetic classification of the first stage, (2) 'facts' concerning the stratigraphy and geologic structure of a region, and (3) the cyclic model. This stage also involved abductive inference, in which the inferred resulting states of affairs (e.g. peneplains, graded rivers, patterns of folds, stratigraphic sequences, etc.) along with the cyclic model, which once again served as inviolate law, were used to infer the controlling state of affairs (sequence of events that produced the current physiography). This two-stage chain of abductive reasoning is common in geohistorical investigations (von Engelhardt and Zimmermann 1988, p. 216).

Although Davis often presented his arguments in deductive form, the objective of his analysis was to determine initial (historical) conditions, not the effect of these conditions. Because he did not have direct knowledge of the initial conditions, his method cannot be properly characterized as deductive. However, Davis had great confidence in his cyclic theory, and in many instances he treated it as certain knowledge. Thus, in contrast to Gilbert, Davis proceeded as if he knew the cause and character of landscape development, and simply had to deduce the effects. This assumption resulted in a circular pattern of reasoning in which Davis deduced effects (resultant landforms) from uncertain knowledge of initial conditions and untested principles of landscape evolution, and then used the perceived match between deduced and observed forms as evidence for the correctness of his postulates (initial conditions combined with cyclic principles) – a problem alluded to by Flemal (1971) and Bishop (1980). However, because cause was inferred, and effect does not guarantee cause, this pattern of reasoning was specious.

The hierarchical chain of abductive reasoning involved in Davisian explanation was only as solid as the initial evidence upon which it was based. This evidence usually consisted of inexact visual information on morphology, which had to be processed through the human mind and therefore had a high potential for subjective bias. Davis was aware of the influence of theoretical preconceptions on visual perception, but apparently felt this influence was an advantage, rather than a problem:

> The more complete the mental scheme by which an ideal system of topography forms is rationally explained, the more clearly can the physical eye perceive the actual features of the land surface (Davis 1894, p. 68).

Davis rarely based his inferences about landscape genesis on any evidence other than subjective, theory-laden visual perceptions of landscape morphology. Thus, the critical independent evidential support so critical to geohistorical inference usually was absent from Davisian analysis. Objectivity suffered as Davis and his followers began to 'see' the landscape in cyclic terms.

The debate about peneplains between Davis and Ralph S. Tarr illustrates rather well the circular nature of Davisian analysis. Tarr (1898) began this debate by expressing concern about the weak evidential basis for identifying peneplains. He clearly recognized that the identification of these landforms was highly inferential and required strong supporting evidence. Tarr (1898, p. 361) felt compelled to argue against the concept of peneplanation, but was frustrated by limited data that made it 'difficult to find positive evidence against this explanation'. Most of his arguments were not evidential in nature, but were based on reasoning that mixed accepted background knowledge of geologic processes, which at the time was inadequate to decide the issue, with various speculative assumptions. Unconvinced, Davis (1899), the master of science by debate, responded in kind to Tarr with equal or greater effectiveness.

Tarr (1898) did, however, present some quantitative evidence to support his argument that the ridge crests in New Jersey and southern New England do not represent a relict peneplain. Because Tarr (1898, p. 356) believed that evidence based on 'appearance to the eye may be most deceptive', he used topographic maps to plot profiles across the regions of supposed accordant summits. These profiles revealed 'a very distinct lack of uniformity of the upland crests, even if all the lower hills are eliminated' (Tarr 1898, p. 356). Local differences in elevation exceeded 500 feet (150 meters) and regional differences were in some cases greater than 1100 feet (335 meters). Tarr (1898) noted that advocates of the peneplanation theory accounted for variations in elevation by proposing that they are the result of regional tilting and the preservation of monadnocks following uplift. He clearly recognized that without independent evidence to support these inferences, the explanation becomes theory-determined:

> I am unable to find that there is any other proof that this interpretation is correct than that which comes from the necessity of such an explanation, made necessary by first accepting the existence of the peneplain (Tarr 1898, p. 358).

Tarr was concerned that a single piece of highly dubious information (observation of pseudo-accordant summits with a regional trend) was being used to support multiple, interrelated inferences (former existence of a peneplain covered with monadnocks, uplift and tilting of the peneplain, preservation of monadnocks following uplift and erosion). Tarr (1898) also used topographic maps to determine the areal extent of summits with elevations within 300 feet (91 meters) of one another in relation to the total area of all summits. This analysis revealed that summits within 300 feet (91 meters) of one another occupy only 10–25% of the total summit area. In other words, 75–90% of the peneplain was being 'filled in' on the basis of reasonable accordance among 10–25% of the summit crests. Based on these results, Tarr (1898, p. 359) concluded that 'there seems to be very little real evidence upon which to construct the ancient peneplain, and I am led to raise the query whether, even granting in its entirety the evidence claimed, we would be warranted in drawing so broad a conclusion from so small a basis of fact'.

Davis's response to Tarr was nothing short of astounding, revealing just how strongly the cyclic theory infected his science. With regard to the topographic evidence on discordance among the summits Davis remarked:

> Considerable as the inequalities of altitude are, frequent study of the maps and repeated views of the uplands from various hill tops impress me much more with the relative accordance of their altitudes than with their diversity. I cannot admit that the appearance of accordance from hill top to hill top is an optical deception. There is an important matter of fact behind the appearance (Davis 1899, p. 210).

To Davis, the eye was more accurate than the surveying instrument on which the map is based! The fragmentary nature of the so-called peneplain did not strike Davis (1899, p. 212) as a 'serious or novel difficulty' because 'geologists are often compelled to work on fragmentary evidence'. Extending a peneplain based on a few fragments was no worse than extending a rock formation based on scattered outcrops. As a rejoinder to Tarr's concern about introducing tilting and monadnocks to explain variations in summit elevations, Davis simply retreated to the cycle to defend the cycle:

> the lack of uniformity of the uplands – a fact perfectly familiar to those who accept the peneplain idea – is partly the result of tilting ... and that for the rest the unevenness of the uplands of to-day is a natural result of imperfect peneplanation followed by submature dissection (Davis 1899, p. 210).

Davis (1899, p. 212) fully agreed with Tarr's assessment that one had to accept the existence of a peneplain with monadnocks on faith and then assume it was tilted after it was uplifted: 'I have repeatedly insisted that it was only by recognizing the existence of a peneplain that uplift or deformation could be determined in certain cases. ...' Arguing that it would be 'as extraordinary to find no slanting peneplains as to find no inclined strata', Davis (1899, p. 213) stated that 'it does not seem warranted to conclude that the peneplain theory is invalidated because certain peneplains are now uplifted on a slant ...'. In these statements, Davis merely reaffirms Tarr's (1898) concern that cyclic explanations are excessively theory-laden.

With regard to the existence of monadnocks, Tarr (1898) challenged Davis to produce independent evidence showing that the most discordant summits are underlain by the most resistant rocks. Davis (1899, p. 219) admitted that he 'has given no particular attention to monadnock rocks; indeed, it has generally seemed to me reasonable to infer their greater resistance on account of their form'. He also stated it is equally encumbent upon Tarr to

produce independent evidence for or against the existence of monadnocks. Davis concluded this discussion by insisting that he is not an advocate of the theory of peneplanation, but in disagreeing with Tarr is merely following the guidance of the best evidence
he can find:

> ... I cannot say too emphatically that the peneplain idea shall find no 'defense' from me. Let
> us all set forth the *pros* and *cons* to the best of our ability, and then the peneplain idea must
> look out for itself, and stand or fall according to its value (Davis 1899, p. 221).

Fortunately for Davis, the fall did not come until after he had died.

The exchange between Davis and Tarr suggests that at least in some instances Davis's
approach to observation encompassed all six forms of theory-ladenness (Table 2.1),
including the extreme forms, TD1 and TD2, that seriously compromise scientific objectivity. Davis's strong adherence to the cycle infected his way of perceiving the landscape;
he 'recognized' relict peneplains as 'an important matter of fact' based on visual perceptions of accordant ridge tops that were not in fact accordant (TD1). He also ignored, or
at least dismissed as unimportant, precise data which showed that these ridge tops were not
accordant (TD2). He was able to reject these data as evidence against his hypothesis
because cyclic explanation did not specify precisely the conditions required to refute the
underlying assumptions upon which it was based. This extreme theory-ladenness resulted
in a substantial degree of incommensurability between Tarr and Davis. Throughout his
response to Tarr, Davis repeatedly retreated to the cycle of erosion to defend peneplanation
and the existence of relict peneplains. Thus, he had difficulty 'interpreting' the evidential
status of information outside of the context of cyclic explanation. This difficulty also led to
a certain amount of exasperation on Davis's part concerning Tarr's inability to 'see' what
Davis saw, an exasperation that is expressed succinctly in the following statement about
whether or not the uplands in Connecticut and Massachusetts are accordant:

> Hence it must be agreed that with the same facts before us, both outdoors and on maps, our
> descriptions and interpretations of them do not correspond; one of us being impressed with the
> diversity of upland altitudes, and the other with their accordance (Davis 1899, p. 211).

Later in his career, this inability to extricate himself from cyclic thought helped to fuel the
famous debate over landscape development between Davis and Walter Penck (Chorley et
al. 1973).

Finally, the constructivist element of Davisian geomorphology should not be under-
estimated. In his reply to Tarr, Davis recounts the following story about a trip to the upper
Mississippi River basin to visit a friend:

> During an excursion to that part of the country, I pointed out the very even skyline of a
> dissected peneplain. My friend dissented, thinking no such special explanation necessary;
> ordinary denudation would suffice, he thought, to produce the observed forms, without
> specification of control by different baselevels. A year later, on meeting the same friend, our
> talk happened to turn on peneplains, and he said: "I should like to show you an excellent
> example of that sort of thing," proceeding to describe the very region we had seen together.
> "How" I asked, "did you come upon that explanation?" "I cannot say precisely how," he
> replied; "it is nothing new" (Davis 1899, pp. 222–223).

Although Davis (1899, p. 223) attributed his friend's change in attitude to 'the unconscience encouragement given to the idea of peneplanation in a quiescent environment', it seems equally, if not more likely that the friend had fallen victim to the influence of a powerful personality. Davis certainly had a deep, profound influence on the inchoate discipline of geomorphology during the first half of the twentieth century (Chorley et al. 1973; Pyne 1980, pp. 254–261; Tinkler 1985, pp. 152–153). Davis's powerful personality, along with the simple elegance of his model, created a situation in which adherents of the cycle of erosion had difficulty viewing the landscape in terms other than those proposed by Davis (Hart 1986, p. 21). In the end, extreme theory-ladenness probably had as much to do with the downfall of Davisian geomorphology as other factors.

Observation in Contemporary Geomorphology

The assessment of observation in Gilbert's and Davis's science provides a context for evaluating the character and epistemic function of observation in contemporary geomorphology. Any attempt to identify the inauguration of contemporary geomorphology necessarily will be subjective. However, one possible point of departure is the benchmark paper by Strahler (1952), in which he announced the need for a process-oriented approach to geomorphic research grounded in the principles of physics, chemistry, and biology.

The 'return to Gilbert'

A careful reading of Strahler (1952) indicates that the primary focus of his paper is the call for a shift in the underlying basis for explanatory theory in geomorphology. Geomorphologists were coming to the conclusion that the Davisian approach was limiting in the sense that it did not readily accommodate background knowledge from other areas of science, particularly physics and chemistry, which was perceived as relevant to geomorphic phenomena. By relating geomorphic phenomena to background knowledge from other areas of science, geomorphologists were infusing the discipline with a large body of well-established theoretical presuppositions that could be used to guide the formulation of geomorphic theories. This background knowledge served several purposes. First, it enhanced the scope of laws and principles used in geohistorical inferences, while at the same time subsuming the Davisian model and highlighting the limitations of this model (e.g. Chorley 1965). Second, it provided a conceptual basis for field and laboratory investigations that seek to disclose the physical, chemical, and biological mechanisms constituting geomorphic processes. Third, in stark contrast to the Davisian scheme, it provided a flexible framework for importing new background knowledge into the discipline. Over the past 45 years this framework has easily accommodated the infusion of a large variety of scientific concepts into geomorphology, including those of systems (Chorley 1962), thermodynamics (Leopold and Langbein 1962), mathematical physics (Scheidegger 1961; Middleton and Wilcock 1994), equilibrium (Thorn and Welford 1994), catastrophe theory (Graf 1979), and nonlinear dynamics (Phillips and Renwick 1992) to name a few. Fourth, because physical and chemical principles have been and continue to be viewed by geomorphologists as 'established', they also play an evidential role in rational decisions among competing geomorphic hypotheses. Geomorphologists, like geologists (e.g. Kitts 1974), do not use physical and chemical laws merely as inference tickets, but proceed on the assumption that these laws constitute inviolate 'truths' about nature; therefore, all geomorphic theories must not only be consistent with these laws, but should embody them. Judgments about the explanatory adequacy of competing theories commonly reflect this underlying sentiment. Geomorphic theories that explicitly and formally include physical or chemical laws are often seen as having greater 'explanatory adequacy' than alternative theories that have greater empirical adequacy, but are only loosely or informally tied to underlying physical/chemical principles.

Strahler (1952) had surprisingly little to say about the implications of his proposed shift in the conceptual basis for geomorphology either for observation or for the manner in which geomorphologists conduct scientific inquiry. However, the discussion near the end of the article, which is littered with terms such as 'data', 'variables', and 'quantitative determinations of landform characteristics and causative factors', provides a hint of what this shift was to imply for observation in geomorphology. Ironically, despite Strahler's (1952) emphasis on implications for explanatory theory, the shift in background knowledge had its greatest immediate impact on the character of geomorphic observation. Initial studies under the new process-based approach focused largely on data-driven analyses (e.g. Leopold and Maddock 1953; Strahler 1957; Melton 1958; Chorley 1966), leading to concern and criticism that the discipline was sliding into raw empiricism (Mackin 1963). These 'empirical studies', however, did not occur in a theoretical vacuum but were performed within the context of statistical theory, systems concepts, and incompletely specified physical theory. It was not until the appearance of Scheidegger's (1961) landmark treatise that physical theory was employed in a formal, mathematical fashion to develop explanatory geomorphic theories in the manner envisioned by Strahler (1952). Since that time there has been steady growth in the infusion of physical, chemical, and biological theory into geomorphology, both at the explanatory theoretical level and the exploratory observational level.

The emergence and rise to dominance of the process-oriented approach to geomorphology over the last 40 years have been referred to within the geomorphic community as a 'return to Gilbert'. This characterization is appropriate not only in the sense that systems concepts and physical principles once again provide the basis for explanatory inference in geomorphology, but also in the sense that geomorphologists now 'observe' geomorphic systems and use their observations in a manner consistent with Gilbert's example. This contemporary approach to observation in geomorphology also is generally consistent with the naturalized view of observation in the philosophy of science.

The character of geomorphic observations

One need only refer to a volume such as *Geomorphological Techniques* (Goudie 1990) to recognize that geomorphologists now largely equate the term 'observation' with quantitative data, rather than visual perception. This characterization is as true for geohistorical investigations as it is for process-oriented studies. The term 'technique' emphasizes the procedural, rule-based nature of contemporary observational methods in geomorphology. These methods range from relatively simple procedures (e.g. taping of distances) to highly sophisticated analytical techniques (e.g. atomic absorption spectrometry, remote sensing of planetary surfaces). The sophistication of observational techniques clearly has increased over the last 45 years and all indications suggest that this trend will only accelerate in the foreseeable future. By embracing a conceptual framework based on established background knowledge in other areas of science, geomorphologists have positioned themselves to take advantage of a smorgasbord of technological developments. Like many other sciences, geomorphology, is being fueled and invigorated by these technological developments.

In accord with the naturalized view of observation, the sophistication of a specific geomorphic technique clearly is not viewed as a deterrent to its reliability. In fact just the opposite is true. Geomorphologists readily adopt sophisticated observational techniques because they view data generated by these techniques as far more reliable than impressions resulting from visual perception. What geomorphologists often fail explicitly to recognize, is that because all observational techniques depend on background theories, the data resulting from them are theory-laden to some extent. Even when data are derived solely from visual perception, such as when sedimentary structures are identified in a depositional sequence, background theory plays an essential role in data production. It is the background theory (in this case from sedimentology) that allows the investigator to convert raw sensory stimuli into meaningful data. Explicit recognition of the theory-ladenness of observation seems to have escaped many geomorphologists, even though this issue has been discussed in relation to geology by Kitts (1974) and to physical geography by Haines-Young and Petch (1986, pp. 37–40).

Observation and objectivity in geomorphology

Given that geomorphic data are theory-laden, a crucial concern is whether this theory-ladenness seriously undermines the objectivity of empirical testing. Through their temporary detour into the world of Davisian science, and their subsequent return to Gilbert, geomorphologists seem to have learned implicitly that at least five factors influence the objectivity, reliability, and certainty of empirical testing:

1. The degree of independence between exploratory and explanatory theories;
2. The availability of independent information on cause and effect (as embodied in scientific arguments);
3. The degree to which information on cause and effect has been modified, altered, or confounded by environmental factors;
4. The fixedness and precision of observational techniques;
5. The fixedness and precision of theoretical arguments with which the empirical evidence is compared.

Although these five components are isolated for the purpose of discussion, in actuality they are often interrelated in specific instances.

The discussion in the first part of this chapter has shown that independence between exploratory theories and explanatory theories contributes to objectivity by helping to ensure that data constitute unbiased evidence, thereby preserving the adjudicatory power of data. In many cases in geomorphology, the theory (theories) underlying an observational technique is (are) independent to some degree from the explanatory theory under test:

independent in the sense that the outcome of data production will not, by necessity, be determined by the explanatory theory alone (i.e. if the explanatory theory is incorrect, the data can be inconsistent with theoretical expectations). For example, a fluvial geo-morphologist who uses an electromagnetic flow sensor to evaluate a hydrodynamic theory of stream channel dynamics will draw upon the explanatory hydrodynamic theory to define the variables to measure and to design the sampling scheme required to produce appropriate spatial and temporal coverage within the stream. It is also true that the operation of electromagnetic current meter depends in part on hydrodynamic theory, a dependency that is evident in calibration experiments designed to assess the sensitivity of the sensor to various types of flow conditions (e.g. Lane et al. 1993). In other words, the data resulting from the measurement program depend at least to some extent on the explanatory theory. Although this dependency partly influences the type of data obtained from the overall measurement program, it does not wholly predetermine individual data values because the operation of the sensor also relies on electromagnetic theory, which is largely independent of hydrodynamic theory. Of course, the role of independence in objectivity is predicated on the sensor's capability to interact causally with the phenom-enon of interest in the real world.

Independence between exploratory and explanatory theory is common in geomorphol-ogy because geomorphologists draw upon a background of established theoretical princi-ples from other areas of science to formulate their theories and to select observational techniques, rather than devising fundamentally new principles or developing observational techniques that involve underlying principles which are poorly understood (as sometimes occurs at the forefront of physics, chemistry, or biology). In most instances, the issue of independence can be decided a priori based on consideration of the types of theoretical knowledge underlying the explanatory theory and a potential observational technique. Such a priori decisions are possible because the issue of independence has largely been resolved by the parent discipline in which the theory underlying the technique originally emerged. For example, an adequately trained fluvial geomorphologist need not deliberate long and hard about testing a hydrodynamic model of channel change with survey data on channel form because the independence of surveying theory and hydrodynamic theory is apparent from his/her store of background knowledge. The lesson to be learned is that geomorphologists should become familiar with a broad spectrum of background knowledge and stay current on recent developments within other areas of science so they put them-selves in an optimal position to employ this knowledge to their full advantage.

Independence between exploratory and explanatory theory helps to ensure that data are reliable evidence; however, the degree of certainty and reliability assigned to empirical testing also hinges on the capacity of an investigation to generate independent evidence for cause and effect. The results of studies that produce multiple types of independent evi-dence both for cause and for effect are viewed as more reliable and certain than results that include only single pieces of evidence for cause and effect, that include evidence for effect only, or that are based on evidence that is known to be highly modified, altered, or otherwise confounded. Many contemporary process-oriented research programs are designed to yield (at least over a series of investigations) various types of independent evidence for cause and effect. For example, an investigation directed toward evaluating a theoretical model of meander bend dynamics may use independent measurements of velocity fields, water surface topography, and patterns of bedload transport to corroborate the existence of predicted hydrodynamic phenomena, and independent measurements of bed topography and grain-size patterns to document how the existence of the predicted phenomena produce certain expected effects (in this case changes in equilibrium channel-bed conditions). If strong agreement is found between model predictions and evidence, the results are accorded a high degree of reliability and certainty.

In geohistorical investigations, data on cause cannot be obtained directly because the cause occurred in the past. Thus, geohistorical inferences inherently have a weaker empirical basis and greater degree of uncertainty than inferences derived from process-oriented studies (Rhoads and Thorn 1993). This characteristic of geohistorical inferences in no way is the result of an inferior methodology or scientific approach; instead it reflects a limitation imposed on such investigations by nature. Kitts (1977) has emphasized the crucial importance in geohistorical investigations of obtaining multiple types of independent evidence on effects to support inferences from effect to cause. Gilbert's (1896) investigation of Coon Butte is a nice example of this approach. On the other hand, the exchange between W.M. Davis and Tarr illustrates how inferring cause (i.e. origin and genesis) from a single piece of evidence on form (accordant summits) can lead to science by debate. The danger involved in basing an inference about the cause of a landform on a single type of evidence, especially perceptual information on morphology, has been recognized for some time. Russell (1949, p. 3) remarked that it makes about as much sense to infer the development of a landform based on form alone as it does to amputate one's leg before entering a footrace.

The exchange between Tarr and Davis also emphasizes that the problem of decisive testing is enhanced when available evidence on effect is weak or ambiguous because it has been highly altered or modified over time. Recent work indicates that explanatory theory is less likely to strongly influence visual perception and interpretation if the incoming sensory information is strong; on the other hand, weak or ambiguous sensory inputs often lead to disagreement, even among individuals with shared background knowledge (Brewer and Lambert 1993). Brown (1995) has shown that even when two individuals have separate conceptual backgrounds clear evidence can prevent incommensurability as long as the two individuals have some shared basis for communication – a situation that certainly exists within geomorphology (or any other scientific discipline) where proponents of different theories share enough knowledge to provide a basis for common ground. The need for multiple types of independent evidence on effects is especially acute when each of the types, considered individually, is ambiguous. Debate about the cause of a particular relict landscape feature can be tentatively resolved if several pieces of weak evidence, when considered in combination, are consistent with only one of several possible explanations.

Another factor that contributes to the objectivity of testing in contemporary geomorphology is the definiteness and precision of observational techniques. Most contemporary techniques involve standardized rule-based procedures. In many cases, geomorphologists have imported techniques from other disciplines and thereby have *de facto* adopted established conventions for implementing the technique. An example is sieve analysis of sediments. Geomorphologists who use sieve analysis are expected to follow certain procedures for the data to be reliable (i.e. unbiased, capable of being reproduced with minimal error). These procedures transcend disciplinary boundaries; they represent standards that apply to the entire scientific community.

Standardized, rule-based observational techniques contribute to scientific objectivity by increasing the precision and intersubjective repeatability of data (or observations). By reducing the potential for different investigators to obtain different information about a specific property of an object of investigation, such techniques minimize the volitional character of the data and improve the basis for consensus about its evidential status. In other words, conventional procedures enhance the reliability of data as evidence for or against an explanatory theory. This concern about reliability of data explains why geomorphologists, along with other scientists, are interested in the capacity of a procedure to isolate information of interest (i.e. its sensitivity to confounding factors or background noise) and to replicate results precisely in calibration experiments and in investigative applications (Woodward 1989). The importance of calibration to objectivity has been demonstrated by Franklin et al. (1989), who argue that even in cases where the technique used to test a theory is an instance of the theory, the technique can still be used objectively if it is calibrated against known information (determined, of course, by an alternative, independent technique). Such cases may occur in geomorphology, but probably are rare. Most geomorphic techniques are drawn from other areas of science; thus, the issue of independence is obvious a priori, based on casual consideration of background knowledge in the area of science from which the technique derives.

Once again the contrast between Davis and Gilbert illustrates these points. Gilbert relied on data derived from precise surveys and compass readings to test his hypotheses concerning the origin of Coon Butte. Both the surveying procedure and use of the compass conform with standardized procedures that would still be used today were the measurements to be repeated. Gilbert also performed calibration experiments to assess the reliability of his data. Davis, on the other hand, used visual perception as his primary source of information on the relative elevations of adjacent ridge tops; moreover, when faced with precise data on these elevations from maps, he rejected it as misleading. The disagreement between Tarr and Davis about the visual appearance of the ridge tops shows that this observational technique produced data that was neither precise nor intersubjectively repeatable. Davis could use perceptual information to his advantage because he did not (probably because he could not) specify standardized rules for determining the elevations of ridge tops visually. This specific example highlights the importance of mensuration as a component of observation. Contemporary geomorphology, perhaps as an unwitting consequence of implicit dissatisfaction with Davis's visual approach to observation, has fully embraced mensuration in an effort to avoid the subjectivity that accompanies testing of inferences with impressionistic evidence.

The complement of definiteness and precision of observational techniques is definiteness and precision of theoretical expectations. The more rigid the limits placed on these expectations, the more restrictive the theory becomes regarding the data. The combined influence of invariable, tightly prescribed observational techniques and theoretical expectations defines the 'fixedness' of the observational setting (Hudson 1994). Assuming a technique has the potential to interact causally with real-world phenomena (i.e. produce involuntary information), the chances of obtaining informative data about the theory under test are greatest when the observational setting has a high degree of fixedness (i.e. a precise theoretical expectation is tested using a standardized, precise observational technique). In fact, Hudson (1994) has argued that as long as the theory provides fixed constraints on expectations, dependence of the observational setting on the theory under test actually enhances, rather than detracts from, objective testing because it is this fixedness that leads to 'unexpected' (i.e. disconfirmatory) results. For example, consider the extreme case where a geomorphologist proposes the theory that all rivers are 3 meters wide, and then tries to prejudice testing of the theory by: (1) developing a measuring device that has a length equal to the distance between the banks of several nearby rivers, all of which happen to have the same absolute width; (2) assigning a value of 3 meters to the length of this device. Clearly here is a case where the observational technique depends incestuously on the theory under test; it has been intentionally rigged to provide only confirmatory evidence for the theory under test. However, despite this dependency, the device can in principle be used to provide disconfirmatory evidence for the theory, provided the observational setting remains fixed. Suppose a flood erodes the banks of the streams, increasing the distance between them by twofold. If the same device is used to measure the widths of the channels after the flood, it will produce disconfirmatory evidence for the theory *as long as the rules governing the observational procedure are not changed arbitrarily (e.g. the value of 3 meters is not arbitrarily reassigned to the new bank-to-bank distance) or the theoretical expectation is not salvaged through the introduction of* ad hoc *auxiliary hypotheses (e.g. the 'real' channel is 3 meters wide, the other 3 meters of width constitute a separate entity called 'excavated space beside the channel following a flood').*

The concept of fixedness draws attention to a fundamental shortcoming of the Davisian scheme of landform analysis. By not specifying precisely theoretical expectations, Davis was able to accommodate virtually any anomalous data, even data generated by standardized rule-based procedures (e.g. Tarr's data on ridge-top elevations), through the introduction of *ad hoc* assumptions designed to preserve the evidential basis for an explanatory theory. In contrast, Gilbert (1896) clearly recognized the importance of a fixed observational setting and did not attempt to salvage the meteorite impact hypothesis for Coon Butte by introducing *ad hoc* qualifications, even though it was probably his preferred explanation.

The emphasis on precise data and quantitative modeling in contemporary geomorphology shows that geomorphologists implicitly recognize the value of fixed observational settings. On the other hand, unlike theorists in foundational sciences, geomorphologists generally treat certain theoretical principles, especially those from physics and chemistry, as sacrosanct. When data obtained within a fixed observational setting do not conform with theoretical expectations based on physical principles, geomorphologists do not discard the physical principles. Instead, they view the data as evidence *against the particular specification of these principles, including the underlying assumptions.* The explanatory theory may have invoked the wrong principles, combined these principles in an incorrect manner, or incorporated too many untested assumptions, but the fundamental validity of physical and chemical laws is unquestionable. In this sense, chemistry and physics provide foundational principles for explanatory theories in geomorphology, much as the cycle of erosion was foundational in Davisian geomorphology. In contrast to Davis, contemporary geomorphologists are willing to view specific formulations of a physically based theory as incomplete or even incorrect based on empirical testing. *A posteriori* qualifications are introduced not to salvage the evidential basis of a fixed explanation, but to modify the explanation so that it yields a new set of precise theoretical expectations that must accommodate the fixed observational data.

The Search for Geomorphic Phenomena

A final issue is the manner in which observation contributes to the general goal of geomorphology: to generate knowledge about landforms and landform-shaping processes on the earth and other planetary surfaces at a variety of temporal and spatial scales. An attribute of explanatory theory that is highly valued among scientists, including geomorphologists, is specification of cause and effect. As discussed in the first part of this chapter, many causal phenomena cannot be perceived directly; instead their existence must be inferred from and justified by data (Bogen and Woodward 1988; Woodward 1989). This situation is common both in process-oriented and geohistorical geomorphologic investigations, but for slightly different reasons in each case.

Process-oriented studies, such as field experiments dealing with stream channel dynamics, are designed to test theories that refer to imperceptible causal processes (phenomena) responsible for landform development and change (e.g. boundary shear stresses associated with a distinctive pattern of fluid motion). Data are gathered both to document the existence of the processes specified by an underlying physical theory as well as to evaluate the impact of the processes on landscape morphology (e.g. structuring or restructuring of the bed and banks of the stream channel into a specific morphology). In such cases, geomorphologists clearly are engaged in a search for causal phenomena. Because the phenomena of interest are assumed to be governed by physical, chemical, or biological principles, explanatory theories commonly include such principles. However, it is the phenomena, not the principles, that are of interest. Geomorphologists are not concerned about testing the validity of physical or chemical principles, rather they are seeking the ways in which various physical and chemical mechanisms combine in nature to constitute distinctive processes that shape planetary landscapes. In some cases, the search for geomorphic processes is guided explicitly by a formal explanatory theory expressed in the language of physics or chemistry. In others, it occurs within an abductive framework where evidence on effects (often morphology or morphologic change) and background knowledge of physical and chemical principles are used to infer the causal mechanisms responsible for landform development and change. In process-oriented investigations, the causal phenomena, because they are believed to be governed by underlying physical or chemical principles, are assumed to be generalizable; therefore, the focus is on testing of generalized theoretical statements about the relationship between cause and effect.

Geohistorical investigations also involve a search for unobservable causal phenomena, in this case historical events that produced relict landscape features or deposits. Here again the search relies heavily on data. The search for geohistorical events is almost always cast within an abductive framework (Rhoads and Thorn 1993) because it is not possible to measure a historical event itself (because it has occurred in the past), only the effect of the event. When reasoning abductively, it is essential to construct a coherent web of evidence, each piece of which provides an independent element of support for a particular explanation, while at least some of the same data serve as disconfirmatory evidence for competing explanations.

The high degree of complexity, and thus the seeming uniqueness of a geohistorical event, makes it difficult to develop generalizations about the event as a *type* of phenomenon (Kitts 1974). Thus, the focus, at least initially, is on determining the character of the event, not on formulating generalizations about the relationship between the event as a type of phenomenon and the effects of this type of phenomenon. However, any geohistorical event can be viewed as a mix of environmental/historical contingencies and general processes governed by physical laws. Because an event consists of a multitude of interacting phenomena manifested at different scales, causal generalizations resulting from process-oriented research programs play an important role in reconstructing the character of a specific event. These physically or chemically based generalizations, by assigning an immanent character to an event, also provide the basis for classifying a seemingly unique event as a type. In cases where a type of event that at first appears to be unique, such as cataclysmic glacial outburst floods, is found to have other instances, the discovery of new instances is guided by generalized background knowledge about the various forms of evidence that an event of that *type* produces (e.g. Rudoy and Baker 1993).

CONCLUSION

This chapter has endeavored to illustrate six main themes with regard to the relationship between observation and theory in geomorphology:

1. That a naturalized philosophical perspective on science, in which accurate depiction of the scientific process takes precedence over prescription or proscription, provides a powerful conceptual tool for understanding the scientific character of contemporary geomorphology.
2. That the naturalized philosophical perspective characterizes scientific observation in a manner that is readily recognizable to practicing scientists, including geomorphologists.
3. That at least since the time of Gilbert and Davis, despite an implicit undercurrent of radical empiricism, geomorphologists have not practiced theory-neutral observation, but have pursued and employed theory-laden observational techniques.
4. That the character of observations in contemporary geomorphology is broadly consistent with naturalized philosophical notions of data or evidence, which despite their theory-ladenness, preserve objectivity of testing.
5. That geomorphology differs from the basic sciences (e.g. physics, chemistry) in that established principles from these sciences are viewed as sacrosanct, regardless of the outcome of empirical testing (i.e. geomorphology is not involved in testing the validity of principles developed in the basic sciences).
6. That it is the search for geomorphic phenomena, not a reliance on novel methodologies or theoretical principles, that distinguishes geomorphology from other sciences.

A quintessential point to emerge from this essay is that observation is theory-dependent, but that this dependence does not necessarily undermine its vital role as the policing agent in geomorphic inquiry. A corollary of this view is that geomorphology will be best served by explicit recognition of the symbiotic relation between fieldwork/experimentation and theorizing. Because geomorphology depends strongly on other sciences, particularly physics, chemistry, and biology, both for its observational techniques as well as for the formulation and justification of its explanatory theories, scientific progress in the discipline is largely a function of how well informed geomorphologists are about theoretical and technological developments in other areas of science. Similarly, this dependency suggests that attempts to explore geomorphology philosophically should at least consider philosophical discussions about other sciences. Seen in this light, it is not a lack of fieldwork that historically has acted as the primary constraint on the discipline's advance, but rather a reluctance to embed geomorphology directly within the theoretical and technological contexts of physics, chemistry, and biology – a reluctance that may have been overcome by a stronger tradition of philosophical debate among geomorphologists.

ACKNOWLEDGEMENTS

Critical comments on an early draft of this chapter by Harold Brown and Richard Chorley are greatly appreciated.

REFERENCES

Achinstein, P. 1968. *Concepts of Science*, Johns Hopkins Press, Baltimore, 266 pp.

Baker, V.R. and Twidale, C.R. 1991. The reenchantment of geomorphology, *Geomorphology*, **4**, 73–100.

Beckinsale, R.P. 1976. The international influence of William Morris Davis, *Geographical Review*, **66**, 448–466.

Bishop, P. 1980. Popper's principle of falsifiability and the irrefutability of the Davisian cycle, *Professional Geographer*, **32**, 310–315.

Bogen, J. and Woodward, J. 1988. Saving the phenomena, *Philosophical Review*, **97**, 303–352.

Bogen, J. and Woodward, J. 1992. Observations, theories and the evolution of the human spirit, *Philosophy of Science*, **59**, 590–611.

Boyd, R., Gaspar, P., and Trout, J.D. (eds) 1991. *The Philosophy of Science*, The MIT Press, Cambridge, Mass., 800 pp.

Bradshaw, M. and Weaver, R. 1993. *Physical Geography: An Introduction to Earth Environments*, Mosby, St Louis, 640 pp.

Brewer, W.F. and Chinn, C.A. 1994. The theory-ladenness of data: an experimental demonstration, in *Proceedings of the Sixteenth Annual Conference of the Cognitive Science Society*, edited by A. Ram and K. Eiselt, Lawrence Erlbaum Associates, Hillsdale, NJ, pp. 61–65.

Brewer, W.F. and Lambert, B.L. 1993. The theory-ladenness of observation: evidence from cognitive psychology, in *Proceedings of the Fifteenth Annual Conference of the Cognitive Science Society*, edited by A. Ram and K. Eiselt, Lawrence Erlbaum Associates, Hillsdale, NJ, pp. 254–259.

Brown, B.W. 1974. Induction, deduction, and irrationality in geologic reasoning, *Journal of Geology*, **2**, 456.

Brown, H.I. 1987a. *Observation and Objectivity*, Oxford University Press, New York, 255 pp.

Brown, H.I. 1987b. Naturalizing observation, in *The Process of Science*, edited by N.J. Nersessian, Martinus Nijhoff, Dordrecht, pp. 179–193.

Brown, H.I. 1993. A theory-laden observation can test the theory, *British Journal for the Philosophy of Science*, **44**, 555–559.

Brown, H.I. 1995. Empirical testing, *Inquiry*, **38**, 353–399.

Carnap, R. 1966. *Philosophical Foundations of Physics*, Basic Books, New York, 300 pp.

Chamberlin, T.C. 1890. The method of multiple working hypotheses, *Science*, **15**, 92–96.

Chamberlin, T.C. 1897. The method of multiple working hypotheses, *Journal of Geology*, **5**, 837–848.

Chao, E.C.T., Shoemaker, E.M. and Madsen, B.M. 1960. First natural occurrence of coesite, *Science*, **132**, 220–222.

Chorley, R.J. 1962. Geomorphology and general systems theory, *United States Geological Survey Professional Paper 500-B*. US Government Printing Office, Washington, DC, 10 pp.

Chorley, R.J. 1965. A re-evaluation of the geomorphic system of W.M. Davis, in *Frontiers in Geographical Teaching*, edited by R.J. Chorley and P. Haggett, Methuen, London, pp. 21–38.

Chorley, R.J. 1966. The application of statistical methods to geomorphology, in *Essays in Geomorphology*, edited by G.H. Dury, American Elsevier, New York, pp. 275–387.

Chorley, R.J., Beckinsale, R.P. and Dunn, A.J. 1973. *The History of the Study of Landforms*. Vol. 2: *The Life and Work of William Morris Davis*, Methuen, London, 874 pp.

Churchland, P. 1988. Perceptual plasticity and theoretical neutrality: a reply to Jerry Fodor, *Philosophy of Science*, **55**, 167–187.

Collins, H.M. 1983. An empirical relativist programme in the sociology of scientific knowledge, in *Science Observed*, edited by K. Knorr-Cetina and M. Mulkay, Sage, London, pp. 85–113.

Davis, W.M. 1885. Geographic classification, illustrated by a study of plains, plateaus, and their derivatives, *Proceedings of the American Association for the Advancement of Science*, **33**, 428–432.

Davis, W.M. 1894. Physical geography in the university, *Journal of Geology*, **2**, 66–100.

Davis, W.M. 1899. The geographical cycle, *Geographical Journal*, **14**, 481–504.

Davis, W.M. 1905. Complications of the geographical cycle, *Report of the Eighth Geographical Congress, Washington, 1904*, pp. 50–163.

Davis, W.M. 1909. *Geographical Essays*, edited by D.W. Johnson, Ginn, Boston, 777 pp.

Davis, W.M. 1911. The Colorado Front Range, *Annals of the Association of American Geographers*, **1**, 21–84.

Davis, W.M. 1912. Relations of geography to geology, *Bulletin of the Geological Society of America*, **23**, 93–124.

Davis, W.M. 1913. Speculative nature of geology, *Bulletin of the Geological Society of America*, **24**, 686–687.

Davis, W.M. 1915. The principles of geographic description, *Annals of the Association of American Geographers*, **5**, 61–105.

Davis, W.M. 1922. The reasonableness of science, *Scientific Monthly*, **15**, 193–214.

El-Baz, F. 1980. Gilbert and the moon, *Geological Society of America Special Paper* **183**, 69–91.

Feyerabend, P. 1958. An attempt at a realistic interpretation of experience, *Proceedings of the Aristotelian Society*, n.s., **58**, 143–170.

Feyerabend, P. 1993. *Against Method*, 3rd edn, Verso, London, 279 pp.

Flemal, R.C. 1971. The attack on the Davisian system of geomorphology: a synopsis, *Journal of Geological Education*, **19**, 3–13.

Fodor, J. 1984. Observation reconsidered, *Philosophy of Science*, **51**, 23–43.

Fodor, J. 1988. A reply to Churchland's 'perceptual plasticity and theoretical neutrality', *Philosophy of Science*, **55**, 188–198.

Franklin, A., Anderson, M., Brock, D., Coleman, S., Downing, J., Gruvander, A., Lilly, J., Neal, J., Peterson, D., Price, M., Rice, R., Smith, L., Speirer, S. and Toering, D. 1989. Can a theory-laden observation test the theory? *British Journal for the Philosophy of Science*, **40**, 229–231.

Gilbert, G.K. 1886. The inculcation of scientific method by example, with an illustration drawn from the Quaternary geology of Utah, *American Journal of Science*, **31**, 284–299.

Gilbert, G.K. 1896. The origin of hypotheses, illustrated by the discussion of a topographic problem, *Science*, n.s., **3**, 1–13.

Gilluly, J. 1963. The scientific philosophy of G.K. Gilbert, in *The Fabric of Geology*, edited by C.C. Albritton, Jr, Freeman, Cooper, Stanford, Calif., pp. 218–224.

Gilman, D. 1992. What's a theory to do...with seeing? or some empirical considerations for observation and theory, *British Journal for the Philosophy of Science*, **43**, 287–309.

Glymour, C. 1980. *Theory and Evidence*, Princeton University Press, Princeton, NJ, 383 pp.

Goudie, A. (ed.) 1990. *Geomorphological Techniques*, 2nd edn, Unwin and Hyman, London, 570 pp.

Graf, W.L. 1979. Catastrophe theory as a model for change in fluvial systems, in *Adjustments of the Fluvial System*, edited by D.D. Rhodes and G.P. Williams, Kendall Hunt, Dubuque, Iowa, pp. 13–32.

Greenwood, J.D. 1990. Two dogmas of neo-empiricism: the 'theory-informity' of observation and the Quine–Duhem thesis, *Philosophy of Science*, **57**, 553–574.

Hacking, I. 1983. *Representing and Intervening*, Cambridge University Press, Cambridge, 287 pp.

Haines-Young, R.H. and Petch, J.R. 1983. Multiple working hypotheses: equifinality and the study of landforms, *Transactions of the Institute of British Geographers*, n.s. **8**, 458–466.

Haines-Young, R.H. and Petch, J.R. 1986. *Physical Geography: Its Nature and Methods*, Harper and Row, London, 230 pp.

Hanson, N.R. 1958. *Patterns of Discovery*, Cambridge University Press, Cambridge, 240 pp.

Hart, M.G. 1986. *Geomorphology Pure and Applied*, George Allen and Unwin, London, 228 pp.

Harvey, D. 1969. *Explanation in Geography*, Edward Arnold, London, 521 pp.

Higgins, C.G. 1975. Theories of landscape development: a perspective, in *Theories of Landform Development*, edited by W.N. Melhorn and R.C. Flemal, State University of New York at Binghamton, Binghamton, NY, pp. 1–28.

Hudson, R.G. 1994. Background independence and the causation of observations, *Studies in the History and Philosophy of Science*, **25**, 595–612.

Johnson, D. 1993. Role of analysis in scientific investigation, *Bulletin of the Geological Society of America*, **44**, 461–493.

Johnson, J.G. 1990. Method of multiple working hypotheses: a chimera, *Geology*, **18**, 44–45.

Judson, S. 1960. William Morris Davis – an appraisal, *Zeitschrift für Geomorphologie*, **4**, 194–201.

Kaiser, M. 1991. From rocks to graphs – the shaping of phenomena, *Synthese*, **89**, 111–133.

Kitts, D.B. 1963. Historical explanation in geology, *Journal of Geology*, **71**, 297–313.

Kitts, D.B. 1974. Physical theory and geological knowledge, *The Journal of Geology*, **82**, 1–23.

Kitts, D.B. 1977. *The Structure of Geology*, SMU Press, Dallas, 180 pp.

Knorr-Cetina, K. 1983. The ethnographic study of scientific work: towards a constructivist interpretation of science, in *Science Observed*, edited by K. Knorr-Cetina and M. Mulkay, Sage, London, pp. 115–140.

Kosso, P. 1989. *Observability and Observation in Physical Science*, Kluwer, Dordrecht, 165 pp.

Kuhn, T.S. 1962. *The Structure of Scientific Revolutions*, University of Chicago Press, Chicago, 172 pp.

Kuhn, T.S. 1970. *The Structure of Scientific Revolutions*, 2nd edn, University of Chicago Press, 210 pp.

Lane, S.N., Richards, K.S. and Warburton, J. 1993. Comparison between high frequency velocity records obtained with spherical and discoidal electromagnetic current meters, in *Turbulence: Perspectives on Flow and Sediment Transport*, edited by N.J. Clifford, J.R. French, and J. Hardisty, Wiley, Chichester, pp. 121–163.

Leopold, L.B. and Langbein, W.B. 1962. The concept of entropy in landscape evolution, *United States Geological Survey Professional Paper 500-A*, US Government Printing Office, Washington, DC, 20 pp.

Leopold, L.B. and Maddock, Jr, T. 1953. The hydraulic geometry of stream channels and some physiographic implications, *United States Geological Survey Professional Paper 252*, Government Printing Office, Washington, DC, 56 pp.

Longino, H. 1990. *Science as Social Knowledge*, Princeton University Press, Princeton, NJ, 262 pp.

McDonald, J. 1992. Is strong inference really superior to simple inference, *Synthese*, **92**, 261–282.

Mackin, J.H. 1963. Rational and empirical methods of investigation in geology, in *The Fabric of Geology*, edited by C.C. Albritton, Jr, Freeman, Cooper, Stanford, Calif., pp. 135–163.

Maxwell, G. 1962. The ontological status of theoretical entities, in *Minnesota Studies in the Philosophy of Science*, vol. 3, edited by H. Feigl and G. Maxwell, University of Minnesota Press, Minneapolis, pp. 3–27.

Melton, M.A. 1958. Correlation structure of morphometric properties of drainage systems and their controlling agents, *Journal of Geology*, **66**, 442–460.

Middleton, G.V. and Wilcock, P.R. 1994. *Mechanics in the Earth and Environmental Sciences*, Cambridge University Press, Cambridge, 459 pp.

Nelson, A. 1994. How *could* scientific facts be socially constructed? *Studies in the History and Philosophy of Science*, **25**, 535–547.

Nickles, T. 1987. 'Twixt method and madness, in *The Process of Science*, edited by N.J. Nersessian, Martinus Nijhoff, Dordrecht, pp. 41–67.

Oliver, J.E. 1991. *The Incomplete Guide to the Art of Discovery*, Columbia University Press, New York, 208 pp.

Parvsnikova, Z. 1992. Is a postmodern philosophy of science possible? *Studies in the History and Philosophy of Science*, **23**, 21–37.

Phillips, J.D. and Renwick, W.H. (eds) 1992. *Geomorphic Systems*, Elsevier, Amsterdam, 487 pp.

Pickering, A. 1990. Knowledge, practice, and mere construction, *Social Studies of Science*, **20**, 658–663.

Platt, J.R. 1964. Strong inference, *Science*, **46**, 347–353.

Popper, K.R. 1968. *The Logic of Scientific Discovery*, Harper and Row, New York, 480 pp.

Putnam, H. 1962. What theories are not, in *Logic, Methodology, and Philosophy of Science*, edited by E. Nagel, P. Suppes, and A. Tarski, Stanford University Press, Stanford, Calif., pp. 240–251.

Putnam, H. 1974. The 'corroboration' of theories, in *The Philosophy of Karl Popper*, edited by A. Schilpp, Open Court Publishing, LaSalle, Ill., pp. 221–240.

Pyne, S.J. 1980. *Grove Karl Gilbert: A Great Engine of Research*, University of Texas Press, Austin, 306 pp.

Rhoads, B.L. and Thorn, C.E. 1993. Geomorphology as science: the role of theory, *Geomorphology*, **6**, 287–307.

Rhoads, B.L. and Thorn, C.E. 1994. Contemporary philosophical perspectives on physical geography with emphasis on geomorphology, *Geographical Review*, **84**, 90–101.

Ritter, D.F. 1986. *Process Geomorphology*, W.C. Brown, Dubuque, Iowa, 579 pp.

Rorty, R. 1979. *Philosophy and the Mirror of Nature*, Blackwell, Oxford, 401 pp.

Rudoy, A.N. and Baker, V.R. 1993. Sedimentary effects of cataclysmic late Pleistocene glacial outburst flooding, Altay Mountains, Siberia, *Sedimentary Geology*, **85**, 53–62.

Russell, R.J. 1949. Geographical geomorphology, *Annals of the Association of American Geographers*, **39**, 1–11.

Sack, D. 1991. The trouble with antithesis: the case of G.K. Gilbert, geographer and educator, *Professional Geographer*, **43**, 28–37.

Scheidegger, A.E. 1961. *Theoretical Geomorphology*, Springer-Verlag, Berlin, 333 pp.

Schumm, S.A. 1991. *To Interpret the Earth: Ten Ways to be Wrong*, Cambridge University Press, Cambridge, 133 pp.

Schumm, S.A., Mosley, M.P. and Weaver, W.E. 1987. *Experimental Fluvial Geomorphology*, Wiley, New York, 413 pp.

Shapere, D. 1982. The concept of observation in science and philosophy, *Philosophy of Science*, **49**, 485–525.

Shapere, D. 1984. *Reason and the Search for Knowledge*, Reidel, Dordrecht, 438 pp.

Shapere, D. 1987. Method in the philosophy of science and epistemology, in *The Process of Science*, edited by N.J. Nersessian, Martinus Nijhoff, Dordrecht, pp. 1–39.

Shoemaker, E.M. 1960. Penetration mechanics of high velocity meteorites, illustrated by Meteor Crater, Arizona, *Proceedings, 21st International Geological Congress*, **18**, 418–434.

Stoddart, D.R. 1966. Darwin's impact on geography, *Annals of the Association of American Geographers*, **56**, 683–698.

Strahler, A.N. 1950. Davis' concepts of slope development viewed in the light of recent quantitative investigations, *Annals of the Association of American Geographers*, **40**, 209–213.

Strahler, A.N. 1952. Dynamic basis of geomorphology, *Bulletin of the Geological Society of America*, **63**, 923–938.

Strahler, A.N. 1957. Quantitative analysis of watershed geomorphology, *Transactions of the American Geophysical Union*, **38**, 913–920.

Summerfield, M.A. 1991. *Global Geomorphology*, Wiley, New York, 537 pp.

Suppe, F. 1977. The search for philosophic understanding of scientific theories, in *The Structure of Scientific Theories*, 2nd edn, edited by F. Suppe, University of Illinois Press, Urbana, Ill., pp. 3–241.

Suppe, F. 1989. *The Semantic Conception of Theories and Scientific Realism*, University of Illinois Press, Urbana, Ill., 475 pp.

Tarr, R.S. 1898. The peneplain, *American Geologist*, **21**, 351–370.

Thorn, C.E. and Welford, M.R. 1994. The equilibrium concept in geomorphology, *Annals of the Association of American Geographers*, **84**, 666–696.

Tinkler, K.J. 1985. *A Short History of Geomorphology*, Croom Helm, London, 317 pp.

Van Fraassen, B.C. 1980. *The Scientific Image*, Clarendon Press, Oxford, 235 pp.

Van Fraassen, B.C. 1989. *Laws and Symmetry*, Clarendon Press, Oxford, 395 pp.

Von Englehardt, W. and Zimmermann, J. 1988. *Theory of Earth Science*, translated by L. Fischer (first published in German in 1982), Cambridge University Press, Cambridge, 381 pp.

Weckert, J. 1985. The theory-ladenness of observations, *Studies in the History and Philosophy of Science*, **17**, 115–127.

Werth, R. 1980. On the theory-dependence of observations, *Studies in the History and Philosophy of Science*, **11**, 137–143.

Wittgenstein, L. 1922. *Tractatus Logico-Philosophicus*, Harcourt, Brace, New York, 189 pp.

Wittgenstein, L. 1969. *On Certainty*, edited by G.E.M. Anscombe and G.H. von Wright, translated by D. Paul and G.E.M. Anscombe, Blackwell, Oxford, 90 pp.

Wittgenstein, L. 1980. *Culture and Value*, edited by G.H. von Wright, translated by P. Winch, University of Chicago Press, Chicago, 94 pp.

Woodward, J. 1989. Data and phenomena, *Synthese*, **79**, 393–472.

Young, R. and McDougall, I. 1993. Long-term landscape evolution: early Miocene and modern rivers in southern New South Wales, Australia, *Journal of Geology*, **101**, 35–49.

3 Hypotheses and Geomorphological Reasoning

Victor R. Baker

Department of Geosciences, University of Arizona

Science does not rest upon rock-bottom. The bold structure of theories rises, as it were, above a swamp, but not down to any natural or 'given' base; and when we cease our attempts to drive our piles into a deeper layer, it is not because we have reached firm ground. We simply stop when we are satisfied that they are firm enough to carry the structure, at least for the time being.

(Popper, 1959, p. 111)

ABSTRACT

Geomorphology is a way of thinking about the surface of planet Earth. Controlled experimentation, in the manner of pure physics, is not possible for most geomorphological concerns. Thus, much of conventional analytical philosophy of science, which is based on the exemplar of experimental physics, fails to portray important aspects of geomorphological reasoning. This is particularly true of hypothesizing, which was recognized by Gilbert, Chamberlin, and Davis as a central methodological concern of geomorphology. Geomorphological reasoning largely relies upon retroductive inference, which Charles S. Peirce described as 'the spontaneous conjectures of instinctive reason'. Because it reasons from real effects to real causes, eventually colligating (bring together) facts under a conceptual scheme (hypothesis), retroduction bridges the gulf between nature and mind. Geomorphological indices, such as landforms and sediments, are signs for which causative processes are inferred retroductively. Though superficially similar to lucky 'guessing', retroductive inference succeeds in generating fruitful hypotheses (some of them outrageous) because the human mind is instinctively attuned to certain aspects of nature. This instinctive propensity in science to 'guess right', which Galileo called *il lume naturale*, may derive from fundamental properties of the universe and mind that modern cosmologists have named the 'anthropic principle'.

The Scientific Nature of Geomorphology: Proceedings of the 27th Binghamton Symposium in Geomorphology held 27–29 September 1996. Edited by Bruce L. Rhoads and Colin E. Thorn. ©1996 John Wiley & Sons Ltd.

INTRODUCTION

Geomorphology is a way of thinking (*Logos*) about the surface (*morphos*) of planet Earth (Gaia). It is also viewed as the body of knowledge or facts about Earth's surface. The distinction in these two meanings has consequences for the conduct of the science. As a body of facts geomorphology assumes monolithic proportions, since the potential facts are presumed to be endless in number and bewildering in complexity. It is obvious to many scientists that the appropriate task in the face of this complexity is to simplify. This can be done by developing a scheme of classifying the numerous individual facts into a smaller number of categories. This is one type, not the only type, of synthesis, whereby entities are combined in wholes, which, in this case, are conceptual categories. Alternatively or subsequently, simplification can arise by establishing a small number of basic principles from which it should be possible, at least in theory, to deduce the details of landform and process complexity on Earth's surface. This activity is analysis, the separation of the extremely complex (presumably intractable) whole into its component parts in order to study them in simplified form and thereby understand.

Inference is the logical process in which some conclusion is reached from a set of statements. Analytical inference derives its conclusion from premises that are assumed to be true. Synthetic inference begins with factual statements and reasons toward the conclusion. Analytical inference corresponds fairly well to the logic of deduction. Much more controversy surrounds synthesis, which many philosophers of science have equated to the logic of induction. Induction is the reasoning from individuals or particulars to general or universal statements about them. This is not the only kind of synthetic reasoning, as will be shown in this chapter.

Closely related to the idea of simplifying complexity is reductionism, or reductivism, which is a philosophical belief in one type of methodology or one science that encompasses the principles applicable to all phenomena. From such principles it should be possible to deduce all observed entities. If geomorphology is viewed as a bewildering complex of facts, an analyst will find great appeal in this philosophy. Thus, a reductionist might conceive of geomorphology as a science for which one should establish simplifying and generalizing principles from which one might then construct model simulations of landscape evolution and/or process operation. This viewpoint also leads to an emphasis upon readily measurable processes, which are those operating over the relatively small spatial and temporal scales that are most easily accessible for the presumed verification or falsification of theoretical models (Baker 1988a).

Modern philosophy of science reflects the general concerns of Western philosophy as a whole. Until about 1960 it would have been rather easy to divide those concerns between analytical philosophy and continental philosophy. Existentialism was the predominant form of the latter, and logical positivism (or logical empiricism) was a predominant form of the former. Since 1960 both these ideologies have fallen into disfavor, but the analytical versus continental distinction has remained. Most philosophy of science is in the analytical tradition, and much of what many scientists themselves know of it is in the older (discredited) logical empiricist form, which held science to deal with problems of fact while philosophy deals with problems of methodology and conceptual analysis. The closely related logical positivist school held to a single, identifiable scientific methodology and to various principles of verification and cognitive meaning, all based on strict rules of logic. It is not difficult to see why the science of physics, particularly classical mechanics, became an exemplar or paradigm for analytical philosophy (Frodeman 1995). Physicists seek the timeless, invariant laws of nature. The laws of physics make predictions that can then be verified against the facts of nature. This verification is possible because nature can be studied objectively in appropriately isolated systems known as 'controlled experiments'.

Since 1960 analytical philosophy has moved in many new directions. Kuhn (1962) shocked the reductive and logical sensibilities of the logical empiricists with his notions of relativism and the role of the community of scientists in determining research paradigms. Popper (1959) and Quine (1953), among others, showed that several basic concepts of logical positivism, including the verification principle, were untenable. Nevertheless, assumptions of the logical positivist program remain, if not universally among analytical philosophers, still rather prominent in the thinking of many scientists who write about the foundations of their subject. For example, highly reductive analytical philosophers have proclaimed a new field of cognitive science that aspires to combine the conceptual skills of philosophy with artificial-intelligence modeling and experimental psychology.

Analytical philosophy of science devotes great emphasis to the methodological treatment of laws, theories, and hypotheses. These three methodological categories can be considered ideas in a hierarchical order in which laws are 'better established' than theories, and theories are better established than hypotheses. Notice that this view preserves the logical empiricist concept of justifying knowledge objectively. In this treatment, laws and theories, which find their most elegant expression in mathematical physics, receive inordinate attention. In contrast, geomorphology, like geology, can be considered to be a science of hypothesis, which simply means to a reductionist that geomorphological theorizing has not advanced to the point where sophisticated theorizing and the establishment of fundamental laws have been achieved. Shamos (1995, p. 96) makes the point this way:

> ... In general, as a science matures, it passes first through a purely descriptive stage, then proceeds to an experimental stage, and finally, as meaningful patterns are seen to emerge, to a theoretical stage, where one usually finds it highly productive to use the language of science (mathematics) to describe natural phenomena and uncover new knowledge. Physics and chemistry are already at this final stage; biology and Earth science, involving as they do far more complex systems, have been slower to mature to this level.

GEOMORPHOLOGY AS A WAY OF THINKING

There is no science in a body of absolute facts. Science is a process of thought and observation directed toward understanding, which is a faculty of mind that allows human beings to grasp reality and thereby cope with the real world. Reductionism, objectivity, and simplification are means for achieving this goal, but they are not the only means, nor do they necessarily have special privilege in this task. Though reductionism can proceed with an especially effective mode of logic, deduction, its tools are highly limited for relating to the messy complexity of the real world, as encountered in sciences like geomorphology for which experimental access to nature's reality is very limited. Indeed, there are sound logical reasons why in Earth science controlled experiments are inadequate to the task of verifying or validating the predictive consequences of theoretical models (Oreskes et al. 1994). More practical issues in regard to controlled experimentation in geomorphology are discussed by Ahnert (1980) and Church (1984).

Thinking about Earth's surface involves a developmental process of growth in that thought. In contrast, reductionism involves a narrowing of thought. If the research aim is reductionistically to deduce or predict certain phenomena from first principles, then the obvious task of the field researcher is to measure those phenomena very carefully in order to check and refine the theory. In this way the complexity of the field problem is reduced to manageable proportions and to a focus on phenomena relevant to key scientific questions. The alternative might be presumed to be a chaotic program of pointless observation, perhaps motivated by a faith that some order will be discerned through induction.

Induction, as noted above, is one type of synthesis in which inference is from particular instances of something to some general or universal statement about those instances. For Francis Bacon, and many other early philosophers of science, induction was a method of scientific discovery. One began with data (particulars in the above definition) and worked toward high-level principles (universal statements about those data). Clearly, science, notably physics, generates such universal statements, and it does so by reference to data through scientific experiments. All science seems to be decidedly inductive. The inductive method was critically assessed by the great empiricist philosopher David Hume. Hume reasoned that it was impossible to ever justify a law by experiment or observation. Just because one sees the sun rise each day does not, of itself, require that it will rise the next day. This is the famous logical problem of induction, which Sir Karl Popper (1959, p. 54) summarized in terms of three seemingly incompatible principles:

> ...(a) Hume's discovery... that it is impossible to justify a law by observation or experiment, since it 'transcends experience'; (b) the fact that science proposes and uses laws 'everywhere and all the time'... To this we add (c) *the principle of empiricism* which asserts that in science, only observation and experiment may decide upon the *acceptance or rejection* of scientific statements, including laws and theories...

Popper claimed to solve the problem of induction via his principle of falsification. This allowed principle (a) to remain compatible with principles (b) and (c). As he states (Popper 1959, p. 54):

> ...the acceptance by science of a law or theory is *tentative only*; which is to say that all laws and theories are conjectures, or tentative *hypotheses* (a position which I have sometimes called 'hypotheticism'); and that we may reject a law or theory on the basis of new evidence, without necessarily discarding the old evidence which originally led us to accept it...

Popper's insight is widely regarded as one of the great achievements for the analytical approach to philosophy of science (Lindh 1993). Popper further reasoned that, since science cannot advance by inductive confirmation, the appropriate mode of advancement is by the imposition of bold conjectures (Popper 1969). These are hypotheses about the world that require completely new models for their scientific exposition. However, as psychological matters of creative thought by individual scientists, the origin of these conjectures lies outside of analytical philosophical discourse. It follows, of course, that such philosophy can tell us little about the growth of geomorphological thought, since that thought depends on the origin of hypotheses, as argued by Gilbert (1886, 1896) and by Chamberlin (1890).

One notes in Popper's work the epistemological distinction between the context of discovery and the context of justifying knowledge, the latter being the sole philosophical concern. However, in the last 25 years there has been a reexamination of discovery by analytical philosophers. Kantorovich (1993) divides current thinking on this matter into two camps. The predominant view is that discovery is not a logical matter (Laudan 1980), as Bacon proposed for induction. Instead, discovery is studied in a historicist/particularist manner, in which individual scientific case studies are examined for their lessons about the scientific process. A great many social and ideological factors are found to play a role. Another, growing, view holds discovery to be a matter for cognitive science, eventually seeking to model the process via computerized artificial intelligence.

Quite peripheral to the mainstream philosophy of science, described above, is the resolution of the problem of induction nearly a century before Popper's work. The resolution is controversial, as is the work of the logician who accomplished it, Charles S. Peirce, who is now widely regarded as 'the one truly universal mind that nineteenth-century America produced' (Dusek 1979). Peirce's life contains many ironies that delayed external recognition of his philosophical importance (Brent 1993), but he was well known to the scientists of his own day (Fisch 1980), and his relevance to our own time is increasingly being recognized (Ochs 1993; Hausman 1993). Peirce is of particular interest because of his direct connection to the late nineteenth-century American geomorphologists G.K. Gilbert, T.C. Chamberlin, and W.M. Davis (Baker in press). All these scientists wrote highly influential papers on the nature of geomorphological reasoning, and these show the possible influence of Peirce's philosophy (Baker in press).

W.H. Davis (1972, p. 34) describes Peirce's resolution of the problem of induction:

> The point [Peirce] made is essentially this: contrary to Hume and contrary to practically every epistemologist since him *scientific reasoning does not depend upon induction at all*! Nor does it depend upon anything so simple as our ability to take habits. The rising of the sun as an example of inductive reasoning is drastically misleading by its simplicity. Scientific reasoning, indeed all of our reasoning, depends upon the mind's ability to have insights, to see things coherently and harmoniously, to see laws and principles, in short, to make up hypotheses. Hume has misled generations of philosophers because he utterly ignored the place of hypothesis in human thinking. Perhaps it is enough that he should have seen the vast importance of the law of association. But when someone grasps the principle behind the workings of some machine or of some feature of nature, he is *not merely being impressed by a succession of regularities*, he is not merely gaining a habit. He is having an insight, seeing principles, grasping interconnections. This is the feature of our mental life which was so wonderfully emphasized by Peirce, but Whewell, long before, saw the same truth.

Peirce credited Cambridge mineralogist William Whewell as one of the few logicians to have properly understood the logic of scientific inference. Whewell wrote extensively on history and philosophy of science in the early nineteenth century (Fisch 1991). His central concern was with the forming of antithetical couplings between (1) the objective facts of nature, and (2) new concepts suggested to scientific minds. Whewell considered this process to be a colligation ('binding together') of existing facts that are unconnected in themselves but get connected through mental concepts. In his treatises on logic, Whewell

(1858, 1860) referred to this process as 'induction', but, as we will see, Charles Peirce has distinguished this form of synthetic inference from the 'induction' of which Hume (and later Popper) are speaking. Peirce accorded it the various names 'hypothesis', 'abduction', 'retroduction', and 'presumption'.

Whewell supplemented his prospective view of hypothesis generation as 'colligation of facts' with a retrospective concept of hypothesis (or theory) appraisal in terms of criteria for 'factual truth'. These included (1) a prospective theory's tendency to converge and to simplify, and (2) the theory's repeated exhibition of explanatory surprise. The latter phenomenon he termed the 'consilience of inductions'. This idea that scientific hypotheses could prospectively approximate truths in nature was criticized vehemently by John Stuart Mill (1846). Whewell tried to support his position with historical case studies (Whewell 1837, 1840), arguing that science in practice works as he described. Mill, in contrast, argued on a detached logical basis that the real work of science lay in the establishment of knowledge by inductive proof. Mill's views were nominalist (discussed in the next section) and empiricist. These were the positions subsequently embraced by Bertrand Russell and other founders of logical empiricism and related strands of modern analytical philosophy. Whewell's ideas were declared defeatist because he made science dependent upon ingenuity and luck. As Wettersten and Agassi (1991, p. 345) observe: '. . . because philosophers ignored him and scientists did not write histories of philosophy, he was forgotten'.

Another unusual aspect of Whewell's philosophy was his interest in geomorphology and the Earth sciences. Whewell's review of Charles Lyell's *Principles of Geology* contains a devastating critique of a metaphysical principle espoused by Lyell and named 'uniformitarianism' by Whewell. Whewell's logic supported the possibility of an alternative position which he named 'catastrophism'. History shows that Lyell, a former lawyer, was the more skillful advocate of his particular philosophy, the defects of which continue to plague the Earth sciences (Baker 1978). One of the great ironies in history must be that the very scholars who have so long ignored Whewell's insights on science and logic bear the name that he first conferred upon them in his remarkable historical studies: 'scientists'.

NOMINALISM VERSUS REALISM

Charles Peirce tied his vision of scientific hypothesizing to a kind of scholastic realism, traceable back to the writings of Aristotle and medieval philosophers, such as Duns Scotus. This view holds that universals (including theories, general concepts, and hypotheses) exist independently of our perceptions of them. The alternative, dominating in much of modern analytical philosophy and modern philosophy of physics, is nominalistism, a doctrine holding that generals do not refer to something real, but rather to the names we attach to things. Peirce resurrected older ideas of scholastic realism because of his mathematical explorations of the continuum concept (Ketner and Putnam 1992). Peirce believed his mathematical conclusions justified the view that nature was infused with a very rich kind of logic. To appreciate Peirce's view, however, we must consider ideas that developed during mainstream philosophy of science in the first part of this century.

Logical positivism, the dominating view for twentieth-century philosophy of science up to about 1960, is founded upon a nominalism espoused by Ernst Mach, Karl Pearson, Henri Poincaré, and Bertrand Russell. The metaphysics of this position is best expressed in how the logical positivists dismissed any element of reality from all notions of theory, including the laws and hypotheses of science. Eisele (1959, p. 461) describes this as follows:

> ... Poincaré felt that science can never reveal any absolute truth concerning nature. It can only set a relationship between a hypothesis and its implication, and the so-called 'laws' are but 'conventions' adopted as a matter of 'convenience' from among many possibilities. In this 'descriptive' theory of science, the natural laws were for Poincaré, as they were for Mach and Pearson, merely fictions created by science to organize sense-impressions, and any order therein is imposed by the mind of man. A hypothesis resulting from observed data has no correlative object in nature. It is an intellectual device for stimulating and directing the discovery of further data, according to Poincaré.

In this conventionalist view, scientific logic can be conceived as a syllogism of the following form:

> Major, premise : All men are mortal.
>
> Minor premise : Socrates is a man.
>
> Conclusion : Socrates is mortal.

In classical physics the major premise, or rule, can be thought nominalistically to correspond to the various conventions, such as Sir Isaac Newton's laws of mechanics. The minor premise, or case, consists of real facts, for this example: the instantaneous relative positions and velocities of all particles at some point in time. Deductive logic allows a conclusion: that various accelerations will follow from the premises.

Poincaré was able to describe the method of physics in these terms. One first sets up an axiomatic system from which consequences can be deduced. These are then judged retroactively by testing (experimentation) against what exists independently of us. Objective experimentation leads to further experimentation. Reality exists only in facts, which are organized through generalization. Every generalization is a hypothesis to be tested, and hypotheses are merely intellectual devices for facilitating the discovery of further data; they do not have corresponding objects in nature. Hypotheses, and the scientific theories and 'laws' to which they lead, are conventions adopted for the convenience of facilitating this scientific process. The order embodied in natural laws is an order imposed by the human mind. This is because statements in logic and mathematics are true by definition and not discovered by examining reality. Their application to natural laws, therefore, is definitional convention applied to reality.

The conventionalism of Poincaré, Mach, and others was key to the revolution in physics during the early twentieth century. Though relativity and Einstein, an admirer of Mach, are most popularly associated with this revolution, it is quantum mechanics that has proven to be the most interesting in its philosophical implications. The generalization of quantum mechanics known as quantum electrodynamics is arguably the most successful theory in physics, judged in terms of its astonishing predictions and their experimental confirmation. This success is readily interpreted nominalistically. For example, one of the founders of quantum mechanics, Niels Bohr, was once asked whether the equations of this theory could be considered to somehow mirror an underlying, real quantum world. Bohr's answer is a succinct statement of the nominalistic character of the philosophy of physics (Petersen 1985, p. 305): 'There is no quantum world. There is only an abstract quantum physical description. It is wrong to think that the task of physics is to find out how nature is. Physics concerns what we can say about nature.'

The nominalistic philosophy of physics, and the logical empiricist scientific philosophy that it inspired, holds hypotheses to be useful fictions. They have no basis in reality. This is not the position of Charles Peirce, who made the bold conjecture that something akin to a logical pattern in nature was far richer than the restrictive deductive scheme given above. In his 1898 Cambridge Conference lecture series (Ketner 1992, p. 161) he wrote:

> What is reality? Perhaps there isn't any such thing at all. As I have repeatedly insisted, it is but a retroduction, a working hypothesis which we try, our one desperate forlorn hope of knowing anything . . . so far as there is any reality, what that reality consists in is this: that there is in the being of things something which corresponds to the process of reasoning, that the world *lives*, and *moves*, and *has its being*, in [a] logic of things.

Peirce's view was unpopular in his own time and it remains unpopular today. Peirce attributed this unpopularity to the pervasive nominalism in the metaphysical presumptions of his contemporaries. He continues in his 1898 lecture (Ketner 1992, p. 161) as follows:

> I point out that Evolution wherever it takes place is one vast succession of generalizations, by which matter is becoming subjected to ever higher and higher Laws; and I point to the infinite variety of nature as testifying to her Originality or power of Retroduction. But so far, the old ideas are too ingrained. Very few accept my message.

Note that the logic which Peirce ascribes to nature is not the familiar induction or deduction discussed in most philosophy of science. He introduces his new term, retroduction, to describe this logic of hypothesizing.

The importance of the realist/nominalist debate can be made clear in regard to Peirce's view that hypothesizing involves reasoning (retroduction) from effect to cause. In the science of the nineteenth century a careful distinction was often made between (1) causes presumed to have a real existence in nature, and (2) figments of mind, presumed to posit such causes (Laudan 1987). Type (1) causes were named *verae causae*, following Sir Isaac Newton's famous 'Rules of Reasoning' in the *Principia*. The first of these 'rules' stated, 'We are to admit no more causes of natural things than such are both true and sufficient to explain their formation'. Type (2) concepts, which might well be nominalistic, were given the name 'hypotheses'. It is the type (2) concepts of which Newton speaks in this famous passage from the *Principia*:

> I feign no hypotheses (*hypotheses non fingo*), for whatever is not deduced from the phenomenon is to be called a hypothesis, and hypotheses, whether metaphysical or physical, whether of occult qualities or mechanical have no place in experimental philosophy. In this philosophy particular propositions are inferred from the phenomena and afterwards rendered general by induction.

The 'hypotheses' described by Newton derive from mind alone; they are not 'inferred from the phenomena'. Peirce recognized Newton's distinction and noted that *verae causae* must be inductions, and are not matters of hypothetic inference. Peirce also agreed with Newton that 'hypotheses' (type 2) are not matters for belief. However, he envisioned scientific hypotheses as part of an inference, retroduction, that transcends the gulf presumed to exist between mind and real causes in nature. This connection was possible because of Peirce's view that all thought is in signs. Peirce's semiotic (theory of signs) is such that a natural continuity exists between the real causes in nature and the interpretations that are eventually made of those causes.

GEOMORPHOLOGICAL SEMIOTICS

Logic is considered to be the study of rules for exact reasoning. For a nominalist logician, like John Stuart Mill, logic merely gives us a system of names for patterns of inference; there is no reality in the names, though the thought systems described by these names may refer to real objects. For Charles Peirce logic is a formal semiotic. It provides valid patterns (forms) for signs, which are the entities of which Peirce believed all thought to be composed. In Peirce's semiotic (scientific study of signs), he has a rather special meaning for the concept of a sign. The usual meaning is dyadic: a sign is anything that stands for something else. Peirce's semiotic involves a triadic definition of sign. His definition (Peirce 1902a, p. 527) is as follows:

> ... Anything which determines something else (its *interpretant*) to refer to an object to which itself refers (its *object*) in the same way, the interpretant becoming in turn a sign, and so on *ad infinitum*.

This definition highlights in striking fashion Peirce's view that thought is continuous. It cannot be arbitrarily divided with absolute distinctions, as nominalism does when distinguishing between particular things and the universals or generals that those particular things share in common. This nominalistic distinction for Peirce is contrary to the spirit of scientific inference, particularly as it seeks to interpret the continuity between the signs of nature with those of which our thought is composed. The continuity is essential in the retroductive phase of inquiry, when reason is creatively employed to generate a candidate best explanation. Nominalism in logic may be motivated by a sense that some sort of objectivity is needed with regard to testing hypotheses, but it is an inappropriate view to take in the creative phase of hypothesis generation.

Earth scientists make common reference to the sign system of nature using metaphors such as 'what the rocks tell us' or 'conversation with the Earth' (Cloos 1953). It is interesting to contrast these language references to those applied to controlled experimentation by Sir Francis Bacon. Objective control means that questions put to nature (experiment) occur as in an interrogation (Keller 1985). Nature is coerced, presumably to reveal her secrets, when data serve the sole function of verifying or falsifying. The metaphorical function of conversation is denied in such interaction. In the Earth sciences, however, controlled experimentation (interrogation) is largely precluded; one must converse with the Earth, metaphorically speaking.

How does one view this semiotically? In Peirce's semiotics the most sophisticated signs are symbols, for which the interpretant is essential. Mathematics can be considered a symbolic language of great power in deductive systems such as the theories of physics. Indices are signs for which the objects are essential and which have existence independent of their interpretants. Landforms, river alluvium, and radiocarbon samples are all indices in that their objects are the processes which cause them and their interpretants are what geomorphological thought makes of them. Thus, geomorphology can be viewed as a sign system involving natural indices. Geomorphological reasoning involves inferences organized through a formal semiotic, that is a logic, that is embedded in these indices. Similarly, mathematics and mathematical physics emphasize symbolic sign systems in their logical inferences.

It was Peirce's view that scientists, like all human beings, reason within thought, which is a system of signs. This contrasts the more common view, traced back to Descartes, that thoughts occur as clear and distinct ideas within individual minds. For Peirce the act of reasoning within collective thought involves other scientists as well as nature. Communication with other scientists is through familiar signs, called symbols, which include the language systems of words, mathematics, and symbolic logic. Communication with nature is less abstract, but it also includes signs; in this case the signs are indices. All signs have objects and interpretants. It is the challenge of the scientist to appreciate the advantages and disadvantages of these different sign systems for the pursuit of their inquiries. Those operating in one sign system, symbols, for example, may erroneously disparage the logic of those operating in another system, indices, for example. This would have analogous validity to a speaker of Latin disparaging a speaker of Chinese for their communication skills. The advocate of scientific symbolic language is likely to share the same appreciation of indices as the Latin advocate would hold for Chinese.

This brings us to the interesting question of the origin of geomorphological hypotheses. Grove Karl Gilbert (1886, 1896) proposed that hypotheses derive from analogy to antecedent phenomena of the real world. Gilbert's invocation of analogy as a source of hypotheses is very puzzling to a modern analytical philosopher. Kitts (1980) simply argued that Gilbert was totally wrong, and that hypotheses derive from theory. Kitts seems to have been so positive on this point because of the nominalistic logic that underpins much of analytical science. In his popular *System of Logic* John Stuart Mill (1865) described analogy as ' . . . reasoning from particulars to particular'. This would certainly not produce theories, which are generals, not particulars.

In conventional logic (Salmon 1973) analogy is treated as a weak form of induction. The word 'weak' here refers to the strength of its explanatory power. It is important to remember that Gilbert was interested in discovery, not explanation, in the origin of hypotheses, not their justification. Analogy, as Gilbert (1886) clearly describes it, relates consequents (e.g. landforms) to their antecedents (e.g. causative processes). This logical manipulation of indices is handled by the cruder reasoning tool of analogy. Because the goal is not so much to generalize, which is the conventional philosophical goal attributed to induction, analogy functions as a combination of retroduction (inference to a hypothesis) with induction (inference from particulars). Thus, in the method of science described by C.S. Peirce and likely familiar to Gilbert (Baker in press), hypotheses are logically inferred when analogy is employed. In this sense, at least, hypotheses can be considered to originate by analogy, exactly as Gilbert (1886, 1896) proposed, but this is merely one example of the richness of potential reasoning that lies in the semiotic world of natural indices.

HYPOTHESES AND SCIENTIFIC EXPLANATION

The view that scientific hypotheses can be derived from existing theories has been criticized by the philosopher Paul Feyerabend. Feyerabend (1975) proposes the advocacy of any alternatives that occur to scientists regardless of how outrageous they might seem. Like Davis (1926), he is concerned that normal scientific practice may result in potentially important hypotheses being rejected because they initially appear contrary to the prevailing theoretical structure. Feyerabend presents his own outrageous hypotheses about the nature of science, comparing its practice to that of religion and arguing that there is no such thing as a scientific method (Feyerabend 1975). These latter claims have led to Feyerabend's status as a maverick among philosophers of science. Nevertheless, many of Feyerabend's scientific critiques of philosophy of science accord well with the attitudes of Earth scientists. He argues that much of analytical philosophy has concentrated on trivial issues of logic and theoretical meanings, such as the nature of realism. Earth science gets on very well without these fine points of logic and theory.

Feyerabend's claim that hypotheses do not derive from theory is an important insight, but his proposed alternative that hypothesis choice is a kind of poetry can be rather easily misunderstood. He claims that scientists should choose hypotheses for the pleasure derived from that choice. This would seem the height of irrationality to analytical philosophers. Indeed, Feyerabend's 'anything goes' anarchical approach to scientific method has been seized upon by postmodernist critics of scientific rationalism and objectivity. If science can be 'what you like', then all forms of knowledge have equal status, including mysticism, witchcraft, and voodoo. Feyerabend's insights suggest that a consideration of outrageous hypotheses in geomorphology may reveal more about the reasoning process than will conventional hypotheses of the type that are often rhetorically involved in published papers to provide the appearance of objective scientific methodology. This point will be developed in the next section of this chapter.

Like Feyerabend, C.S. Peirce is also critical of positivists and their logical empiricist successors for arbitrarily separating the context of discovery from the context of justification. Unlike Feyerabend, however, Peirce argued for a rationalist approach, but one in which the human mind is instinctively attuned to nature. Peirce supported his position, much as Whewell had before him, through extensive reference to the history of science. His triadic view of the sign relationship derived from a triadic phenomenology that underpins much of his philosophy. This distinguishes his view of scientific explanation from that commonly assumed in analytical philosophy of science. The prevailing nominalistic, dualistic approach to scientific explanation employs the 'deductive–nomological' model in which explanatory arguments subsume some resulting state of affairs (called the 'explanandum') under some covering laws (called 'explanans'), which act as the controlling state of affairs (Hempel 1966). The perusal of most scientific papers will confirm to their readers that this mode of logic prevails in published scientific explanation. On the other hand, Sir Peter Medawar (1991) argued that scientific papers were nearly all fraudulent. This rather forceful statement derives from the fact that it is a rare paper indeed that conveys the true process by which the science is actually done. Scientific papers provide dualistic models of scientific explanation because they leave out the third element, involving discovery of causative aspects of reality, that was essential to the whole scientific process. Describing the commonsense, acritical logic of scientific discovery is not a socially acceptable mode for presentation in 'scientific' papers.

Peirce (in Hartshorne and Weiss, 1931, p. 36) expresses his triadic notion of explanation by reference to the syllogistic model, noted above:

> ...An explanation is a syllogism of which the major premise, or rule, is a known law or rule of nature, or other general truth; the minor premise, or case, is the hypothesis or retroductive conclusion, and the conclusion, or result, is the observed (or otherwise established) fact.

This approach to science may have influenced the major writers on late nineteenth century method of geomorphology, particularly G.K. Gilbert, T.C. Chamberlin, and W.M. Davis (Baker in press). In relation to the deductive–nomological model, noted above, the explanations would have two parts: laws of nature (the major premise) and statements about causative controlling states of affairs (Peirce's 'hypotheses' or retroductive conclusions as the minor premise). This is how von Engelhardt and Zimmermann (1988) introduce Peirce's logic as a major element in Earth science reasoning (though they fail to develop the realism/nominalism issue in regard to this reasoning). It is also interesting to note that retroductive inference, long ignored by analytical philosophers of science, has now emerged as one of the 'hottest topics' in artificial intelligence (Peng and Reggia 1990) and cognitive science research (Josephson and Josephson 1994).

Before providing examples of geomorphological hypothesizing it is important to be clear on Peirce's meaning for retroduction, which he also called abduction. This inference of creative discovery is often confused with induction, so Peirce's distinction of the two is important (in Burks 1958, pp. 136–137):

> ...Nothing has so much contributed to present chaotic or erroneous ideas of the logic of science as failure to distinguish the essentially different characters of different elements of scientific reasoning; and one of the worst of these confusions, as well as one of the commonest, consists in regarding abduction and induction taken together (often mixed also with deduction) as a simple argument. Abduction and induction have, to be sure, this common feature, that both lead to the acceptance of a hypothesis because observed facts are such as would necessarily or probably result as consequences of that hypothesis. But for all that, they are the opposite poles of reason... The method of either is the very reverse of the other's. Abduction makes its start from the facts, without, at the outset, having any particular theory in view, though it is motivated by the feeling that a theory is needed to explain the surprising facts. Induction makes its start from a hypothesis which seems to recommend itself, without at the outset having any particular facts in view, though it feels the need of facts to support the theory. Abduction seeks a theory. Induction seeks for facts. In abduction the consideration of the facts suggests the hypothesis. In induction the study of the hypothesis suggests the experiments which bring to light the very facts to which the hypothesis had pointed. The mode of suggestion by which, in abduction, the facts suggest the hypothesis is by *resemblance*, – the resemblance of the facts to the consequences of the hypothesis. The mode of suggestion by which in induction the hypothesis suggests the facts is by *contiguity*, – the familiar knowledge that the conditions of the hypothesis can be realized in certain experimental ways.

The construction of hypotheses is a matter of scientific attitude. It does not derive from some body of theory, nor does it follow some rigorous method. However, Peirce was nearly unique among philosophers in noting that a method could be discerned in hindsight. His use of the word 'abductive' was a recognition of Aristotle's use of that word in the *Organon*, though Peirce believed a mistranslation of the original document confused logicians on this point for all later history. His description continues (in Burks 1958, pp. 137–138) as follows:

> ... that the matter of no new truth can come from induction or from deduction, we have seen. It can only come from abduction; and abduction is, after all, nothing but guessing. We are therefore bound to hope that, although the possible explanations of our facts may be strictly innumerable, yet our mind will be able, in some finite number of guesses, to guess the sole true explanation of them. *That* we are bound to assume, independently of any evidence that it is true. Animated by that hope, we are to proceed to the construction of a hypothesis.
>
> Now the only way to discover the principles upon which anything ought to be constructed is to consider what is to be done with the constructed thing after it is constructed. That which is to be done with the hypothesis is to trace out its consequences by deduction, to compare them with results of experiment by induction, and to discard the hypothesis, and try another, as soon as the first has been refuted; as it presumably will be.

The last sentence of this quote describes the essence of scientific method according to Peirce. The goal of discarding hypotheses would seem to be a severe requirement. Consider the dilemma for modern computer modelers. If we hold that models are really hypotheses (Baker 1985), then the task is not so much to make a beautifully elegant theoretical model. Rather, it is to falsify models. This is abuse indeed of our theoretical constructs!

Peirce (in Burks 1958, p. 138) goes on to list the considerations that determine the choice of a hypothesis:

> ... In the first place, it must be capable of being subjected to experimental testing. It must consist of experiential consequences with only so much logical cement as is needed to render them rational. In the second place, the hypothesis must be such that it will explain the surprising facts we have before us which it is the whole motive of our inquiry to rationalize.... In the third place, quite as necessary a consideration as either of those I have mentioned, in view of the fact that the true hypothesis is only one out of innumerable possible false ones, in view, too, of the enormous expensiveness of experimentation in money, time, energy, and thought, is the consideration of economy.

Peirce (1902b, pp. 761–672) also notes that the testing process involves some special considerations in regard to hypothesis choice:

> ... It is desirable to understand by a verifiable hypothesis one which presents an abundance of necessary consequences open to experimental tests, and which involves no more than is necessary to furnish a source of those consequences. The verification will not consist in searching the facts in order to find features that accord or disagree with the hypothesis. That is to no purpose whatsoever. The verification, on the contrary, must consist in basing upon the hypothesis predictions as to the results of experiments, especially those of such predictions as appear to be otherwise least likely to be true, and in instituting experiments in order to ascertain whether they will be true or not.

Thus, it is the surprising results of experiments, surprising except in terms of the hypothesis being tested, that are most important to the verification process. While experiments need not be elaborate or many, Peirce holds it of great importance that they be independent. This means that their results should be capable of explanation only through the hypothesis under investigation and not through some other hypothesis. To achieve this seemingly difficult independence of experiments Peirce directed attention to his famous 'economy of research' (Peirce 1879; Cushen 1967).

REALITY IN OUTRAGEOUS HYPOTHESES

As stated thus far, geomorphology is considered to be a way of thinking about Earth's surface and its processes, involving signs or indices of those real-world phenomena for which probable causes and/or explanatory concepts (that are real in Peirce's sense of the word) are inferred (through Peirce's 'retroduction') in such a way as to fit formerly known but seemingly unrelated phenomena (Whewell's 'colligation of facts'). Though super-ficially similar in appearance to lucky guessing, this retroductive process leads uncannily to hypotheses that display the repeated exhibition of explanatory surprise in regard to natural observation of experiments (Whewell's 'consilience of inductions'). The claim has been made that this involves the 'logic' of geomorphology as science. I now want to describe some important examples of geomorphological thinking to illustrate the process.

I wish my examples to involve scientific discovery, since that is important to the advancement of science, though hypotheses also obviously aid in the study of what is already known. Kantorovich (1993) notes that discovery can occur either by exposure or by generation. In exposure, some new phenomenon is observed, or information hidden in a set of statements is revealed (as in prediction from a new theory). Hypotheses are most important in discovery by generation, in which new theoretical concepts are developed. New theoretical concepts must replace old ones, but presumably the old concepts are adhered to for good reasons. How does one stimulate progress, getting past what most of the scientific community considers to be good working principles at some moment in time?

Late in life William Morris Davis reflected on this problem (Davis 1926, p. 464):

> ... Are we not in danger of reaching a stage of theoretical stagnation, similar to that of physics a generation ago, when its whole realm appeared to have been explored? We shall be indeed fortunate if geology is so marvelously enlarged in the next thirty years as physics has been in the last thirty. But to make such progress, violence must be done to many of our accepted principles; and it is here that the value of outrageous hypotheses, of which I wish to speak, appears. For inasmuch as the great advances of physics in recent years and as the great advances of geology in the past have been made by outraging in one way or another a body of preconceived opinions, we may be pretty sure that the advances yet to be made in geology will be at first regarded as outrages upon the accumulated convictions of to-day, which we are too prone to regard as geologically sacred.

Davis was sufficiently motivated by this view that, against the judgment of most of his contemporaries, he even proposed a serious contemplation of 'the Wegener outrage of wandering continents'. Wegener's hypothesis was well known to have so many flaws that it was considered untenable by many geologists in the 1920s. The flaws included the physical process whereby it operated and many errors of geological detail that were advanced by its meteorologist-inventor. However, this hypothesis also colligated numerous otherwise inexplicable facts, including the 'fit' of the continents and the distribution of Permian glacial deposits of the Southern Hemisphere. The problems were eventually resolved through the concepts of sea-floor spreading and plate tectonics, the latter being arguably the most successful conceptual scheme to have emerged in the Earth sciences (LeGrand 1988; Menard 1986; Stewart 1990). Wegener's concept was clearly a lucky guess inspired by an array of facts. It was motivated by the need for a theory to explain otherwise unrelated facts, but it flew in the face of existing theory.

As Davis's 1926 paper on 'the value of outrageous geological hypotheses' was being published, another outrage was being perpetrated by J Harlen Bretz, who had hypothesized a catastrophic flood origin for the Channeled Scabland region of the northwestern United States (Bretz, 1923, 1928). Bretz challenged a prevailing uniformitarian view that extraordinary causes, unlike those in operation today, should be precluded from hypothetical consideration. This view derived from the misconception that uniformitarianism constituted a fundamental principle of nature rather than a regulative guide to fallible human reasoning about nature. More will be said about such regulative principles in relation to hypothesizing in a subsequent section of this chapter. It is important here to note that the rationale for serious consideration of cataclysmic flooding as an explanation was based solely on field relationships on the Columbia Plateau (Baker 1978, 1995). Arguments against it were partly regulative and partly theoretical in that it was presumed that such immense flood flows were not physically reasonable and that there was no reasonable source for the vast quantities of water required (Baker and Bunker 1985). Once the field relationships were generally accepted by the scientific community, however, the debate was followed in short order by the demonstrated physical consistency of dynamical cataclysmic flood processes with scabland landforms and sedimentary sequences (Baker 1973a, b).

The reality of the Wegener and Bretz outrageous hypotheses seems to be clear in hindsight. How does one view such hypothesizing as it is unfolding? This may be a more relevant example in terms of geomorphological reasoning in relation to the continuity of thought envisioned by Charles Peirce. My candidate for a presently developing outrageous geomorphological hypothesis, true to the Davis (1926) definition, is the proposal by John Shaw (1994) that a very interesting assemblage of late Pleistocene glacial erosional and depositional features, including tunnel channels, Rogen moraine, and drumlins, may be indicative of massive subglacial flooding. It has long been known that a variety of landforms, involving both erosion and deposition, form beneath thick continental glaciers. These landforms have generally been explained by various subglacial deformation processes. However, modern glaciers do not provide satisfactory analogs to the ancient ice sheets that produced these landforms. In Shaw's hypothesis catastrophic subglacial flooding has been posed to explain the spatial and temporal associations of these features. Large-scale, subglacial meltwater floods are inferred to be responsible for certain drumlins (Shaw 1983; Shaw and Sharpe 1987), tunnel channels (Brennand and Shaw 1994), bedrock erosional marks (Shaw 1994), and Rogen moraine (Fisher and Shaw 1992).

The problem of subglacial cataclysmic flooding provides an excellent example of the methodological/philosophical quandaries faced in geomorphology and geology. There are no modern process equivalents to the gigantic late Pleistocene warm-based ice sheets of

North America and Eurasia. There is no possibility of controlled experiments scaled to the dynamics of such ice sheets. Nevertheless, there is a detailed assemblage of landforms and sediments (Shaw, 1994) for which we have no modern analogs. How is science to be done in such circumstances? Theories can predict the landforms by deduction, but which theories apply? Effects can be classified for extrapolation to theoretical generalization, but such verification of theory is logically flawed (Popper 1959). The solution is found in a process of generating hypotheses, as classically argued in geology (Baker 1988b). The goal is to infer cause from effect, or 'consequent from antecedent' as Gilbert (1886) so aptly described it. The logic of this reasoning is neither deductive nor inductive. It is retroductive or abductive (Baker in press), a reasoning central to geology (von Engelhardt and Zimmermann 1988) but greatly misunderstood by the advocates of deductive/inductive approaches in science, who use the exemplar of experimental/theoretical physics to justify their arguments.

THE LOGIC OF HYPOTHESES IN THE SCIENTIFIC PROCESS

If hypotheses violate the usual conventions of science, why should they be pursued? Presumably such hypotheses would have a very low probability of confirmation against future experimental tests. There are an immense number of possible outrageous hypotheses. If hypotheses are merely convenient fictions, and do not have some connection to reality, why do outrageous hypotheses, like those described in the previous section, prove very fruitful for further scientific investigation? One might think that there is some probability of the hypothesis being true that recommends itself to the hypothesizer. Tables for hypothesis probability have even been suggested (Strahler 1987).

Peirce's view on the probability of hypotheses is consistent with his view of retroduction as a unique process of discovery, and not something subject to a frequency of confirmation or disconfirmation. A retroductive hypothesis is accepted through a directed scientific process of inquiry involving a great complex of evidence. The process is not the trivial logical comparison of some statement of cases for true/false testing according to a nominalistic logic. This is simply not what scientists do; rather, it is something idealized by philosophers inexperienced in actual scientific practice. As Peirce stated it (in Hartshorne and Weiss 1931, p. 78): 'It is nonsense to talk of the probability of a law, as if we could pick universes out of a grab-bag and find in what proportion of them the law held good.' Peirce's frequentist views of probability were even extended to inductive scientific inference, a view that would be quite controversial to modern philosophers of probability.

Peirce's emphasis on how scientists actually reason can be reconciled with the view that logic is a normative science, referring to how reasoning ought to be done. For Peirce this normative function is divided between a formulated, scientific, and critical logic, which he called the *logica docens*, and an implicit and acritical logic, the *logica utens*, that is instinctive and part of the commonsense reasoning of the inquirer. Peirce believed that the *logica utens* is antecedent to systematic reasoning. Because the instinctive *logica utens* does not serve on all occasions, one must embark upon the *logica docens*. As will be seen below, Peirce hypothesized that the *logica utens* has a very important function in science, allowing the scientist to 'guess right' in the formulation of hypotheses. Guessing right does not mean that one derives the absolutely correct answer by retroduction. It does mean that one reasons to a hypothesis that is fruitful and productive on the path of scientific inquiry. Because it involves reason, Peirce considers these issues to be matters of logical concern. This is the sense in which he views the logic of scientific discovery. Although a few modern analytical philosophers, notably Hanson (1958), tried to argue some aspects of Peirce's logical position, they have generally not accepted his instinctive basis of the *logica utens*. Most philosophical opinion denies that a logic of discovery is possible (Kantorovich 1993), but this opinion generally denies this from the position of nominalistic *logica docens*.

It is interesting that when Peirce considered the eventual testing of hypotheses, not to be confused with the retroductive phase of inquiry, his view came to a position remarkably similar to that espoused many years later by Sir Karl Popper. As published by Hartshorne and Weiss (1931, p. 48), Peirce's view is as follows:

> It is a great mistake to suppose that the mind of the active scientist is filled with propositions which, if not proved beyond all reasonable cavil, are at least extremely probable. On the contrary, he entertains hypotheses which are almost wildly incredible, and treats them with respect for the time being. Why does he do this? Simply because any scientific proposition whatever is always liable to be refuted and dropped at short notice. A hypothesis is something which looks as if it might be true and were true, and which is capable of verification or refutation by comparison with facts. The best hypothesis, in the sense of the one most recommending itself to the inquirer, is the one which can be the most readily refuted if it is false. This far outweighs the trifling merit of being likely. For after all, what is a *likely* hypothesis? It is one which falls in with our preconceived ideas. But these may be wrong. Their errors are just what the scientific man is out gunning for more particularly. But if a hypothesis can quickly and easily be cleared away so as to go toward leaving the field free for the main struggle, this is an immense advantage.

The issue here is one of hypothesis selection. Reasoning to a best hypothesis must involve the 'one most recommending itself to the inquirer'. If we view this selection process purely objectively, in which there is not necessarily any immediate connection to nature's reality (though testing may yet reveal one), then one can come to the position that the creative process might well generate an unlimited number of explanatory hypotheses. This multiple generation process is quite valueless. The valuable activity is selection of the most fruitful combinations of explanatory possibilities. The mathematical physicist Henri Poincaré argued that such selection occurs according to some inner sense of the inquirer, such as the 'mathematical beauty' or 'symmetry' concepts that are often described in the published reminiscences of famous mathematicians and physicists. This inner sense may be combined with a psychological interplay between the conscious and subconscious. This psychological process involves a period of deep and intense thought which fails to solve an especially difficult problem. The scientist then pursues some activity totally unrelated to the problem, usually something leisurely that involves little thought. During this activity a flash of illumination strikes the mind, resolving the original problem. Poincaré (1914) provides a striking example of this in an anecdote about his inability to resolve a mathematical problem. The solution came to him while engaged in a geological excursion!

The selection process for mathematical discovery involves reasoning in symbols that are the most formal and detached from natural connections of any in science. When complex signs, such as the indices relating to real causes in nature, enter the reasoning process, it is not likely that the creative process is exactly the same as used by a mathematical genius. Presumably there are selection criteria available to any scientist that will apply in hypothesis selection.

Parsimony and Simplicity

The two regulative principles generally invoked for Earth science hypothesizing are uniformitarianism and evolutionism (von Engelhardt and Zimmermann 1988). Evolutionism will receive some consideration under the topic of naturalism. The concern here will begin with uniformitarianism, one of the few Earth science issues to have received appreciable philosophical attention. Like all regulative principles, uniformitarianism has a methodological, or 'weak' form that specifies procedures for reasoning about the Earth, especially hypothesizing. It also has a substantive, or 'strong' form that makes ontological claims about how the Earth actually behaves. Though long recognized (Gould 1965, 1987), the various types of uniformitarianism continue to cause confusion for practicing Earth scientists (Shea 1982). The Channeled Scabland controversy involved a mistaken application of substantive uniformitarianism by those who criticized Bretz for advocating catastrophism (Baker 1981). Ironically Bretz's hypothesis was formulated according to a regulative rule of simplicity (Baker 1987), which is now recognized as the methodological manifestation of uniformitarianism (Goodman 1967; Gould 1987).

The principle of simplicity (methodological uniformitarianism) serves as a means for promoting hypothesis selection from among multiple causal possibilities. This has logical status, as promoted by the famous nominalist medieval logician William of Ockham. The principle may be stated in modern terms as follows (Jeffreys and Berger 1992, p. 64): Among competing hypotheses favor the simplest one. This is also known as 'Ockham's razor' or the 'principle of scientific parsimony'. If logic is totally divorced from connection to real entities, as nominalists presume, then this principle can be globally applied to adjudicating between competing hypotheses, no matter what the causes to which those hypotheses refer. However, the alternative, that the presumption of simplicity implies something real about the nature of the world, does not have to be as drastic as substantive uniformitarian claims that catastrophic processes are precluded. Such an ontological claim can be as simple as our supposition that the laws of mechanisms apply both on Earth and at the orbit of Jupiter. While the success of spacecraft launches attests to this claim, it does not prove it. On the other hand, if miracles can occur, allowing the laws of mechanics to be different at different points in the solar system, a great deal of science will be precluded.

The full argument that simplicity may imply some subtle ontological claims is beyond the scope of this chapter. Sober (1988) provides a good summary of these arguments in relation to issues in evolutionary biology. The key point I want to make here is that regulative principles are not necessarily purely methodological. Arguments that they are so may be developed in the crisp logic of deduction. They are far more difficult to make for the logic of induction, to which David Hume applied an early formulation of uniformity of nature: the presumption that the future (or past) will be like the present. If simplicity or parsimony applies to inductive inference, what principle applies to retroduction, the notion of reasoning to a best hypothesis?

The Light of Nature

Charles Peirce came to a remarkable conclusion about hypothesis selection. Moreover, if we agree with Peirce that hypothesizing through retroduction is the whole basis of a natural science like geomorphology, then this conclusion also applies to the basis of that science. Here is the statement as Peirce presented it in his 1898 lectures (Ketner 1992, pp. 176–177):

> The only end of science, as such, is to learn the lesson that the universe has to teach it. In Induction it simply surrenders itself to the force of facts. But it finds, at once, – I am partly inverting the historical order in order to state the process in its logical order, – it finds I say that this is not enough. It is driven in desperation to call upon its inward sympathy with nature, its instinct for aid, just as we find Galileo at the dawn of modern science making his appeal to *il lume naturale*. But insofar as it does this, the solid ground of fact fails it. It feels from that moment that its position is only provisional. It must then find confirmations or else shift its footing. Even if it does find confirmations, they are only partial. It still is not standing upon the bedrock of fact. It is walking upon a bog, and can only say, this ground seems to hold for the present. Here I will stay till it begins to give way. Moreover, in all its progress science vaguely feels that it is only learning a lesson. The value of *Facts to it*, lies only in this, that they belong to Nature; and Nature is something great, and beautiful, and sacred, and eternal, and real, – the object of its worship and its aspiration.

Note that Peirce's metaphor of the bog is exactly the same as that quoted by Sir Karl Popper at the beginning of this chapter. The metaphysics of science, accounting for its foundations, is immensely difficult in philosophical terms. What has been remarkable, however, and a great puzzle to philosophers (Gjertsen 1989), is that scientists are able to be amazingly successful in explaining nature without ever resolving these foundational difficulties. This is what is revealed so well in historical study of scientific inquiry.

Peirce was greatly influenced by his historical study of Galileo. He particularly noted Galileo's appeal to the light of nature (*il lume naturale*) as the means of retroductive success, or guessing right (Eisele 1979). Peirce (1891, p. 165) writes:

> A modern physicist on examining Galileo's works is surprised to find how little experiment had to do with the establishment of the foundations of mechanics. His principal appeal is to common sense and *il lume naturale*. He always assumes that the true theory will be found to be a simple and natural one.

Peirce also found that the appeal to nature was often confused with the notion of simplicity. As a logician Peirce must have been interested in this confusion, as he had been in Newton's use of the word 'hypothesis'. Peirce (1908, p. 104) wrote the following description of simplicity as a criterion for hypothesis selection:

> Modern science has been builded after the model of Galileo, who founded it on *il lume naturale*. That truly inspired prophet had said that, of two hypotheses, the *simpler* is to be preferred; but I was formerly one of those who, in our dull self-conceit fancying ourselves

more sly than he, twisted the maxim to mean the *logically* simpler, the one that adds the least to what has been observed, in spite of three obvious objections: first, that so there was no support for any hypothesis; secondly, that by the same token we ought to content ourselves with simply formulating the special observations actually made; and thirdly, that every advance of science that further opens the truth to our view discloses a world of unexpected complications. It was not until long experience forced me to realise that subsequent discoveries were every time showing I had been wrong, while those who understood the maxim as Galileo had done, early unlocked the secret, that the scales fell from my eyes and my mind awoke to the broad and flaming daylight that it is the simpler Hypothesis in the sense of the more facile and natural, the one that instinct suggests, that must be preferred; for the reason that unless man have a natural bent in accordance with nature's, he has no chance of understanding nature at all ... I do not mean that logical simplicity is a consideration of no value at all, but only that its value is badly secondary to that of simplicity in the other sense.

This tendency of the human mind to 'have a natural bent in accord with nature's' is an unusual form of naturalism, much as Peirce's realism is not the usual form of that doctrine. Peirce's ascribing of a major role of Galileo's scientific reasoning to *il lume naturale* (the light of nature) is a very striking statement. Indeed, an anonymous reviewer of an early version of this essay (clearly a philosopher and historian of science) wrote:

I know Galileo's writings fairly well, along with a good bit of the scholarly literature. I cannot think of any place where such a notion appears, let alone plays a prominent role. A review of several relevant texts has not provided me with any evidence that such a notion played an important role in Galileo's understanding of science.

The review statement is reminiscent of the disbelief some modern historians of science accorded Galileo's writings in which he commonly invokes reasoning *ex suppositione* (hypothesizing from effects to causes). Galileo's logic in such writings seems to come from his interest in Aristotelian and Thomistic writings, rejecting those of Plato and various nominalists, such as William of Ockham. Wisan (1978, p. 47) expressed her disbelief as follows: ' ... today, of course, everyone knows that one cannot argue rigorously from effects to causes ... '.

Of course, Galileo, like Peirce, was not interested in *ex suppositione* reasoning for its value in rigorous argument. He was interested in discovering new explanations, and his remarkable success in that endeavor led Peirce to take Galileo's methodological writings very seriously. McMullin (1978) also recognized the retroductive character of Galileo's inferences, and Wallace (1981) extensively explored the logical basis of his thought in the Aristotelian tradition.

The phrase *il lume naturale* appears prominently in Galileo's most mature work, *Discourse Upon Two New Sciences*, published in 1638. Drake's (1974) well-known English version gives the bizarre translation 'my good sense', but Drake (1974, p. 162) places the original in parentheses, thereby highlighting its unusual invocation. Wallace (1992a) recently traced the source directly, and his translation of Galileo's logical treatises (Wallace 1992b) contains the multiple references to *il lume naturale*. Wallace (1992a) even related how and why Galileo's logical treatises, written early in his career, were ignored and not even translated by modern scholars. Of course, Charles Peirce, who explored the history of logic for planned but uncompleted books, would have been desperately interested in all of Galileo's logical pronouncements. He recognized their importance long before modern scholars did. This failure of modern scholars, until very recently, to understand the Aristotelian realism of Galileo's logic provides a very instructive example of how modern beliefs impact historical inquiry. Peirce would probably have attributed the oversight to the blinding nominalism of the scientific reasoning process presumed by most philosophers and historians of science.

NATURALISM AND HYPOTHESES

Naturalism is a philosophical position that holds scientific explanations to arise through methods that are continuous from the natural domain of objects and events. Naturalism has become important in various evolutionary epistemologies, such as David Hull's (1988) modeling of the development of science on the natural selection process. Kantorovich (1993) expands Hull's descriptive thesis into a full explanatory theory of science based on naturalistic evolutionary criteria.

The concern here will be more limited, confined to finding the natural basis for *il lume naturale*, the instinct that Peirce claims to provide the basis for retroduction ('guessing right'). Peirce's naturalism has been labeled 'ecstatic' by Corrington (1993). Particularly in his later writings Peirce sees in nature not so much something to be described in an objective sense, as the reductionist program of some scientists would have it, but rather an entity with which the human mind has an affinity that leads to self-understanding. Peirce's emotional arguments in this regard were much criticized during the heyday of logical positivism (Buchler 1939; Goudge 1950). However, the utter purposelessness implied by a reductive scientific program has recently produced a new philosophical debate about science and human values. On one hand, there are critics of science's cold austerity and objectivity who claim that these qualities in our modern technological world are destroying humankind's aesthetic sense (Appleyard 1992). On the other hand, certain scientists are espousing a scientism that denies the commonsense origins of knowledge (Cromer 1993) and glorifies the unnatural nature of science (Wolpert 1992). Much of this polarization centers around the view of science as objective knowledge, theories, and facts about the world, rather than science as a human process, naturally attuned to nature's inspiration. The natural connection in science may be more effectively conveyed by considering the origin of hypotheses, rather than the explanatory power of scientific 'laws'. Sciences of natural indices, like geomorphology, may be more appropriate for its exposition than sciences of detached symbols, like mathematical physics.

Charles S. Peirce made the astonishing proposal that the human mind is naturally adapted to 'guess right', that is, to form highly probable causal hypotheses based on the experience of mental interaction with the world. This mental interaction involves relating to the world in a system of signs. Of course, more sophisticated interactions take place through language, which is itself a system of signs. The role of language is so important that some constructivist philosophers argue that scientific reasoning is necessarily embedded in language. A corollary to this view is the relativist position that the language world of every person precludes any truly objective access to reality. Peirce would vehemently disagree with many current relativist, idealist views of science. He strongly supported the experimental approach in which inductive measurement is matched against deduced consequences of theories. Much more importantly, however, Peirce held that reality influences causal hypothesizing.

The idea of an instinctive basis for a type of inference (retroduction) would be denied by behavioral psychologists, such as Pavlov and Skinner, who argued that young children and animals are conditioned in their responses to the world. Repeated associations of phenomena may give some appearance of causation to sophisticated human adults, as the philosopher David Hume argued, but the ability to infer is presumed to be a later development. Animals and young children, according to the behavioralists, are incapable of hypothetical inferences as Peirce envisioned them. In contrast, however, recent work in cognitive psychology is more consistent with Peirce's naturalistic view of human thought. In experiments with very young infants, Leslie (1982) and Leslie and Keeble (1987) demonstrate that even children at age four-and-a-half months perceive interactions between objects in causal terms. This understanding of causality proceeds in parallel to the acquisition of language skills, and is very advanced before children learn to speak. The understanding of causality is developed to a high degree in play and in pretending.

Even animals may have a natural reasoning instinct, contrary to behavioralist assumptions (Dickinson 1980). In a remarkable set of experiments on laboratory rats, Holland and Straub (1979) show that rats can apply a logic of inference that links two sets of associations to infer a third relationship. While this may fall short of strict causality, it is much more sophisticated than the simple associative reasoning envisioned by Pavlov. Cheney and Seyfarth (1990) document even more sophisticated hypothetical inference by vervet monkeys. Their experiments document true 'if..., then...' inferences by these animals. The implications of these studies are that the human mind has evolved to the present ability to conduct scientific inquiry, and that this process of inquiry is strongly rooted in nature itself. This hypothesis has the merit of explaining a remarkable paradox: How it is that science has been so successful in explaining the natural world. Could its success derive from it being the only method that has emerged (evolved?) from that world?

Socioevolutionary theories of science are currently popular in philosophy of science. For example, Kantorovich (1993) argues that the growth of science is analogous to an evolutionary process in which there is a coevolution of human action (analogous to the evolution of sensorimotor organs) and human understanding (analogous to the brain). He adds to this a shock effect caused by unintended or serendipitous discoveries. One might well extend this evolutionism to Peirce's problem of *il lume naturale*. A candidate for this is gene-cultural evolution, the sociobiological thesis that culture is elaborated under the influence of hereditary learning propensities, while the genes prescribing the propensities are spread in a cultural context (Lumsden and Wilson 1985). An example is the socio-biological argument for an instinctive human fear of snakes. Wilson (1994) notes that phobias are rarely acquired for the really dangerous items of modern life, including guns and speeding automobiles. These have not threatened the human species long enough for acquisition of predisposing genes that will ensure avoidance. Snakes, however, have a long history of close association with humans with likely adverse consequences. Phobia as a means of snake avoidance may have conveyed survival advantages, and gene-based evolution is at least a plausible explanation for the development of that phobia.

Presumably the survival value of 'guessing right' is at least as good a candidate for sociobiological explanation as snake phobia. On the other hand, I doubt that the extreme reductionism of the sociobiologists would have been very appealing to Peirce, who entertained a very different metaphysical view of evolution than these modern treatments.

An alternative to the evolutionary model has been offered by Thomas Nagel (1986), who treats the problem as one of self-understanding. As he defines this (Nagel 1986, p. 78), self-understanding is ' . . . an explanation of the possibility of objective knowledge of the real world which is itself an instance of objective knowledge of that world and our relation to it'. Nagel is critical of any evolutionary explanation, which he considers (Nagel 1986, p. 78) to be ' . . . an example of the pervasive and reductive naturalism of our culture'. He notes that Darwinian natural selection explains selection among generated organic possibilities, but that it does not explain the possibilities themselves. As he puts it (Nagel 1986, p. 78), 'It may explain why creatures with vision or reason will survive, but it does not explain how vision or reason are possible.'

Nagel's language is reminiscent of Peirce in how he marvels at the abilities of human beings to understand the world. He comes to the conclusion that scientific theorizing and its success must involve some essential or fundamental quality of the universe. Evolution might be a method of bringing this forth, but its potentiality must already exist (Nagel 1986, p. 81):

> I don't know what an explanation might be like either of the possibility of objective theorizing or of the actual biological development of creatures capable of it. My sense is that it is antecedently so improbable that the only possible explanation must be that it is in some way necessary. It is not the kind of thing that could be either a brute fact or an accident . . . the universe must have fundamental properties that inevitably give rise through physical and biological evolution to complex organisms capable of generating theories about themselves and it.

Nagel's concern with the possibilities of self-understanding is quite analogous to Peirce's concern with the possibility of scientific success through retroduction. Peirce's metaphysics, which is mostly beyond the scope of this chapter, envisions a phenomenology of pure categories that provide the structure through which all perceptions and experiences come to be understood. The first of these categories involves the individual being of things and includes a kind of chance. The chance of this Firstness category is not entirely uncaused and irregular. Instead, it is a spontaneity that is some degree regular and reflects a kind of reason. This is also an absolute chance in which the universe finds beginning; Peirce called it 'tychism'. Peirce would argue that this tychism is what gives reason its tendency toward better explanation. But reason is also tied together by a continuity that involves connection through the other categories of understanding. This continuity he called 'synechism', and he believed it to be displayed in commonsense reasoning, as in the *logica utens*. Because it connects the relating tendencies (Secondness) and the generalizing or law-establishing tendencies (Thirdness) to pure potentiality for developing laws (Firstness), this synechism is exactly with what the mind is presumed to be attuned in hypothesizing.

The above may well confirm the reader of G.K. Gilbert's description of Peirce in a letter cited by Davis (1927, p. 269): ' . . . a man so metaphysical I should never had tho't of going to him with a practical question'. Nevertheless, it is also fascinating that current biological research into self-organization and complexity holds that a kind of order emerges naturally from the most complex of systems (Kauffman 1995). This order presumably derives from forces that operate right at the edge of chaotic behavior. Is this a mathematical manifestation of Peirce's tychism?

The cosmological implications of these principles also have their counterparts in current physical theories of the universe. Among modern cosmologists there is a split between those who see the universe as totally devoid of purpose and those who see it filled with causal necessity. Typical of the former camp is Weinberg's (1993, p. 154) conclusion, '...the more the universe seems comprehensible, the more it seems pointless'. In complete contrast is Dyson's (1979) argument of universal growth in richness and complexity. Wheeler (1983) develops similar arguments from the requirement that observers participate in the development of the universe, a creative process that Barrow and Tipler (1986) have named the 'Participatory Anthropic Principle'.

The Anthropic Cosmological Principle was postulated to deal with the physics of explaining the universe and its origins (Hawking 1988). It holds that the universe can in some ways be explained that it must be such as to contain people. Its weak form is similar to a methodological rule (Carter 1974, p. 291): '...what we can expect to observe must be restricted by the conditions necessary for our presence as observers'. This is not controversial. The strong version, however, makes a substantive statement about the universe (Carter 1974, p. 294): '...the Universe (and hence the fundamental properties on which it depends) must be such as to admit the creation of observers within it at some stage'.

The tentative conclusion is that the importance of the Anthropic Principle for cosmology (Barrow and Tipler 1986) is paralleled by the importance to science of the principle that our reasoning (particularly retroductive hypothesizing) is a part of the nature we seek to fathom. In its strong form, the ontological version, our minds are attuned to nature in some manner, as hypothesized by Peirce, and worthy of further scientific and philosophical inquiry.

DISCUSSION AND CONCLUSIONS

Modern philosophers of 'science' have generally used the experimental, conceptual sciences as their ideal for discussing the justification of knowledge. Much current interest is devoted to distinguishing the predominant philosophy of physics from the newly emergent philosophy of biology. The Earth sciences, outside of minor interest in the plate-tectonic 'revolution' and the medieval metaphysics of uniformitarianism, have not figured prominently in philosophy of 'science'.

Controlled experiments, at least for the most interesting phenomena, are not possible in much of the Earth sciences, including geomorphology. Instead of the detached objectivity of sterile laboratories, the geomorphologist must be concerned with the open problem of the field, in which nature is 'taken as it is'. Great geomorphologists of the past, including Chamberlin, Gilbert, and Davis, championed the use of hypotheses for understanding the field problem. However, their concept of hypotheses is very different than that of physicists like Henri Poincaré, who viewed them as theoretical instruments. The hypothesis derived from theory alone, while fine for a controlled experimental physics of Earth's surface (geomorphysics?), is not appropriate for a holistic understanding of that surface. Hypotheses must have a connection to the reality that they purport to represent, and this connection must be achieved at their inception, not just through some later correspondence between theoretical prediction and objective measurement. This position was argued effectively a century ago by the famous logician C.S. Peirce, who advanced the notion that our minds are constituted in some way to be continuous in thought with aspects of nature. Peirce's hypothesis about hypothesizing is not in itself a complete explanation for how reasoning in a natural science like geomorphology is both possible and successful. Rather, it is a recognition of real world coincidences that cries out for an explanation. In this regard it is a manifestation of what Whewell termed 'consilience' in science.

The guessing instinct that regulates our scientific hypothesizing does not reveal the truth of things; it merely places the inquirer on a fruitful path of investigation. That path leads to the deduction of consequences and their experimental testing. However, the nominalistic logic that was developed to be effective in these latter activities may not be a good model for the retroductive reasoning that underpins hypothesizing. If we agree that scientific reasoning is a continuous process of inference connecting the real world and concepts of that world, one must be careful with logical tools like Ockham's razor. J Harlen Bretz expressed the commonsense methodology of geomorphologists aptly in a 1978 communication to me: 'I have always used Chamberlin's method of multiple working hypotheses. I applied Ockham's razor to select the most appropriate hypothesis, but always with due regard for possible dull places in the tool.'

What would a science be like that consisted only of theory, or systems of laws about the physical world, plus stark facts about the world's reality? Some analytical philosophies of science seem to imply that these are the sole elements of science conducted objectively within individual scientific minds. The sole scientific communication with nature consists of tests against the real world, and these well-tested theories build up an objective base for coping with the world. There are examples of human beings that function in much this way. These individuals operate within individual minds and are unable to view nature from nature's point of view, nor are they able to see the world from the points of view of other people. These people are able to meet their individual needs and function in realistic terms, but their relation to the collective human mind is severely handicapped. Perhaps the analogy to this tragic human condition, known as autism, should not be stretched too far. The warning it implies to analytical philosophers of science is that science has human, social, and naturalistic elements, if only because it is conducted by human beings. Denial of these elements may be as unhealthy to the human enterprise of science as such denial is to the mental health of individual humans.

Much as Peirce's views were criticized by philosophers earlier in this century, one can argue that the proponents of naturalism and realism in the hypothetical reasoning underpinning geomorphological inference (Gilbert 1886, 1896; Chamberlin 1890; Davis 1926) were philosophically naive (Kitts 1980). Of course, the analytical philosophy of which these geomorphologists were innocent had largely not been formulated when they wrote their seminal papers. Another explanation is that the late nineteenth century geomorphologists were operating in a different philosophical tradition. Frodeman (1995) recently proposed that this tradition may have elements which we can recognize in modern continental philosophy. Alternatively, as argued here, the tradition may relate to positions espoused by William Whewell and especially by Charles Peirce, who envisioned all thought as a system of signs, including a continuity to nature that includes an instinctive capability for producing fruitful hypotheses in scientific investigation. Whatever the view, it is clear that the question of the philosophical foundations to geomorphology is not so clear-cut as some would have us believe. Those foundations do lie upon a bog or swamp of metaphysical concerns, as described by both Peirce and Popper. However, just because we have not yet successfully probed the depths of that swamp does not mean that we should ignore its presence. All science relies upon regulative principles and other metaphysical notions. It is precisely because we know so little of them that they should be thoroughly criticized and alternatives proposed that may seem more in accord with our actual practices of reasoning.

ACKNOWLEDGEMENTS

I thank Mike Church and an anonymous reviewer for their stimulating criticism of an earlier version of this chapter. This chapter is AUMIN Contribution No. 9.

REFERENCES

Ahnert, F. 1980. A note on measurements and experiments in geomorphology, *Zeitschrift für Geomorphologie*, Supplementband **35**, 1–10.

Appleyard, B. 1992. *Understanding the Present: Science and the Soul of Modern Man*, Doubleday, New York, 269 pp.

Baker, V.R. 1973a. *Paleohydrology and Sedimentology of Lake Missoula Flooding in Eastern Washington*, Geological Society of America Special Paper 144, Boulder, Colo., 79 pp.

Baker, V.R. 1973b. Erosional forms and processes for the catastrophic Pleistocene Missoula floods in eastern Washington, in *Fluvial Geomorphology*, edited by M. Morisawa, Publications in Geomorphology, State University of New York, Binghamton, New York, pp. 123–148.

Baker, V.R. 1978. The Spokane Flood controversy and the Martian outflow channels, *Science*, **202**, 1249–1256.

Baker, V.R. (ed.) 1981. *Catastrophic Flooding: The Origin of the Channeled Scabland*, Dowden, Hutchinson and Ross, Stroudsburg, Pa., 360 pp.

Baker, V.R. 1985. Models of fluvial activity on Mars, in *Models in Geomorphology*, edited by M. Woldenberg, Allen and Unwin, Boston, pp. 287–312.

Baker, V.R. 1987. The Spokane Flood debate and its legacy, in *Geomorphic Systems of North America*, edited by W.L. Graf, Geological Society of America Centennial Vol. 2, Boulder, Colo., pp. 416–423.

Baker, V.R. 1988a. Cataclysmic processes in geomorphological systems, *Zeitschrift für Geomorphologie*, Supplementband **67**, 25–32.

Baker, V.R. 1988b. Geological fluvial geomorphology, *Geological Society of America Bulletin*, **100**, 1157–1167.

Baker, V.R. 1995. Joseph Thomas Pardee and the Spokane Flood controversy, *GSA Today*, **5**, 169–173.

Baker, V.R. in press. The pragmatic roots of American Quaternary geology and geomorphology, *Geomorphology*.

Baker, V.R. and Bunker, R.C. 1985. Cataclysmic late Pleistocene flooding from glacial Lake Missoula: a review, *Quaternary Science Reviews*, **4**, 1–41.

Barrow, J.D. and Tipler, F.J. 1986. *The Anthropic Cosmological Principle*, Oxford University Press, Oxford, 706 pp.

Brennand, T.A. and Shaw, J. 1994. Tunnel channels and associated landforms, south-central Ontario: their implications for ice-sheet hydrology, *Canadian Journal of Earth Science*, **31**, 505–522.

Brent, J. 1993. *Charles Sanders Peirce: A Life*, Indiana University Press, Bloomington, Ind., 388 pp.

Bretz, J H. 1923. The Channeled Scabland of the Columbia Plateau, *Journal of Geology*, **31**, 617–649.

Bretz, J H. 1928. The Channeled Scabland of eastern Washington, *Geographical Review*, **18**, 446–477.

Buchler, J. 1939. *Charles Peirce's Empiricism*, Columbia University Press, New York, 275 pp.

Burks, A.W. (ed.) 1958. *Collected Papers of Charles Sanders Peirce*, Vol. 7, Harvard University Press, Cambridge, Mass., 415 pp.

Carter, B. 1974. Large number coincidences and the Anthropic Principle in cosmology, in *Confrontation of Cosmological Theories with Observational Data*, edited by M.S. Longair, Reidel, Dordrecht, pp. 290–298.

Chamberlin, T.C. 1890. The method of multiple working hypotheses, *Science*, **15**, 92–96.

Cheney, D.L. and Seyfarth, R.M. 1990. *How Monkeys See the World*, University of Chicago Press, Chicago, 377 pp.

Church, M. 1984. On experimental methodology in geomorphology, in *Catchment Experiments in Fluvial Geomorphology*, edited by T.P. Burt and D.E. Walling, Geobooks, Norwich, pp. 563–580.

Cloos, H. 1953. *Conversation with the Earth*, Alfred A. Knopf, New York, 413 pp.

Corrington, R.S. 1993. *An Introduction to C.S. Peirce: Philosopher, Semiotician, and Ecstatic Naturalist*, Rowan and Littlefield, Lanhan, Md., 227 pp.

Cromer, A. 1993. *Uncommon Sense*, Oxford University Press, Oxford, 240 pp.

Cushen, W.E. 1967. C.S. Peirce on benefit–cost analysis of scientific activity, *Operations Research*, **14**, 641–648.

Davis, W.H. 1972. *Peirce's Epistemology*, Martinus Nijhoff, The Hague, 163 pp.

Davis, W.M. 1926. The value of outrageous geological hypotheses, *Science*, **63**, 463–468.

Davis, W.M. 1927. Biographical Memoir – Grove Karl Gilbert, 1843–1918, *National Academy of Sciences, Biographical Memoir v. 21, 5th Memoir*, Washington, DC, 303 pp.

Dickinson, T. 1980. *Contemporary Animal-Learning Theory*, Cambridge University Press, Cambridge, 177 pp.

Drake, S. 1974. *Galileo Galilei: Two New Sciences*, University of Wisconsin Press, Madison, 323 pp.

Dusek, V. 1979. Geodesy and the Earth sciences in the philosophy of C.S. Peirce, in *Two Hundred Years of Geology in America*, edited by C.J. Schneer, University Press of New England, Hanover, NH, pp. 265–276.

Dyson, F.J. 1979. Time without end: physics and biology in an open universe, *Review of Modern Physics*, **51**, 447–460.

Eisele, C. 1959. Poincaré's positivism and Peirce's realism, in *Actes du IXe Congrès International d'Histoire des Sciences*, Asociación para la Historia de la Ciencia Española Universidad de Barcelona, Barcelona, pp. 461–465.

Eisele, C. 1979. The influence of Galileo on Peirce, in *Studies in the Scientific and Mathematical Philosophy of Charles S. Peirce: Essays by Carolyn Eisele*, edited by R.M. Martin, Mouton, The Hague, pp. 169–176.

Feyerabend, P. 1975. *Against Method*, New Left Books, London, 279 pp.

Fisch, M.H. 1980. The range of Peirce's relevance, *Monist*, **63**, 269–276.

Fisch, M. 1991. *William Whewell: Philosopher of Science*, Oxford University Press, Oxford, 220 pp.

Fisher, T.G. and Shaw, J. 1992. A depositional model for Rogen moraine, with examples from the Avalon Peninsula, Newfoundland, *Canadian Journal of Earth Science*, **29**, 669–686.

Frodeman, R. 1995. Geological reasoning: geology as an interpretive and historical science, *Geological Society of America Bulletin*, **107**, 960–968.

Gilbert, G.K. 1886. The inculcation of scientific method by example, *American Journal of Science*, **31**, 284–299.

Gilbert, G.K. 1896. The origin of hypotheses illustrated by the discussion of a topographic problem, *Science*, **3**, 1–13.

Gjertsen, D. 1989. *Science and Philosophy*, Penguin Books, London, 296 pp.

Goodman, N. 1967. Uniformity and simplicity, *Geological Society of America Special Paper*, **89**, 93–99.

Goudge, T.A. 1950. *The Thought of C.S. Peirce*, University of Toronto Press, Toronto, 360 pp.

Gould, S.J. 1965. Is uniformitarianism necessary?, *American Journal of Science*, **263**, 223–228.

Gould, S.J. 1987. *Time's Arrow, Time's Cycle: Myth and Metaphor in the Discovery of Geological Time*, Harvard University Press, Cambridge, Mass., 222 pp.

Hanson, N.R. 1958. *Patterns of Discovery*, Cambridge University Press, Cambridge, 241 pp.

Hartshorne, C. and Weiss, P. 1931. *Collected Papers of Charles Sanders Peirce*, Vol. 2, Harvard University Press, Cambridge, Mass., 486 pp.

Hausman, C.R. 1993. *Charles S. Peirce's Evolutionary Philosophy*, Cambridge University Press, Cambridge, 230 pp.

Hawking, S.W. 1988. *A Brief History of Time*, Bantam, New York, 198 pp.

Hempel, C.G. 1966. *Philosophy of Natural Science*, Prentice-Hall, Englewood Cliffs, NJ, 116 pp.

Holland, P.C. and Straub, J.J. 1979. Differential effects of two ways of devaluing the unconditioned stimulus after Pavlovian appetitive condition, *Journal of Experimental Psychology: Animal Behavior Processes*, **5**, 65–78.

Hull, D. 1988. *Science as a Process*, University of Chicago Press, Chicago, 586 pp.

Jeffreys, W.H. and Berger, J.O. 1992. Ockham's razor and Bayesian analysis, *American Scientist*, **80**, 64–72.

Josephson, J.R. and Josephson, S.G. 1994. *Abductive Inference: Computation, Philosophy, Technology*, Cambridge University Press, Cambridge, 306 pp.

Kentorovich, A. 1993. *Scientific Discovery: Logic and Tinkering*, State University of New York Press, Albany, 281 pp.

Kauffman, S. 1995. *At Home in the Universe*, Oxford University Press, Oxford, 336 pp.

Keller, E.F. 1985. *Reflections on Gender and Science*, Yale University Press, New Haven, 193 pp.

Ketner, K.L. (ed.) 1992. *Reasoning and the Logic of Things: Charles Sanders Peirce*, Harvard University Press, Cambridge, Mass., 297 pp.

Ketner, K.L. and Putnam, H. 1992. Introduction: the consequences of mathematics, in *Reasoning and the Logic of Things: Charles Sanders Peirce*, edited by K.L. Ketner, Harvard University Press, Cambridge, Mass., pp. 1–54.

Kitts, D.B. 1980. Analogies in G.K. Gilbert's philosophy of science, *Geological Society of America Special Paper*, **183**, 143–148.

Kuhn, T.S. 1962. *The Structure of Scientific Revolutions*, University of Chicago Press, Chicago, 210 pp.

Laudan, L. 1980. Why was the logic of discovery abandoned?, in *Scientific Discovery, Logic and Rationality*, edited by T. Nickles, Reidel, Dordrecht, pp. 173–183.

Laudan, R. 1987. *From Mineralogy to Geology*, University of Chicago Press, Chicago, 278 pp.

LeGrand, H.E. 1988. *Drifting Continents and Shifting Theories*, Cambridge University Press, Cambridge, 313 pp.

Leslie, A.M. 1982. The perception of causality in infants, *Perception*, **11**, 173–186.

Leslie, A.M. and Keeble, S. 1987. Do six-month-old infants perceive causality?, *Cognition*, **25**, 265–288.

Lindh, A.G. 1993. Did Popper solve Hume's problem?, *Nature*, **366**, 105–106.

Lumsden, C.J. and Wilson, E.O. 1985. The relation between biological and cultural evolution, *Journal of Social and Biological Structure*, **8**, 343–359.

McMullin, E. 1978. The conception of science in Galileo's work, in *New Perspectives on Galileo*, edited by R.E. Butts and J.C. Pitts, Reidel, Dordrecht, pp. 209–257.

Medawar, P. 1991. *The Threat and the Glory*, Oxford University Press, Oxford, 291 pp.

Menard, H.W. 1986. *The Ocean of Truth*, Princeton University Press, Princeton, 353 pp.

Mill, J.S. 1846. *A System of Logic: Ratiocinative and Inductive*, Longmans and Green, London, 860 pp.

Nagel, T. 1986. *The View from Nowhere*, Oxford University Press, New York, 244 pp.

Ochs, P. 1993. Charles Sanders Peirce, in *Founders of Constructive Postmodern Philosophy*, State University of New York Press, Albany, pp. 43–87.

Oreskes, N., Shrader-Frechette, K. and Berlitz, K. 1994. Verification, validation, and confirmation of numerical models in the Earth sciences, *Science*, **263**, 641–646.

Peirce, C.S. 1879. Note on the theory of the economy of research, in *Report to the Superintendent of the United States Coast Survey*, Government Printing Office, Washington, DC, pp. 197–201.

Peirce, C.S. 1891. The architecture of theories, *The Monist*, **1**, 161–176.

Peirce, C.S. 1902a. Sign, in *Dictionary of Philosophy and Psychology*, Vol. 2, edited by J.M. Baldwin, Macmillan, New York, p. 527.

Peirce, C.S. 1902b. Verification, in *Dictionary of Philosophy and Psychology*, Vol. 2, edited by J.M. Baldwin, Macmillan, New York, pp. 761–762.

Peirce, C.S. 1908. A neglected argument for the reality of God, *Hibbert Journal*, **78**, 90–112.

Peng, Y. and Reggia, J.A. 1990. *Abductive Inference Models for Diagnostic Problem-Solving*, Springer-Verlag, New York, 284 pp.

Petersen, A. 1985. The philosophy of Niels Bohr, in *Niels Bohr: A Centenary Volume*, edited by A.P. French and P.J. Kennedy, Harvard University Press, Cambridge, Mass., pp. 299–310.

Poincaré, H. 1914. *Science and Method*, T. Nelson, London, 248 pp.

Popper, K. 1959. *The Logic of Scientific Discovery*, Basic Books, New York, 479 pp.

Popper, K. 1969. *Conjectures and Refutations*, Routledge and Kegan Paul, London, 412 pp.

Quine, W.V.O. 1953. *From a Logical Point of View*, Harvard University Press, Cambridge, Mass., 184 pp.

Salmon, W.C. 1973. *Logic*, Prentice-Hall, Englewood Cliffs, NJ, 150 pp.

Shamos, M.H. 1995. *The Myth of Scientific Literacy*, Rutgers University, New Brunswick, NJ, 261 pp.

Shaw, J. 1983. Drumlin formation related to inverted meltwater erosional marks, *Journal of Glaciology*, **29**, 461–479.

Shaw, J. 1994. A qualitative view of sub-ice-sheet landscape evolution, *Progress in Physical Geography*, **18**, 159–184.

Shaw, J. and Sharpe, D.R. 1987. Drumlin formation by subglacial meltwater erosion, *Canadian Journal of Earth Sciences*, **24**, 2316–2322.

Shea, J.H. 1982. Twelve fallacies of uniformitarianism, *Geology*, **10**, 455–460.

Sober, E. 1988. *Reconstructing the Past: Parsimony, Evolution, and Inference*, MIT Press, Cambridge, Mass., 265 pp.

Stewart, J.A. 1990. *Drifting Continents and Colliding Paradigms: Perspectives on the Geoscience Revolution*, Indiana University Press, Bloomington, 285 pp.

Strahler, A.N. 1987. *Science and Earth History: The Evolution–Creation Controversy*, Prometheus Books, Buffalo, NY, 552 pp.

Von Engelhardt, W. and Zimmermann, J. 1988. *Theory of Earth Science*, Cambridge University Press, Cambridge, 381 pp.

Wallace, W.A. 1981. *Prelude to Galileo*, Reidel, Dordrecht, 369 pp.

Wallace, W.A. 1992a. *Galileo's Logic of Discovery and Proof*, Reidel, Dordrecht, 323 pp.

Wallace, W.A. 1992b. *Galileo's Logical Treatises*, Reidel, Dordrecht, 239 pp.

Weinberg, S. 1993. *The First Three Minutes*, Basic Books, New York, 203 pp.

Wettersten, J. and Agassi, J. 1991. Whewell's problematic heritage, in *William Whewell: A Composite Portrait*, edited by M. Fisch and S. Schaffer, Oxford University Press, Oxford, pp. 345–369.

Wheeler, J.A. 1983. Law without law, in *Quantum Mechanics and Measurement*, edited by J.A. Wheeler and W.H. Zurek, Princeton University Press, Princeton, NJ, pp. 200–210.

Whewell, W. 1837. *History of the Inductive Sciences*, Cass, London, 3 vols.

Whewell, W. 1840. *The Philosophy of the Inductive Sciences, Founded Upon Their History*, Cass, London, 2 vols.

Whewell, W. 1858. *Novum Organon Renovatum*, J.W. Parker, London, 476 pp.

Whewell, W. 1860. *On the Philosophy of Discovery*, J.W. Parker, London, 531 pp.

Wilson, E.O. 1994. *Naturalist*, Island Press, New York, 380 pp.

Wisan, W.L. 1978. Galileo's scientific method: a reexamination, in *New Perspectives on Galileo*, edited by R.E. Butts and J.C. Pitts, Reidel, Dordrecht, pp. 1–57.

Wolpert, J. 1992. *The Unnatural Nature of Science*, Harvard University Press, Cambridge, Mass., 191 pp.

4 Fashion in Geomorphology

Douglas J. Sherman

Department of Geography, University of Southern California

ABSTRACT

Geomorphology is a philosophically sedate discipline. It might be argued that this philosophical and methodological quietude represents the progression of Kuhn's 'normal'
science. However, this same torpor can be interpreted as indicative of a discipline waiting
for a fashion leader. Sperber (1990) adopted the concept of fashion change from the
design and arts disciplines to explain one means of controlling the developments and
directions of science. He contends that changes in the goals, subjects, methods, philosophies, or practice of science can often be attributed to the emergence of an opinion (or
fashion) leader, pointing toward a different path – setting out the new fashion. The fashion
process relies upon fashion dudes to advance their disciplines (and their careers). This
chapter examines the applicability of the fashion process to geomorphology through an
evaluation of eight criteria established by Sperber as common to fashion-dominated disciplines. It is concluded that the fashion process is applicable to geomorphology, and
examples of past leaders are presented. Some fashion futures, and their implications, are
considered. Whether the fashion process has been a positive or negative force in geomorphology remains to be assessed, but its recognition as an agent of change is critical for
a maturing discipline.

INTRODUCTION

Fortunately, practice never quite fulfills the expectations of prescription
(Leighly 1955, p. 317).

There are more geomorphologists alive today than at any time in the past. More geomorphological research is being conducted, more papers are being published and
presented, and there is more general interest in the discipline than in any previous era. It is
an exciting, challenging, and promising time for us. It is also a season appropriate for

*The Scientific Nature of Geomorphology: Proceedings of the 27th Binghamton Symposium in Geomorphology held 27–29
September 1996.* Edited by Bruce L. Rhoads and Colin E. Thorn. © 1996 John Wiley & Sons Ltd.

some disciplinary introspection – an inventory and appraisal. This is an activity that most of us can (and do) avoid for most of the time, but there are occasions when serious contemplation is called for. This is such an occasion. Recent debates concerning the place and status of geomorphology as a science betray a fundamental insecurity regarding the intellectual viability of our discipline. These debates also provide the opportunity to move away from the orthodoxy of tradition toward the establishment of a vibrant geomorphology; an opportunity that requires critical introspection.

The theme of this volume is 'the scientific basis of geomorphology', as designated by Bruce Rhoads and Colin Thorn to include consideration of the historical, methodological, and philosophical issues surrounding our discipline. This chapter will touch on aspects of all three of these realms as it reconsiders the developmental history of geomorphology through lenses ground with Sperber's (1990) concept of 'science as a fashion process'. The discussion is structured around five themes:

1. Pertinent aspects of the relationships between philosophy, science, and geomorphology;
2. Science as a social process;
3. The applicability of Kuhn's model of scientific revolutions to geomorphology;
4. The concept of fashion in science as described by Sperber (1990);
5. The propriety of a model of geomorphology as a discipline developing according to the fashion process.

It is not my purpose to start from first arguments and principles here. Instead I will rely on summary arguments directly related to the contention that geomorphology has been, and remains, a fashion science. The recognition that we, as a scientific community, are influenced by social forces is not new. But the manners through which fashion processes control our disciplinary development are worthy of exploration. Categorical denial of the power of the fashion process in shaping geomorphology is retrogressive.

There are several steps necessary to get to point 5, each of which involves a substantial literature. Examples of these literatures are cited, but the reader who is interested or skeptical will need to read beyond these pages. I have also relied to a substantial degree on direct quotation from pertinent sources. There are two reasons for this – first, I find paraphrasing tedious and unrewarding, with a nontrivial risk of misstating the original author's intent, and second, in most cases the thoughts of the respective authors are best presented in their own words. I have tried diligently to avoid divorcing these quotes from their context. Bryan (1950, p. 206) chided: 'Direct quotation and copious footnotes used merely to convince the reader of the scientific verity of statements should be avoided.' I have avoided footnotes entirely (unless Bryan was really referring to references), and this is not an effort to convince the reader of this piece about its scientific verity.

PHILOSOPHY, SCIENCE, AND GEOMORPHOLOGY

The reason for the neglect of philosophical issues by geomorphologists is unclear. Perhaps it reflects a basic sense of philosophical security

(Rhoads and Thorn 1994, p. 91).

Geomorphology is a philosophically sedate discipline. This is not to suggest that there are not philosophers among us – there are. And it is not to suggest that there are not critical philosophical underpinnings to our discipline – there are. It is to contend that most of us are psychologically unconcerned or academically unprepared to initiate or entertain philosophical debate about the nature of our science. It is to contend that this disconcern represents a surrender of disciplinary responsibility to a small cadre of opinion leaders, and a complacency to the fact that we are governed by trends, or fashions, rather than by logic or theory. It might be argued that philosophical training is superfluous to the education of a geomorphologist, or it is at least unimportant relative to more immediately germane subjects. Philosophical issues remain largely innate in our research, and mainly ignored, except in occasional explanatory retrospection. We can rationalize this fact by agreeing with the statement of Einstein (1936, p. 349) 'It has often been said, and certainly not without justification, that the man of science is a poor philosopher.' Indeed, why should geomorphologists, in their roles as scientists, engage in philosophical considerations anyway? Are we not able to avoid philosophical issues by saying 'not my job'? Is it not sufficient to do 'good science' and get on with it? The answers to these questions, and the motivations behind the questions, are central to an understanding of the internal and external processes that guide geomorphology. In a broader context, the importance of these questions (and some partial answers) was also underscored by Einstein (1936) when he continued his discussion to stress the necessity that physicists not leave the contemplation of the metaphysical foundations of their discipline to philosophers. The involvement of the scientist in disciplinary self-examination is especially critical during times of fundamental uncertainty or change, '. . . for, he himself knows best, and feels more surely where the shoe pinches' (Einstein, 1936, p. 349). Recent developments suggest that geomorphology may be entering (or perhaps continuing) a period of fundamental unrest. In fact the calling of this symposium seems highly diagnostic of such conditions. Few of us would admit to being part of a disciplinary proletariat that is chained to an intellectual hegemony forged by a few academic power brokers more than a half century ago. But most of us are. And most of us remain curiously silent concerning the scientific and social foundations of the discipline that we embrace.

The recognition that there tends to be a philosophical *laissez-faire* (at least) in geomorphology is not new. Most of us have some personal experience or views along this vein, and these attitudes are often noted in the literature (e.g. Chorley 1978; Schumm 1991; Rhoads and Thorn 1993; Bassett 1994; Sherman 1994). Ironically, of course, virtually every pronouncement on the paucity of philosophical discourse in geomorphology is prologue to such discourse. It is no different here, and it is in this spirit that I offer another perspective on where we stand. My interests, perhaps my temerity, in undertaking this chapter stem from the quasi-coincidences of reading Sperber (1990), Baker and Twidale (1991), Yatsu (1992), and Rhoads and Thorn (1993). This was coupled with teaching a graduate seminar with a prototypical postmodernist, Michael Dear (e.g. Dear and Wassmansdorf 1993), and the preparation of an essay on social relevance and geomorphology (Sherman 1994). These events made me aware of how much is taken for granted concerning the development of our science, especially differences between the explanations of how science should work and the histories of how science does work. This symposium affords an opportunity to offer an interpretation of how our discipline, its ideas, theories, methods, and philosophies, appears to evolve, rather than how it is supposed to behave.

Wittgenstein (1921, 4.111) wrote: 'Philosophy is not one of the natural sciences. (The word "philosophy" must mean something whose place is above or below the natural sciences, not beside them)' and (4.113 and 4.14): 'Philosophy sets limits to the much disputed sphere of natural science. It must set limits to what can be thought; and, in doing so, to what cannot be thought.' If we allow, *prima facie*, that geomorphology is a science and that Wittgenstein was correct, then we are obligated to treat our discipline as a core of science sandwiched by philosophies that must condition our practices. Speculation concerning such science (the task essayed here) must be tied to a speculation concerning the philosophy of that science. We can analogize the Kantian metaphysical dyad of ontology and epistemology (wherein the latter must flow from the former) to represent the bread of our sandwich. The upper crust, ontology, conditions our broader beliefs in the organizational structure that governs the behavior of geomorphological systems. The lower crust, epistemology, conditions our rules for obtaining, evaluating, and accepting knowledge concerning the behavior of geomorphological systems. In reality, of course, our sandwich is crushed to an extent that although we can still recognize a 'top' and 'bottom', the filling leaks here and there, and there is substantial ambiguity about where the metaphysical bread and the scientific filling meet.

It is apparent that although most of us are quite content with the filling of our geomorphological sandwich, our diets are not complete without the bread. Further, a well-constructed sandwich keeps our hands clean and enhances the taste (thus our enjoyment) of the filling. Just as this metaphorical sandwich represents an entity, we must also recognize that science, philosophy, and landforms must be conceptually costructured to represent the entity of geomorphology. The distinctions between science and philosophy must be recognized as the organizational (disciplinary) conveniences that they are meant to represent. It is critical to recognize that science and philosophy (just like our sandwich) are human endeavors; created, shaped, and practiced by humans. Past, present, and future philosophies and sciences must be considered through the sociologies that produced them, use them, or desire to change them. It is how we think about and use science and philosophy that is important (Longino 1990). If the subjective foundations of philosophy and science are accepted, then a stage is prepared for the interpretation of scientific development as a social process. And, if we are to consider the development of the discipline of geomorphology as occurring in some manner other than that of the orderly, sterile (mythical?) scientific method, then we must search for, and appraise, alternative models. There are many such models to choose from (including the model of no models!) but only the models of Kuhn (1970) and Sperber (1990) are contrasted here because they seem best to describe the practices and developments of geomorphology. And this comparison first requires some support from a discussion of social influences and controls on science.

SCIENCE AS A SOCIAL PROCESS

> ...*authoritarianism, heroism and idolatry are deeply buried in the mental substrata of geomorphologists' communities which strictly prohibit libertarian activities in geomorphology*
>
> (Yatsu 1992, p. 115).

Science is a social process. It depends upon contracts, contacts, and communication. Its practice depends upon funding, publication, and communication. It is organized into disciplines, with boundaries defined and drawn through tradition or argument. And within these disciplines we have the 'invisible colleges' where like-minded scientists group together to attack particular problems (Crane 1969, 1972). According to Price and Beaver (1966, p. 1011) 'The basic phenomenon seems to be that in each of the more actively pursued and highly competitive specialties in the sciences there seems to exist an "in group"... Since they constitute a power group... they might... control the administration of research funds and laboratory space. They may also control personal prestige and the fate of new scientific ideas...' Coinage of the term 'invisible college' is credited to the origins of the Royal Society of London (Price and Beaver 1966). We can recognize that there are invisible colleges, of different degrees of organization and formality, in geomorphology. These groups respond to external and internal social and political pressures and also exert social pressures on their nonmembers. Invisible colleges work as part of the selection process for symposia such as this one. Invisible colleges determine who is funded, who is published, who is in, and who is out. J. T. Jutson, for example, was judged 'out' by his contemporaries (according to Brock and Twidale 1984), and is, therefore, near-forgotten by his discipline. And Davis was judged to be 'in'.

There is nothing inherently wrong with the practice of science being a social process. There is still a requirement that the quality of science be judged according to agreed-upon rules of evidence. Longino (1990) compared the so-called objectivity of science with similar standards of appraisal in the arts and in philosophy. She wrote (Longino 1990, pp. 74–75): 'Objectivity... is a characteristic of a community's practice of science rather than of an individual's...' and 'Scientific knowledge is, therefore, social knowledge. It is produced by processes that are intrinsically social, and once a theory, hypothesis, or set of data has been accepted by a community, it becomes a public resource.' Feyerabend (1975, p. 309) was, as we might expect, a little more blunt in this regard: 'Science itself uses the method of ballot, discussion, vote, though without a clear grasp of its mechanism, and in a heavily biased way.' But a key requirement is that the community somehow agrees on standards for the appraisal of work, that there is some method at least to decide which issues are on Feyerabend's ballot. We must also recognize that the community tacitly allows power (responsibility) to concentrate into the hands of a few. According to the findings of Price and Beaver (1966, p. 1017) '... the research front is dominated by a small core of active workers and a large and weak transient population of the collaborators...'.

The manner in which these standards are set, and the criteria that they embody, must reflect the central concepts held by that community, or at least those of the invisible college claiming proprietary rights. Latour and Woolgar (1979) have likened this process to a rhetorical contest, where the audience is most likely to be persuaded by the scientist(s) bringing the greatest resources to bear on the problem. They include the mustering of references as one means of fighting this battle. Certainly that represents a principal methodological tool in chapters like this one! But the purposes motivating the struggle, providing the impetus for struggle, are often omitted from the literature describing the sociologies of science. A recent review of this literature (Golinski 1990) reveals very little about what motivates scientists to adopt a particular position, or why they might refuse to adopt other positions. It is the issue of motivation, however, that must underlie the explanations of human agencies for disciplinary change. Hull (1988) is bold enough to write that science only works to the degree that it does because scientists are motivated by, and receive, individual credit for their product – knowledge. Because crediting is a social process, and because the judgment of credit due is a public process, we then have a portrait of scientists moved to please the social circle represented by their invisible colleges.

In the pre-Kuhnian era, it was commonly believed, or at least commonly pretended, that science was practiced according to some objective (e.g. Baconian) standards. Logical positivism provided the most common philosophical stance, at least for those that were trying to pay attention. It was sciencey, it was safe, it was acceptable, and it represented dispassionate objectivity. The scientist was emotionless, clean, fair, and motivated by a love of truth and the desire to improve the human condition. Kuhn suggested otherwise, with arguments founded on the subjectively derived rules developed by scientific communities. There is now a large literature deconstructing the practice and aims of science (e.g. Miller 1987; Hull 1988; Chalmers 1990; or Wolpert 1992). Many historians, sociologists, and philosophers concerned with science now subscribe to the notion of science as a relativistic (subjective) process. Although this perspective is popular, it is not universal. Popper, for example, remained unconvinced by much of this. As reported in an interview by Horgan (1992, p. 42), 'Popper is repelled by the currently popular view that science is driven more by politics and social custom than by a rational pursuit for truth. He blames this attitude on a plot by social scientists . . .'. There is not even agreement on how to disagree!

The attack on rationalism reached new levels with the publication of Feyerabend's (1975) *Against Method*. Feyerabend tried to apply concepts of political revolution and anarchism to the pursuit of science, claiming (Feyerabend 1975, p. 23) 'The only principle that does not inhibit progress is: *anything goes.*' It is readily apparent that the invisible colleges, or other forms of social organizations within scientific communities, do not, and cannot, allow anything to go. Although the argument that anything goes can be used to substantiate any particular stance embraced by the leadership of an invisible college, this posture might also rob them of authority and eliminate many of their gatekeeper functions (this may, in fact, pose one of the fundamental dilemmas confronting postmodernists – how can they denigrate positivism while advocating a methodological postmodernism?). We must explore models that do describe how change can be accomplished for the development of scientific disciplines. For nearly three decades, the paradigm for scientific development and the evolution of disciplines has been Thomas Kuhn's vision of 'scientific revolution'. Of interest here is the frequent application of Kuhn's model, implicitly or explicitly, to describe progress in geomorphology. If it is correct to propose a geomorphology governed by fashion processes as described by Sperber, then the viability of Kuhn's model must be considered first.

KUHN'S MODEL OF SCIENTIFIC REVOLUTION

. . . Kuhn tried to say something about the way science is created . . .

(Thorn 1988, p. 20).

Any modern attempt to explain disciplinary change must at least consider the applicability of Kuhn's model for scientific revolutions. His model is founded first upon the concept that a discipline follows the tenets of 'normal science'. According to Kuhn (1970, p. 10), '...normal science means research firmly based upon one or more past scientific achievement, achievements that some particular scientific community acknowledges for a time as supplying the foundation for its further practice'. Normal science is characterized by the establishment of paradigms. Although Kuhn has been criticized for his relatively loose use of this term (e.g. Shapere 1964), he sets forth a basic definition – 'A paradigm is what members of an academic community share, *and*, conversely, a scientific community consists of men who share a paradigm' (Kuhn 1970, p. 176). This is pretty vague, but Kuhn (1970, p. 182) clarifies the term as being equivalent to signifying '"disciplinary matrix": "disciplinary" because it refers to the common possession of the practitioners of a particular discipline; "matrix" because it is composed of ordered elements of various sorts, each requiring further specification'. Normal science progresses under the constraint and opportunities presented by a discipline's paradigm. During this progression there will arise 'anomalies', or instances where '...nature has somehow violated the paradigm-induced expectations that govern normal science' (Kuhn 1970, pp. 52–53). The scientific community will attempt to adjust their paradigm to accommodate the anomaly. However, where accommodations cannot be made (or forced), and where the anomaly (or set of anomalies) is of sufficient importance that it cannot be ignored, there must be a paradigm change. The new paradigm should be capable of assimilating the facts of the old paradigm and its anomalies while allowing considerable freedom for additional discovery.

Kuhn's concept of paradigm change and scientific revolution was directed to the development of mature disciplines, and geomorphology may not fit this description. However, geomorphologists have not been shy in their attempts to interpret the development of geomorphology from a Kuhnian perspective. For example, it is commonly accepted that the paradigm governing the discipline through the early decades of the twentieth century was the Davisian model of the geographical cycle (e.g. James 1972, p. 350; Sack 1992; Strahler 1992; Rhoads and Thorn 1993, among many others). Certainly this was the first general landscape model to receive widespread, international acceptance (e.g. Chorley et al. 1973; Beckensale 1976; Tinkler 1985) and it provided the touchstone for geomorphological development and debate for decades. It is also commonly accepted that this paradigm was rejected sometime around the middle of the century and replaced by process (or quantitative/dynamic/systematic) geomorphology (e.g. Vitek and Ritter 1989; Sack 1992; Strahler 1992). This is a North American perspective, as other regions embraced, at least temporarily, what Beckensale and Chorley (1991) have categorized as 'historical geomorphology' and 'regional geomorphology'. The latter was defined to include morphoclimatic geomorphology. Others can and do argue that there are other, mainstream, paradigms adopted by geomorphologists. In the modern era we can most easily see paradigmatic regimes categorized as 'scientific' (i.e. nomothetic or process-based) or 'nonscientific' (i.e. idiographic or historical), as discussed by Jennings (1973), Chorley (1978), Baker and Twidale (1991), Schumm (1991), or Yatsu (1992), among many. There is relatively little 'crossover' between the geomorphological communities practicing in these divergent camps, and there may occur substantial tension between these, and other, camps as each makes proprietary claims on the discipline. We see aspects of this type of struggle in the continuing efforts of geologists to disenfranchise geomorphologists trained in the process-oriented geographical tradition (e.g. Baker and Twidale 1991). Bauer (Chapter 16, this volume) addresses this issue in a carefully considered essay. This is a disciplinary power struggle in the most basic sense, clothed in a guise of intellectualism. It could be considered to be a paradigmatic struggle according to the Kuhnian model, if that model actually described the way geomorphology works.

THE FAILURE OF THE KUHNIAN MODEL IN GEOMORPHOLOGY

> *It now seems certain that the philosophical vacuum created by the disenchantment with Davisian geomorphology was not replaced by another general theory of landscape development*
>
> (Ritter 1988, p. 165).

Kuhn's model of scientific revolution through paradigm failure cannot be applied to geomorphology as a coherent discipline, and an alternative explanation for disciplinary evolution must be sought. My argument to support this contention is relatively simple and is posed through a consideration of geomorphology's prototypical paradigm, Davis's geographic cycle, under the light of Kuhn's protocol. The most rational conclusion that can be obtained is that the geographical cycle was not a paradigm and geomorphology is not a 'normal science'.

We must recognize that geomorphology did not exist as a distinct intellectual enterprise prior to the latter half of the nineteenth century (if it has even attained such a status today). What we now recognize as the roots of our discipline were grounded in physics, chemistry, theology, engineering, geography, geology, and medicine, among other disciplines (Chorley et al. 1964; or Bauer, Chapter 16 this volume). When Davis began the proclamations of his model toward the end of the nineteenth century (e.g. Davis 1889), geomorphology had no paradigms that would distinguish it from other disciplines. It had no foundational claims for recognition. Therefore it could not have been a 'normal science'. In Chapter II of his book, *The Route to Normal Science*, Kuhn (1970) notes that normal science is usually preceded by a period of fact-collecting. Preparadigms arise through the efforts to organize these facts and to build theories to accommodate them. Without a paradigmatic context, it is difficult to evaluate the worth of facts, and it is difficult to prescribe a particular research agenda to seek more facts. Therefore preparadigms are usually founded upon existing facts or those easily obtained. 'The resulting pool of facts contains those accessible to casual observations...' (Kuhn 1970, p. 15). The development of the model of the geographical cycle was perforce predicated upon explaining just such a pool of facts. Further, in the absence of a belief system by which to address the pool of facts, such a system '...must be externally supplied, perhaps by a current metaphysic, by another science, or by personal or historical accident' (Kuhn 1970, p. 17). We can presume that Davis's body of belief derived from geology and biology. We see in his theory a preparadigm that comprises Chamberlin's notions of 'old' and 'new' valleys in the Driftless Area (according to Peattie, 1950) and Darwinian evolution. Thus Davis derived a theoretical framework for explanation of landscape evolution that accommodated the facts at hand rather neatly. Divergent views, or alternative preparadigms (and there were several), were largely ignored or swept away as the geographical cycle and its champion won new adherents.

The new discipline of geomorphology was dominated by the Davisian theory, and in that sense the theory might be considered a paradigm according to Kuhn's definitions. Indeed, the acceptance of the model was so pervasive that the geomorphological community might have earned the application of Kuhn's (1970, p. 22) words, 'They had . . . achieved a paradigm that proved able to guide the whole group's research. Except with the advantage of hindsight, it is hard to find another criterion that so clearly proclaims a field a science.' Of course we do have the advantage of hindsight with which to judge the success of the geographical cycle as a paradigm for geomorphology.

The key evidence against the Davisian model as a paradigm for geomorphology, and thus geomorphology as a normal science, is based on Kuhn's discussion of *The Response to Crisis* (his Chapter VIII). There are two key statements in this chapter that underpin the argument made here. First, Kuhn (1970, p. 77) states that ' . . . once it has achieved the status of a paradigm, a scientific theory is declared invalid only if an alternative candidate is available to take its place'. Second, he continues (on that same page) with a second circumstance of scientific revolution, 'The decision to reject one paradigm is always simultaneously the decision to accept another, and the judgement leading to that decision involves the comparison of both paradigms with nature *and* with each other.' Note especially Kuhn's use of 'only if' in the first statement, and 'always' in his second. The contentious discussions over the content and practice of geomorphology, continuing to this symposium, are ample evidence that our discipline is not presently united by a paradigm, or even a set of paradigms. Nor has it been for more than half a century (e.g. Mosley and Zimpfer 1976). And certainly no serious student of the recent history of geomorphology would argue that the geographical cycle was refuted, in the manner required by Kuhn, in favor of an alternative paradigm. St Onge (1981) complained in fact that *vis-à-vis* a general theory (paradigm), geomorphology was in a conceptual vacuum.

There were several contemporary challenges to the viability of the geographical cycle, but these had minimal impact (e.g. discussion by Bishop 1980; or Tinkler 1985). Penck's (1924) alternative slope model, perhaps the most serious competitor, would not qualify as the alternative paradigm. Although its worldview was quite different from that of Davis, Penck's model did not receive the acceptance necessary to reach paradigm status. Through the middle part of this century, many geomorphologists were content to discard much or all of both models and muddle along by retreating to the safer havens offered by the paradigms of geology or geography. For example, at the Association of American Geographers Symposium on Penck's Contribution to Geomorphology, published in volume 30 of the *Annals*, little credence was paid to the viability of Penck's model, especially as a successor to the Davisian. John Leighly (1940, p. 225) notes that 'The great forward step beyond Davis that Penck took is not to be comprehended as a system that replaces even a part of Davis's work . . .'. In that same volume, Douglas Johnson (1940, p. 231) called Penck's slope model ' . . . one of the most fantastic errors ever introduced into geomorphology'. And Kirk Bryan (1940) approved of neither approach as a general theory.

A decade after the Penck Symposium, the Association of American Geographers held a symposium to commemorate the 100th anniversary of Davis's birth, again marked by a special issue of the *Annals*. Once more, investigation of that literature fails to reveal a scientific revolution. O. D. von Engeln (1950, p. 177) notes that 'The geographers, though not disowning Davis, on the other hand, have more and more abandoned his ways and approaches.' Baulig (1950, p. 195) concludes his essay with 'Whatever success may attend the study of processes, the interpretation of forms will remain the ultimate aim of geomorphology. To this end, the Davisian method has not, thus far, been superseded.' Even the quantitative slope studies of the young lion, Strahler, could not produce anomalous results. In summarizing his own results, Strahler (1950, p. 212) allows that 'This suggests strongly that slopes recline in angle with time, unless constantly refreshed by stream corrosion. Davis's concept of reclining slopes in the erosion cycle may thus to some extent be confirmed.' Gregory (1985) discusses extensions, alternatives, and additions to the Davisian model, but not substitutes. There are many other accounts to confirm that the geographical cycle was never refuted, but rather ' . . . many of Davis's ideas simply hang in suspended animation; they are largely in disfavor, but have never actually been addressed comprehensively, let alone disproven' (Thorn 1988, p. 132).

So if geomorphology has any claim to disciplinary status, we are stuck when we try to apply Kuhn's model. In appraising our present situation we are left to accept one of two options: (1) to be a normal science we need a paradigm. Davis's model has been purported to be such a paradigm. This paradigm has not been simultaneously replaced in a scientific revolution. Therefore modern geomorphology would still operate under the aegis of the geographical cycle as a marcescent paradigm. The only rational alternative is that (2) geomorphology does not fit the model for normal science. From this stance, it can be argued that geomorphology has never been governed by a distinct paradigm, although it has been influenced substantially by the elbows of geological and geographical paradigms. Davis's model, and the general models of Penck (1924) or King (1953), and the general methodologies of Quaternary, process, and climatic geomorphologies (among others) all represent competing preparadigms. This option is reinforced by the frequent mention of the geomorphologists who were contemporaries of Davis but who did not accept the general or specific applicability of the geographical cycle (see discussions in Chorley et al. 1973). Sack (1992, p. 254) noted that there were several ' . . . investigators who worked outside the dominant geomorphic paradigm during the Davisian era . . . '. According to Kuhn's definition of paradigm, quoted above, these investigators could not have been members of the geomorphological community if they did not accept the geomorphological paradigm. This conclusion is unambiguous. If we accept the paradigmatic concept for geomorphology, we must then argue that Gilbert and Bryan and Salisbury, and many other scientists pursuing similar research agenda, were not geomorphologists. This is an unreasonable corollary.

It is more reasonable, and certainly more satisfying intellectually, to conclude that the second option is the more viable explanation. The postmodern scramble of competing worldviews may represent nothing more than preparadigms struggling for domination. It is to be expected and encouraged as diagnostic of a healthy and vigorous discipline, but it should not be accepted without a critical awareness of what the associated social processes are. This perspective also leaves us without Kuhn's model for scientific revolution as a framework for describing the development of geomorphology. There are other models that might be applicable to fill this void. But the purpose here is to assess geomorphology in the context of Sperber's model for disciplinary development as a result of fashion processes.

THE FASHION MODEL

The game of science is, in principle, without end

(Popper 1934, p. 53).

Irwin Sperber (1990) attempted to explain episodes of irregular or irrational behavior in the scientific community by explicitly acknowledging the importance of a discipline's social structure and the vulnerability of that structure to the play of fashion. This approach grows out of the general body of literature produced by historians, sociologists, and philosophers of science that recognizes the importance of social controls on the development of disciplines (e.g. Livingstone 1984). Sperber acknowledges more directly the influences of Georg Simmel and Alfred Kroeber, social scientists who published some of the initial explorations of the interplay of fashion with modern society. In his book, Sperber addresses a fundamental question: 'How do the rules of Popper's game of science change?' His argument is based on the notion that the fashion process is pervasive but unacknowledged (even denied) in the scientific community, that the process is a dominant factor in the development of many scholarly disciplines. The fashion process operates in a scientific community to cause '...styles or models of scientific thought {to} rise to and fall from prominence just as shifting hemlines and chrome fenders...' (Sperber 1990, p. ix). He defines the fashion process as

> ...a form of collective behavior marked by a series of normative preoccupations: keeping in step with the times, with the latest of developments; following the examples of prestigious opinion leaders who 'keep their ears to the ground' and articulate the shared and implicit sentiments of the public; admiring proposals for adoption when they are in good taste and new, discarding them when they are in bad taste and old; dismissing the weight of tradition while rediscovering and repackaging old proposals as though they were unprecedented, exciting, and modern; ignoring or downgrading explicit criteria by which competing proposals can be evaluated (Sperber 1990, p. x).

Several aspects of this definition appear immediately pertinent to the discipline of geomorphology as it constitutes a scientific community as defined by Sperber (1990, p. 6): '...any discipline, profession, institution, or network of organizations regarded as both responsible for producing and conveying ideas about a given order of social or physical phenomena and the source of those particular ideas defined as *authentic*, *certified*, and *valid* descriptions or explanations of such phenomenon'. Sperber claims that the adoption of various modes of scientific inquiry and the recognition of disciplinary leaders both replicate the processes associated with the rise and fall of ideas and leaders in the fashion and design industries. The disciplinary leaders are also variously described as fashion leaders, fashion dudes, or opinion leaders. His formal definition (Sperber 1990, p. 6) '...is any scholar whose work is regarded as highly influential and prestigious by his or her peers, reasonably representative of the most important and best research in a given area of study, and generally successful in "staying on top" (at least in the short run) despite the entry of rival designers into the selection process'. He then argues vigorously that the fashion process is an apt model for the behavior of many scientific communities. His particular concern is with the progressions of social science, but his model can be, and should be, held up against the natural sciences for appraisal because 'the delights of nature and the play of fashion in science seem to be "made" for each other' (Sperber 1990, p. 249).

The fashion process plays to the socially acceptable side of a scientific community, especially that of its leadership, a side that longs for status, popularity, conformity, and stability. The fashion process also degrades the importance of logical progress in the development of knowledge unless that mode of the development is in vogue. Most importantly, the fashion process recognizes that scientists are people, and that science is a social activity, and that any social community will have leaders. Even though the term 'fashion process' may have negative connotations for many of us, it is not intrinsically a bad thing. The value of the fashion process depends upon the selection and action of particular opinion leaders. The operation of the fashion process in a discipline may either allow – indeed encourage – the adoption of radical new approaches, or it may provide the appearance of change through a continuous reshuffle of existing fashions while minimizing true change. The process can be used as a funnel, a sledgehammer, a door, or a window. The recognition of the fashion process by a disciplinary mainstream, however, does provide a mechanism to maximize the potential benefits by increasing the participation in decision-making.

Throughout his book, Sperber posits his arguments as direct refutation of several other models of scientific development, especially Kuhn's descriptions of revolutions as the means by which science progresses. Indeed Kuhn, according to Sperber (e.g. Sperber 1990, pp. 209–210), was also involved as part of a fashion process, and he became a fashion leader because his work spoke to the desires of both his academic and the popular community. The logical, scholarly components of his work were of less direct impact in fostering an acceptance of his model, than was his timely critique of the way science works. The approaches of Kuhn and Sperber are not entirely contradictory, as they both rely upon a recognition of the critical roles played by social processes in directing scientific communities. Their models differ instead through the treatment of the motivations in the communities that result in change. Kuhn's model suggests that scientific communities change their paradigms when they cannot avoid doing so (this is a very simplified statement). Sperber's model suggests that scientific communities change their focus when it becomes fashionable, through any number of reasons.

It was argued above that the Kuhnian model is not generically applicable to geomorphology as a scientific community. It is now argued that Sperber's fashion model is applicable to, and important for, the development of geomorphology. We must recognize and embrace the importance of collective behavior in setting the geomorphological agenda. If we acknowledge the influences of the fashion process and the roles of opinion leaders in directing our discipline, we appropriate the ability to control the process in a critical and beneficial manner. This will, in turn, empower our community in the development of a vibrant and progressive science. Denial of the fashion process leaves our community vulnerable to the repressive aspects of fashion.

GEOMORPHOLOGY AND THE FASHION MODEL

Geomorphic fashions have swung from the qualitative statements of early observers to the highly mathematical treatments . . . and back again to inductive studies derived from field measurement

(Carson and Kirkby 1972).

The term 'fashion' can be found frequently in the literature of geomorphology. It is usually employed in a manner that is at least mildly derogatory. Use of the word in a discipline's literature, however, does not provide a sufficient basis to conclude that the fashion model is operative. A more rigorous standard must be applicable. To meet this need, Sperber has defined eight criteria that are common to (and indicative of) scientific communities governed (at least partly) by fashion processes. An appreciation and evaluation of each criterion are fundamental to any attempt to link the fashion process to a science, and every criterion must be satisfied in order to conclude that a discipline is governed by the play of fashion. This is desirable information because the recognition of the potential roles of fashion in the development of a scientific discipline, and a subsequent acknowledgement of their importance, reflects a laudatory professional self-awareness. Sperber claims that this appraisal can lead to an objective decision concerning the discipline's status *vis-à-vis* the fashion process. The intent in this section of the chapter is to assess the applicability of each of these criteria to geomorphology in the context of the discipline's development and present status. If a relevant example can be described for each criterion (and I have tried to provide multiple examples for each), then we may conclude that the fashion process represents, at a minimum, a suitable hypothesis for exploring the way the discipline evolves. The criteria, as presented here, are paraphrased and abridged from Sperber's original.

Criterion 1

The most recognized research of a discipline's leadership may be seriously flawed in logic, evidence, or conclusions.

Can such an example be identified for geomorphology? It is often difficult to point to an individual publication as representing a scientist's 'most recognized work'. It must also be a subjective appraisal, and therefore open to debate. To simplify (but not necessarily relax) Sperber's criterion, it will be defined here as representing an oft cited or influential, substantive piece of work by a recognized leader (or leaders) of the community of geomorphologists. And from this basis many examples can be found.

It is prudent to avoid the litany of now humorous explanations of landform development that were offered prior to the twentieth century. The reader is directed to accounts in Chorley et al. (1964), Davies (1969), or Tinkler (1985) for compendia of examples. But there remain several notable examples from more recent literature. First and foremost can be offered Davis's suite of publications on the geographical cycle.

There is little debate that the major, internal aspects of Davis's model are logically consistent. Indeed Bishop (1980) has made a credible argument that Davis constructed a system that was internally consistent to the degree that it is not falsifiable in Popper's sense. Because his system was deductively contrived (e.g. Thorn 1988), such internal consistency should not be surprising. This reflects nothing more than the inherent emptiness of deduction (e.g. Reichenbach 1951). However, logical consistency does not necessarily imply scientific credibility. Deduction based upon faulty premises must lead to faulty conclusions. Bishop (1980) concludes that Davis's model probably is not a scientific theory because it cannot be falsified. One might argue that this conclusion is valid only to the extent that Popper's view of science is valid. However, it may be more powerfully argued that the Davisian model is not scientific because, as posed by Davis, it cannot lead to new knowledge. Therefore, because the prototypical geomorphological theory (albeit a theory in current disrepute) is not scientific, geomorphology, at least near the turn of this century, satisfies Sperber's first criterion. But the efforts of Davis do not stand alone. Lustig (1967, p. 5), discussing desert geomorphology, states ' ... much of the litera-ture ... as well as current research effort, is inextricably enmeshed in basic concepts of debatable validity ... '. And our arid lands colleagues are not unusual in being faced with this situation.

The rate of production of illogical, nonscientific literature by leaders of the community of geomorphologists seems to be accelerating. To mention a few examples, and without detailing particular correspondence relative to criterion 1, we can consider the subset of fashionist propaganda embodied in the prescriptive lists of concepts, canons, principles, error or propositions, and rules, constraints or cautions, as published by Strahler (1952), King (1953), Thornbury (1954), Chorley (1978), Schumm (1985, 1991), Scheidegger (1987), Brunsden (1990), and Baker and Twidale (1991). These targets are easy because many of these authors making fashion statements were writing in essay modes, and it is important to note that they were not necessarily concerned with logic, evidence, science, or quality of conclusions. However, all of these examples represent larger bodies of research and the writings are diagnostic of the geomorphological fashions embraced by their respective authors. The authors making fashion statements must be held to high standards of accountability because their works (beliefs) are expounded at conferences, appear in journals and books aimed at broad audiences, and their publications all address the metaphysics, methods, or goals of geomorphology. All of these publications aim to control the practice of geomorphology through the explication of beliefs concerning what geomorphology does or what it should look like. Because few of these publications are based upon logical consideration of all pertinent evidence, they must be held as evidence validating criterion 1 for geomorphology. For the reader concerned with more substantive examples of persistent error impacting conclusions, the case of frequent and continuing misused regression analyses of velocity profiles (discussed in Bauer et al. 1992) serves as one specific example.

Criterion 2

These flaws stem unintentionally from an embrace of a fashion stance empowered by the social context of the discipline. Sperber (1990, p. 7) notes, as examples of such stances, ' ... what is new or modern is necessarily best', or 'the scholar who works alone is irresponsible and his findings are suspect', or 'good taste, congeniality, and moderation are preferable to heated debate, political conflict and disruption of the professional status quo'.

There are several examples where we can see clear evidence of fashion stances in the community of geomorphologists. The focus here will be on the specific cases arising from the notion that ' ... what is new or modern is necessarily best' (Sperber 1990, p. 7). The importance of this attitude in geomorphology is best exemplified by Jennings' (1973) classic, '"*Any millenniums today, lady?" The geomorphic bandwaggon parade.*' This chapter alone substantiates the role of fashion in the practice of geomorphology. In Jennings' (1973, p. 115) own words, ' ... I propose to muse about fashions in geomor-phology on a thread of personal experience.' And then Jennings dictated his list of the fashions that had influenced the development of geomorphology. He singles out for mention general systems theory, climatic geomorphology, a revitalized historical geo-morphology exploiting breakthroughs in dating methods, the 'new morphometry', process geomorphology, quantification, and a new structural geomorphology. Jennings (1973, p. 128) semi-concludes with '...new ideas and approaches must always be welcomed, though it does not follow that everybody should busy themselves with them'. This is a critical notion. The fashion process embraces what is new for the sake of its newness. As scholars, we have a responsibility to be able to critically evaluate, accept, or discard what is new. This can be accomplished within the context of a fashion model, but it cannot be done with innocence.

Other examples of 'if it's new it's good' are found abundantly, if implicitly, in efforts to import innovation to geomorphology. In the last three (or so) decades we have seen claims for entropy (Leopold and Langbein 1962), general systems theory (Chorley 1962), probability theory (Scheidegger and Langbein 1966), allometry (Bull 1975), catastrophe theory (Graf 1979), chaos, or nonlinear dynamics (Phillips 1993), and fractal physio-graphy (Outcalt et al. 1994). Each of these concepts has seen a brief flurry of attention, and most are mentioned in recent reviews of developments in geomorphology. But none have prospered as general approaches after the novelty wore off (although it is too early to extend this verdict to nonlinear dynamics and fractal physiography). And it is critical to note that the ephemeral attention paid each of these 'fashion flashes' reflects but little upon the potential utility of a given theory – just its popularity. Discussion of the implications of different temporal scales of fashion churning in geomorphology is beyond the scope of this chapter. However, the occurrence of these flashes validates Sperber's second criterion for geomorphology.

Criterion 3

There are many models competing for recognition, and the scientific community recog-nizes successful models through the perspectives of criteria 1 and 2, and through the results of a disciplinary 'popularity contest'.

It is possible to identify a population of models competing for domination of our discipline, and the competition does resemble a popularity contest run according to the rules of criteria 1 and 2. A straightforward demonstration of the existence of many geomorphological models can be found in the compendium published by the Japanese Geomorphological Union (1989) describing the history of geomorphology in 32 countries. The reader of these accounts will be struck by similarities and differences in the devel-opment of national geomorphologies. That there is not one common history, and that there persist different geographical patterns of present-day approaches is confirmatory of the competing models condition. More specific evidence for this competition can be found in the discussions of Butzer (1973) and the polemics of Baker and Twidale (1991). After his discussion of four major research streams in contemporary geomorphology, Butzer (1973, p. 41) notes '...it becomes apparent that geomorphology has unusual methodological problems' but continues (p. 43) 'Diversity is a source of strength but it can only be so when pluralism is accepted and tolerated.' This open-handed stance toward competing models is an interesting contrast to the closed-mindedness of Baker and Twidale's (1991, p. 84) plaint concerning '...the current infatuation with theory in Geo-morphology'. Further, I would offer Baker and Twidale (1991) and Rhoads and Thorn

(1993) in their (continuing?) debate over the role of theory in our discipline as examples of geomorphologists pitting wits in a contest that can only be resolved as a popularity contest because their arguments are incommensurate from any other perspective. This example alone is ample confirmation of criterion 3 for geomorphology.

Criterion 4

The process of selection among competing models is governed more by matters of taste and consensus than matters of logic or empiricism.

It is possible to identify evidence indicating that competing geomorphological models are selected by a fashion process rather than through rigorous considerations of their logical or empirical qualities. The annals of geomorphology are ripe with such examples, many of them found in the 'prehistory' of geomorphology. For example, Tinkler (1985, p. 50) describes hostile reactions to Hutton's work stemming from the misled beliefs that surface erosion cannot produce substantial landform change. Empiricism certainly would have to (and, in fact, does) weigh in for Hutton. Tinkler (1985, p. 230) also makes the eloquent point: 'It is not always the case that the most visible authors in an era are those who, potentially and from the point of view of posterity, ought have had the most influence.' And Higgins (1975) argues that criterion 4 processes (not using this terminology, of course) contributed substantially to the widespread and rapid acceptance of the geographical cycle (see also Thorn's 1988 discussion).

But it is more instructive to consider this criterion in the light of more modern evidence. And this is also abundant. Twidale (1977, p. 85) considers this issue in some detail and offers evidence to support his contentions that '...regrettably, too many geomorphological explanations and hypotheses which either in brief retrospect or even at the time of their announcement appeared untenable, have nevertheless achieved considerable acceptance' and '...some of the answers produced in geomorphology are unacceptable, not because they have been shown to be wrong or unfashionable but because they are so obviously lacking in logic and testing'. These statements are intriguing. Twidale says that acceptability of an answer should be based more upon its logic and testing than upon the degree to which it might be correct. These are curious words from a geomorphologist who later argues for the importance of serendipity and outrageous hypotheses to the development of geomorphology, who espouses a 'maverick geomorphology', and who clearly recognizes 'The distinction between justification and process of discovery' (Baker and Twidale 1991, p. 87). Despite Twidale's earlier opposition to fashion in geomorphology, his later writings indicate a substantial embrace of that very approach.

A more specific example concerns the continued use of Bagnold's (1936) model for aeolian sand transport despite there being little empirical evidence to indicate that it is superior to any of a number of competing models. Indeed, most objective comparisons of aeolian sand transport models, based upon field experiments, find that other models work better (e.g. Berg 1983 or Sarre 1988). The persistent use of Bagnold's model is a reflection of taste and consensus substituting for logic and empiricism, although the model itself may indeed be worthy. We do not know, but it does not seem to matter. The embrace of Bagnold's model is diagnostic of fashion operating in our discipline, and substantiation of criterion 4 in geomorphology.

Criterion 5

Leading scholars in a scientific community are not recognized primarily because of their scholarship, but instead because their work epitomizes the fashion tastes of their discipline and because of the resulting popularity of their research results and perspectives. Rival work is not objectively evaluated against that of the leaders.

It has been argued and demonstrated that the community of geomorphologists recognizes some of its disciplinary leaders more because of their fashion statements than because of their scholarship. This is partially the basis of the evaluation of criterion 1, above. Further, Yatsu (1992) has argued strongly that this has too often been the case for geomorphology, and his essay alone provides a basis for accepting this criterion. He is amusingly critical of many of geomorphology's leaders, and allows special attention to Chorley, claiming (Yatsu 1992, p. 101): 'It seems to the present author that Chorley is not a geomorphologist at all, but probably an exponent of the enlightenment.' Yatsu continues by noting Chorley's misapprehension of the differences between closed and isolated systems. Despite these (and other) flaws, there can be little doubt that Chorley enjoyed a leadership position in geomorphology for more than a decade.

Other examples are abundant, and probably comprise much of the generic material used in geomorphological thought (concepts/principles/theory/methods) courses. Certainly more geomorphologists have read and appreciated Strahler's (1952) process manifesto than any of his landform studies. Strahler (1992) claims that it is his most frequently cited publication (although the work of Bodman 1991, suggests that this is not so recently). I presume that a similar state maintains for Baker (although I am less convinced about Twidale), and Rhoads, Jennings, and Dury. Twidale (1977, p. 93) explicitly acknowledges his perception of the reality of criterion 5 conditions when he discusses science as a game '... in which, unfortunately, personalities, plausibility, and intelligibility are ephemerally as important as the fundamental worth of the ideas propounded'. Whether it is unfortunate or ephemeral or not, it is true in instances common enough to validate criterion 5.

Criterion 6

The disciplinary status quo, or orthodoxy, is defended by the scientific community against unwashed rabble, and this defense may include personal attacks on the outsiders who are 'out of step' with the governing fashion.

There is evidence for such a geomorphological orthodoxy being protected by the community (especially through the actions of the fashion leaders) using these defenses. This is perhaps why Rhoads and Thorn (1994, p. 100) felt constrained to write '... philosophical introspection provides an excellent antidote to scientists who wish to divide their colleagues into winners and losers on the basis of methodological preferences'. This may explain why Baker and Twidale (1991) decided that it was appropriate to include gratuitous attacks on human geography in their plea for a geomorphology that is connected with nature. These cases represent both indirect and direct evidence that validates the application of criterion 6 to geomorphology. The first case is a response to the problem, the second is an example of the problem. But issues concerning disciplinary turf protection are not a modern development. They have colored debate in geomorphology for more than a century. Even Davis was moved to complain to W. Penck (in a 1921 letter, published in Chorley et al. 1973, pp. 547–551) about the conservative nature of their discipline's earlier leadership: 'They objected, as many still object, to the use of explanatory methods of description, because of their danger, because of their use of deduction, etc. etc. In a word, they wished to remain purely observational, purely inductive. Indeed many were so prejudiced against deduction that they decried it in others, even when they used it themselves.'

From the many references we have that relate to Davis's harsh treatment of his contemporaries (e.g. the numerous examples provided by Chorley et al. 1973) we might conclude that he prided himself as a keeper of the geomorphological flame of righteousness. Thus Jones, in the 1950 Association of American Geographers Davis Symposium, after dismissing the validity of the geographical cycle, claims that he (Jones 1950, p. 179) was still '. . . a great admirer of Davis for other of his ideas, not the least of which was heckling presenters of papers at AAG meetings'. Of course Davis was not spared contemporary or posthumous payment in kind. Daly, for example (as quoted in Chorley et al. 1973), made the exclamation concerning some of Davis's work, 'Excellent illustrations, but not a word of truth in it.' Strahler (1950) also had some none too gentle criticisms.

There are numerous examples to substantiate the operation of criterion 6 in geomorphology. I would like to address briefly one specific orthodoxy that is protected by at least part of the geomorphological community. This orthodoxy can only be supported as a fashion, because it fails in aspects of logic, science, and practice. This is the myth of multiple working hypotheses; the notion that a 'good' earth scientist must hold simultaneous, parallel beliefs concerning the explanation of the topic of investigation. The original concept is usually credited to Chamberlin (1897), although Baker and Pyne (1978) assert that the idea was really Gilbert's, and there are many modern practitioners advocating that this method be used by geomorphologists (e.g. Haines-Young and Petch 1983; Schumm 1991). This myth could have served as a case study for Brush's (1974) essay, 'Should the history of science be rated X?' The concept of multiple working hypotheses is commonly taught to earth science students (the primary target audience of Schumm's 1991 book, for example). Yet it is not really a viable methodology and cannot be so. Hull (1988), for example, notes that there are an infinite number of hypotheses available to explain any natural phenomenon, and that the only way to winnow such a spectrum is to rely on one's vision of the relationships of the problem at hand. Although there may be several options remaining after such a process, it is difficult to see the simultaneous evaluation of parallel possibilities. This myth has been critiqued in the essay of Johnson (1990), but it is still advocated as the 'right' approach. It is difficult to interpret whether Baker and Twidale (1991) are supporting or rebuking the concept. But if they are supporting it, then their advocacy of pursuing the 'outrageous hypothesis' (and, of course, there must be more than one 'outrageous hypothesis' per problem) creates a scenario that would grind scientific activity to a halt if taken literally.

The gap between what is practiced and what is taught is extremely large in this instance, and the divergence goes right back to the beginnings of the mythology. Gilbert, claimed as the champion of the method of multiple working hypotheses, did not always practice that method in his own research. An example from Baker and Pyne's (1978, p. 102) essay on Gilbert illustrates this point (and, in fact, a host of others about the practice of science): 'As he approached the Henry Mountains on August 23, he had already formulated a conception of their structure. His field notebook for that day contains the following entry:

"*My idea of yesterday in regard to H.M. are confirmed by this view...*". In short, Gilbert had conceived the structure of the mountains before he ever actually visited the scene.'

Criterion 7

The orthodoxy, despite any appearance of invincibility, represents a fashion that has replaced an earlier model, and is, in turn, subject to future replacement. Present and past fashions are subject to rediscovery as new and daring fashions in the future.

There is every reason to believe that our present orthodoxy is subject to replacement as part of a fashion cycle, although the issue is a little sticky *vis-à-vis* geomorphology, because it avoids the preliminary question of whether there is a dominant orthodoxy. For the moment, and for the sake of this discussion, let us presume both that there is such an orthodoxy, and that it is process geomorphology. I do not hold this belief, but it is commonly subscribed to by many. For example, we must believe that it is indicated as so by the dedication of the fourth (final?) volume of the series *The History of the Study of Landforms, or the Development of Geomorphology* to tracking the development of process geomorphology, primarily subsequent to World War II (promised in Beckensale and Chorley 1991). Vitek and Ritter (1989) and Strahler (1992) claim that it is so. Kennedy (1992) seems to suggest that at least many take it as so. Sack (1992) says that it is so. Chorley (1978) says that it is so and Phillips (1992) writes like he believes it. And I must presume that Baker and Twidale (1991) believe it or else they would not have bothered with their essay. Yatsu (1992) may (p. 109 and p. 112) or may not (p. 105) think that process geomorphology is the reigning orthodoxy.

If this supposition is accepted (and it costs little to do so), then we see that in almost every example that discusses the ascension of the process balloon there is a direct acknowledgement of its replacement of the geographical cycle, or historical approaches (e.g. Carson and Kirkby 1972). Similarly the last decade has seen increased criticism aimed at the practice, if not the aims, of a process-based science as the core of our discipline (e.g. Baker and Twidale 1991). And given the fact that most geomorphologists do not in fact practice a genuine form of process geomorphology, this domination must be considered nothing more than (perhaps) wishful thinking or popular mythology. Certainly the bulk of recent publications, as reported by Marston (1989) do not support the process fashion (was Marston's listing of *Earth Surface Forms and Processes* in his Table 1 Freudian?), and there are increasing numbers of articles that criticize the utility of process studies, both from a scientific and social perspective. There are even more words that seem to imply a psychic yearning for the gentler, simpler times when historical and regional studies were the orthodoxy. Certainly Baker and Twidale appear to long for those days when storytelling sufficed for science. They have forgotten, or have failed to realize, that science must be founded upon theory. We can pay heed to a reminder from a historian of science, Miller (1987, p. 381): 'In modern physical science, there is not a single empirical principle which is firmly believed by all without crucial reliance on a theoretical underpinning.'

It is extremely revealing that in the closing sections of Baker and Twidale (1991, pp. 93–96) entitled 'Possible futures' and 'Proposed action' the authors are caught taking a hard look backward. The geomorphologists that they cite in those pages are, with one exception, long dead. The earth science citations they chose averaged more than 70 years of age in 1991 (and this excludes the 1864 reference to Marsh). And in these sections we read phrases such as 'In returning to its common-sense roots, geomorphology...'. Wolpert (1992, pp. xi–xii) argues convincingly that science is unnatural, 'Scientific ideas are, with rare exceptions, counter-intuitive... doing science requires a conscious awareness of the pitfalls of "natural" thinking. For common sense is prone to error when applied to problems requiring rigorous and quantitative thinking; lay theories are highly unreliable.' We have seen the play of fashion as indicated by criterion 7, and it is us.

Criterion 8

The disciplinary proletariat are under constant pressure to conform to the fashion dicta of opinion leaders, even as those fashions change.

There is evidence to support the premise that the 'working stiffs' of the geomorphological community are under constant pressure to conform to changing fashions as dictated by our leadership, although once again it is difficult to demonstrate this through direct example because seldom will we read an 'or else' statement. But one exemplar is an older case, as described by Butzer (1973, p. 42), 'The academic intolerance of Davis and his followers between World War I and II has become proverbial and can still be readily savored in the editorial policies implicit in the defunct *Journal of Geomorphology*.' There are, of course, few greater strangleholds on a discipline than to control the editorial reins of a central journal. This condition remains as an often unspoken, but frequently felt, pressure to conform. Similar, but more insidious pressures come from the panels that meet to evaluate and fund (or not fund) research proposals. For how could these groups comprise any who do not satisfy some measure of orthodoxy? The publication and granting processes must be controlled largely by members of our own invisible colleges because in these cases our community would not (could not) tolerate judgment from outsiders. Further, the repetitive publications telling us what we should study, and how, are all applying pressure, especially if they have passed through the process of affirmation known as refereeing. If the new 'Ten Commandments' are published in *Zeitschrift für Geomorphologie*, does that not indicate that the hierarchy has stamped their approval on the product? Finally, there is an orthodoxy represented in the collective opinions of the senior faculty (and other professionals) who are called upon to evaluate junior colleagues for tenure and promotion. Therefore, although generational tensions must and should exist, these tensions must also be modulated by issues of taste and consensus – witness the plight of Bretz. Certainly there has been, and continues to be, pressure to conform to an orthodoxy. Or to change it.

I have little doubt that geomorphology behaves as a fashion science according to the criteria proposed by Sperber. The examples provided here may not be the most compelling that exist for our science, but this evaluation requires only adequate examples, and these we have. Geomorphology is a fashion science.

FASHIONS FOR A DEAD MILLENNIUM

An investigator may be likened to a hunter in desperate search for food. He is prepared to shoot whatever appears

(Conant 1967, p. 312).

Fashion Dudes in Geomorphology

Sperber (1990, p. 220) defines the fashion, or opinion leaders of a discipline as ' . . . not merely exemplars and trend setters: these opinion leaders are in reality the collectively perceived heroes of their day'. As a graduate student, especially, I held Strahler, Chorley, Leopold, Bagnold, and Scheidegger in esteem to the extent that they represented the (process-oriented) geomorphological heroes that I aimed to emulate (aside from my graduate advisor and dissertation committee members, of course). I am hard-pressed to nominate one candidate for our present fashion hero. The lack of a clear leader is tied to the lack of a dominant fashion as we turn toward the next millennium. Our present state of methodological and theoretical tension dictates that there are multiple, competing fashions, each with a leadership in quest of power over the direction for future geomorphologies. None has dominion now. But we can identify some of the fashion leaders (and near leaders), and their research streams, that have set the stage for the present era, and we can speculate on directions for the pursuit of our science. The list presented here is merely one subset from geomorphology's pantheon. The purpose is to illustrate fashion dudes, and then ferret their commonalities. A recognition of common traits of some of the past masters should illuminate the characteristics to be seen in present contenders.

Davis as the prototype geomorphological fashion dude

It is easy to support an argument for William Morris Davis as the prototype fashion dude for geomorphology. The literature (including the biography by Chorley et al. 1973) discussing the man and his science is replete with mention of attributes suited to a fashion leader. He was a propagandist and heckler. He was a prolific writer and public speaker, and he was a powerful political advocate for his own ideas and for the stature of our discipline. He founded and led the invisible college of his time, and then took it semi-public as manifested in the Association of American Geographers. His leadership position also transferred status to a series of disciples, acolytes, and partial believers: Baulig, Wooldridge, Cotton, Linton, and Johnson among several others. Davis was a one-man invisible college (for a while), and remains the standard against which fashion competitors must be measured.

Gilbert as an underappreciated fashion dude

There is little doubt that Gilbert was and is highly regarded for his contributions to geomorphological science. In many instances his approaches to problems were decades (or more) ahead of those used by his contemporaries. He remains a true, albeit dusty, hero of geomorphology, and he has been usurped by recent generations as a philosophical and methodological ancestor. But although he was a respected member of his scientific community during his professional career, he was not a fashion leader of his time. There are numerous explanations for why this might have been (e.g. Sack 1991), but they are unimportant here. The pertinent issue is that Gilbert failed to establish geomorphological fashions among his peers. His work was admired, but not immediately emulated. If it had been, we might presume that there would have been much less of Davis and his disciples to occupy our attention.

Strahler as a faded dude

Strahler accomplished an exquisite, fashion power-move. As a relative newcomer to the discipline of geomorphology, and with a fashion vacuum in the discipline, he took it upon himself to outline a plan for the restructuring of the science in a manner sure to upset the complacency of the postwar community. He used polemic and exhortation as policy tools to promote his vision (e.g. Strahler 1950, 1952). He obtained and maintained a leadership position largely as a result of the stimulation generated by his 1952 call to arms, through the success of a cadre of his students, and, perhaps most significantly, through his recognition as 'the man who wrote the book'. His retreat from the research frontier created a power void in the invisible college, an opportunity for new leadership. Strahler's long absence from the cutting edge precludes his status as a current fashion leader. He can, however, be cast in the role of an esteemed forerunner to a modern leader. Strahler was especially fortunate that Richard Chorley went to Columbia, and was willing to take up the mantle of leadership.

Chorley as a science dude

From the process geomorphology perspective on our history, Chorley followed Strahler (perhaps to an extent by pushing him) at the fashion helm of our discipline. He recognized that changes in geomorphology seemed to occur very slowly relative to related disciplines, and that there was much to be gained by vigorously expropriating theory and methods. Chorley has been often criticized for his multiple approaches to the practice of geomorphology, but his versatility was one key to his fashion success. In this sense he epitomizes one aspect of the successful opinion leader – he keeps just ahead of the pack. When the fashion becomes accessible to the masses, the fashion must change. But Chorley made the bold moves, he argued strongly and convincingly for changes aimed to bring respectability to geomorphology, and he published extensively on issues associated with a broad spectrum of geomorphological topics. Finally, his contributions to the *History of the Study of Landforms* (not to minimize the roles of Beckensale and Dunn) guaranteed his position as one of our fashion leaders. But it may well be argued that Chorley's day is past. And, just as we might see Chorley as descended from Strahler, we might ask if the succession now continues through a subsequent anointing of Goudie.

Some common threads

These four brief sketches are not meant to summarize the careers of these geomorphologists. Neither is this short list meant to be representative of the history of our discipline, except from a process-oriented position. But these cases are sufficient to demonstrate some of the threads common to fashion leaders in any aspect of geomorphology, and to contrast the fashion leaders (Davis, Strahler, and Chorley) with the one non-leader example (Gilbert). There are at least six characteristics that are shared by these (and perhaps all) geomorphological fashion leaders. First, all of the successful leaders wrote voluminously, and their most recognized works appealed to a large cross-section of their scientific community, rather than just one or two narrow subdisciplines. It was therefore possible for a large number of their contemporary colleagues to participate in consensus building for the new orthodoxy. Second, there is a strong indication that all of our successful fashion leaders have written text (or text-like) books, that have been widely used to educate nascent geomorphologists. This is the case with all of the examples noted above. Third, forceful presentations at professional meetings, promoting the new fashion and often denigrating opposing viewpoints, seem to be requisite. Fourth, long-term positions in academia seem to be a desirable foundation for fashion setting. Fifth, all of the fashion leaders seem to possess an academic or professional heritage that placed them at or near the center of the leadership of an existing invisible college. Sixth, all of the fashion dudes seem to obtain recognition relatively early in their careers, and are widely acclaimed by their peers, either directly through the bestowal of awards, or indirectly through citations in the literature. We can recognize all six of these traits as common to Davis, Strahler, and Chorley, but also recognize that Gilbert's profile differs substantially. At least four of these characteristics (i.e. the second, third, fourth, and fifth) do not seem to apply to Gilbert's career. Ability is never sufficient (and perhaps only partly necessary in some instances) to become a fashion leader! This summary avoids the potentially critical roles of nationality and national influences because I remain uncertain concerning their importance. There is no doubt that there are national and local fashion leaders, but it is unclear whether they comprise a pool of candidates for the broader geomorphological population, or part of a fashion pyramid.

Consideration of these points might also provide explanations of why it is difficult to decipher the present fashion dude(s). For example, although contemporary geomorphology boasts several prolific writers, most of them confine the bulk of their productivity to a preferred research subdiscipline. A more central (popular) position is desirable for a fashion dude, because it must be extremely difficult to move the discipline from the edges under conditions where most colleagues will not read or hear fashion statements. Another example is that few of the present generation of geomorphologists receive general recognition from the discipline as being fashion leaders. This is implied in the examination of citation patterns published by Bodman (1991) that suggest our discipline is dominated by an entrenched seniority. However, my interpretation is that these leaders are really place holders, as there is no agreement about where we are, or where we (as a discipline) are going. Notable in Bodman's list is the rapidly rising influence of Goudie (again, based upon the citation indicator).

Looking for leadership

There is a struggle going on for the leadership of geomorphology. The struggle is about power, prestige, and control. Geomorphologists are pitting wits, spending resources, risking egos, and buying ballots in attempts to set the agenda for geomorphology in the next millennium. The leaders of our invisible college are reviewing candidates constantly. The tempo of debate and discussion concerning philosophies, methodologies, and the scientific nature of geomorphology is accelerating. This level of interest and attention is diagnostic either of a discipline experiencing healthy growing pains or of a discipline in need of rejuvenated leadership. Given the nature of much of the recent rhetoric, I fear our condition is the latter. Clearly we are not in the throes of scientific revolution. That term is too grandiose a description for our paradigmless discipline. We are instead in the preliminary stages of reconstructing geomorphology for the next generation. It is tempting to ignore the struggle using our own research agenda as justification, or to sit back and enjoy the spectacle and cheer for an underdog in order to prolong the fight. But we are witnessing continuing struggles between factions of particularism and pluralism; between science and history; between geography, engineering, and geology; between the past and the future. It is a contest between rival invisible colleges, and there will be winners and losers. We can accept the fashions that we are presented, or we can be critics and contenders; assessing, constructing, and refusing to accept a shoddy product. We do not have the luxury of watching the procession unless we are willing to bet that the 'good guys' will win, or unless we are willing to concede that we are part of a disciplinary proletariat lined up and waiting to be told what the next fashion is. We need to work to make sure that geomorphology, in a vital and relevant form, wins. And because there is this struggle, there is a correspondingly long list of those who are intentional or accidental competitors for status as our contemporary fashion leaders. The list will shorten substantially only after the initial sparring has selected some finalists.

One reason for this attempt to establish the importance of fashion in shaping the development of geomorphology is that such a disciplinary self-awareness will allow us to appropriate the fashion process in a critical and therefore useful manner. For although we cannot all be fashion dudes, we all have the option (perhaps the obligation) to be fashion critics. In the preceding section, I have outlined some of the characteristics that we might recognize in a leader, and we can employ these traits in an examination of conditions conducive to the establishment of a new fashion leader. The first goal of this process might be to reduce the present number of competing fashions because this will reduce the length of our list of potential fashion contenders. Although this may not be a desirable process from a disciplinary perspective, it is eminently fashionable! Appraisals of potential semifinalists would be based upon current or potential stature. Alternatively, we might justify the search for a larger population of fashions from which to choose. This can be rationalized through an argument that none of the present contenders are truly worthy.

It must be assumed that the successful opinion leaders will be able to transcend their subdisciplines, either by intentionally writing for a larger audience or because their specific research leads to a discovery that is widely applicable. Examples of broad topical areas that seem to be of modern geomorphological interest and where many subdisciplines can participate include applied geomorphology (e.g. Nordstrom and Renwick 1984 or Sherman 1989), natural hazards (e.g. Gares et al. 1994), public policy (e.g. Graf, Chapter 18 this volume), or environmentalism (e.g. the new text series being edited by Goudie and Viles).

We might also agree on several assumptions about what character traits and professional attributes are desirable (acceptable) for election as a fashion leader. The contenders must have reasonable scientific credentials, they must be in mainstream or near mainstream branches of geomorphology (I do not know if we would line up behind a karst geomorphologist – maybe). They must be somewhat politicized and 'publicly active'. They must be willing to stick their necks out to compete for a leadership role, because it is presumed that no one can become and remain a leader unintentionally. They must be generationally young enough to at least give the appearance of being (post)modern and progressive, a substantive fashion requirement. Finalists in the fashion competition will probably embrace one or more aspects of public policy/social relevance as a campaign theme.

The fashion dude for geomorphology at the dawn of the twenty-first century must be clever and wise enough to manage the power and authority necessary for the position. This person must be willing to serve as the president or director of our invisible college. He or she will probably need to write and speak concerning important issues of methodology and philosophy, and be willing to take unpopular stances and convince our community that the position is fashionable (because it is important). The new leader will be a public face for geomorphology and a defender of the discipline. Our leader will not hide behind the anonymity of review panels and editorial boards, although he or she may manipulate those bodies. That is part of the inevitable role of the invisible college. The leader will not hesitate to establish and maintain rules, and change rules as the moment (fashion) demands.

The operation of the fashion process in geomorphology frees us to embrace new theory, method, philosophy. We are able to skip Baconian generations in advancing our discipline, because the fashion model encourages creative ahistoricism. We can enjoy a more critical perspective on debate and power struggles. The sociologists, historians, and philosophers of science, are necessarily preoccupied with the past, at least in terms of the evidence and events that govern their interpretations of science. We cannot rely upon these scholars for the appraisals of geomorphology past, or for predictions of future geomorphologies. We need to be in an informed position to make conscious decisions about the future of our science. We can recognize that the quest for power or prestige moves geomorphology as much as (or more than) some scientific process. We can identify potential leaders and appraise them according to guidelines of our own fashion. It is destructive and foolish to deny the role of fashion in geomorphology, or to pretend that it is unimportant. Geomorphology as a science and as a discipline can benefit from the innovative qualities of the fashion process, if we choose to participate. The decision is ours. Geomorphology belongs to us.

ACKNOWLEDGEMENTS

This chapter has benefited substantially from the helpful comments of several colleagues – Bernie Bauer, Michael Dear, and Curt Roseman; the symposium organizers and editors of this volume – Bruce Rhoads and Colin Thorn; and the referees – Scott Morris, Irwin Sperber, and some nameless human geographer. I am grateful for their input and am responsible for the mistakes.

REFERENCES

Bagnold, R.A. 1936. The movement of desert sand, *Proceedings, Royal Society of London, Series A*, **157**, 594–620.
Baker, V.R. and Pyne, S. 1978. G.K. Gilbert and modern geomorphology, *American Journal of Science*, **278**, 97–123.
Baker, V.R. and Twidale, C.R. 1991. The reenchantment of geomorphology, *Geomorphology*, **4**, 73–100.
Bassett, K. 1994. Comments on Richards: the problems of 'real' geomorphology, *Earth Surface Processes and Landforms*, **19**, 273–276.

Bauer, B.O., Sherman, D.J. and Wolcott, J.F. 1992. Sources of uncertainty in shear stress and roughness length estimates derived from velocity profiles, *Professional Geographer*, **44**, 453–464.

Baulig, H. 1950. William Morris Davis: master of method, *Annals, Association of American Geographers*, **30**, 188–195.

Beckensale, R.P. 1976. The international influence of William Morris Davis, *Geographical Review*, **66**, 448–466.

Beckensale, R.P. and Chorley, R.J. 1991. *The History of the Study of Landforms*, Vol. 3, *Historical and Regional Geomorphology 1890–1950*, Routledge, London, pp. 496.

Berg, N.H. 1983. Field evaluation of some sand transport models, *Earth Surface Processes and Landforms*, **8**, 101–114.

Bishop, P. 1980. Popper's principle of falsifiability and the irrefutability of the Davisian cycle, *Professional Geographer*, **32**, 310–315.

Bodman, A.E. 1991. Weavers of influence: the structure of contemporary geographic research, *Transactions Institute of British Geographers*, NS, **16**, 21–37.

Brock, E.J. and Twidale, C.R. 1984. J.T. Jutson's contributions to geomorphological thought, *Australian Journal of Earth Sciences*, **31**, 107–121.

Brunsden, D. 1990. Tablets of stone: toward the ten commandments of geomorphology, *Zeitschrift für Geomorphologie*, Supplementband **79**, 1–37.

Brush, S.G. 1974. Should the history of science be rated X?, *Science*, **183**, 1164–1172.

Bryan, K. 1940. The retreat of slopes, *Annals, Association of American Geographers*, **30**, 254–267.

Bryan, K. 1950. The place of geomorphology in the geographic sciences, *Annals, Association of American Geographers*, **40**, 196–208.

Bull, W.B. 1975. Allometric change of landforms, *Geological Society of America Bulletin*, **86**, 1489–1498.

Butzer, K.W. 1973. Pluralism in geomorphology, *Proceedings, Association of American Geographers*, **5**, 39–43.

Carson, M.A. and Kirkby, M.J. 1972. *Hillslope Form and Process*, Cambridge University Press, Cambridge, 475 pp.

Chalmers, A. 1990. *Science and its Fabrication*, University of Minnesota Press, Minneapolis, 142 pp.

Chamberlin, T.C. 1897. The method of multiple working hypotheses, *Journal of Geology*, **5**, 837–848.

Chorley, R.J. 1978. Bases for theory in geomorphology, in *Geomorphology: Present Problems and Future Prospects*, edited by C. Embleton, D. Brusden, and D.K.C. Jones, Oxford University Press, Oxford, pp. 1–13.

Chorley, R.J. 1962. Geomorphology and general systems theory, *US Geological Survey Professional Paper*, 500B.

Chorley, R.J., Beckensale, R.P. and Dunn, A.J. 1973. *The History of the Study of Landforms*, Vol. 2, *The Life and Work of W.M. Davis*, Methuen, London, 874 pp.

Chorley, R.J., Dunn, A.J. and Beckensale, R.P. 1964. *The History of the Study of Landforms*, Vol. 1, *Geomorphology before Davis*, Methuen, London, 678 pp.

Conant, J.B. 1967. Scientific principles and moral conduct, *American Scientist*, **55**, 311–328.

Crane, D. 1969. Social structure in a group of scientists: a test of the 'invisible college' hypothesis, *American Sociological Review*, **34**, 335–352.

Crane, D. 1972. *Invisible Colleges: Diffusion of Knowledge in Scientific Communities*, University of Chicago Press, Chicago, 213 pp.

Davies, G.L. 1969. *The Earth in Decay*, Macmillan, London, 390 pp.

Davis, W.M. 1889. The rivers and valleys of Pennsylvania, *National Geographic*, **1**, 183–253.

Dear, M. and Wassmandorf, G. 1993. Postmodern consequences, *Geographical Review*, **83**, 321–325.

Einstein, A. 1936. Physics and reality, *Journal of the Franklin Institute*, **221**, 349–382.

Feyerabend, P. 1975. *Against Method: Outline of an Anarchistic Theory of Knowledge*, Verso (1978 edn), London, 339 pp.

Gares, P.A., Sherman, D.J. and Nordstrom, K.F. 1994. Geomorphology and natural hazards, *Geomorphology*, **10**, 1–18.

Golinski, J. 1990. The theory of practice and the practice of theory: sociological approaches in the history of science, *Isis*, **81**, 492–505.

Graf, W.L. 1979. Catastrophe theory as a model for change in fluvial systems, in *Adjustments of the Fluvial System*, edited by D.D. Rhodes and E.J. Williams, Allen & Unwin, London, pp. 13–32.

Gregory, K.J. 1985. *The Nature of Physical Geography*, Edward Arnold, London, 262 pp.

Haines-Young, R.H. and Petch, J.R. 1983. Multiple working hypotheses: equifinality and the study of landforms, *Transactions, Institute of British Geographers*, **8**, 458–466.

Higgins, C.G. 1975. Theories of landscape development: a perspective, in *Theories of Landform Development*, edited by W.N. Melhorn and R.C. Flemal, Allen & Unwin, London, pp. 1–28.

Horgan, J. 1992. Karl Popper: the intellectual warrior, *Scientific American*, **267**, 38–44.

Hull, D.L. 1988. *Science as a Process: An Evolutionary Account of the Social and Conceptual Development of Science*, University of Chicago Press, Chicago, 586 pp.

James, P.E. 1972. *All Possible Worlds: A History of Geographical Ideas*, Odyssey, Indianapolis, 622 pp.

Japanese Geomorphological Union 1989. 10th anniversary issue: history of geomorphology, *Transactions, Japanese Geomorphological Union*, **10**-B.

Jennings, J.N. 1973. 'Any millenniums today, lady?' The geomorphic bandwaggon parade, *Australian Geographical Studies*, **11**, 115–133.

Johnson, D. 1940. Comments, *Annals, Association of American Geographers*, **30**, 228–232.

Johnson, J.G. 1990. Method of multiple working hypotheses: a chimera, *Geology*, **18**, 44–45.

Jones, W.D. 1950. Remarks, *Annals, Association of American Geographers*, **40**, 179.

Kennedy, B.A. 1992. Hutton to Horton: views of sequence, progression and equilibrium in geomorphology, *Geomorphology*, **5**, 231–250.

King, L.C. 1953. The canons of landscape evolution, *Bulletin, Geological Society of America*, **64**, 721–752.

Kuhn, T.S. 1970. *The Structure of Scientific Revolutions*, 2nd edn, University of Chicago Press, Chicago, 210 pp.

Latour, B. and Woolgar, S. 1979., *Laboratory Life: The Construction of Scientific Facts*, Sage, Beverly Hills, 272 pp.

Leighly, J. 1940. Comments, *Annals, Association of American Geographers*, **30**, 223–228.

Leighly, J. 1955. What has happened to physical geography? *Annals, Association of American Geographers*, **45**, 309–318.

Leopold, L.B. and Langbein, W.B. 1962. The concept of entropy in landscape evolution, *US Geological Survey Professional Paper*, 500A.

Livingstone, D.N. 1984. The history of science and the history of geography: interactions and implications, *History of Science*, **22**, 271–302.

Longino, H.E. 1990. *Science as Social Knowledge: Values and Objectivity in Scientific Inquiry*, Princeton University Press, Princeton, NJ, 262 pp.

Lustig, L.K. 1967. *Inventory of Research on Geomorphology and Surface Hydrology of Desert Environments*, Office of Arid Lands Research, University of Arizona, Tucson, 189 pp.

Marston, R.A. 1989. Geomorphology, in *Geography in America*, edited by G.L. Gaile and C.J. Wilmott, Merrill, Columbus, Ohio, pp. 70–94.

Miller, R.W. 1987. *Fact and Method: Explanation, Confirmation and Reality in the Natural and the Social Sciences*, Princeton University Press, Princeton, NJ, 611 pp.

Mosley, M.P. and Zimpfer, G.L. 1976. Explanation in geomorphology, *Zeitschrift für Geomorphologie*, NF, **20**, 381–390.

Nordstrom, K.F. and Renwick, W.H. 1984. A coastal cliff management district for protection of eroding high relief coasts, *Environmental Management*, **8**, 197–203.

Outcalt, S.I., Hinkel, K.M. and Nelson, F.E. 1994. Fractal physiography? *Geomorphology*, **11**, 91–106.

Peattie, R. 1950. Remarks, *Annals, Association of American Geographers*, **30**, 178–179.

Penck, W. 1924. *Die Morphologische Analyse*. Engelhorn, Stuttgart, 283 pp.

Phillips, J.D. 1992. Nonlinear dynamical systems in geomorphology: revolution or evolution, *Geomorphology*, **5**, 219–229.

Phillips, J.D. 1993. Instability and chaos in hillslope evolution, *American Journal of Science*, **293**, 25–48.

Popper, K.R. 1934. *The Logic of Scientific Discovery*, Harper (1959 edn), New York, 480 pp.

Price, D.J. and Beaver, D. deB. 1966. Collaboration in an invisible college, *American Psychologist*, **21**, 1011–1018.

Reichenbach, H. 1951. *The Rise of Scientific Philosophy*, University of California Press, Berkeley, 333 pp.

Rhoads, B.L. and Thorn, C.E. 1993. Geomorphology as science: the role of theory, *Geomorphology*, **6**, 287–307.

Rhoads, B.L. and Thorn, C.E. 1994. Contemporary philosophical perspectives on physical geography with emphasis on geomorphology, *Geographical Review*, **84**, 91–101.

Ritter, D.F. 1988. Landscape analysis and the search for geomorphic unity, *Bulletin, Geological Society of America*, **100**, 160–171.

Sack, D. 1991. The trouble with antithesis: the case of G.K. Gilbert, geographer and educator, *Professional Geographer*, **43**, 28–37.

Sack, D. 1992. New wine in old bottles: the historiography of a paradigm change, *Geomorphology*, **5**, 251–263.

St Onge, D.A. 1981. Presidential address: theories, paradigms, mapping and geomorphology, *Canadian Geographer*, **25**, 307–315.

Sarre, R.D. 1988. Evaluation of aeolian sand transport equations using intertidal zone measurements, Saunton Sands, England, *Sedimentology*, **35**, 671–679.

Scheidegger, A.E. 1987. The fundamental principles of landscape evolution, *Catena*, **10**, 199–210.

Scheidegger, A.E. and Langbein, W.B. 1966. Probability concepts in geomorphology, *US Geological Survey Professional Paper*, 500C.

Schumm, S.A. 1985. Explanation and extrapolation in geomorphology: seven reasons for geologic uncertainty, *Transactions, Japanese Geomorphological Union*, **6**, 1–18.

Schumm, S.A. 1991. *To Interpret the Earth: Ten Ways to be Wrong*, Cambridge University Press, Cambridge, 133 pp.

Shapere, D. 1964. The structure of scientific revolutions, *Philosophical Review*, **73**, 383–394.

Sherman, D.J. 1989. Geomorphology: praxis and theory, in *Applied Geography: Issues, Questions, and Concerns*, edited by K.S. Kenzer, Kluwer, Dordrecht, pp. 115–131.

Sherman, D.J. 1994. Social relevance and geographical research, *Geographical Review*, **84**, 336–341.

Sperber, I. 1990. *Fashions in Science: Opinion Leaders and Collective Behavior in the Social Sciences*, University of Minnesota Press, Minneapolis, 303 pp.

Strahler, A.N. 1950. Davis' concepts of slope development in the light of recent quantitative investigations, *Annals, Association of American Geographers*, **30**, 209–213.

Strahler, A.N. 1952. Dynamic basis of geomorphology, *Bulletin, Geological Association of America*, **63**, 923–938.

Strahler, A.N. 1992. Quantitative/dynamic geomorphology at Columbia 1945–60: a retrospective, *Progress in Physical Geography*, **16**, 65–84.

Thorn, C.E. 1988. *An Introduction to Theoretical Geomorphology*, Unwin Hyman, London, 247 pp.

Thornbury, W.D. 1954. *Principles of Geomorphology*, John Wiley, New York, 618 pp.

Tinkler, K.J. 1985. *A Short History of Geomorphology*, Barnes and Noble, Totowa, NJ, 317 pp.

Twidale, C.R. 1977. Fragile foundations: some methodological problems in geomorphological research, *Revue de Géomorphologie Dynamique*, **26**, 84–95.

Vitek, J.D. and Ritter, D.F. 1989. Geomorphology in the United States, *Transactions, Japanese Geomorphological Union*, **10**, 225–234.

Von Engeln, O.D. 1950. Remarks, *Annals, Association of American Geographers*, **30**, 177–178.

Wittgenstein, L. 1921. *Tractatus Logico-Philosophicus* (1961 translation), Routledge & Kegan Paul, London, 166 pp.

Wolpert, L. 1992. *The Unnatural Nature of Science*, Harvard University Press, Cambridge, Mass., 191 pp.

Yatsu, E. 1992. To make geomorphology more scientific, *Transactions, Japanese Geomorphological Union*, **13**, 87–124.

5 Toward a Philosophy of Geomorphology

Bruce L. Rhoads and Colin E. Thorn

Department of Geography, University of Illinois at Urbana-Champaign

ABSTRACT

Few attempts have been made to examine philosophically the scientific nature of geomorphology. The reluctance of geomorphologists to engage in philosophical analysis reflects, at least in part, a widespread skepticism of nonempirical forms of inquiry among practicing scientists. This perspective is an outgrowth of the a priori prescriptive nature of traditional philosophy of science. Contemporary philosophers of science have responded to the skepticism of practicing scientists by developing naturalized philosophies that illuminate the complexity of scientific inquiry through direct examination of scientific practice.

The objective of this chapter is to illustrate the potential for philosophical analysis to strengthen the intellectual foundation of geomorphology by providing insight into the scientific nature of the discipline. Several issues are introduced that have relevance for understanding geomorphology as a science, including classification, laws and causality, theory and models, discovery, gender issues, and applied studies. The discussion calls attention to unexamined aspects of these issues in geomorphology and briefly reviews contemporary perspectives on them in the philosophy of science. The purpose of the discussion is not to provide a penetrating philosophical investigation of each issue, but to establish an informative framework for future analysis.

INTRODUCTION

Geomorphologists generally have not exhibited much enthusiasm for engaging in philosophical introspection. Whereas the mention of theory commonly elicits the proverbial reaction of reaching for one's soil auger (Chorley 1978), the mention of philosophy is

The Scientific Nature of Geomorphology: Proceedings of the 27th Binghamton Symposium in Geomorphology held 27–29 September 1996. Edited by Bruce L. Rhoads and Colin E. Thorn. ©1996 John Wiley & Sons Ltd.

perhaps the surest way to get a geomorphologist into the field posthaste! The reasons for this aversion to philosophy are probably manifold (Rhoads and Thorn 1994), but no doubt it stems in part from an inherent skepticism about nonempirical forms of inquiry among practicing scientists. Philosophy often is largely ignored by scientists until a period of intradisciplinary dissension arises, whereupon forays into philosophy are conducted in an effort to provide external standards for resolving internal disputes (e.g. Sloep 1993). This book is partly the product of such a situation; it emerged in response to the recent spate of books and articles on philosophical and methodological issues in geomorphology (Douglas 1982; Starkel 1982; Church et al. 1985; Haines-Young and Petch 1986; Baker 1988, 1993; Ritter 1988; Thorn 1988; Brunsden 1990; Richards 1990, 1994; Baker and Twidale 1991; Montgomery 1991; Schumm 1991; Yatsu 1992; Rhoads and Thorn 1993, 1994; Bassett 1994; Rhoads 1994), which collectively suggest that the discipline currently is, if not in crisis, at least experiencing acute growing pains.

One reason why scientists in general, including geomorphologists, have viewed philosophy with a jaundiced eye is that traditional philosophy of science, particularly logical empiricism, has been highly normative or prescriptive in nature, a characteristic many scientists find annoying. No one likes to be told how to do their job better by someone who has not actually performed the tasks involved. This problem has been accentuated by the analytical nature of logical empiricism, which holds that the knowledge providing the basis for epistemic norms in science can be grasped a priori (i.e. through nonempirical reflection on the meaning of certain propositions). In other words, the philosophical program to understand science is independent of any specific scientific results, beliefs, or methods.

The nonempirical foundation of logical empiricism clearly conflicts with the empirical *modus operandi* of scientists. Contemporary philosophers of science fully recognize the need to grapple with this problem, and over the past 30 years (i.e. since the demise of logical empiricism) have focused their efforts on developing naturalized philosophical perspectives that attempt to capture the knowledge-producing potential of science as it is actually practiced. As noted by Shapere (1987, p. 10):

> Not only has traditional epistemology failed to provide the 'analyses' it promises; it turns out to have been misguided in principle in its methodological approach. For an understanding of the nature of knowledge – of the knowledge-seeking and knowledge-acquiring enterprise – can only be obtained through a study of the knowledge we have actually attained, of how we have attained it, and of how the goal of knowledge itself has been constructed and altered in that process.

Naturalized epistemology is largely descriptive or empirical, and in many ways constitutes a 'science of science'.

> Philosophy of science does not exist and function on a level above and independent of the substantive content of scientific beliefs; it is integrally and inseparably linked to that content, and its methods and conclusions must rest on the results of the very science with which it is concerned (Shapere 1987, p. 24).

This trait has led to criticisms that science is now being used to evaluate the knowledge claims of science, obviously a circular analysis, and that abandonment of a traditional role of philosophy, to provide an independent meta-narrative on science, threatens to obviate

Table 5.1.   Main tasks of contemporary philosophy of science (after Shapere 1987)

Critical function
- Continue its traditional task of exposing confused or mistaken interpretations of science

Overview function
- Provide an overview of the rationale (or lack thereof) of scientific change
- Determine how specific beliefs develop and change
- Ascertain how certain beliefs are considered knowledge (i.e. are judged free from specific and compelling doubt
- Formulate conceptions of scientific reasoning and knowledge

Detailing function
- Conduct detailed studies of science, including case studies within specific disciplines, to determine how important presuppositions, beliefs, methods, criteria, goals, and so forth have developed and changed and to demonstrate important commonalities and differences among these factors across the various domains of science.

philosophy of science (Trigg 1993). Debate about the extent to which science subsumes epistemology or epistemology subsumes science continues within the philosophy of science (Maffie 1990a, b, 1993; Axtell 1993), but many contemporary philosophers feel that by developing more accurate depictions of the ways in which science (or various branches of science) 'successfully' pursues certain types of goals, even the most extreme forms of naturalized epistemology can maintain a normative role in relation to science (Brown 1989).

A naturalized approach does not require that philosophy abandon its meta-narrative role in relation to science. Complete separation between philosophy and science is necessary only when one is attempting to avoid absolute skepticism about the possibility of scientific knowledge, not when one is interested in determining how particular processes within science produce certain *scientific* beliefs that are free from specific and compelling doubt (Shapere 1987; Nickles 1987). In this sense, naturalized philosophy still has important tasks to undertake in relation to science (Table 5.1).

What is the relevance of these developments in philosophy of science for geomorphology? Of course, the response to such a question ultimately must be a matter of opinion. Nevertheless, opinions will be based in part on whether persuasive reasons can be given for the value of a particular intellectual pursuit. The purposes of this chapter are to show how contemporary philosophical inquiry promises to illuminate the scientific nature of geomorphology and, in the process, contribute to the intellectual depth of the discipline. The discussion briefly identifies a broad range of important topics, each of which is deserving of penetrating analysis in the future. The intent is to provide another step forward toward a comprehensive philosophy of geomorphology.

DEFINITION OF GEOMORPHOLOGY

In preparing this book, a reviewer opined that a volume of this sort should provide a definitive definition of geomorphology. At first glance, such a request seems quite reasonable; should not a volume on the scientific nature of the discipline define geomorphology once and for all? Moreover, such a request does not seem to be too difficult to accommodate; one has simply to consult various introductory texts for cursory definitions that can provide the basis for a more elaborate definition (Table 5.2). Although some may be disappointed, such a definition will not be provided here. In keeping with naturalized approaches to the study of science, geomorphology is viewed as historically dynamic, having the potential to change character as it evolves through time. Although the discipline may eventually either cease to change or cease to exist, thereby either allowing for a stable definition or obviating the need for such a definition, the evolution of geomorphology cannot be determined a priori. All that can be done is to identify the current state of affairs and to speculate about the implications of emerging trends for the future of the discipline.

Textbook definitions of geomorphology have much in common and appear to be adequate for identifying the fundamental core, or *domain* (Shapere 1974), of contemporary geomorphology. These definitions suggest that few geomorphologists would disagree with the claim that the aim of the discipline is to investigate the surface forms and processes on the terrestrial portion of the earth. At present, inclusion of the morphology of the ocean floor or the study of the surfaces of other planets within the domain

Table 5.2.   Definitions of geomorphology

- '...geomorphology is...devoted to the explanation of the earth's surface relief and to an understanding of the processes which create and modify landforms' (Bridges 1990, p. vii)
- 'Geomorphology is the study of landforms, and in particular their nature, origin, processes of development and material composition' (Cooke and Doornkamp 1990, p. 1)
- 'Geomorphology is the study of the surface of the Earth. Classically, geomorphologists have studied landforms, which are shapes that have been categorized or named by geomorphologists or other Earth scientists' (Mayer 1990, p. 1)
- 'Geomorphology...is the scientific study of the geometric features of the earth's surface. Although the term is commonly restricted to those landforms that have developed at or above sea level, geomorphology includes all aspects of the interface between the solid earth, the hydrosphere and the atmosphere. Therefore, not only are the landforms of the continents and their margins of concern, but also the morphology of the sea floor. In addition, the close look at the moon, Mars and other planets provided by spacecraft has created an extraterrestrial aspect to geomorphology' (Chorley et al. 1984, p. 1)
- 'Geomorphology is best and most simply defined as the study of landforms. Like most simplistic definitions, the actual meaning is somewhat vague and open to interpretation' (Ritter et al. 1995, p. 3)
- 'Geomorphology is the study of the origin and evolution of topographic features by physical and chemical processes operating at or near the earth's surface.... the study of surface processes and landforms relies heavily on geologic principles. Yet, like other sciences, geomorphology also depends on the application of basic principles of physics, chemistry, biology, and mathematics to natural systems' (Easterbrook 1993, p. 2)
- '*Structures, materials, processes*, and the *history* of changing landforms, are the four essential components of a study of the nature and origin of the modern land surface...' (Selby 1985, p. 1)
- '...the systematic description, analysis, and understanding of landscapes and the processes that change them...the description, analysis, and understanding of landforms...' (Bloom 1991, p. 1)
- 'Geomorphology is the science concerned with the form of the landsurface and the processes which create it. It is extended by some to include the study of submarine features, and with the advent of planetary exploration must now incorporate the landscapes of the major solid bodies of the Solar System. One focus for geomorphic research is the relationship between landforms and the processes currently acting on them' (Summerfield 1991, p. 3)
- '...geomorphology is broadly defined as the study of past, present, and future landforms, landform assemblages (physical landscapes), and surficial processes on the earth and other planets' (Rhoads and Thorn 1993, p. 288)

of geomorphology may be more hopeful than well-founded. Few scientists that study deep ocean basins or the surfaces of planets such as Jupiter or Saturn probably consider themselves geomorphologists. One might qualify the extended definition by restricting it to near-coastal submarine forms or to the surfaces of solid planets, but this type of restriction begins a slide down the slippery slope of additional qualifications (e.g.. solid planets with atmospheres similar to the Earth's atmosphere). In the end, this problem merely points out that whereas a core of geomorphology can be identified, its periphery is rather fuzzy, a trait that characterizes many fields of knowledge (Shapere 1974).

NATURAL KINDS

One way in which philosophical analysis can contribute to geomorphology is to help unify the discipline by identifying bases for common ground among a diverse group of scientists all of whom consider themselves geomorphologists. Whereas some geomorphologists have viewed methodology as a potential unifier (Haines-Young and Petch 1986; Richards 1990, 1994), Rhoads and Thorn (1993, 1994) have argued that methodology is a healthy source of diversity, rather than unity in the discipline. The chapters in this volume reinforce this perspective and also highlight the contributions that diverse methodologies have made to geomorphological knowledge. This situation suggests that any sense of disciplinary unity is best achieved by focusing on some aspect of geomorphologic inquiry *other than* methods of inquiry. An obvious alternative is the objects of inquiry.

An important component of any science, including geomorphology, is classification. Geomorphologists have approached the study of the Earth's surface by classifying it into discrete categories, or taxa, known as landforms. The ubiquitous reference to landforms in definitions of geomorphology (Table 5.2) clearly demonstrates the centrality of this concept in the discipline. Landforms provide the basis for process and geohistorical investigations of the Earth's surface. In turn, landform taxonomy often is refined based on the results of such investigations. An example is recent work in fluvial geomorphology, which has identified anastomosing and wandering gravel-bed rivers as distinct types that differ fundamentally from meandering, braided, or straight rivers (Church 1983; Knighton and Nanson 1993).

The centrality of the concept of 'landform' in geomorphology raises several important epistemological and metaphysical issues that have not been adequately addressed by geomorphologists. Despite the frequent mention of landforms in definitions of geomorphology, none of these definitions specifies what a landform is. Only Mayer (1990) addresses this issue directly and his explication is largely unhelpful. Because the concept of 'landform' underpins geomorphologic classification, which in turn provides the basis for much geomorphologic inquiry, it is important to consider the metaphysical nature and epistemic utility of this concept. In particular, geomorphologists have not adequately grappled with questions such as: How is 'landform' to be defined in geomorphological taxonomy? What is the epistemic purpose of landform classification? What standards or criteria provide the basis for defining types of landforms? Are these standards or criteria consistent with the epistemic purpose of classification? Does landform classification have an objective basis in nature or is it merely an artifact of human comprehension? Answers to these questions are important because they can help clarify whether the epistemic basis of geomorphological classification is consistent with the overall epistemic goals of geomorphological inquiry and whether the science of geomorphology can be justified on ontological grounds.

Philosophical analysis of scientific classification has centered on the problem of natural kinds. The pivotal idea behind the concept of natural kinds is that individual objects in the world are naturally divided into distinct classes, or *kinds* of entities by virtue of certain shared intrinsic properties. Natural kinds define the basic types of objects that exist in the world. Categories that do not have an objective basis in nature, but that have been developed for human purposes only, are artificial or nominal kinds (Schwartz 1980). The notion that a goal of science is to discover natural kinds and develop scientific explanations for the existence of these kinds constitutes an implicit, widely held conviction among practicing scientists. The search for natural kinds can be viewed as epistemologically privileged in relation to other types of human inquiry because it represents an attempt to uncover the true way in which the world is structured independently of human thought. It represents an attempt to 'carve the world at its joints'. Natural kinds are fundamental in science because they serve as the foci of theoretical generalizations (e.g. Rothbart 1993).

The concept of natural kinds has a long and controversial history within the philosophy of science (Hacking 1991), dating to Aristotle (Granger 1989; Suppe 1989, pp. 204–205). Arguments for and against the concept have been best developed for biology, especially with regard to whether or not species constitute natural kinds (e.g. Kitts and Kitts 1979; Dupre 1981, 1989, 1994; Fales 1982; Kitcher 1984; Ruse 1987; Wilkerson 1988, 1993; Stanford 1995), but the concept has also been applied in physics (Quine 1992), chemistry (van Brakel 1986), and even economics (Nelson 1990). Much debate has centered around the basis on which natural kinds should be identified (cf. Quine 1969). Perhaps the most controversial aspect of contemporary philosophical debate about natural kinds is the notion that any object belongs to an unambiguously discoverable natural kind on the basis of a certain essential (necessary and sufficient) property or set of properties. Often the properties that determine the real 'essence' of a kind are viewed as underlying causal mechanisms, powers, or processes (e.g. the molecular structure of water (Putnam 1975) or the genetic structure of living organisms (Wilkerson 1988)). The implication of such a view is that any object belongs to a natural category *independent of the context of inquiry*, that this category is determined by some shared 'real' essence among certain objects, and that it is the goal of science to discover these 'hidden' or 'theoretical' real essences, thereby revealing the true structure of the natural world. Such a view, with its emphasis on 'hidden', 'internal', 'microscopic' causal powers has metaphysical connotations, providing the basis for many realist perspectives on science (Boyd 1991). It also has reductionist implications for scientific inquiry (Meyer 1989). For example, if relations among microscopic physical particulars constitute the real essences (causal mechanisms) of macroscopic phenomena (i.e. these relations determine the macroscopic properties of macroscopic phenomena), then all generalizations about macroscopic phenomena can be at least quasi-reduced to fundamental physics because the structure and function of macroscopic phenomena are ontologically (and possibly epistemically) quasi-reducible to microscopic physical entities and causal relations (e.g. Melnyk 1995).

Not all philosophers agree that natural kinds are determined by essential, underlying causal powers or mechanisms. Many hold that the definition of a natural category, even one that is theoretically based, depends on the context of inquiry (De Sousa 1984; van Brakel 1992; Dupre 1993; Shain 1993), or that the concept of essentialism, as outlined by its proponents (e.g. Putnam 1975; Kripke 1980; Leplin 1988), cannot be sustained when examined philosophically or within the context of actual scientific practice (Mellor 1977; Nersessian 1991; Shapere 1991; Stroll 1991; Li 1993). The challenge posed by context-dependent kinds is that such a view threatens to undermine the epistemically privileged status of scientific inquiry (i.e. the search for the 'essential' set of natural kinds that exist in the world).

Are landforms natural kinds? It is beyond the scope of this chapter to analyze this question in detail. However, future analysis of the question may be fruitful given the centrality of classification in geomorphology. Such analysis may yield insights about the epistemological role of classification in the discipline and whether or not such classification is merely epistemically convenient (as argued for geology by Watson 1966) or has a justifiable ontological basis. A starting point for addressing this question would be to determine whether specific types of landforms are fixed by a necessary and sufficient property or set of properties. Morphologic properties alone cannot provide necessary and sufficient conditions because these properties vary in detail among individual landforms of the same type. Moreover, the concept of equifinality suggests that similar morphologic properties can be produced by different causal mechanisms, calling into question (at least from an essentialist perspective) the appropriateness of existing landform categories. A possible solution could be to revise existing categories based on reductionist analyses of underlying mechanisms, but such an endeavor could have important implications for the ontological status of landforms. For example, Wilkerson (1988), a leading advocate of essentialism (and thus realism), argues that geological and geographical kinds (e.g. cliffs, beaches, mountains, valleys, volcanoes, rivers, glaciers) do not have real essences and thus are unlikely to yield theoretical generalizations, a claim that may be difficult to refute given the current status of theory development in geomorphology. To Wilkerson, these features are 'superficial' kinds, not natural kinds. Scientific inquiry in fields such as geology or geography is possible only because the superficial kinds of interest are composed of physical and chemical constituents (i.e. natural kinds) that do have real essences. Such a view does not exclude an epistemic role for geomorphology (i.e. to uncover underlying relations among physical/chemical kinds responsible for similarities among superficial properties of the landscape that lead us to classify it into landforms), but it does sharply reinforce the popular, implicit ideology that geomorphology is nothing more than applied physics and chemistry (because landforms have no ontological status apart from their physical/chemical constituents and properties). Such a perspective stands in stark contrast to Dupre's (1993) claim that the conventionality of classification at all levels of scientific inquiry undermines essentialism and reductionism, and supports a form of ontological pluralism he calls promiscuous realism.

The problem of natural kinds also raises the issue of whether classification of the physical landscape into categories known as landforms is a nominalistic exercise that is providing a misleading theoretical picture about the 'real' structure of the landscape. When classification is applied to a landscape (as in geomorphological mapping), it requires the imposition of boundaries, boundaries that often are useful from a practical or methodological perspective, but that may not exist in nature. Gould (1987, pp. 160–161) argues that although discrete 'islands of form' can sometimes be identified in nature, 'we must accept shadings and continua as fundamental'. This alternative perspective on the natural world provides ontological and epistemological support for holistic landscape analyses that treat planetary surfaces as continua (e.g. fractal analyses of planetary terrains; physically based models based on continuum concepts), rather than as assemblages of individual landforms. On the other hand, sophisticated conceptions of natural kinds recognize and attempt to accommodate intraclass variation and indistinct boundaries between natural categories (Boyd 1989; Suppe 1989). Under these conceptions, the 'essence' of the kind (with the kind being characterized by a variable but clustered set of properties) will consist in the complete catalogue of laws or causal mechanisms responsible for the clustering of properties. In any case, a naturalized view of geomorphology maintains that philosophical arguments for or against the existence of natural geomorphologic kinds will be adjudicated over the long term by the relative empirical adequacy and explanatory power of geomorphological theories that posit the existence of such kinds versus those that do not.

LAWS, CAUSALITY, AND CAUSAL EXPLANATION

Geomorphology is a science that deals with complex, dynamic natural systems consisting of physical, chemical, and biological constituents and attributes. The questions arise whether it is possible for sciences of this type to develop their own laws and if such sciences can successfully employ laws of the basic sciences to explain (and possibly predict) the natural phenomena with which they are concerned. In part, the answers to these questions depend on how one defines the concept of law and the role that one assigns to laws in scientific explanation.

According to the logical empiricist conception, laws of nature express empirical regularities. As noted by Carnap (1966, p. 3) 'if a certain regularity is observed at all times and all places, without exception, then the regularity is expressed in the form of a "universal law"'. This perspective implies that the cognitive content of a law consists in a predicted pattern of perceptual observations and that the evidence for a law consists in a set of observations that instantiate this pattern (Boyd 1985). Of course, logical empiricists recognized that not all empirical regularities constitute laws, but in keeping with the empiricist aversion to metaphysical commitment, they approached this issue as a linguistic problem about laws as *universal statements*. In other words, distinguishing law-like generalizations from accidental generalizations should be based on the syntactic properties of particular statements expressing these generalizations (cf. Lambert and Brittan 1970, pp. 37–45).

Geomorphologists, like most scientists, are greatly concerned with explaining the natural phenomena they study. The concept of explanation is itself deserving of further exploration, but many practicing scientists value greatly *causal* explanations, or those that identify causes of empirical phenomena (Dilworth 1994; Barnes 1995). The empiricist conception of causal explanation, the covering-law model, is derivative from the view of laws as regularities:

The type of explanation which has been considered here so far is often referred to as causal explanation. If E describes a particular event, then the antecedent circumstances described in the sentences $C_1, C_2, \ldots C_k$ may be said jointly to 'cause' that event, in the sense that there are certain empirical regularities, expressed by the laws $L_1, L_2, \ldots L_r$, which imply that whenever conditions of the kind indicated by $C_1, C_2, \ldots, C_k$ occur, an event of the kind described in E will take place. Statements such as $L_1, L_2, \ldots, L_r$ which assert general and unexceptional connections between specified characteristics of events, are customarily called causal, or deterministic, laws (Hempel and Oppenheim 1948, p. 139).

The covering-law model associates causal explanation with subsumption of an event to be explained (E) under deterministic laws, where deterministic refers to exceptionless generalizations. It is clearly an attempt to analyze causal relations in terms of laws. However, because the empiricist concept of a law consists in nothing more than reference to regular patterns in observable data, the covering-law model of causal explanation does not permit a metaphysical interpretation. In other words, it implies that *causation consists in nothing more than regularities in the behavior of observables.* This strategy represents an attempt to reduce the concept of causation to an empirical interpretation (Tooley 1990) – laws are merely summaries of what is observed (Boyd 1985). Such a perspective opposes the intuitive understanding of causation as involving a precipitating event, a resulting event, and a *causal process that connects the two events by propagating a causal influence from one space-time locale to another* (Salmon 1984, p. 155).

The empiricist conception of laws is highly controversial. Realist philosophers have challenged this conception by attempting to provide laws with an ontological status. These alternative conceptions are worth examining given the recent concern about realism in geomorphology (Richards 1990, 1994; Bassett 1994; Rhoads 1994). Realist views on laws consist of two types of claims: (1) that laws describe *necessary relations* between universal properties associated with certain objects (e.g. Dretske 1977; Armstrong 1983; Tooley 1987), and (2) that laws describe manifestations of *causal powers, capacities, or dispositions* possessed by certain objects or classes of objects (Cartwright 1989; Bigelow et al. 1992; Woodward 1992). The difference between these two positions is subtle but important. The first treats laws as universals, i.e. relations that exist separately from objects and govern regular behavior among objects. These universals are treated as irreducible primitives whose ontological status is unanalyzable and simply must be accepted. The second conception does not assign an independent existence to laws; instead laws derive from the capacities of kinds of objects to effect change. The difference between the two conceptions can be captured by an analogy to a chess game. In the first case, laws represent rules that govern the movements of specific types of pieces; these rules exist independently of the board and pieces in the form of a rule book. Thus, if the rule book was written differently (i.e. allowing rooks to move diagonally), the types of pieces could and would move differently. According to the second view, the rules (laws) arise from the capacity of the types of pieces themselves to move only in specific patterns. The rules are what they are because of the capacities of the types of pieces; the rules are prescribed by these capacities and are not a contingent matter. In other words, causal capacities are fundamental relative to natural laws.

One implication of the view that laws ensue from causal capacities is that the concepts of causation and causal law can be divorced from the concept of regularity (Cartwright 1989; Woodward 1992). The causal capacities of particular kinds of objects may manifest themselves differently (i.e. produce different outcomes) depending on the specific context in which the cause operates. Ascriptions of capacities define the range of possible outcomes a kind of object can cause, but are too general to be used for precise predictions. Causal lawfulness, on the other hand, is based on the criterion of *invariance of causal relations*, rather than solely on regularities in empirical data (Woodward 1992). Invariance refers to functional stability of a causal relation as initial conditions change over a specified range in a constrained setting. Because causal laws define causal relations that obtain *in specific situations*, they can be used for prediction. The need to specify the circumstances within which a particular causal relation obtains, however, implies that all causal laws are *ceteris paribus* generalizations (Lange 1993; Cartwright 1995). The manifestation of causal capacities in complex systems, such as those studied in geomorphology, will vary, depending on the nature of interactions among various capacities within a specific context. Regularities will emerge only when a capacity or set of interacting capacities is triggered repeatedly within uniform settings (i.e. those that appropriately shield the capacity from interacting with specific features of a new situation). Thus, the study of complex phenomena poses a problem both for inductively establishing the existence of underlying causal laws based on regularity principles and for identifying possible combinations of underlying causal laws by combining mathematical formulations of these laws in predictive models. This situation may account for the fact that geomorphology has not been very successful at developing its own laws, or in using simple models that combine a few basic physical laws to predict the form and dynamics of specific landforms – a topic which has received considerable attention elsewhere in this volume. It also points out the need both for detailed experimental work, in the field and in the lab, and for large-sample investigations (e.g. Richards, Chapter 7 this volume). In experimental work, specific conditions can either be created artificially or at least precisely documented so that particular manifestations of causal capacities can be deduced from data or specific claims derived from causal laws can be evaluated from patterns of data (Peakall et al., Chapter 9 this volume). On the other hand, large-sample investigations are useful for isolating statistically causal capacities that operate irregularly within uniform settings (and thus underlie probabilistic causal laws) or that operate regularly, but whose effects are readily confounded by interaction with other causal capacities in nonuniform settings (Woodward 1992; Dupre 1993, pp. 194–217).

Another challenge to the empiricist account of laws comes from the realm of nonlinear dynamics, a topic that is beginning to have an impact on geomorphology (Phillips, Chapter 13 this volume). Sensitive dependence of outcomes on initial conditions greatly complicates efforts to inductively derive or test the functional form of an underlying nonlinear law based on regularities in patterns of data, especially for complex natural systems in which initial conditions are likely to exhibit considerable variability (Holt and Holt 1993). This empirical problem is also in part a problem for the realist account of laws, but, by embracing the evidential role of accepted background knowledge on causal capacities, which is held to be at least approximately true, the realist has additional epistemic resources for evaluating the validity of competing theoretical models, all of which may be 'equifinal' in the sense of having similar predictive accuracy (e.g. Beven, Chapter 12 this volume). To the realist, the evaluation of a theoretical model is based not only on empirical adequacy, but also on how well the model is grounded in accepted background knowledge. To take full advantage of these additional epistemic resources, nonlinear geomorphological models should be explicitly and unambiguously linked to known causal mechanisms. However, at present, many nonlinear models in geomorphology are based on simple mathematical functions (ordinary differential equations) that include aggregated variables, each of which may subsume a complex amalgamation of physical, chemical, or biological mechanisms. Thus, qualitative stability analysis of such models cannot yield explanations that specify the role that underlying mechanisms play in system response; instead, one must assume that the aggregated variables can effect change in the manner specified by the structure and functional form of the equations. In this sense, many nonlinear models, like multivariate statistical models, are 'black box' models.

An obvious question is: can philosophical discussion about the nature of laws be adjudicated in any way by an analysis of scientific practice? The answer is – to some extent. The logical empiricist perspective in large part arose from the fact that theoretical laws in physics often do not specify causes: 'the reason why physics has ceased to look for causes is that, in fact, there are no such things' (Russell 1917, p. 174). However, more recent examinations of physics (Cartwright 1981, 1983) and other areas of science (Cartwright 1989) seem to indicate that the empiricist claim that scientists have a greater epistemic commitment to laws than to causes is flawed. Not only do scientists attempt to identify causes, but they often treat ascriptions of causal capacities as more fundamental than theoretical laws. Whether or not geomorphologists conform with this assessment will not be considered here, but recent concern about the search for causal mechanisms in geomorphology (e.g. Richards, Chapter 7 this volume) suggests that this issue at least is worthy of further exploration. A concern with the nature of laws and causal explanation is important in geomorphology because it is intimately linked to scale-related issues, especially the potential for the character of geomorphological methods and explanations to vary over the temporal and spatial range of inquiry (e.g. Rhoads and Thorn 1993; Church, Chapter 6 this volume). The recent trend toward reductionist analyses based on the principles of mechanics may be motivated not only by pragmatic problems related to the development of empirical laws for complex, evolving natural systems (e.g. van der Steen and Kamminga 1991), but by a fundamental concern about the causal relevance of macrolevel phenomena – a worry generated by proponents of reductionism (e.g. Kim 1989). The widespread adoption of mechanics in geomorphology suggests that causal powers may lie in physical entities such as forces (e.g. Tuchanska 1992; Cartwright 1995), but geomorphologists should be cautioned that the ontology of forces and other entities that populate physics is far from clear (Bigelow et al. 1988; Jones 1991). On the other hand, those geomorphologists with an antireductionist bent (e.g. Haff, Chapter 14 this volume) may find solace in recent philosophical work on macrolevel causation (e.g. Henderson 1994) and on the value of case studies for deriving and evaluating causal explanations about complex, seemingly unique phenomena (Shrader-Frechette 1994). In any case, the problems of the existence, relevance, and spatial–temporal variation of causal agents in geomorphology are *a posteriori* theoretical issues that can only be adjudicated on the basis of how various theory-directed research programs fare in competition with one another. Realists will view such evolution, should it occur, as the triumph of truth over falsity (e.g. Richards 1990), relativists will see it as the triumph of particular research styles (Vicedo 1995; Osterkamp and Hupp, Chapter 17 this volume) or fashions (Sherman, Chapter 4 this volume), pragmatists will hail it as the triumph of practical utility and societal relevance (Baker 1994), and empiricists will proclaim it as the triumph of empirical adequacy (Beven, Chapter 12 this volume).

THEORY AND MODELS

Theory is a central concept in science. Most scientific activity centers around the development and testing of theory. Given that geomorphologists generally consider their discipline a science, it is not surprising that they have expressed an interest in the role of theory in geomorphology (Baker and Twidale 1991; Rhoads and Thorn 1993). Contemporary philosophical analysis suggests that theories are not merely storehouses for scientific knowledge, but that they also have a pervasive methodological influence on scientific inquiry (Brown, Chapter 1 this volume). Most observational procedures are now viewed as theory-dependent (at least to some extent) – a perspective that appears to apply to geomorphology (Rhoads and Thorn, Chapter 2 this volume). More analysis is required to determine the extent to which methodological procedures are infused with theory in geomorphology and to examine the epistemological implications of the relationship between theory and observation in specific instances.

Another type of philosophical investigation that may prove fruitful is formal analysis of the structure of geomorphological theories. The concept of a theory is certainly a fuzzy one within the philosophy of science, within science in general, and within geomorphology in particular. A continuum of perspectives on theory has emerged from the philosophy of science, ranging from the simple notion of a theory as a hypothetical claim with empirical content (Popper 1965, p. 115) to the sophisticated, logical–analytic view of theories as axiomatized, hierarchical systems of deductively connected statements (Feigl 1970). Most scientists, including earth scientists, tend to make an implicit distinction between hypotheses and theories. In general, a theory is viewed as more comprehensive and reliable than a hypothesis (von Engelhardt and Zimmermann 1988, p. 234). Similarly, philosophers interested in formal analysis of theory structure usually examine theories that consist of more than a singular hypothetical claim.

The view of theories as axiomatized systems of statements is commonly referred to as the Received View. Many geomorphologists may be familiar with the basic tenets of the Received View through their training in geography (Harvey 1969; Amedeo and Golledge 1975) or geology (Kitts 1963; von Engelhardt and Zimmermann 1988). This perspective emerged from logical empiricism and incorporates many of its basic tenets, including the observational/theoretical distinction, knowledge empiricism, and the verifiability theory of meaning (Rhoads and Thorn 1994). It is beyond the scope of this chapter to review in detail the Received View (for comprehensive overviews see Suppe 1977a, pp. 6–61 and 1989, pp. 39–62). The important point here is that this view has never been popular among scientists given that its main focus is to provide an artificial reconstruction of existing theories within an explicitly characterized formal language, rather than to provide an accurate depiction of how scientists actually construct and use theories (Feigl 1970). In particular, the emphasis on theories as linguistic entities fails to adequately capture the pervasive use of models in science.

Over the past 30 years, considerable effort has been devoted to an alternative to the Received View known as the Semantic or Model-Theoretic View (MTV) of theories (Suppe 1977a, pp. 221–230, also 1989; van Fraassen 1980, pp. 41–69, also 1987; Giere 1988, pp. 62–91). According to MTV, a theory is specified by defining a family of abstract structures, i.e. its models. Because models are nonlinguistic entities, they can be characterized in many ways using many different languages. In other words, although an expression of a theory may include statements in a specific language (including equations), the use of this language is not fundamental because the same class of models could be described in other languages as well. Specification of a class of models involves theoretical definitions that draw upon the laws or postulates of the theory (Giere 1979, pp. 63–83). Thus, the relation between the models and the underlying theory is unproblematic; the models are, by definition, true representations of the theory. However, the goal of scientific inquiry is not to study relations between theories and their models, but to examine relations between theories and some real-world phenomena (i.e. the intended scope of the theory). The link between the models, or idealized abstract systems representing the theory, and some identified class of real phenomena is achieved by specifying theoretical hypotheses. These linguistic statements make claims about the world in relation to the model, usually of the type that the phenomena would be as the model prescribes if all of the idealized conditions specified in the model had actually obtained. One implication of this view is that the evaluation of hypotheses is not performed by comparing statements about phenomena with direct observations (sensory perceptions) of phenomena, but rather by comparing theoretical hypotheses with data about phenomena (Figure 5.1). The production of data draws upon various types of auxiliary theories, including those governing data collection

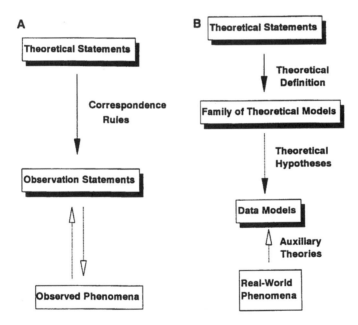

Figure 5.1. Contrasting philosophical perspectives on the structure of scientific theories: (A) the Received View. Theoretical statements are connected to observational statements via correspondence rules (explicit or partial definitions). Observational statements are directly testable (verified or refuted) by comparing these statements with observations. (B) The Model-Theoretic View. Basic theoretical statements define families of theoretical models that represent abstract, idealized representations of some domain of real-world phenomena. Raw data on real-world phenomena along with auxiliary theories governing data reduction and analysis are used to develop data models, which provide the basis for evaluating theoretical hypotheses that individually make claims about the real-world system being a system of the type defined by the theory

(experimental design, instrumentation, selection of certain types of information as opposed to others), data reduction and analysis (statistics, processing routines, presentation methods), and data interpretation (inferences about extant patterns and their relation to attributes of underlying phenomena). The outcome of this process is the construction of a data model. The purpose of creating a data model is to ensure compatibility between the data gathered from the real world and the form of information specified by the theoretical model. In other words, a theoretical model specifies the pattern of data a phenomenon or set of phenomena should generate under a particular set of idealized conditions, and the data model reveals whether this pattern of data is, in fact, present in the data collected from the real-world system.

The advantages of MTV are several. First, in characterizing theory, it puts models, rather than hierarchical systems of statements expressed in a formal language, on center stage – a portrayal of theory that accords well with actual scientific practice. The large number of chapters on models and modeling in this volume suggests that geomorphology conforms at least to some extent with this characterization. Second, MTV, by emphasizing the nonlinguistic, abstract nature of theories, easily accommodates the notion that theories are conceptual devices (Suppe 1977b, pp. 706–716) or mental representations (Giere 1992, 1994) that can be expressed mathematically, qualitatively, or iconically (e.g. Da Costa and French 1990). This aspect of MTV holds promise for geomorphology, which includes qualitative conceptual models as well as various types of quantitative models. Also, the implication that, as idealizations, theories always fall short of completely capturing the full complexity of real-world phenomena should appeal to geomorphologists who believe that the greatest asset of theory is its fallibility, i.e. its role in highlighting anomalous data and in promoting the search for new theories. The search for phenomena through the development of data models can serve as the impetus for theory change as new phenomena are revealed that cannot be accommodated by any possible model of an accepted theory (van Fraassen 1987). Third, MTV emphasizes that the relation between theoretical propositions and the real world is not direct, as implied by the Received View, but instead is mediated through abstract idealizations, i.e. models (Figure 5.1). Current attempts to apply physical theory to geomorphic phenomena appear to be broadly consistent with this perspective. The general hypothesis that a particular feature of the landscape is a type of natural mechanical system is not evaluated by attempting to assess directly basic propositions from classical mechanics (e.g. Newton's Laws of Motion); instead, the validity of this general hypothesis is assessed by formulating a class of models (expressed mathematically) and testing specific claims about the relation of the predictions of a particular model to data on real-world phenomena. Fourth, MTV stresses the important role of hypotheses in science. This aspect should have an intuitive appeal for geomorphologists, who traditionally have embraced hypotheses and hypothesizing as an important component of geomorphologic inquiry (Schumm 1991; Baker, Chapter 3 this volume). Fifth, MTV has been applied to the formalization of theories not only in physics and chemistry (Suppe 1989), but also in biology (Thompson 1983; Beatty 1987; Lloyd 1989; Sintonin 1991), a discipline which, like geomorphology, is characterized by qualitative and quantitative models that are based on a mixture of principles from within the discipline and from other disciplines. One particularly promising application of MTV in biology suggests that it can accommodate seemingly nonuniversal theories characterized by *ceteris paribus* conditions that include entities at different levels of aggregation (i.e.

different scales) (Schaffner 1980). Such versatility is probably essential for efforts to formalize geomorphologic theories, given the centrality of environmental contingency and scale issues in geomorphological explanations.

DISCOVERY

A criticism of philosophy of science frequently specified by scientists, including geomorphologists (Baker and Twidale 1991), is that most philosophical analysis focuses on justification, rather than on discovery of scientific knowledge, even though the latter, rather than the former, is the most vital dimension of science. Such perceptions reflect the pervasive influence that logical empiricism, with its distinction between the contexts of discovery and justification (e.g. Reichenbach 1938), has had on scientists' views of the philosophy of science. In contrast, nineteenth-century philosophers of science, such as John Herschel, William Whewell, and Charles Peirce, devoted considerable attention to the problem of scientific discovery, especially with regard to the problem of how hypotheses are formulated and evaluated, a topic of special interest in geomorphology (Baker, Chapter 3 this volume). Hanson (1958) and Kuhn (1970) initiated a renewed interest in the problem of discovery and the literature on this topic has increased rapidly over the past 25 years in conjunction with the emergence of postpositivist interpretations of science (Nickles 1980, 1985, 1990; Kelly 1987; Lai 1989; Kantorovich 1993).

Discovery in science generally involves at least one of the following: (1) the invention of new, reliable concepts or ideas about the world (i.e. hypothesis–theory formulation), and (2) the encounter with novelties (anomalous data) in scientific investigations. Scientists may view either or both of these factors as necessary for discovery, depending on the particulars of a specific situation. Whereas the second factor often, but not always (e.g. Brewer and Chinn 1994), provides an impetus for the first (e.g. Darden 1992), the first factor can occur independently of the second (e.g. Chi 1992; Gooding 1992). Exactly where discovery stops and justification begins is viewed as a moot issue in contemporary philosophy of science. Instead, discovery is now seen as a continuum that includes both the invention of new ideas and the validation of these ideas. Thus, discovery and justification are intertwined, rather than discrete components of scientific inquiry.

Recent work has not focused on the development of a formal logic of discovery, such as the logical analysis empiricists developed for justification, but on efforts to determine the heuristics, or general rule-based strategies, that underlie aspects of scientific inquiry that lead to discoveries (Kleiner 1993). The most formal analyses of this type have been conducted by those interested in artificial intelligence, some of whom have developed computational algorithms of discovery (Simon 1973, 1977; Langley et al. 1987; Kulkarni and Simon 1988; Thagard 1988; Shrager and Langley 1990).

Other work on heuristics has been of a less formalized nature, attempting only to identify components of scientific inquiry that appear to play a role in discovery in at least some instances. A general theme that has emerged from this type of analysis is that most discoveries occur within a context of background information, including those based on serendipity (e.g. Kantorovich and Ne'eman 1989; Kleiner 1993, p. 308), analogical and imagistic reasoning (e.g. Nersessian 1992), or abductive inference (e.g. Kapitan 1992), all of which have been discussed by geomorphologists interested in the discovery process

(Gilbert 1886, 1896; Baker and Twidale 1991; Rhoads and Thorn 1993). Background information includes deeply ingrained metaphysical beliefs about the nature of reality (e.g. Dilworth 1994), presuppositions that influence decisions about the 'best' way to explain this reality (e.g. Barnes 1995), and accepted scientific knowledge (Kleiner 1993, pp. 59–86). This information guides discovery by constraining the range of possibilities when formulating or entertaining new theories or hypotheses. In the earth sciences, principles from physics, chemistry, and biology, as well as accepted knowledge within a specific discipline, often provide relevant background information for constraining the range of possible explanations of new situations (Kitts 1982; Bardsley 1991; Rhoads and Thorn, Chapter 2 this volume). The extent to which a particular item of background information constrains possible new hypotheses is usually in direct proportion to the extent to which that item is held to be true. Through detailed examination of investigations that have led to discoveries in geomorphology, the discipline may develop a better understanding of the reasoning processes that underlie geomorphologic inquiry and the relationship of these processes to those in other areas of science.

GENDER ISSUES

Prior to 1950 geomorphology was almost exclusively a male preserve. Over the past 45 years the number of women geomorphologists has increased dramatically. Despite this increase, women are still a distinct minority in the discipline; geomorphology remains a male-dominated scientific field. In this sense, the gender structure of geomorphology is similar to that in other physical-science and engineering-related disciplines (Sonnert 1995).

The increasing presence of women in science at large has raised some practical issues related to gender, such as concern about equality of access for women with regard to educational opportunities (Matyas and Dix 1992; Ginorio 1995) and science-related careers (Vetter 1992; Rayman and Brett 1993; National Research Council 1994), of which geomorphologists should be made aware. The effort to include female authors in this volume serves as an example of the practical challenges facing women scientists today. Two women who agreed to contribute chapters had to cancel their commitments immediately prior to the deadline for submission because of unanticipated personal situations which required their immediate and full attention. These situations were not, by necessity, uniquely female, but they were ones in which women often are expected to assume a disproportionate share of responsibility relative to men. Although social expectations based on gender are less prevalent today than in the past, they still exist to some extent. Men in a male-dominated discipline must not only contribute to the dismantling of such stereotypes, they must also be sensitive to the ways in which the persistence of such stereotypes can obligate women differently from men.

The increasing infusion of women into science has also had an influence on the philosophy of science through the development of gendered and feminist perspectives on contemporary science (Keller 1985, 1992; Bleier 1986; Harding 1986, 1991; Haraway 1989; Tuana 1989; Code 1991; Alcoff and Potter 1993; Shepard 1993; Rose 1994; Spanier 1995) and on the history of science (Benjamin 1991; Scheibinger 1993). Much gender-oriented philosophical analysis is highly naturalized and postmodern, drawing upon work in behavioral psychology, Marxist theory, and sociology of science for its inspiration. Most of this work is directed at offering constructive critiques of contemporary science, focusing in particular on the influence of masculine ideology on the scientific enterprise, with the hope of illuminating how such ideology can obstruct scientific objectivity. Feminist epistemology takes the critique to the next level by arguing that women, as marginalized persons in science, are in a privileged epistemic position to recognize, expose, and correct male-related bias in scientific inquiry, thereby enhancing the objectivity of this inquiry (Hartsock 1983; Rose 1986; Haraway 1988; Harding 1993). Although this claim is controversial (e.g. Pinnick 1994), it appears to be relevant in at least some scientific contexts, particularly in studies in which the subject matter is directly or indirectly related to women or gender (Sismondo 1995). Other feminist philosophy of science has employed psychoanalytic theory on personality development in an attempt to understand the nature of science as a human endeavor (e.g. Keller 1987). According to this perspective, science can be seen as a dialectic between the desires for mastery over, and union with, nature. On the one extreme, knowledge obtained through detached, objective scientific analysis that emphasizes the use of quantitative models and empirical data can be pursued to gain some measure of power or control over the world. This image of science reflects aggressive, autonomous human behavior. At the other extreme, scientific research may employ qualitative methodologies in an effort to converse with nature or to let the data suggest the answer to a problem. The goal here is simply the pleasure associated with knowing the world so that we might better appreciate it and our place within it – an image that reflects romantic predilections. According to Keller (1987) examples of these dialectic elements can be found throughout science, both now and throughout the history of science. The contrast between certain chapters in this volume suggests that this dialectic also exists in geomorphology. Although the components of the dialectic may be somewhat exaggerated, stereotypes about scientific style, whether perceived or real, can have an important influence on how attractive a particular discipline is to individuals with particular types of personality traits. Such stereotypes are shaped not only through exposure to actual research in the discipline, but also through educational experiences.

The emergence of gender-related issues in science has not escaped the attention of some women geomorphologists, especially those affiliated with the discipline of geography, where interest in feminist approaches to human geography has exploded over the past several years (e.g. Hanson 1992; McDowell 1992; Rose 1993; Monk 1994). An open forum on physical geography entitled 'Is Gender an Issue?' was conducted by the Women in Geography Study Group (WGSG) in September 1995 at the Royal Geographical Society, London, UK. The goal of the forum was to initiate discussion and exchange ideas about women, gender, and physical geography (Joanna Bullard, 1995, personal communication). Several questions served as an impetus for discussion (Table 5.3). Although no serious problems related to gender were identified at the forum, several concerns were raised (Table 5.3). This initial meeting did not consider in depth the possible role of feminist theory in physical geography, but the nature of the concerns identified at the meeting suggests that further exploration of the relevance of feminist epistemology and methodology to geomorphology may be worthwhile. For example, the concern about possible overemphasis on quantitative methods and empirical analysis in physical geography (Table 5.3) is consistent with many feminist epistemologies. A possible point of

Table 5.3. Focus questions for the open forum on physical geography and gender-related concerns emerging from the forum

---

Focus questions:
Can we attract more women to undergraduate courses in physical geography?
Why do so few female physical geographer postgraduates go on to become lecturers?
Is there a role for a feminist physical geography?
What is the role of the WGSG in physical geography?

Concerns raised at the forum:
Female role models are important for attracting female undergraduates, but the most important attributes of the teacher/lecturer are enthusiasm for the subject and the ability to make course material interesting.
On undergraduate field trips the presence of a female staff member is important in moderating the often male-dominated environment, but so too is a balance between male and female students.
At graduate or postgraduate level there is sometimes the expectation that women participants on a field trip will cook and clean up.
Physical geography is seen as a discipline in which quantitative data and statistical analysis are valued and the more philosophical side of the discipline is undervalued. One must first prove one's ability as a 'hard' scientist before engaging broad philosophical issues.
Questions asked of applicants for undergraduate degree programs and academic jobs are often inappropriate in the sense that they are gender-specific. An example is concern about the physical capabilities of women in performing fieldwork.
Teaching of physical geography at the undergraduate level often emphasizes a 'continual onslaught of equations' in the context of an insipid style of presentation. This approach to teaching is inaccessible and unappealing.

---

departure for future debate about this issue is the controversy over the role that quantification should play in feminist-oriented research in human geography (Mattingly and Falconer-Al-Hindi 1995; McLafferty 1995; Moss 1995; Lawson 1995; Rocheleau 1995).

APPLIED STUDIES

One of the most fundamental changes that has occurred in geomorphology over the past several decades is the dramatic increase in the number of studies with an applied dimension. The original purpose of geomorphology, to provide knowledge of the evolutionary history of landscapes, has been supplemented by the goal of explaining and predicting landscape dynamics for societal benefit. This supplemental focus implies that geomorphological studies conducted on human time scales no longer need to be justified solely on the basis of their contribution to the goal of understanding geologic-scale landscape evolution. It also raises some interesting philosophical and ethical issues.

The distinction between basic and applied science is common, but ambiguous. Although most scientists distinguish between basic and applied research, this distinction rests mainly on an intuitive foundation. The tension involved in drawing the distinction is reflected in the well-known cliché that all scientific knowledge has practical value, albeit perhaps in a highly indirect and unforeseeable manner. Nevertheless, because the distinction between basic and applied research often enters into political policy decisions that directly affect science (i.e. government decisions about science policy and funding) (Graf, Chapter 18 this volume), consideration of this issue is important, if not for intellectual reasons, then for pragmatic purposes.

Traditional philosophy of science has been largely unhelpful in illuminating the philosophical basis for the basic versus applied distinction because it has focused mainly on what both philosophers and scientists would characterize as basic research (in the intuitive sense). Recently, however, some philosophers of science have begun to direct attention toward this issue, especially with regard to the distinction between science and technology (e.g. Bunge 1985; Kroes 1989; Ihde 1991). While recognizing that scientific research forms a broad continuum that does not allow for dichotomous categorization, this work does maintain that distinctions can be made between characteristics of research in disparate portions of the continuum. Much of this work has challenged the standard conception that science and technology are connected in a directed, linear manner with technological knowledge consisting in nothing more than applications of scientific knowledge within specific contexts.

One useful way to characterize the distinction among basic science, applied science, and technology is on the basis of differences in the utilities that define the aims of inquiry and in the structure of scientific statements (Niiniluoto 1993). The primary aim of basic research is generally viewed as the purist ideal of science: to accurately explain and understand reality. Thus, the epistemic utilities of truth, knowledge, and explanatory power serve as cognitive virtues at this level. At the other extreme, the aim of technology is to produce artifacts that create new possibilities of action for humans in their interaction with the world. The utility underlying this aim is a practical one: the effectiveness of the artifact relative to its intended application. In between, the aim of applied science is not to produce artifacts, but to produce information that can be used either for prognostication or for enhancing the effectiveness of human actions. The utilities of applied science often are a mix of the epistemic utilities of basic science and the practical utility defined by the value of the information in relation to human concerns or goals.

Differences in the structure of scientific statements can also be identified among various types of scientific research (Table 5.4). Descriptive statements are stated in language that is meant to be value-neutral (factual statement in indicative mood), although this trait does not guarantee that the statement is completely value-free in the sense that the selection of the scientific problem underlying the claim and the process of establishing the content of the claim do not involve value judgments on the part of scientists. When explanation is the aim, statements that specify cause usually are preferred, whereas when the aim is predictive power, statements that identify a reliable (but not necessarily causal) relation between two variables are desired. Technical norms, on the other hand, combine a categorical normative statement (you ought to do X) with a statement about want or preference. Although technical norms include an explicit statement about valuation, usually of an ethical nature (i.e. you want or desire A), such statements are still capable of being analyzed scientifically once this valuation is fixed by some process external to science (e.g. Hempel 1960). In other words, scientific testing can still reveal whether doing X achieves A.

The criteria of utility and type of scientific statement can be combined to identify four different types of scientific research along a continuum ranging from descriptive basic science to technology (Table 5.4). Specific examples from fluvial geomorphology illustrate how this characterization of science and technology applies to geomorphology. Most work in geomorphology to date generally falls within the categories of basic, descriptive science or applied, predictive science. The increasing visibility of geomorphologists in

Table 5.4. A scheme for characterizing basic science, applied science, and technology in geomorphology (adapted from Niiniluoto 1993)

	Primary utilities defining aim of scientific research	Structure of scientific statements	Example from fluvial geomorphology
Basic, descriptive science	Epistemic: truth, knowledge, explanatory power	Descriptive, causal [X causes A in situation B, or X tends to cause (with probability p) A in situation B]	Research on meandering e.g. Dietrich and Smith (1983)
Applied, predictive science	Mix of epistemic and practical with emphasis on predictive power of scientific information	Descriptive, relational [X is related to A (with probability p) in situation B]	Threshold functions of human-induced channel instability e.g. Brookes (1987)
Applied, design science	Mix of epistemic and practical with emphasis on usefulness of information for developing technological artifacts	Technical norm [if you want A, and believe you are in situation B, then you ought to do X]	Development of river restoration/management guidelines based on geomorphologic principles e.g. Brookes (1995)
Technology Professional problem solving	Practical: effectiveness of technological artifacts for achieving human goals	NA	Implementation of restoration schemes based on geomorphological principles e.g. Brookes (1990), Rhoads and Herricks (in press)

professional roles has also initiated the development of applied design science components in some areas of the discipline. This type of science, which draws upon principles derived from descriptive basic and applied sciences, directly supports the development of problem-solving tools and skills needed to support a professional practice. However, the relationship among these components is not a unidirectional, linear path from basic science to technology. In some cases, existing scientific knowledge or information to support a particular technology may be poorly developed (Figure 5.2) and the constraints of time or money may not allow for testing of normative statements to determine their validity. In such cases, a 'trial and error' approach may be adopted at the technological level, which in turn may lead to the formulation of technical norms. Moreover, because these norms include statements about relations between variables found in descriptive scientific statements, information generated by 'trial and error' at the technological level occasionally may be relevant to the evaluation of knowledge claims in basic, descriptive science.

This depiction of the relation between basic science, applied science, and technology suggests that the growth of geomorphology as a practical profession requires that geomorphologists continue to devote effort to developing and refining a design science to

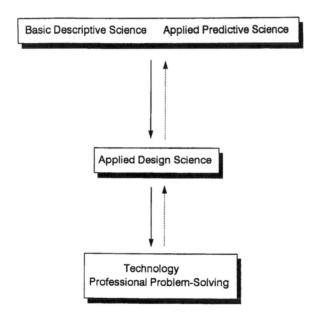

Figure 5.2. Relationships among basic science, applied science, and technology. Knowledge and information from basic descriptive science and applied predictive science provide support for an applied design science, which in turn sustains some technology or profession (practical problem-solving). On the other hand, trial and error efforts (usually in situations for which the design science provides little guidance) at the professional or technological level can lead to improvements in the design science, which in turn can contribute to knowledge in basic descriptive science. For example, trial and error attempts at stream-channel restoration could generate information that proves useful for improving restoration guidelines (technical norms) (e.g. Table 5.4). Moreover, this information may have value for basic descriptive science and applied predictive science if it leads to improved understanding of the basic mechanisms governing river-channel dynamics, thereby allowing for more accurate predictions of river response to human disturbance

support this profession. Such a design science can provide the basis for professionalization of the discipline by codifying a body of information, tools, and skills for licensing or certification programs. Establishment of a design science, however, represents only one aspect of the effort to professionalize geomorphology. Because geomorphologists involved in applied design science and professional geomorphologists both deal with technical norms, which are explicitly value-laden, the issues of moral responsibility and professional ethics become a concern. As noted by Niiniluoto (1993), a person who implements technical norms or who helps to establish ways of attaining these norms is morally responsible for helping to effect the valuations stated in these norms. Little or no attention has been given to ethical issues in geomorphology. Although most applied work in the discipline appears to rest on an underlying environmental ethic, the exact nature of this ethic and its potential implications for a professional code of conduct have yet to be explored in detail (Pierce and VanDeVeer 1995).

CONCLUSION

This chapter has introduced a variety of philosophical topics, the exploration of which may shed light on the scientific nature of geomorphology and the ways in which geomorphology is similar to and different from other scientific disciplines. The choice of topics has been selective, and a host of other issues in the philosophy of geomorphology may also be worthy of investigation. Some geomorphologists may remain unconvinced that philosophical analysis has anything useful to offer to the discipline. For them, it is time to head to the field. On the other hand, the chapters in this volume show that some geomorphologists are interested in, and maybe even concerned about, exploring more fully the scientific nature of geomorphology. Certainly, this chapter should not be viewed as a call for a vast number of geomorphologists to become philosophers. As scientists, geomorphologists should primarily practice science, not philosophy. On the other hand, given the extent to which philosophical discussion of geomorphology has been avoided in the past, a small to moderate dose of philosophy will probably not hurt us too much, and, who knows, it may contribute something of genuine value, not only to geomorphology, but to philosophy of science as well. Now where is that soil augur?

ACKNOWLEDGEMENTS

A special thank you to Harold Brown for providing critical comments on a preliminary draft of this chapter.

REFERENCES

Alcoff, L. and Potter, E. (eds) 1993. *Feminist Epistemologies*, Routledge, New York, 312 pp.
Amedeo, D. and Golledge, R.G. 1975. *An Introduction to Scientific Reasoning in Geography*, Wiley, New York, 431 pp.
Armstrong, D.M. 1983. *What is a Law of Nature?* Cambridge University Press, Cambridge, 180 pp.

Axtell, G. 1993. Naturalism, normativity, and explanation: the scientific biases of contemporary naturalism, *Metaphilosophy*, **24**, 253–274.

Baker, V.R. 1988. Geological fluvial geomorphology, *Geological Society of America Bulletin*, **100**, 1157–1167.

Baker, V.R. 1993. Extraterrestrial geomorphology: science and philosophy of Earthlike planetary landscapes, *Geomorphology*, **7**, 9–36.

Baker, V.R. 1994. Geological understanding and the changing environment, *Transactions of the Gulf Coast Association of Geological Societies*, **XLIV**, 1–8.

Baker, V.R. and Twidale, C.R. 1991. The reenchantment of geomorphology, *Geomorphology*, **4**, 73–100.

Bardsley, W.E. 1991. Some thoughts on the automated generation and selection of hypotheses in the earth sciences, *Mathematical Geology*, **23**, 241–256.

Barnes, E. 1995. Inference to the loveliest explanation, *Synthese*, **103**, 251–277.

Bassett, K. 1994. Comments on Richards: the problems of 'real' geomorphology, *Earth Surface Processes and Landforms*, **19**, 273–276.

Beatty, J. 1987. On behalf of the semantic view, *Biology and Philosophy*, **2**, 17–23.

Benjamin, M. 1991. *Science and Sensibility: Gender and Scientific Inquiry, 1780–1945*, Blackwell, Oxford, UK, 295 pp.

Bigelow, J., Ellis, B. and Pargetter, R. 1988. Forces, *Philosophy of Science*, **55**, 614–630.

Bigelow, J., Ellis, B. and Lierse, C. 1992. The world as one of a kind: natural necessity and laws of nature, *British Journal for the Philosophy of Science*, **43**, 371–388.

Bleier, R. (ed.) 1986. *Feminist Approaches to Science*, Pergamon, New York, 212 pp.

Bloom, A.L. 1991. *Geomorphology: A Systematic Analysis of Late Cenozoic Landforms*, Prentice-Hall, Englewood Cliffs, NJ, 532 pp.

Boyd, R. 1985. Observations, explanatory power, and simplicity: toward a non-Humean account, in *Observation, Experiment, and Hypothesis in Modern Physical Science*, edited by P. Achinstein and O. Hannaway, MIT Press, Cambridge, Mass., pp. 47–94.

Boyd, R. 1989. What realism implies and what it does not, *Dialectica*, **43**, 5–29.

Boyd, R. 1991. Realism, anti-foundationalism, and the enthusiasm for natural kinds, *Philosophical Studies*, **61**, 127–148.

Brewer, W.F. and Chinn, C.A. 1994. Scientists' responses to anomalous data: evidence from psychology, history, and philosophy of science, in *PSA 1994*, Philosophy of Science Association, East Lansing, Mich., pp. 304–313.

Bridges, E.M. 1990. *World Geomorphology*, Cambridge University Press, Cambridge, 260 pp.

Brookes, A. 1987. The distribution and management of channelised streams in Denmark, *Regulated Rivers: Research and Management*, **1**, 3–16.

Brookes, A. 1990. Restoration and enhancement of engineered river channels: some European experiences, *Regulated Rivers; Research and Management*, **5**, 45–56.

Brookes, A. 1995. Challenges and objectives for geomorphology in UK river management, *Earth Surface Processes and Landforms*, **20**, 593–610.

Brown, H.I. 1989. Normative epistemology and naturalized epistemology, *Inquiry*, **31**, 53–78.

Brunsden, D. 1990. Tablets of stone: toward the ten commandments of geomorphology, *Zeitschrift für Geomorphologie*, Supplementband **79**, 1–37.

Bunge, M.A. 1985. *Philosophy of Science and Technology*, Reidel, Dordrecht, 2 vols.

Carnap, R. 1966. *Philosophical Foundations of Physics*, Basic Books, New York, 300 pp.

Cartwright, N. 1981. The reality of causes in a world of instrumental laws, in *PSA 1980*, Vol. 2, edited by P. Asquith and R. Giere, Philosophy of Science Association, East Lansing, Mich., pp. 38–48.

Cartwright, N. 1983. *How the Laws of Physics Lie*, Oxford University Press, New York, 221 pp.

Cartwright, N. 1989. *Nature's Capacities and their Measurement*, Clarendon Press, Oxford, 268 pp.

Cartwright, N. 1995. Précis of *Nature's Capacities and their Measurement, Philosophy and Phenomenological Research*, **55**, 153–156.

Chi, M.T.H. 1992. Conceptual change within and across ontological categories: examples from learning and discovery in science, in *Cognitive Models of Science*, edited by R.N. Giere,

Minnesota Studies in the Philosophy of Science, Vol. 15, University of Minnesota Press, Minneapolis, pp. 129–186.

Chorley, R.J. 1978. Bases for theory in geomorphology, in *Geomorphology: Present Problems and Future Prospects*, edited by C. Embleton, D. Brunsden and D.K.C. Jones, Oxford University Press, Oxford, pp. 1–13.

Chorley, R.J., Schumm, S.A. and Sugden, D.E. 1984. *Geomorphology*, Methuen, New York, 605 pp.

Church, M. 1983. Pattern of instability in a wandering gravel bed channel, in *Modern and Ancient Fluvial Systems*, edited by J.D. Collinson and J. Lewin, International Association of Sedimentologists, Special Publication 6, Blackwell, Oxford, pp. 169–180.

Church, M., Gomez, B., Hickin, E.J. and Slaymaker, O. 1985. Geomorphological sociology, *Earth Surface Processes and Landforms*, **10**, 539–540.

Code, L. 1991. *What Can She Know? Feminist Theory and the Construction of Knowledge*, Cornell University Press, Ithaca, New York, 349 pp.

Cooke, R.U. and Doornkamp, J.C. 1990. *Geomorphology in Environmental Management*, Clarendon Press, Oxford, UK, 410 pp.

Da Costa, N.C.A. and French, S. 1990. The model-theoretic approach in the philosophy of science, *Philosophy of Science*, **57**, 248–265.

Darden, L. 1992. Strategies for anomaly resolution, in *Cognitive Models of Science*, edited by R.N. Giere, Minnesota Studies in the Philosophy of Science, Vol. 15, University of Minnesota Press, Minneapolis, pp. 251–273.

De Sousa, R. 1984. The natural shiftiness of natural kinds, *Canadian Journal of Philosophy*, **14**, 561–580.

Dietrich, W.E. and Smith, J.D. 1983. Influence of the point bar on flow through curved channels, *Water Resources Research*, **19**, 1173–1192.

Dilworth, C. 1994. Principles, laws, theories and the metaphysics of science, *Synthese*, **101**, 223–247.

Douglas, I. 1982. The unfulfilled promise: earth surface processes as a key to landform evolution, *Earth Surface Processes and Landforms*, **7**, 101.

Dretske, F.I. 1977. Laws of nature, *Philosophy of Science*, **44**, 248–268.

Dupre, J. 1981. Natural kinds and biological taxa, *Philosophical Review*, **90**, 66–90.

Dupre, J. 1989. Wilkerson on natural kinds, *Philosophy*, **64**, 248–251.

Dupre, J. 1993. *The Disorder of Things: Metaphysical Foundations of the Disunity of Science*, Harvard University Press, Cambridge, Mass., 308 pp.

Dupre, J. 1994. The philosophical basis of biological classification, *Studies in the History and Philosophy of Science*, **25**, 271–279.

Easterbrook, D.J. 1993. *Surface Processes and Forms*, Macmillan, New York, 520 pp.

Fales, E. 1982. Natural kinds and freaks of nature, *Philosophy of Science*, **49**, 67–90.

Feigl, H. 1970. The 'orthodox' view of theories: remarks in defense as well as critique, in *Analysis of Theories and Methods of Physics and Psychology*, edited by M. Radner and S. Winokur, Minnesota Studies in the Philosophy of Science, Vol. 4, University of Minnesota Press, Minneapolis, pp. 3–16.

Giere, R.N. 1979. *Understanding Scientific Reasoning*, Holt, Rinehart and Winston, New York, 371 pp.

Giere, R.N. 1988. *Explaining Science: A Cognitive Approach*, University of Chicago Press, Chicago, 321 pp.

Giere, R.N. 1992. Introduction: cognitive models of science, in *Cognitive Models of Science*, edited by R.N. Giere, Minnesota Studies in the Philosophy of Science, Vol. 15, University of Minnesota Press, Minneapolis, pp. xv–xxviii.

Giere, R.N. 1994. The cognitive structure of scientific theories, *Philosophy of Science*, **61**, 276–296.

Gilbert, G.K. 1886. The inculcation of scientific method by example, with an illustration drawn from the Quaternary geology of Utah, *American Journal of Science*, 3rd series, **3**, 284–299.

Gilbert, G.K. 1896. The origin of hypotheses, illustrated by the discussion of a topographic problem, *Science*, NS 3, **53**, 1–13.

Ginorio, A.B. 1995. *Warming the Climate for Women in Academic Science*, Association of American Colleges and Universities, Washington, DC, 38 pp.

TOWARD A PHILOSOPHY OF GEOMORPHOLOGY 139

Gooding, D. 1992. The procedural turn; or, why do thought experiments work?, in *Cognitive Models of Science*, edited by R.N. Giere, Minnesota Studies in the Philosophy of Science, Vol. 15, University of Minnesota Press, Minneapolis, pp. 45–76.

Gould, S.J. 1987. *The Flamingo's Smile*, Norton, New York, 476 pp.

Granger, H. 1989. Aristotle's natural kinds, *Philosophy*, **64**, 245–252.

Hacking, I. 1991. A tradition of natural kinds, *Philosophical Studies*, **61**, 109–126.

Haines-Young, R.H. and Petch, J.R. 1986. *Physical Geography: Its Nature and Methods*, Harper and Row, New York, 230 pp.

Hanson, N.R. 1958. *Patterns of Discovery*, Cambridge University Press, Cambridge, 240 pp.

Hanson, S. 1992. Geography and feminism: worlds in collision?, *Annals of the Association of American Geographers*, **82**, 569–586.

Haraway, D.J. 1988. Situated knowledges: the science question in feminism and the privilege of partial perspective, *Feminist Studies*, **14**, 575–609.

Haraway, D.J. 1989. *Primate Visions: Gender, Race, and Nature in the World of Modern Science*, Routledge, New York, 486 pp.

Harding, J. 1986. *Perspectives on Gender and Science*, Falmer Press, London, 217 pp.

Harding, S. 1991. *Whose Science? Whose Knowledge?: Thinking from Women's Lives*, Cornell University Press, Ithaca, NY, 319 pp.

Harding, S. 1993. Rethinking standpoint epistemology: what is 'strong objectivity'? in *Feminist Epistemologies*, edited by L. Alcoff and E. Potter, Routledge, New York, pp. 49–82.

Hartsock, N.C.M. 1983. The feminist standpoint: developing a ground for historical materialism, in *Feminist Perspectives on Epistemology, Metaphysics, Methodology and Philosophy of Science*, edited by S. Harding and M. Hintikka, Reidel, Dordrecht, pp. 282–311.

Harvey, D.W. 1969. *Explanation in Geography*, Edward Arnold, London, 521 pp.

Hempel, C.G. 1960. Science and human values, in *Social Control in a Free Society*, edited by R.E. Spiller, University of Pennsylvania Press, Philadelphia, pp. 39–64.

Hempel, C.G. and Oppenheim, P. 1948. Studies in the logic of explanation, *Philosophy of Science*, **15**, 135–175.

Henderson, D.K. 1994. Accounting for macro-level causation, *Synthese*, **101**, 129–156.

Holt, D.L. and Holt, R.G. 1993. Regularity in nonlinear dynamical systems, *British Journal for the Philosophy of Science*, **44**, 711–727.

Ihde, D. 1991. *Instrumental Realism: the Interface between Philosophy of Science and Philosophy of Technology*, University of Indiana Press, Bloomington, Ind., 159 pp.

Jones, R. 1991. Realism about what? *Philosophy of Science*, **58**, 185–202.

Kantorovich, A. 1993. *Scientific Discovery: Logic and Tinkering*, State University of New York Press, Albany, NY, 281 pp.

Kantorovich, A. and Ne'eman, Y. 1989. Serendipity as a source of evolutionary progress in science, *Studies in the History and Philosophy of Science*, **20**, 505–529.

Kapitan, T. 1992. Peirce and the autonomy of abductive reasoning, *Erkenntnis*, **37**, 1–26.

Keller, E.F. 1985. *Reflections on Gender and Science*, Yale University Press, New Haven, Conn., 193 pp.

Keller, E.F. 1987. Feminism and science, in *Sex and Scientific Inquiry*, edited by S. Harding and J. O'Barr, University of Chicago Press, Chicago, pp. 233–246.

Keller, E.F. 1992. *Secrets of Life, Secrets of Death: Essays on Language, Gender, and Science*, Routledge, New York, 195 pp.

Kelly, K.T. 1987. The logic of discovery, *Philosophy of Science*, **54**, 435–452.

Kim, J. 1989. The myth of nonreductive materialism, *American Philosophical Association Proceedings*, **63**, 31–47.

Kitcher, P. 1984. Species, *Philosophy of Science*, **51**, 308–333.

Kitts, D.B. 1963. The theory of geology, in *The Fabric of Geology*, edited by C.C. Albritton, Freeman and Cooper, Stanford, Calif., pp. 49–68.

Kitts, D.B. 1982. The logic of discovery in geology, *History of Geology*, **1**, 1–6.

Kitts, D.B. and Kitts, D.J. 1979. Biological species as natural kinds, *Philosophy of Science*, **46**, 613–622.

Kleiner, S.A. 1993. *The Logic of Discovery: A Theory of the Rationality of Scientific Research*, Kluwer, Dordrecht, 334 pp.

Knighton, A.D. and Nanson, G.C. 1993. Anastomosis and the continuum of channel pattern, *Earth Surface Processes and Landforms*, **18**, 613–625.

Kripke, S. 1980. *Naming and Necessity*, Harvard University Press, Cambridge, Mass., 172 pp.

Kroes, P. 1989. Philosophy of science and the technological dimension of science, in *Imre Lakatos and Theories of Scientific Change*, edited by K. Gavroglu, Y. Goudaroulis and P. Nicolacopoulos, Kluwer, Dordrecht, pp. 375–382.

Kuhn, T.S. 1970. *The Structure of Scientific Revolutions*, University of Chicago Press, Chicago, 210 pp.

Kulkarni, D. and Simon, H. 1988. The processes of scientific discovery: the strategy of experimentation, *Cognitive Science*, **12**, 139–175.

Lai, T. 1989. How we make discoveries, *Synthese*, **79**, 361–392.

Lambert, K. and Brittan, G. 1970. *An Introduction to the Philosophy of Science*, Prentice-Hall, Englewood Cliffs, NJ, 113 pp.

Lange, M. 1993. Natural laws and the problem of provisos, *Erkenntnis*, **38**, 233–248.

Langley, P., Simon, H.A., Bradshaw, G.L. and Zytkow, J.M. 1987. *Scientific Discovery: Computational Explorations of the Creative Process*, MIT Press, Cambridge, Mass., 357 pp.

Lawson, V. 1995. The politics of difference: examining the quantitative/qualitative dualism in post-structuralist feminist research, *Professional Geographer*, **47**, 449–457.

Leplin, J. 1988. Is essentialism unscientific? *Philosophy of Science*, **55**, 493–510.

Li, C. 1993. Natural kinds: direct reference, realism and the impossibility of necessary a posteriori truth, *Review of Metaphysics*, **47**, 261–276.

Lloyd, E. 1989. The semantic approach and its application to evolutionary theory, in *PSA 1988*, Vol. 2, edited by A. Fine and J. Leplin, The Philosophy of Science Association, East Lansing, Mich., pp. 278–285.

McDowell, L. 1992. Doing gender: feminism, feminists, and research methods in human geography, *Transactions of the Institute of British Geographers*, **17**, 399–416.

McLafferty, S.L. 1995. Counting for women, *Professional Geographer*, **47**, 436–442.

Maffie, J. 1990a. Naturalism and the normativity of epistemology, *Philosophical Studies*, **59**, 333–349.

Maffie, J. 1990b. Recent work on naturalized epistemology, *American Philosophical Quarterly*, **27**, 281–294.

Maffie, J. 1993. Realism, relativism, and naturalized meta-epistemology, *Metaphilosophy*, **24**, 1–13.

Mattingly, D.J. and Falconer-Al-Hindi, K. 1995. Should women count?: a context for the debate, *Professional Geographer*, **47**, 427–435.

Matyas, M.L. and Dix, L.S. (eds) 1992. *Science and Engineering Programs: On Target for Women?* National Academy Press, Washington, DC, 216 pp.

Mayer, L. 1990. *Introduction to Quantitative Geomorphology*, Prentice-Hall, Englewood Cliffs, NJ, 380 pp.

Mellor, D.H. 1977. Natural kinds, *British Journal for the Philosophy of Science*, **28**, 299–312.

Melnyck, A. 1995. Two cheers for reductionism: or, the dim prospects for non-reductive materialism, *Philosophy of Science*, **62**, 370–388.

Meyer, L.N. 1989. Science, reduction and natural kinds, *Philosophy*, **64**, 535–546.

Monk, J. 1994. Place matters: comparative international perspectives on feminist geography, *Professional Geographer*, **46**, 277–288.

Montgomery, K. 1991. Methodological and spatio-temporal contexts for geomorphological knowledge: analysis and implications, *The Canadian Geographer*, **35**, 345–352.

Moss, P. 1995. Embeddedness in practice, numbers in context: the politics of knowing and doing, *Professional Geographer*, **47**, 442–449.

National Research Council. 1994. *Women Scientists and Engineers Employed in Industry: Why So Few?* National Academy Press, Washington, DC, 130 pp.

Nelson, A. 1990. Are economic kinds natural? in *Scientific Theories*, edited by W. Savage, Minnesota Studies in the Philosophy of Science, Vol. 14, University of Minnesota Press, Minneapolis, pp. 102–135.

Nersessian, N.J. 1991. A method to 'meaning': a reply to Leplin, *Philosophy of Science*, **58**, 678–686.

Nersessian, N.J. 1992. How do scientists think? Capturing the dynamics of conceptual change in science, in *Cognitive Models of Science*, edited by R.N. Giere, Minnesota Studies in the Philosophy of Science, Vol. 15, University of Minnesota Press, Minneapolis, pp. 3–44.

Nickles, T. (ed.) 1980. *Scientific Discovery, Logic, and Rationality*, Reidel, Dordrecht, 385 pp.

Nickles, T. 1985. Beyond divorce: current status of the discovery debate, *Philosophy of Science*, **52**, 177–206.

Nickles, T. 1987. 'Twixt method and madness, in *The Process of Science*, edited by N.J. Nersessian, Martinus Nijhoff, Dordrecht, pp. 41–67.

Nickles, T. 1990. Discovery logics, *Philosophica*, **45**, 7–32.

Niiniluoto, I. 1993. The aim and structure of applied research, *Erkenntnis*, **38**, 1–21.

Pierce, C. and VanDeVeer, D. (eds) 1995. *People, Penguins, and Plastic Trees: Basic Issues in Environmental Ethics*, Wadsworth, Belmont, Calif., 485 pp.

Pinnick, C. 1994. Feminist epistemology: implications for philosophy of science, *Philosophy of Science*, **61**, 646–657.

Popper, K.R. 1965. *Conjectures and Refutations, The Growth of Scientific Knowledge*, 2nd edn, Basic Books, New York, 412 pp.

Putnam, H. 1975. The meaning of 'meaning', in *Philosophical Papers*, Vol. II, Cambridge University Press, Cambridge, pp. 215–271.

Quine, W.V.O. 1969. Natural kinds, in *Ontological Relativity and Other Essays*, Columbia University Press, New York, pp. 114–138.

Quine, W.V.O. 1992. Structure and nature, *The Journal of Philosophy*, **89**, 5–9.

Rayman, P. and Brett, B. 1993. *Pathways for Women in the Sciences*, Wellesley College Center for Research on Women, Wellesley, Mass., 177 pp.

Reichenbach, H. 1938. *Experience and Prediction*, University of Chicago Press, Chicago, 410 pp.

Rhoads, B.L. 1994. On being a 'real' geomorphologist, *Earth Surface Processes and Landforms*, **19**, 269–272.

Rhoads, B.L. and Herricks, E.E. in press. Naturalization of headwater agricultural streams in Illinois: challenges and possibilities, in *River Channel Restoration*, edited by A. Brookes and D. Shields, Wiley, Chichester.

Rhoads, B.L. and Thorn, C.E. 1993. Geomorphology as science: the role of theory, *Geomorphology*, **6**, 287–307.

Rhoads, B.L. and Thorn, C.E. 1994. Contemporary philosophical perspectives on physical geography with emphasis on geomorphology, *Geographical Review*, **84**, 90–101.

Richards, K. 1990. 'Real' geomorphology, *Earth Surface Processes and Landforms*, **15**, 195–197.

Richards, K. 1994. 'Real' geomorphology revisited, *Earth Surface Processes and Landforms*, **19**, 277–281.

Ritter, D.F. 1988. Landscape analysis and the search for geomorphic unity, *Geological Society of America Bulletin*, **100**, 160–171.

Ritter, D.F., Kochel, R.C. and Miller, J.R. 1995. *Process Geomorphology*, Wm. C. Brown, Dubuque, Iowa, 546 pp.

Rocheleau, D. 1995. Maps, numbers, text, and context: mixing methods in feminist political ecology, *Professional Geographer*, **47**, 458–466.

Rose, G. 1993. *Feminism and Geography: The Limits of Geographical Knowledge*, University of Minnesota Press, Minneapolis, 205 pp.

Rose, H. 1986. Beyond masculinist realities: a feminist epistemology for the sciences, in *Feminist Approaches to Science*, edited by R. Bleier, Pergamon, New York, pp. 57–76.

Rose, H. 1994. *Love, Power, and Knowledge: Towards a Feminist Transformation of the Sciences*, Indiana University Press, Bloomington, Ind., 326 pp.

Rothbart, D. 1993. Discovering natural kinds through inter-theoretic prototypes, *Method and Science*, **26**, 171–189.

Ruse, M. 1987. Biological species: natural kinds, individuals or what? *British Journal for the Philosophy of Science*, **38**, 225–242.

Russell, B. 1917. *Mysticism and Logic*, Doubleday, New York, 234 pp.

Salmon, W.C. 1984. *Scientific Explanation and the Causal Structure of the World*, Princeton University Press, Princeton, NJ, 305 pp.

Schaffner, K. 1980. Theory structure in the biomedical sciences, *Journal of Medicine and Philosophy*, **5**, 331–371.

Scheibinger, L. 1993. *Nature's Body: Gender in the Making of Modern Science*, Beacon Press, Boston, 289 pp.

Schumm, S.A. 1991. *To Interpret the Earth: Ten Ways to be Wrong*, Cambridge University Press, Cambridge, 133 pp.

Schwartz, S.P. 1980. Natural kinds and nominal kinds, *Mind*, **89**, 182–195.

Selby, M.J. 1985. *Earth's Changing Surface: An Introduction to Geomorphology*, Clarendon Press, Oxford, UK, 607 pp.

Shain, R. 1993. Mill, Quine, and natural kinds, *Metaphilosophy*, **24**, 275–292.

Shapere, D. 1974. Scientific theories and their domains, in *The Structure of Scientific Theories*, edited by F. Suppe, University of Illinois Press, Urbana, Ill., pp. 518–565.

Shapere, D. 1987. Method in the philosophy of science and epistemology, in *The Process of Science*, edited by N.J. Nersessian, Martinus Nijhoff, Dordrecht, pp. 1–39.

Shapere, D. 1991. Discussion: Leplin on essentialism, *Philosophy of Science*, **58**, 655–677.

Shepard, L.J. 1993. *Lifting the Veil: The Feminine Face of Science*, Shambhala, Boston, 329 pp.

Shrader-Frechette, K. 1994. Applied ecology and the logic of case studies, *Philosophy of Science*, **61**, 228–249.

Shrager, J. and Langley, P. (eds) 1990. *Computational Models of Scientific Discovery and Theory Formation*, Morgan Kaufmann, San Mateo, Calif., 497 pp.

Simon, H. 1973. Does scientific discovery have a logic? *Philosophy of Science*, **40**, 471–480.

Simon, H. 1977. *Models of Discovery*, Reidel, Dordrecht, 456 pp.

Sintonin, M. 1991. How evolutionary theory faces the reality, *Synthese*, **89**, 163–183.

Sismondo, S. 1995. The scientific domains of feminist standpoints, *Perspectives on Science*, **3**, 49–65.

Sloep, P.B. 1993. Methodology revitalized? *British Journal for the Philosophy of Science*, **44**, 231–249.

Sonnert, G. 1995. *Gender Differences in Science Careers: The Project Access Study*, Rutgers University Press, New Brunswick, NJ, 187 pp.

Spanier, B. 1995. *Im/partial Science: Gender Ideology in Molecular Biology*, Indiana University Press, Bloomington, Ind., 207 pp.

Stanford, P.K. 1995. For pluralism and against realism about species, *Philosophy of Science*, **62**, 70–91.

Starkel, L. 1982. The need for parallel studies on denudation chronology and present-day processes, *Earth Surface Processes and Landforms*, **7**, 301–302.

Stroll, A. 1991. Observation and the hidden, *Dialectica*, **45**, 165–179.

Summerfield, M.A. 1991. *Global Geomorphology*, Longman Scientific and Technical, Essex, UK, 537 pp.

Suppe, F. 1977a. The search for philosophic understanding of scientific theories, in *The Structure of Scientific Theories*, edited by F. Suppe, University of Illinois Press, Urbana, Ill., pp. 3–232.

Suppe, F. 1977b. Afterword – 1977, in *The Structure of Scientific Theories*, edited by F. Suppe, University of Illinois Press, Urbana, Ill., pp. 617–730.

Suppe, F. 1989. *The Semantic Conception of Theories and Scientific Realism*, University of Illinois Press, Urbana, Ill., 475 pp.

Thagard, P. 1988. *Computational Philosophy of Science*, MIT Press, Cambridge, Mass., 240 pp.

Thompson, P. 1983. The structure of evolutionary theory: a semantic approach, *Studies in the History and Philosophy of Science*, **14**, 215–219.

Thorn, C.E. 1988. *An Introduction to Theoretical Geomorphology*, Unwin Hyman, Boston, 247 pp.

Tooley, M. 1987. *Causation: A Realist Approach*, Clarendon Press, Oxford, 360 pp.

Tooley, M. 1990. Causation: reductionism versus realism, *Philosophy and Phenomenological Research*, **50**, 215–236.

Trigg, R. 1993. *Rationality and Science: Can Science Explain Everything?* Blackwell, Oxford, UK, 248 pp.

Tuana, N. (ed.) 1989. *Feminism and Science*, Indiana University Press, Bloomington, Ind., 249 pp.

Tuchanska, B. 1992. What is explained in science? *Philosophy of Science*, **59**, 102–119.

Van Brakel, J. 1986. The chemistry of substances and the philosophy of mass terms, *Synthese*, **69**, 291–324.

Van Brakel, J. 1992. Natural kinds and manifest forms of life, *Dialectica*, **46**, 243–261.

Van der Steen, W.J. and Kamminga, H. 1991. Laws and natural history in biology, *British Journal for the Philosophy of Science*, **42**, 445–467.

Van Fraassen, B.C. 1980. *The Scientific Image*, Clarendon Press, Oxford, 235 pp.

Van Fraassen, B.C. 1987. The semantic approach to scientific theories, in *The Process of Science*, edited by N.J. Nersessian, Martinus Nijhoff, Dordrecht, pp. 105–124.

Vetter, B.M. 1992. *What is Holding Up the Glass Ceiling?: Barriers to Women in the Science and Engineering Workforce*, Commission on Professionals in Science and Technology, Washington, DC, 26 pp.

Vicedo, M. 1995. Scientific styles: toward some common ground in the history, philosophy, and sociology of science, *Perspectives on Science*, **3**, 231–254.

Von Engelhardt, W. and Zimmermann, J. 1988. *Theory of Earth Science*, Cambridge University Press, 381 pp.

Watson, R.A. 1966. Explanation and prediction in geology, *Journal of Geology*, **77**, 488–494.

Wilkerson, T.E. 1988. Natural kinds, *Philosophy*, **63**, 29–42.

Wilkerson, T.E. 1993. Species, essences and the names of natural kinds, *The Philosophical Quarterly*, **43**, 1–19.

Woodward, J. 1992. Realism about laws, *Erkenntnis*, **36**, 181–218.

Yatsu, E. 1992. To make geomorphology more scientific, *Transactions, Japanese Geomorphological Union*, **13**, 87–124.

METHODOLOGICAL ISSUES

There are a number of sources of friction within the geomorphological community which the discipline can ill afford given the small number of practitioners. Perhaps foremost among these is methodology and/or technique. In an era of rapidly burgeoning techniques, there is a large and increasing burden on the research geomorphologist to master new methods of inquiry and analysis. Many of these methods are quite sophisticated, requiring a considerable investment of time and effort to develop the requisite level of expertise for conducting meaningful scientific research. The net result is that specialists proliferate, generalists are disparaged, and professional exchange within the discipline withers. The personal investment in mastering individual research techniques is now so substantial that loyalty to them is necessarily great and the tendency to view other techniques as flawed appealing. It is, perhaps, worth noting the parallel between this situation and the original thrust of the multiple working hypothesis concept with respect to ruling hypotheses (e.g. Chamberlin 1897).

Another important source of both confusion and conflict within geomorphology is the scale of interest. Scale, both spatial and temporal, pervades geomorphology and challenges geomorphologists like few other issues. Successful scale-linkage is a strong candidate for the Holy Grail of geomorphology. In a discipline where it is perfectly acceptable to investigate regional landscapes that are millions of years old or to monitor the response of sand grains on a beach to individual wind gusts, the diverse problems associated with profound scale changes tend to be a source of professional divergence because it is quite apparent that the discipline's body of theory does not reach seamlessly across the entire range of legitimate scales of interest (Rhoads and Thorn 1993). Quite obviously reductionist approaches that provide a fundamental explanation that may be multiplied or scaled up as needed have been attempted. However, such approaches generally have generated more questions than solutions. Geomorphologists are discovering that new conceptual issues emerge at every scale of interest. Faced with this situation, they have begun to recognize the full complexity of the scale problem.

The discipline would benefit greatly if its practitioners had a more comprehensive and profound appreciation of the concepts held, and obstacles faced, by all geomorphologists regardless of specific interests or methodologies. To steal an analogy from Stephen J. Gould's characterization of evolution, we need to view the discipline not as a tree (or alternatively a ladder) with inherent notions of inferiority and superiority, but rather as a shrub with many shoots. In short, we are all entitled to be an individual shoot, but we need to recognize that we all share the same root stock and that different is not inherently inferior or superior. Hopefully, as geomorphologists of different scale and methodological 'persuasions' enter into active collaborations with one another, all will become more appreciative of the advantages and disadvantages of different types of geomorphological research.

Michael Church examines the fundamental relationships between scale and theory building in geomorphology. From the small to the large, he sees (1) at very small scales, stochastic processes being described statistically, (2) at the scale of classical mechanics, description by deterministic theories, (3) at even larger scales, systems which are still described deterministically but exhibit 'contingent endogenous effects' that make non-linear dynamical models appropriate, (4) at the very largest scales, narrative, particularist descriptions of purely contingent landscape evolution.

Keith Richards takes up the scale issue by contrasting research methods which are extensive (large-N) with those that are intensive (small-N). Sampling theory addresses large-N studies comprehensively, but small-N studies lack a comparably well-developed underpinning. In geomorphology, the seeming weakness of 'uniqueness' in detailed (small-N) case studies may be offset by creating a 'web of research' in which at-a-site cross-referencing of many components within a well-established theoretical framework substitutes for the robustness supplied by replication in large-N studies.

Ron Dorn confronts the scale issue in terms of resolution. He takes as his case study the relationship between established geomorphic models of alluvial-fan development in Death Valley and recent changes in the way Quaternary climatic changes are viewed. While this case study focuses on the use of optical varnish microlaminae in relating fan building to such climatic indicators as 'Heinrich events', he raises the much broader issue of the quality of resolution that can be achieved in paleoreconstructions and the ability of geomorphologists to link mixed signals from different environments in a manner that permits acceptable testing of hypotheses.

Jeffrey Peakall, Phil Ashworth, and Jim Best survey the rigorous demands of attempting to construct physical scale models, focusing upon flume experiments ranging from 1:1 replicas, through Froude scale models, to distorted or analog models. Throughout their review they focus on the underlying principles that determine how well the physical model accurately replicates the corresponding real-world prototype.

REFERENCES

Chamberlin, T.C. 1897. The method of multiple working hypotheses, *Journal of Geology*, **5**, 837–848.
Rhoads, B.L. and Thorn, C.E. 1993. Geomorphology as science: the role of theory, *Geomorphology*, **6**, 287–307.

6 Space, Time and the Mountain – How Do We Order What We See?

Michael Church

Department of Geography, The University of British Columbia

ABSTRACT

The spatial and temporal scales of the principal perceived phenomena have severely constrained the development of geomorphology as a science. The scales of ordinary human perception at which nineteenth-century naturalists attempted to understand the landscape gave way only after about 1950 to the scales of classical mechanics. This chiefly meant constraining space scales to make them commensurable with observable time scales, so that the subject in recent decades has been dominated by considerations of 'dynamical' or 'process' geomorphology. There has been no confrontation of the basic question of scales at which we may expect to observe consistent patterns in data, how these patterns may be expressed, and what – consequently – constitute useful modes of explanation. In this chapter, I attempt to develop this theme. I recognise four distinctive modes of theory construction. At small space and time scales, phenomena are recorded in sequences which describe very large numbers of characteristic events. Descriptions are statistical, and processes are considered to be stochastic. At the scales of classical mechanics, deterministic theories are sustainable. At still larger scales, system evolution reveals contingent endogenous effects which cannot be predicted, even though the system remains deterministic. Nonlinear dynamical models, expressing chaotic behaviour, are appropriate. At the largest scales of space and time, landscape evolution is entirely contingent, and we adopt a narrative, particularistic model of explanation. Each level of theory construction must be consistent with the others if the subject is to present a viable construction of nature, but it is not obvious that phenomena described at each scale can be derived from theory at different scales. Scales are relative and are set by the resolution of measurements and by material virtual velocities. At high frequencies and high resolution, it is not clear that all information possesses coherent patterns of geomorphological

The Scientific Nature of Geomorphology: Proceedings of the 27th Binghamton Symposium in Geomorphology held 27–29 September 1996. Edited by Bruce L. Rhoads and Colin E. Thorn. © 1996 John Wiley & Sons Ltd.

interest. At low frequencies (those over which landscapes evolve), the information available nearly always is highly censored. Significant elements of geomorphological patterns may not be decipherable.

INTRODUCTION

> We ... naturally hope that the world is orderly. We like it that way.... All of us, including those ignorant of science, find this idea sustaining. It controls confusion, it makes the world seem more intelligible. But suppose the world should happen in fact to be *not* very intelligible? Or suppose merely that we do not know it to be so? Might it not then be our duty to admit these distressing facts?
>
> Mary Midgley, in *Science as Salvation: A Modern Myth and its Meaning*, quoted in *Science*, **269** (28 July 1995): 567.

Whatever the truth about the intrinsic orderliness of the world, it is obvious that the order scientifically imposed upon the world is a human invention, a means by which we sustain ourselves. Science is a product of the way we see the world. Our perceptions of the world around us are fundamentally constrained by the dimensions of space and time within which we inhabit the world. These constraints ultimately affect all of science, but they are perhaps most immediately evident within the geographical and historical sciences that directly describe the condition and history of the natural world. Striking evidence for these claims is available in the modern history of geomorphology. The evidence is arresting because of a significant mismatch between the spatial and temporal scales of ordinary human experience of the landscape, and those of geological processes which have created it. Despite the familiar nature of the principal features of the landscape, a virtually unprecedented range of scales must be invoked in order to explain them.

In this chapter, I propose to examine ways in which space and time scales have conditioned our construction of a scientific world-view, using the example of geomorphology to illustrate the discussion. I shall argue that scientific theories are essentially constrained by their associated scales of space and time, and that different kinds of theories are appropriate to describe phenomena at different scales.

An important *desideratum* in science is that theories at different scales be mutually consistent – at least if they are to be viewed as equally valid parts of the description of nature. But it is not obvious that the conceptual foundations for a theory at one scale must be entirely manifest in theories at other scales. Indeed, such often is not the case. An elementary reason for this is information loss through shifts of the scale limits of resolution, but there appear to be deeper reasons residing in the ways that theories are constructed.

I shall commence my argument by endeavouring to illustrate the issues from the history of geomorphology. Almost any other field science could serve as well. My examples will be drawn nearly entirely from the field of fluvial geomorphology; that is merely because it is the topic that I study. In what follows, the specifications of scale magnitude should be understood to represent the scale of resolution, that is, the lower end of the range of scales under consideration.

A SCANDALOUSLY BRIEF HISTORY OF GEOMORPHOLOGY

Chapter 1: Landscape History

Modern geomorphology was born just two centuries ago in the Huttonian revolution of geological thought. James Hutton was a Scottish doctor, landowner and natural philosopher who forcefully established the physical bases of modern earth science. The perceptually evident spatial scale at which Hutton and his successors attempted to understand the landscape was that of ordinary human perception – on the order of kilometres. (Which is not to say that they did not examine closely phenomena at smaller scales for what they could reveal about the landscape and its history.) Hutton's claim was that landscapes develop not by uniquely exceptional cataclysms, but by the same incremental processes of weathering, erosion and sedimentation that we observe today. Unfortunately, the time scale for significant changes in the landscape in this mode is many milleniums; well beyond the ordinary human sense of time. The consequence was an inferential, largely hypothetical (or speculative) discipline preoccupied with attempting to explain the historical particularities of specific landscapes. The underlying purpose was to extract a temporal pattern from such studies which could serve as a template for the interpretation of landscape. The theory, and the quintessential science, lay in the template.

This development is not surprising. To a greater or lesser extent it occurred in all of the nineteenth-century natural sciences, and it persisted until instruments were developed by which we could magnify or telescope our sense of scale. Arguably the most revolutionary development in all of nineteenth-century science is the theory of evolution, and it was subject to just the same temporal constraint as nineteenth-century theories of landscape. Both Huttonian landscape theory and Darwinian evolutionary theory challenged the credulity of scholars no less than lay people by stretching the bounds of time beyond all common reason. ·

Within geomorphology, the archetypal achievement was the 'geographical cycle' of W.M. Davis, published in 1899. For more than half a century it remained the dominant template for landscape interpretation. It portrayed landscape as a staged sequence of erosional transformations of an initially elevated landmass. (A brief description of the geographical cycle is given in Chorley et al. 1984, pp. 17–22; notes from a lecture course given by Davis, compiled and annotated by King and Schumm 1980, give a detailed outline and commentary.) Davis's model incorporates a number of classical hallmarks of scientific theories. It postulates artificially simple initial conditions; tectonic uplift followed by essential stability. (So far as I can tell, isostatic compensation of erosion was practically ignored, even though it presents no significant complications.) In its abstract form, the theory adopts relatively simple boundary conditions, even though the influence of geological structure is prominently acknowledged as one of the principal determinants of landform. This is exemplary (and perfectly reasonable) reductionism. The theory also is remarkably sophisticated in some respects. It recognises contingency in the form of climatic 'accidents', in the possibility for tectonic rejuvenation, and in its acknowledgement of multicyclic landscapes. In seeking evidence to support his theory, Davis 'solved' the time warp by conflating contemporary conditions in different landscapes, by making a loose sort of 'space-for-time' substitution (see Paine 1985, for a modern discussion of this procedure).

Within the attempt to reconcile spatial scales of the order of kilometres with time scales of the order of many milleniums, Davis's theory is a notable achievement. But it lacks a sense of the mechanics of landform development – of the specific physical processes by which erosion and sedimentation proceed. That is because the paradigm of classical mechanics, the ostensible basis for describing and analysing the displacement of earth materials, is also constructed upon the ordinary scales of human perception, and consequently does not readily admit this combination of space and time scales. Yet classical mechanics, which exemplifies the apparently deeply ingrained human wish to find a rationally ascribable cause (an action) to cover every observed result (reaction), is the foundation for our sense of order of the ordinarily observed world.

It is possible to quantify Davis's theory, but Davis did not proceed to that. He probably was too honest (or perhaps merely too blinkered by the conventions of his day) to admit the necessarily massive parameterisation. Only recently have such parameterisations begun to appear in numerical models of landscape development. I expect that the qualitative character of Davis's theory is the reason why it fell not merely into obsolescence, but into positive disrepute after the middle of this century.

The glacial theory – another nineteenth-century theory which originated in geomorphology – illustrates in a different way the problems of spatiotemporal constraints. After about 1800, the (then) entirely hypothetical concept of former glaciers of semi-continental extent became a contender to explain the widespread stony soils – termed 'drift' – of north-west Europe. Although Hutton himself expressed some inklings about glacial action, the theory triumphed only in the years after 1840, when Louis Agassiz published an influential paper proposing a former great ice sheet in northern Europe. (A useful brief account of the origins of the glacial theory is given by Flint (1971, pp. 11–20).) Agassiz had been shown the efficacy of glacial action in high alpine valleys by colleagues in the Society of Natural History at Lucerne. A mechanism was thus demonstrated within commensurable space and time scales which could be invoked – if one possessed sufficient uniformitarian faith to make dramatic extrapolations in both space and time – to explain the widespread drift. Presumably, farmers in the Alps had known the mechanism for centuries, but they lacked the knowledge and motivation to make such unreasonable extrapolations. The theory continued to meet resistance for many years, and never completely captured the allegiance of Charles Lyell, the great nineteenth-century publicist of Huttonian earth science. The difficulty to envisage concepts outside the familiar spatiotemporal range in which they are mechanistically grounded is probably the reason for that. The subsequent history of the glacial theory has continued to be plagued by this problem. Today we remain uncertain about the mechanics of the unstable ice sheets of the Pleistocene Northern Hemisphere (cf. Clark 1994), and therein lies the crux of some of the more tantalising current problems of late Pleistocene earth history (cf. Bond et al. 1992; Lehmann and Keigwin 1992; MacAyeal 1993).

Chapter 2: Functional Geomorphology

In geomorphology, the nineteenth century ended in about 1950. The turning-point apparently was exasperation with the evident impossibility to reconcile the time scale of observable and mechanically explicable processes with the spatial scale of the classically considered landscape (Strahler 1952). A substantial consensus emerged, instead, to constrain the spatial scale of enquiry to match the time scales of observable processes. It probably is no accident that the most influential student of the new paradigm (L.B. Leopold, a hydrologist in the United States Geological Survey) was trained as an engineer. Considering practical problems of land management and public safety, engineers had been studying such landscape processes as field erosion, hillslope failure and river erosion at these scales since the time of Leonardo. Refocusing on new and *commensurable* space and time scales permitted fresh theoretical grounding to enter geomorphology. 'Commensurable scales' are ones which permit landscape phenomena to be described and ordered in terms of some more fundamental knowledge or familiar experience. In the present case, the methods and paraphernalia of applied physics (or engineering science, if you prefer) could now be deployed to measure landform geometry, applied forces and material transfers on time scales from seconds out to decades. The commensurable space scales are on the order of metres.

There has, in succeeding decades, been a preoccupation with 'dynamical' or 'process' geomorphology (influential textbooks include those of Leopold et al. 1964; Carson and Kirkby 1972; Ritter 1978; revised by Ritter et al. 1995; Chorley et al. 1984). Process geomorphology can be defined as the study of 'the erosional and depositional processes that fashion the landform, their mechanics and their rates of operation' (Chorley et al. 1984, p. 3). Major effort is directed towards analysing the equilibrium between the strength of earth materials and the applied erosional stresses on supposedly stable land-forms. This focus of attention is dictated directly by Newtonian mechanical principles (cf. Ritter et al. 1995, pp. 7ff). In the field, observations are conducted at very local sites. There is even the possibility to move important observations into the laboratory. Within this paradigm, and the theories of landform development to which it has given rise, the evolving landscape of the nineteenth-century geomorphologists became a part of the fixed boundary conditions. Theories to explain large-scale landscape development have been quietly ignored.

A representative topic within process geomorphology is the theory of sediment transport in rivers. Achievement of a closed theory, it has been supposed, will lead to mechanical understanding of how the river shapes its channel in the short term and, since rivers are supposed (mainly incorrectly) to form the valleys in which they flow, of the erosional development of the landscape in the long term. Within the theory, it is supposed that the river moves loose granular sediment over the bed of the stream in proportion to the shearing force applied by the flow at the bed:

$$g_b = f[(\tau - \tau_0)/D] \qquad (1)$$

in which g_b is the bedload (traction load) transported per unit width of channel, $\tau = \rho g R S$ is the shear stress (tractive stress) imposed on the bed by a uniform flow of water, ρ is the density of water, g is the acceleration of gravity, τ_0 is the threshold stress for sediment movement, and D is the diameter of the transported sediment. The sediment transport relation is nonlinear, very strongly so near the threshold for motion (Figure 6.1). This approach to sediment transport was initiated by European river engineers towards the end of the chronological nineteenth century, and it has endured ever since. Its appearance in geomorphology was accompanied by a host of other engineering results on the hydraulics of deformable channels which have proven useful to develop a level of understanding about rivers. Similar developments have informed almost every other topic in geomor-

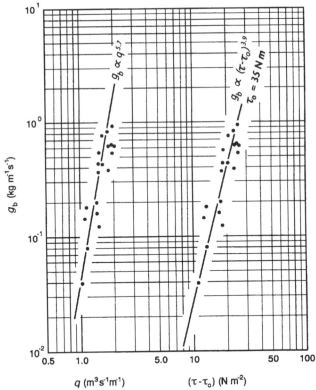

Figure 6.1   The relation between bedload sediment transport and hydraulic quantities in Elbow River at Bragg Creek, Alberta, a cobble-gravel stream. Data of A.B. Hollingshead (1971) obtained using basket traps and infilling of bed excavations, analysed in the theory-oriented paper of Parker et al. (1982b). On the right, the relation between sediment transport and effective shear stress at the bed. The shear stress is based upon section averaged flow depth. The transport rates are low and the relation is very sensitive to changes in shear stress. The sensitivity of the relation is based upon the rapidly increasing proportion of the bed that takes part in the sediment exchange process as the flow increases above the nominal threshold for motion. On the left, the relation between sediment transport and specific discharge. This relation is as good as the last one, although it is very sensitive. The mutual correlations amongst discharge, shear stress and sediment transport indicate that these relations should be regarded as scale relations of the flow

phology. Concomitantly the subject has adopted, in conformity with the rest of physical science, a thoroughly quantitative character. Perhaps the desire to conform drove much of this development. Certainly the desire for orderly and, within the scales, more precise explanation did.

Withal, an interesting feature of the tractive force approach to sediment transport is that it is not mechanically rigorous at all. Some heroic attempts have been made to place the equation on a completely rational foundation (cf. in particular, Yalin 1972; more recently, work by Parker and colleagues, beginning with Parker et al. 1982a, b), but it remains stubbornly empirical. Whilst the formula above has the appearance of a classical Newtonian force–response equation, it must be recognised that conditions at the stream bed are much too complex to admit more than an empirical correlation at the specified scale of examination. The channel-scale measurements that underlie the assessment of shear stress are substantially averaged in time and, often, in space. The simple description of the sediment boundary ignores important structural characteristics of the sediment surface. It is a highly parameterised result. This is true of a substantial range of the results that have been imported into or developed within the 'process' geomorphology of recent decades. Many of the results are, indeed, no more than empirical scale relations (Church and Mark 1980; see also Figure 6.1). This has led to the criticism that process geomorphology represents merely a kind of functionalist thinking that – whatever its merit in the engineering arena – is not good science at all because it does not approach the 'true' phenomena.

I think that this criticism of functionalism is not very helpful. It ignores the fundamental constraint posed by the space and time scales at which observations, and the consequent theories, are pitched. Important mechanical constraints upon the nature of the sediment transport process occur at scales that are below the resolution of most of the observations made in geomorphology until very recently, and well beyond the descriptive capacity of simple mechanical models of the kind that underlies the engineering approach to sediment transport in rivers. The Japanese geomorphologist Eiju Yatsu – who dealt with the difficult topics of weathering and erosion – more than 30 years ago recognised the general problem that is represented here (cf. Yatsu 1966), but his teaching has been generally ignored. The space and time scales of observation constrain the structure and physical content of functionalist theories through their control of the resolution of information in the theory. Our theoretical construction of order in nature is bound by the tyranny of the scales. (Whether a particular scale of enquiry is enlightening or practically helpful is quite another question.) This, of course, opens the possibility that a coherent theory pitched at one scale may be subject to fundamental criticism in light of criteria derived at different scales (Montgomery 1991). Such was the criticism of nineteenth-century landscape science raised by functional geomorphology, and such is the criticism of functionalist geomorphology raised by the realist school.

We are here broaching philosophical issues which run very deep in science. I suppose that the archetypal theory in all of science is Newton's theory of gravitation. Yet for more than two centuries it provoked nagging uncertainty in thinkers who were predisposed not to be satisfied with the appearance of mere functional order in the cosmos. How in heaven could celestial bodies separated by vast gulfs of space influence each other's motion in the formulated manner? Surely explicable order implied more than Newton's sleight-of-mind. It took Einstein, building upon the insights of James Clark Maxwell, to provide an answer. But none of this invalidates the practical utility nor destroys the intellectual satisfaction that Newton's achievement has provided during these last three centuries. Nor does it provide a basis to dismiss the description of fluvial sediment transport achieved within the bounds of tractive force theory.

Chapter 3: Competing Paradigms and Competing Scales

We must recognise that it is perfectly reasonable for more than one spatiotemporally delimited paradigm to be pursued within a science at any given time. A signal example occurs within geomorphology. At the same time that Davis was propounding and refining his geographical cycle, G.K. Gilbert was engaged in investigations, some of them not surpassed for nearly a century, which we would today recognise as quintessentially functional geomorphology. More than most investigators, Gilbert was conscious, as well, of the necessity for theories to be mutually consistent across different scales. That is the most startling feature of *The Geology of the Henry Mountains* (1877). Why, one asks, did it require another 75 years before a functionalist paradigm came to the fore in geomorphology? The answers that have been offered usually have referred to the character and situation of these major actors. But a thorough ransack of the literature reveals a steady production of functional studies from the mid-nineteenth century on. The spirit of the times and the general education of practitioners and public seem to have been much more important. Education, in particular, serves to define for most individuals what is a satisfactory standard for order in their world.

The last 20 years have witnessed the widest proliferation of programmes for study in the history of the subject. One must be careful about such claims. It is the propensity of every generation to declare that it is the grand historical exception. The basis for my claim is that geomorphology appears at present to be entertaining work on a wider range of space and time scales and – accordingly – within a wider range of paradigms than ever before. There appear to be several reasons for this. An important one in the present argument is that geomorphologists have lately become very good at adopting and importing into the field the advanced tools of physical science. One encounters everything from the more exotic atomic microscopes and mass spectrometers at the molecular end, to the satellites and sensors of space science at the global end. This has produced a dramatic expansion in the range of spatiotemporal scales that, through instrumental resources, are more or less directly available for study. Even geologically deep time is forced to reveal some of its secrets to modern absolute dating techniques based on isotope chemistry. Another reason is the increasingly diverse range of educational backgrounds of students drawn to the subject. These immigrants bring with them analytical tools with their own spatiotemporal scales and apply them to order geomorphological phenomena in original ways. Ultimately, the shear size of the discipline promotes diversity.

We may briefly illustrate the claim of diversity by referring again to sediment transport. In order to approach more closely the supposed real mechanics of sediment transport, geomorphologists have made studies of turbulent shear flows using high-frequency velocity probes (see Clifford et al. 1993, for a review). The underlying expectation is that sediment is actually entrained from the stream bed when high-velocity threads of the flow impinge upon it. They have also used advanced flow visualisation techniques in an attempt to establish this claim directly. At the other extreme, geomorphologists have used satellite images of the Amazon – the largest river in the world – to observe and attempt to understand aspects of sediment diffusion and sedimentation (Mertes 1994), not previously accessible to practical observation, that clarify the interaction between the river and its serially reconstructed floodplain and floodplain vegetation (Mertes et al. 1995).

The theoretical constructs into which these disparate observations lead have quite different foundations. Turbulent flows have classically been thought of as random phenomena, as processes with no evident coherent structure. More recently, the picture has been modified to recognise the occurrence of randomly recurrent structures within the turbulent flow. Whilst the phenomenological picture over boundaries of high roughness (i.e. river beds) remains decidedly murky, there appears to be little doubt that recurrent events of some description are centrally implicated in sediment entrainment. The time and space scales of turbulent motions and turbulent structures are seconds and millimetres.

The theoretical constructs are generic and *essentially* statistical. (I mean 'generic' in the sense that characteristic events are defined which are supposed to be repeated infinitely many times with only minor variation; I mean 'essentially statistical' in the sense that there is no means to specify unequivocally an individual event within the characteristic range.) Studies of regions as large as the Amazon floodplain reintroduce contingency. Contingency in this case is represented by the observed pattern of channel and floodplain configuration which is created by the erosion and sedimentation. The scales are decades and tens of kilometres, or greater, and these are consistent with the virtual velocity (i.e. the time-averaged rate of displacement) of sediment movement in the system, hence are a commensurable set. The features observed have individually distinctive histories and contexts which are recognisable at ordinary scales of perception (which does not prevent summary for certain purposes of sets of features or events by *convenient* statistics). These two sets of scales lead to theories quite different in their character, and different again from those associated with functional investigations at the scale of ordinary human perception.

At that scale, recent work has attempted to understand the pattern of channel shifts and sediment storage in river channels. Channel shifting is a consequence of sediment transport and storage. To the trained observer, there is a clear pattern in the bar and channel structure in a braided channel, and in a meandered channel the pattern is evident to almost anyone. But what is the pattern of modification of the channel? Murray and Paola (1994) have created a cellular model of the process which generates a developing braided pattern with statistics similar to those of real braided channels. The model is based on a nonlinear sediment transport rule similar to that given in Figure 6.1. It is entirely deterministic and there is statistical stability in the characteristics of the pattern, but the developing configuration of the channels is unpredictable. Accordingly, the behaviour is chaotic (cf. Turcotte 1992, for a simple description of chaotic processes).

Chaotic behaviour is radically different than classical mechanical theories admit. It incorporates an unpredictable sort of deterministic process. That is, the short-term and local trajectory of the system is clear enough, if sufficient information is collected about it (in the stream channel model referred to above, it is specified), but the overall structure of the system is sufficiently complex that predictive ability more or less rapidly disappears as the system develops. It is difficult to see whether the statistical aspect of such theories is essential or convenient. Systems subject to chaotic behaviour present, perhaps, one of the clearest examples of the scale-bound nature of theories. The pattern of behaviour that is evident in braided channel switching (or in loop development in a meandered channel) may not be derivable at all from characterisation of the microscale sediment transport events. Nor is it evident in the deterministic, mechanical description of channel hydraulics. In a sense, channel switching is 'emergent behaviour'. In fact, in the river channel example, the behaviour arises from the effect that the pattern of sediment storage has on the subsequent sediment transport. For the same reason, one expects – contrary to classical dogma – that the construction of the Amazon floodplain is not derivable from integration of sediment transport mechanics, even though viable theories at both scales must accommodate each other.

It is important to recognise that the theoretical constructs themselves are not tied essentially to scales. We can study eddies in classical mechanical terms; we can study ensembles of large landforms statistically. Theory selection follows from the way we isolate or aggregate events when we have freedom so to do within the scale metric we have chosen. But at the limit of resolution set by the metric, scale may indeed constrain the theoretical possibilities.

Whilst geomorphologists have embraced a very wide range of space and time scales to order – thence 'explain' – observations, there has not been (to my knowledge) any organised attempt to construct the subject about those scales in conscious recognition that the character and quality of explanation will thereby be systematically affected. It is my claim that this is, indeed, what happens. In the next section, I shall attempt to demonstrate this by deliberate recapitulation of a linked set of topics; the flow of water and sediment in stream channels, and the fluvial development of landscape. I shall introduce arbitrary adjustments of scale in order to emphasise how that controls feasible explanation.

A SPATIOTEMPORAL HIERARCHY OF EXPLANATION

Consider the flow of water in a river channel. At the scale of casual observation, we observe a mass of water flowing downstream with some assignable mean depth and mean velocity. A closer inspection, still at the scale of ordinary perception, reveals the existence of eddies, swirling masses of water moving across the main flow in more or less organised and briefly persistent cells. If we abandon ordinary perception and obtain a photograph at an instant in time we observe a highly complicated field of fluid motions. How can we describe it?

Physicists and others, including geomorphologists, have placed velocity sensors with resolution of order 10^{-3} m 10^{-2} s in the flow field. What is observed is a record of apparently random fluctuations about the mean. The recent development of high-resolution sensors for sediment concentration has revealed similar characteristics in sediment flux (Figure 6.2). A.N. Kolmogorov's analysis of such signals represents the classical statistical characterisation of turbulence. The character of the motion appears to remain similar over a substantial range of scales. Instruments have improved to the point that physicists have identified coherent structures in laboratory flows over boundaries of low roughness (cf. Robinson 1991). Evidence has been sought for similar structures in highly sheared flows over rough boundaries (that is, in rivers) because of the conviction that herein lies the mechanistic key to sediment entrainment. The observations remain decidedly equivocal (Clifford and French 1993), probably because the locally conditioned eddy structure over boundaries of high roughness replaces the spontaneously generated structures found in the laboratory. In any event, the description of such phenomena remains essentially statistical and simple patterns of cause and effect cannot be traced. The classical equations of fluid motion have customarily been applied to such flows only through averaging procedures and arbitrary linearisation.

But suppose we accelerate our own time scale by several orders of magnitude and correspondingly shrink the spatial scale of our ordinary perception. (Or, alternately, suppose we magnify the scales of the flow dramatically. Readers unable to suspend reality in this way may imagine, instead, that they are swimming in the Gulf Stream.) The random velocity fluctuations of our turbulent flow will now appear to be well-defined eddies. They will persist for a substantial period (in our accelerated frame), and – with instruments of correspondingly increased resolution – we will be able to measure internal characteristics

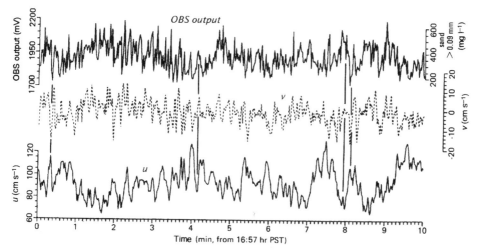

Figure 6.2   Ten-minute time sequence of downstream and vertical velocity components and optical backscatter (OBS) record of water turbidity at 1m above the sand bed of Fraser River, Near Mission, British Columbia, to illustrate the relation between suspended sediment flux and turbulent velocity scales in the flow. Peaks in OBS turbidity are characteristically associated with turbulent peaks in v and troughs in u. Peaks in v and in OBS turbidity are correlated with $r = 0.31$ (in a 2.2 hour record), a level that is typical of turbulent transfer of scalar properties. The displayed record also shows clear fluctuations lasting several minutes which can be interpreted as patchiness in the higher frequency regime. Data of 13 June 1986, recorded by M.F. Lapointe and presented in Lapointe (1992, Figure 8)

of the motion. These characteristics might include an Eulerian advection velocity, a rotation rate, radial velocity gradient, momentum, vorticity of the motion, and so on. A classical mechanical description of the flow becomes locally possible. Given some properties of the motion, we can predict others. If we know something about the neighbourhood of our eddy, we may also be able to calculate some features of its evolution, but we will not be able to extend our predictions for arbitrarily lengthy periods. This is, of course, just a thought experiment, except in large-scale geophysical flows. (The results in large-scale flows are distorted in comparison with those we would see in our river because intrinsic properties of the flow, such as viscosity of the fluid, will not be scaled, and certain exogenous parameters, such as the acceleration of gravity, will remain constant so long as we constrain our imaginations to the Earth's surface.)

Expand our scales by some orders of magnitude again. We are now embedded well inside our eddy, the dimensions of which are substantially larger than our 'ordinary' perception. We will perceive a large, slowly developing system within which we can track a specific history. To achieve that we will have to take into account the evolution of the system in the context of the even larger field of motion around it, with which there are continual momentum and energy exchanges. We will notice the contingencies that govern the evolution of our large eddy. We may calculate the fluid motion at places within the eddy in accordance with classical mechanics, and we may now be able to assume equilibrium conditions for substantial periods within our interval of observation. This, of course, is routinely done for Gulf Stream rings, and for large atmospheric disturbances.

Subject to some constraints about approximate similarity of the phenomena (which need not be very strict) our thought experiment has demonstrated how we may describe a geophysical phenomenon on quite different theoretical bases, depending upon the scales (read resolution) of the enquiry. It also hints at how reasonable assumptions and approximations allow us to embed one theoretical description within systems of phenomena drawn on significantly different scales. A key to understand the changing theoretical basis is to recognise that resolution governs what we can observe about structures in the system and what we can record about the evolution of those structures. When we have many, more or less rapidly evolving structures we know relatively little about each one – certainly too little to appreciate the individual nuances of its development and to assign specific antecedent causes. We adopt a statistical characterisation of what we observe. At the other extreme, we may observe only an interval in the evolution of a major structure. Our description becomes highly contingent. We observe a different sort of order at different scales.

It is useful to remind ourselves that, whilst at each step in the foregoing sequence of scale transitions new explanatory modes become feasible, previously sketched modes remain accessible. The essential constraint upon the character of information resides in the interaction between the resolution of the observations and the information requirements for the particular mode of explanation.

Let us return to our river and perform the scale transformation instead by focusing upon larger scale features of the flow. At scales of 10^0 m 10^0 s (1 metre, 1 second), we again observe eddies. But this time, particular eddy configurations persistently recur. Using appropriate velocity meters, hydrologists accordingly measure secondary currents in the river (Figure 6.3). Our spatial scale is now well within an order of magnitude of that of the channel itself, and we observe the shaping effect of the channel upon the flow. If the boundary is compliant (alluvial or, at least, erodible by the ambient currents), the configuration of these flows will eventually reshape the channel. This is what we observe in an evolving meander bend, or in the successive zones of flow convergence and divergence in a braided channel. This is the scale at which classical sediment transport theory, as exemplified in equation (1), is straightforwardly applied. We obtain measurements at individual points on the bed of velocity, shear stress, and sediment flux over the adjacent boundary. The *eddy scale* phenomenon varies sufficiently slowly to permit classical mechanical descriptions based on averaged quantities, even though the individual flow structures remain transient. We do not resolve the turbulent scales, but this does not entitle us to ignore them conceptually since sediment fluxes depend upon the turbulent-scale correlation of sediment concentration and flow velocity. At the large eddy to channel scale, we also observe the contingencies of the recent history of flows and of the morphological development governing eddy production and the evolution of the averaged secondary flows. We incorporate these effects into the mechanical description in the form of initial and boundary conditions. Our ability to predict their further evolution remains, however, limited. The use of different explanatory modes to order phenomena which interact across a broad range of scales – hence the necessity for mutual consistency between explanations at different scales – is particularly well illustrated here.

Let us expand the space and time scales again to order 10^1 m 10^4 s (about 3 hours). Much of the functionalist geomorphology of recent decades resides in these scales of normal human perception (at least, of rivers). We now consider the mean flow and

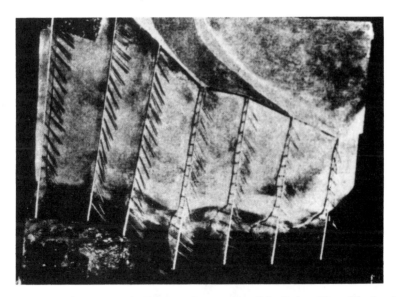

Figure 6.3   Illustration of secondary currents in a section of the Dniepr River, Ukraine, based on measurements of N. de Leliavsky. Flow is left to right (Figure 33 in Leliavsky 1959). De Leliavsky's measurements, taken around 1890, represented the first measurements of secondary currents. Leliavsky (1959, p. 98) notes of these data, 'the principle of non- parallelism of the flow lines in natural rivers, is not merely a matter of turbulent disorder.... It refers to temporal average velocities ... the general pattern of which was capable of being interpreted as and consistent with, an original scour theory.' Leliavsky is asserting that a distinctive theory emerges at this scale, and with the achieved resolution, which reflects the observed persistence of the flow configuration

sediment transport in a reach. This is the scale of usual application of hydrological and hydraulic measurements. At this scale, we are close to a mechanistic view of river channel evolution, a result of intense geomorphological interest. It is also the scale at which most attempts have been made to apply classical sediment transport theory to understand river channel evolution. Moreover, it is now possible to obtain information about the sediment transport process by examining sequential changes in channel morphology (Ashmore and Church in press: see Figure 6.4). This may be much more relevant, geomorphologically, than direct flux measurements. However, the further averaging that is inherent in the observations can introduce bias into the results if they are viewed only at this scale, since we no longer see the mechanistically conceived transport process. Nesting of measurements in adjacent scales is a means to minimise this problem which represents an important connection between scales of enquiry. The averaging arises from the space and time limits of resolution; the bias may arise when changes beyond the limit of resolution are not reflected in the average. An example may be compensating scour and fill in a river bed.

If we again expand our scale of attention, we begin to consider the river as an extended system, in which locally contingent events are happening at many places. Consider scales between the channel scale, above, and 10^4 m 10^9 s (the latter is 30 years). At these scales, styles of explanation and the structure of theories can be seen clearly to depend upon the way in which information is marshalled. A major achievement of L.B. Leopold and his

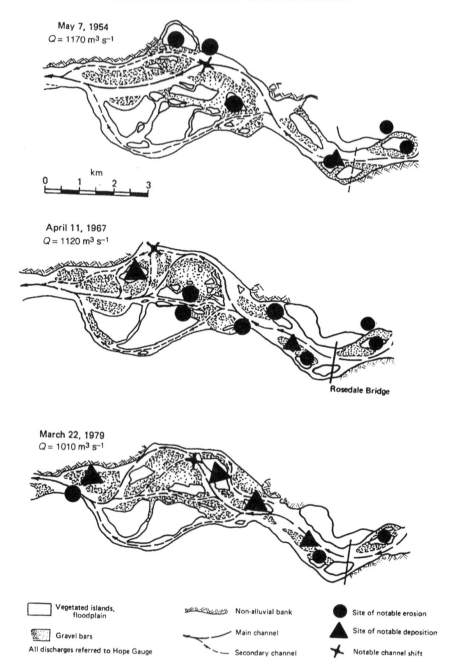

May 7, 1954
$Q = 1170 \text{ m}^3 \text{ s}^{-1}$

km
0   1   2   3

April 11, 1967
$Q = 1120 \text{ m}^3 \text{ s}^{-1}$

Rosedale Bridge

March 22, 1979
$Q = 1010 \text{ m}^3 \text{ s}^{-1}$

Vegetated islands,
floodplain

Gravel bars

All discharges referred to Hope Gauge

Non-alluvial bank

Main channel

Secondary channel

Site of notable erosion

Site of notable deposition

Notable channel shift

Figure 6.4

associates was to notice that if measurements of mean flow and geometry in river channels are averaged and compared over time and throughout the river system (that is, over the current scales), then a functional description of mean behaviour of the river system – known as the hydraulic geometry – becomes available (Leopold and Maddock 1953). These scales are identical with engineering regime scales, and so the coincidence of hydraulic geometry with engineering regime theory of unlined canals (cf. Blench 1957) is not surprising. If, on the other hand, we choose to examine the historical sequence of channel development within these scales, we observe contingent behaviour of two kinds. In the first kind, developments at one place in the system are constrained by endogenous developments elsewhere in the system. The result is the kind of deterministic but unpredictable development that is represented, for example, by braid switching or by meander loop extension and cut-off. In the second kind, exogenous constraints, such as the configuration of the even larger-scale landscape, impose conditions which remain constant (hence, trivially predictable). Climate, the forcing function for runoff and so for river hydrology, imposes an exogenous control that is particularly interesting because climate itself fluctuates significantly (and not yet predictably) on time scales similar to those of the river.

At the very largest scales (up to 10^6 m 10^{12} s), we are within the realm of development of those exogenous constrains. Macklin et al. (in press) have shown that at time scales between 30 and 9000 years a river is a complex system subject to both endogenous and exogenous controls. Short-term observations of channel regime may yield a quite mis-leading picture of long-term river behaviour because of instabilities associated with longer-term trends or a protracted period of relaxation after a major perturbation (see also Church 1981).

Within the last 30 years, some notable attempts have been made to model landscape geometry and development which provide insight into scale constraints. I shall consider two problems. Modern approaches to the description of the drainage network began with R.E. Horton (1945). The most prominent contribution was made by R.L. Shreve (1966, 1969), and some recent developments are presented by Stark (1991) on evolution and by Peckham (1995) on structure. There is essentially a single thread of development. The drainage network is modelled statistically on the basis of the theory of rooted tree graphs. The aim of modelling is not to describe the history of configuration of any particular drainage network, nor to predict the development of particular drainage networks, but to reproduce salient network characteristics of many or all such networks as means to test our

---

Figure 6.4 (*opposite*) Sequential maps of the channel configuration in Fraser River near Agassiz, British Columbia. Bed material transport has been estimated by measuring the volumetric changes in the channel over a number of years. The calculations yield a highly averaged view of river channel changes. Comparison of the results with bedload transport measurements at the Rosedale Bridge shows that the morphological changes estimate the bed material transfer very well. The observations are unbiased in this river because it is very large, so that episodes of erosion or deposition persist at individual sites along the river for a number of years and there is little compensating scour and fill (see Church and McLean 1994, for more details). Although consistency can be demonstrated between the sediment transport and channel evolution, it would not be possible to predict the long-term evolution of this particular channel from sediment transport theory, since details of erosion and deposition would remain inaccessible. Short-term predictions may be accessible via numerical models of flow and sediment transport. The problem is entirely analogous with the weather prediction problem

understanding of how they might have arisen in general. So it is a reduced model. An interesting issue in the present context is that drainage networks are not particularly microscale phenomena in the landscape. None the less, they represent a (practically) undenumerable phenomenon at regional landscape scale and they extend to rather small scales. Statistical representation seems inevitable. This constrains the nature of the predictions that are to be had. This example shows very clearly that there are no absolute scale limits associated with theoretical representation of the landscape. But an important constraint compounded of scale and information is effective: we can practically analyse far too little information about individual drainage basins to permit more detailed, mechanistic modelling.

Models of the evolution of the entire fluvial landscape encounter the same problem. Early models (cf. Ahnert 1976, 1987, and Kirkby 1986) tended to be explicitly mechanistic. By this, I mean that they incorporate statements about the supposed actual driving forces (soil creep; slopewash; mass failures, and so on) and material resistances. Realistically, such models are constrained to simulate landform changes at the synoptic scales of these processes. Beyond that, they might indicate idealised landscape developments. More recent models have recognised the need to parameterise most sediment transfer processes in some manner appropriately generalised to cover integral effects which occur over the long time spans during which landscape development actually occurs. Topographic gradient-driven diffusion processes on hillslopes and scale correlations (cf. Figure 6.1) for fluvial sediment transfers are the usual models. Because of the difficulty to match development scales for slopes and rivers, slope–base sediment storage commonly has been ignored: the models are, in effect, debris supply limited. No doubt problems such as this will be resolved nearly as rapidly as they are properly defined (see Howard et al. 1994, for a recent discussion which appeals to physical principles). The possibility to model the development of particular landscapes appears nevertheless to remain remote. Uncertainty about driving forces in the long range, and the impossibility to reconstruct endogenously emergent events defeat the issue. The purpose of modelling remains similar to that of modelling the drainage network. The problem that can be tackled is not unlike that faced by W.M. Davis. The useful purpose of model development is to simplify consideration of mechanics which occurs on more local scales than that of the overall system in order that features of landscape system development can be clearly defined and studied. But it is not clear, at the scale of landscape history, that useful simplification is to be had. Landscape development appears to be largely a matter of cumulated history. We have arrived back at the scale of nineteenth-century landscape science.

SOME CHARACTERISTICS OF SCALE-DELIMITED THEORY

In the foregoing sections, the relation between scale and resolution has been emphasised as a control upon the feasible representation of order in the landscape. In this section, this relation is explored more systematically and some additional characteristics of observations are introduced that further affect our choice of explanatory mode.

The foregoing paragraphs have essayed a systematic view of geomorphological scales on the basis of the fluvial system. I introduced a range of scales which I have called 'commensurable scales'. Beyond indicating that these are scales at which theory construction is feasible, I have not explicitly defined what commensurable scales might be. They are scales at which, within the resolution set by the dimensions and by our observing methods, information transfer can be detected within the landscape (or, more generally, 'within the system under study'). It is upon the basis of observed information transfer that theory can be constructed about the behaviour of the system, that causes and effects can be assigned, and that we can detect satisfactory order.

For geomorphological systems, information transfer is synonymous with the transfer of earth materials and, to a lesser degree, with the transfer of certain kinds of energy. Commensurable space and time scales are matched by consideration of the velocity for material transfer in the landscape. In the fluvial system at very local scales, this is the characteristic velocity of water (and entrained sediment), about 1 m s^{-1}. But even at the channel scale (the scale of 'ordinary' perception), the velocities of water and sediment have diverged significantly. Since sediments spend most of their time in storage – during which they constitute the visible morphology of landscape which we endeavour to explain – the 'virtual velocity' of the earth material is the critical scale-matching velocity. The virtual velocity is the average transfer rate for material on the time scale of resolution. (For landforms which can be defined by the linear dimension along which sediment transfer occurs – such as a river channel – virtual velocity is equivalent to linear dimension/residence time.) Virtual velocities rapidly decline as we move to larger spatial scales, whence commensurable time scales expand even more quickly. Within stream channel virtual velocities on the order 10^{-4}–10^{-7} m s^{-1} are typical for bed material (these values are the same as 1 m to 1000 m yr^{-1}).

It is probable that a good deal that happens on shorter time scales and more restricted space scales than those of sediment virtual velocity is not of direct interest geomorphologically. For example, Macklin et al. (in press) have shown that the space and time scales of long-term sediment transfer and storage hold the key to understanding river morphology. But most of the change actually happens within a relatively short time, so the relevant time scale for observing geomorphological *processes* is that of ordinary perception of events, 10^1 m 10^4 s. From the geomorphological viewpoint, shorter-term phenomena that occur within such human synoptic scales are effectively averaged. They may, of course, remain intensely interesting in the context of environmental physics, and they may hold the key to mechanistic understanding of the system (as in the appearance that turbulent scale phenomena control sediment entrainment). At these shorter scales statistical modes of explanation dominate. Geomorphologically, there are no coherent patterns of essential interest. Event patterns are random with respect to the geomorphological structure of the landscape and averaged summary quantities are, for strictly geomorphological purposes, appropriate.

Conversely, at low frequencies and over large areas, the customary resolution of measurements (which usually is set by the resources available to collect information within the domain of interest, and by ability to analyse it) creates a more or less highly censored view of geomorphologically significant events. This is especially true at the scale of landscape history, which requires a long retrospective view that is often recovered only from the surviving sedimentary records. Significant elements of geomorphological event sequences remain inaccessible, especially in the temporal scale. In such circumstances, conjectural reconstruction of history, using inference as the evidence permits, is unavoidable. Significant elements of geomorphological patterns may remain inaccessible. Our characterisation will remain particularistic until we have gained sufficient experience to identify a representative class of phenomena, when we seek a mechanistic explanation (that is, one which maximises our appreciation of order in the most parsimonious way). At the landscape scale, models – both conceptual models such as that of Davis and more recent numerical models – serve this purpose.

Between these extremes lies the domain of classically ordered, mechanically tractable phenomena. Provided our time frame is sufficiently short, we use linear or linearised arguments to obtain relatively robust explanations. But on the outer margin of this realm we find a class of phenomena in which nonlinearities come to dominate the mechanistic behaviour and predictive ability decays rather quickly (Figure 6.5). This is the class of so-called chaotic processes. Chaos occurs in a wide range of mechanical systems and is detected when the observing time exceeds some characteristic event time by a substantial margin. Event times in geomorphological systems may, then, define the scale range for chaotic processes. River systems are driven at synoptic or seasonal scales by runoff, so the domain for chaotic behaviour of the channel may reasonably be expected to fall in the range of years to decades.

A significant feature of geomorphological systems is the occurrence of substantial, yet clearly bounded, accumulations of material in the form of sediment deposits. These are generated or consumed over many events. Such stores are repositories of information about the history of the system. Stores introduce significant nonlinearities into system behaviour by modifying force–flux relations (such as equation (1)). They can achieve this by modifying the bounding geometry of the system. Processes are then controlled by the current geometry of the system. This is nothing new. It appears, however, that very detailed features of the geometry, such as sedimentary structure, are of major importance in determining the further development of the sedimentary system. Sedimentary systems are extremely sensitive to current configuration. This claim is new. The phenomenon is a source of deterministic, unpredictable behaviour. The time scale for such behaviour to

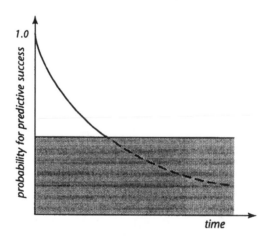

Figure 6.5 Illustration of the deterioration of predictive capability in systems subject to chaotic dynamics. The stippled zone represents probabilities not usefully different from zero

become manifest is consistent with the estimates given above and Figure 6.4 shows an example in which the effect occurs.

The passage of material through stores also creates persistence in the record of material flux (Klemes 1974; Kirkby 1987). Persistence provides an effective short-term memory for the system. The effect of persistence is to suppress the full range of variability in relatively short records of a process. In view of the discussion in the last paragraph, this seems to be a paradox. The limit of practical predictability in chaotic systems is, however, a way of viewing persistence. Persistence identifies a scale within which the controlling conditions remain sufficiently consistent to permit useful predictions to be made. Processes subject to storage effects are ultimately dominated by a sequence of increasingly rare, extreme events which are apt to be revealed only by taking a very long view of the process. It follows that the window of available observations may yield a quite misleading picture of the long term. What constitutes a long view, however, is determined by the commensurable scales of space and time. Intuitive recognition of this circumstance (or, at least, intelligent speculation upon the nature of the evidence left behind by dominant events) has probably influenced our tolerance for particularistic and narrative (hence contingent) 'theories' of large-scale processes, even though our sense of order is best served by reductionist, mechanistic theories of local processes, and may be provisionally satisfied by statistical (and still reductionist) theories of microscale processes.

Patchiness is a phenomenon in space which is homologous with persistence in time. Just as we must have a long view in time in order to detect persistence, in order to detect patches, we must enjoy a view that is far larger than the characteristic dimension of a patch. So we are most apt to detect patchiness in very local phenomena, and to detect it only when the resolution is very much finer than the domain of the study. We can find examples in the river. In an advecting system (such as flow down a river), persistence and patchiness are the same phenomenon. Low-frequency variance in turbulent signals (which is characteristic in rivers; cf. Figure 6.2) indicates the presence of patchiness. But a limit is imposed on the scale range of patchiness at one end by the size of the container (the channel) and at the other end by the onset of dissipative effects – in this case because of viscosity. Sediment accumulations in stream beds exhibit patchiness at the channel scale. The range of patches is delimited by the channel and by the elementary character of the sediments. Another way of viewing alternate theoretical frameworks is to consider that if we possess the means to survey the range of patch characteristics, then we are apt to construct a statistical theory as the only tractable way to digest the quantity of information and to appreciate its essential structure. At the other extreme, if we can scarcely see the structure of the patch system, we are apt to focus on local and particular elements and to construct mechanistic or particular explanations.

The asymmetrical structure of space and time influences how these concepts inform our knowledge. Whilst we have no evident limits in the time dimension – hence the chance always to detect larger-scale persistence – the space domain on the surface of the earth is limited. The domains of specific processes may be even more severely delimited, as the examples of channel-scale phenomena given above demonstrate. For large-scale features of the landscape, we may never arrive at representative domains, and our explanation of the landscape is apt, then, to remain particularistic. What is 'large scale' must be interpreted in terms of the resolution with which we view the world. At turbulent scales, large eddies are large-scale phenomena. One might suppose that absence of representative domains leaves no constraint upon interpretations in the time domain, but the temporal limits for humans to collect information become very effective domain delimiters. Since information is far less readily accessible through time (especially deep geological time) than it is over space, particularistic explanation continues to play a dominating role at the large scales in time as well. In short, history matters.

A final important feature of landscape which must be acknowledged is that different sedimentary systems operate with different virtual velocities and with distinctive phenomenological scales. Furthermore, the relation between phenomenological scales and human scales appears to exert a substantial influence over our construction of order in the system. The contrast between the hillslope system and the fluvial system provides an obvious example. Most hillslopes have quite limited space scales and very small characteristic virtual velocities, which is to say that time scales are very long. Hillslopes are also very sticky systems (friction is high), so that events are exceedingly episodic. On hillslopes, patchiness very quickly becomes evident both spatially and temporally. Interestingly, classical mechanistic theories almost completely dominate work on hillslope development. I guess that is because of the extreme difficulty to observe very local processes in any consistent way (not least because they are apt to be very boring for protracted periods), and – since they are largely erosional systems – because we usually cannot recover information to characterise the behaviour of hillslopes through very long periods, when changing climate and cumulative weathering of earth materials are apt to be dominant considerations.

SUMMARY

In this chapter, I have attempted to argue that our construction of rational order in the world around us is essentially constrained by the scales of space and time within which we examine the world. That is a remarkably large theme, of which the chapter presents only a sketch. I have tried to show that an important element of the constraint is the information that is accessible via the observing techniques at our disposal. Classically, two important domains of theory were recognised. The first summarises human experience in terms of a spatiotemporal narrative in which contingency plays a dominant role. Orderly explanation is couched in terms of recognised contingencies. Space and, particularly, time scales may be large. The second recognises the recurrence of characteristic, classifiable events which are subject to general, mechanistic explanations.

It has been common, since the advent of classical mechanical science, to deny that the first domain is even 'scientific'. So far as the earth sciences are concerned, at any rate, this claim scarcely seems tenable; in the end, we have only one earth to consider. But it appears, more generally, that the scales of enquiry determine the most appropriate mode of explanation, and its seems unreasonable to limit science only to that which is accessible to mechanistic theory. Within contingent explanation, the canons of science are observed in terms of the phenomena that may be admitted, and their expected behaviour at smaller, embedded scales (see Simpson 1963). I have endeavoured to show that the switch in modes of explanation might occur over a wide range of scales, depending upon the resolution of the observations.

This century has seen the accession of two new modes of theoretical organisation of phenomena. Around the turn of the century, statistical explanation entered the purview of science. Whether statistical abstractions of phenomena are a matter of convenience or a reflection of essentially stochastic processes remains a matter of controversy, at least in macroscopic science (see Smart 1979). There is no doubt, however, that the mode of explanation to which they give rise is distinctive in that the quality of the information that is summarised in theory is different than is found in classical or historical modes of organising knowledge. In statistical explanations, we at best know some information about a class of phenomena, within which we may be able to assign probabilities for the appearance of particular outcomes.

The final mode of theoretical organisation is very new, although its foundations also were laid around the turn of this century. It arises in the zone between mechanistic and contingent explanation; it may even emerge as a way to subsume contingency more acceptably into scientific method. It describes the phenomena that emerge in ostensibly well-behaved systems when sufficient time elapses for information to accumulate from remote parts of the system, or for significant information stores to experience system-modifying changes. In this circumstance, highly novel and unpredictable developments may occur which we characterise as 'chaotic behaviour'. In geomorphology, this appears to occur in landform systems over time scales for significant changes in material storage, hence in the configuration of the system.

The domain of relevance for each mode of explanation provides a fundamental connection amongst them. I find it useful to consider these domains in relation to the virtual

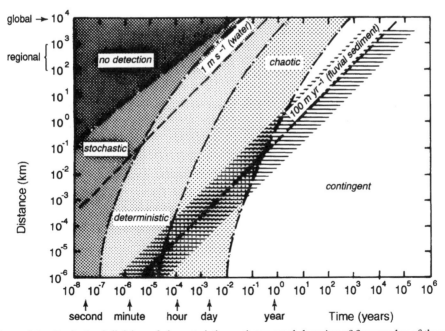

Figure 6.6 Conjectural division of characteristic spatiotemporal domains of four modes of theory construction, specified for fluvial system virtual velocities

velocity for material (or information) transfers in the system under consideration (Figure 6.6). This appears to set scales for theory, because it sets the information requirements to be able to make coherent statements about nature. However, it appears that feasible modes of explanation may change as instruments (hence, the resolution of measurements) and analytical capabilities change. Thus, numerical methods and large-scale automatic computation render a much wider range of phenomena open to classical mechanical description today than was possible only a few decades ago. Preferred modes of explanation appear, however, to be systematically related to customary human scales of perception of the world.

These modes of explanation appear to be sufficient for us to construct an orderly picture of nature. It is not clear to me whether or not there is, in addition, an absolute order in nature which establishes as necessary ones the distinctions I have drawn. I am not sure it matters. Either way, we arrive at the possibility to ground theories of landscape (and, I would claim, of all else) in some concept of order at various distinct scales. This is what humans seek.

ACKNOWLEDGEMENTS

I thank two apparently perplexed and disappointingly anonymous referees for their persistence with my text, and editors Rhoads and Thorn for their patience with my procrastination. I hope that the chapter will eventually repay their investments of good will and good advice.

REFERENCES

Ahnert, F. 1976. Brief description of a comprehensive three-dimensional process–response model of landform development, *Zeitschrift für Geomorphologie*, Supplementband, **25**, 29–49.

Ahnert, F. 1987. Process–response models of denudation at different spatial scales, *Catena Supplement*, **10**, 31–50.

Ashmore, P. and Church, M. In press. Sediment transport and river morphology: a paradigm for study, in *Gravel-bed Rivers in the Environment*, edited by P.E. Klingeman, R.L. Beschta, J. Bradley, and P.D. Komar, Proceedings of the 4th International Workshop on Gravel-bed Rivers, Wiley, Chichester.

Blench, T. 1957. *Regime Behaviour of Canals and Rivers*, Butterworths, London, 138 pp.

Bond, G. and 11 co-authors 1992. Evidence for massive discharges of icebergs into the North Atlantic Ocean during the last glacial period, *Nature*, **360**, 245–249.

Carson, M.A. and Kirkby, M.J. 1972. *Hillslope Form and Process*, Cambridge University Press, London, 475 pp.

Chorley, R.J., Schumm, S.A. and Sugden, D.E. 1984. *Geomorphology*, Methuen, London, 607 pp.

Church, M. 1981. Records of recent geomorphological events, in *Timescales in Geomorphology*, edited by R.A. Cullingford, D.A. Davidson and J. Lewin, Wiley, Chichester, pp. 13–29.

Church, M. and McLean, D.G. 1994. Sedimentation in lower Fraser River, British Columbia: implications for management, in *The Variability of Large Alluvial Rivers*, edited by S.A. Schumm and B.R. Winkley, American Society of Civil Engineers (ASCE Press), New York, pp. 221–241.

Church, M. and Mark, D.M. 1980. On size and scale in geomorphology. *Progress in Physical Geography*, **4**, 342–390.

Clark, P.U. 1994. Unstable behavior of the Laurentide ice sheet over deforming sediment and its implications for climate change, *Quaternary Research*, **41**, 19–25.

Clifford, N.J. and French, J.R. 1993. Monitoring and analysis of turbulence in geophysical boundaries: some analytical and conceptual issues, in *Turbulence*, edited by N.J. Clifford et al., Wiley, Chichester, pp. 93–120.

Clifford, N.J., French, J.R. and Hardisty, J. (eds) 1993. *Turbulence: Perspectives on Flow and Sediment Transport*, Wiley, Chichester, 360 pp.

Flint, R.F. 1971. *Glacial and Quaternary Geology*, Wiley, New York, 892 pp.

Gilbert, G.K. 1877. *Report on the Geology of the Henry Mountains*, United States Geographical and Geological Survey of the Rocky Mountain Region, United States Department of the Interior, pp. 18–98.

Hollingshead, A.B. 1971. Sediment transport measurements in gravel river, *American Society of Civil Engineers, Proceedings, Journal of the Hydraulics Division*, **97**, 1817–1834.

Horton, R.E. 1945. Erosional development of streams and their drainage basins: hydrophysical approach to quantitative morphology, *Geological Society of America Bulletin*, **56**, 275–370.

Howard, A.D., Dietrich, W.E. and Seidl, M.A. 1994. Modeling fluvial erosion on regional to continental scales, *Journal of Geophysical Research*, **99**, 13 971–13 986.

King, P.B. and Schumm, S.A. (compilers) 1980. *The Physical Geography (Geomorphology) of William Morris Davis*, Geobooks, Norwich, 174 pp.

Kirkby, M.J. 1986. A two-dimensional simulation model for slope and stream evolution, in *Hillslope Processes*, edited by A.D. Abrahams, Allen and Unwin, Boston, pp. 203–222.

Kirkby, M.J. 1987. The Hurst effect and its implications for extrapolating process rates, *Earth Surface Processes and Landforms*, **12**, 57–67.

Klemes, V. 1974. The Hurst phenomenon: a puzzle? *Water Resources Research*, **10**, 675–688.

Lapointe, M.F. 1992. Burst-like sediment suspension events in a sand bed river, *Earth Surface Processes and Landforms*, **17**, 253–270.

Lehman, S.J. and Keigwin, L.D. 1992. Sudden changes in North Atlantic circulation during the last deglaciation, *Nature*, **356**, 757–762.

Leliavsky, S. 1959. *An Introduction to Fluvial Hydraulics*, Constable, London, 257 pp. (reprinted by Dover Publications, New York, 1966).

Leopold, L.B. and Maddock, T. Jr 1953. The hydraulic geometry of stream channels and some physiographic implications, *United States Geological Survey Professional Paper 252*, 56 pp.

Leopold, L.B., Wolman, M.G. and Miller, J.P. 1964. *Fluvial Processes in Geomorphology*, W.H. Freeman, San Francisco, 522 pp.

MacAyeal, D.R. 1993. Growth/purge oscillations of the Laurentide ice sheet as a cause of the North Atlantic's Heinrich events, *Paleoceanography*, **8**, 775–784.

Macklin, M.G., Passmore, D.G. and Newson, M.D. In press. Controls of short and long term river instability: processes and patterns in gravel-bed rivers, the Tyne basin, northern England, in *Gravel-bed Rivers in the Environment*, edited by P.E. Klingeman, R.L. Beschta, J. Bradley, and P.D. Komar, Proceedings of the 4th International Workshop on Gravel-bed Rivers, Wiley, Chichester.

Mertes, L.A.K. 1994. Rates of flood-plain sedimentation on the central Amazon River, *Geology*, **22**, 171–174.

Mertes, L.A.K., Daniel, D.L., Melack, J.M., Nelson, B., Martinelli, L.A. and Forsberg, B.R. 1995. Spatial patterns of hydrology, geomorphology, and vegetation on the floodplain of the Amazon River in Brazil from a remote sensing perspective, *Geomorphology*, **13**, 215–232.

Montgomery, K. 1991. Methodological and spatio-temporal contexts for geomorphological knowledge: analysis and implications, *The Canadian Geographer*, **35**, 345–352.

Murray, A.B. and Paola, C. 1994. A cellular model of braided rivers, *Nature*, **371**, 54–57.

Paine, A.D.M. 1985. 'Ergodic' reasoning in geomorphology, *Progress in Physical Geography*, **9**, 1–15.

Parker, G., Dhamotharan, S. and Stefan, S. 1982a. Model experiments on mobile, paved gravel bed streams, *Water Resources Research*, **18**, 1395–1408.

Parker, G., Klingeman, P.C. and McLean, D.G. 1982b. Bedload and size distribution in paved gravel-bed stream, *American Society of Civil Engineers Proceedings, Journal of the Hydraulics Division*, **108**, 544–571.

Peckham, S.D. 1995. New results for self-similar trees with applications to river networks, *Water Resources Research*, **31**, 1023–1029.

Ritter, D.F. 1978. *Process Geomorphology*, William C. Brown, Dubuque, Iowa, 603 pp.

Ritter, D.F., Kochel, R.C. and Miller, J.R. 1995. *Process Geomorphology*, 3rd edn, William C. Brown, Dubuque, Iowa, 538 pp.

Robinson, S.K. 1991. Coherent motions in the turbulent boundary layer, *Annual Reviews of Fluid Mechanics*, **23**, 601–639.

Shreve, R.L. 1966. Statistical law of stream numbers, *Journal of Geology*, **74**, 17–37.

Shreve, R.L. 1969. Stream lengths and basin areas in topologically random channel networks, *Journal of Geology*, **77**, 397–414.

Simpson, G.G. 1963. Historical science, in *The Fabric of Geology*, edited by C.C. Albritton, Jr, Freeman, Cooper, Stanford, Calif., pp. 24–48.

Smart, J.S. 1979. Determinism and randomness in fluvial geomorphology, *Eos*, **60**, 651–655.

Stark, C.P. 1991. An invasion percolation model of drainage network evolution, *Nature*, **352**, 423–425.

Strahler, A.N. 1952. Dynamic basis of geomorphology, *Geological Society of America Bulletin*, **63**, 923–938.

Turcotte, D.L. 1992. *Fractals and Chaos in Geology and Geophysics*, Cambridge University Press, 221 pp.

Yalin, M.S. 1972. *Mechanics of Sediment Transport*, Pergamon, Oxford, 290 pp.

Yatsu, E. 1966. *Rock Control in Geomorphology*, Sozosha, Tokyo, 135 pp.

7 Samples and Cases: Generalisation and Explanation in Geomorphology

Keith Richards

Department of Geography, University of Cambridge

ABSTRACT

Research in geomorphology employs a range of strategies, but one useful distinction is between extensive research methods based on large-N samples, and intensive methods employing small-N case studies. The former may lead to generalisation by empirical statistical means, while the latter generalise by theoretical reasoning. This chapter explores the relationship between these approaches, and some implications of adopting the latter. The case of the historical development of views on the nature of river meandering is used to illustrate that a broad shift occurs from extensive to intensive research as understanding improves. However, this necessitates a more detailed assessment of the rules governing case-study research than has hitherto taken place; while sampling theory for large-N studies is well established, theoretical bases for the selection of field areas for case studies are much less evident. In a case study, it is essential to identify the boundary conditions provided by the field location, in order that generalisation can proceed of the mechanisms inferred from observation. This in turn places great emphasis on the development of new methods for describing those local conditions in time and space. These general considerations are illustrated with examples drawn from fluvial geomorphology, glacial hydrology and slope hydrology.

INTRODUCTION

Discussion of methodology in most disciplines, and geomorphology is no exception, is contentious because it often involves semantic devices of more or less dubious validity. For example, critics of 'positivism' often conveniently ignore the large number of its

The Scientific Nature of Geomorphology: Proceedings of the 27th Binghamton Symposium in Geomorphology held 27–29 September 1996. Edited by Bruce L. Rhoads and Colin E. Thorn. ©1996 John Wiley & Sons Ltd.

variants (up to 12 have been identified; Outhwaite 1987). Instead, they judiciously select the most convenient straw man for their particular purpose. A given position can most easily be defended (or attacked) by deploying arguments about one of its properties, when a different property has been the subject of attack (or defence). Phillips (1992) illustrates this by showing that 'naturalism' in the social sciences involves several distinct properties, and that refuting only one cannot undermine the whole notion. In geomorphology, as in other disciplines, methodological debate may resort to the tactic of reducing a complex issue to a dichotomous variable, with the most familiar being the evolution : equilibrium dichotomy. This particular case (of a dichotomy) has been entrenched metonymically, by symbolic representation through the attachment of the names of claimed or supposed historic authorities (in this case, Davis and Gilbert). As Sack (1992) has shown, detailed textual analysis often reveals that the symbolic representative provides no evidence of having been a true proponent of the methodological position assigned to, or chosen by, him or her.

These debating strategies often result in polarisation of views, and this in turn encourages belief in the reality of the various, often opposing, positions held. This is unhelpful, because it results in methodological statements being treated normatively, as a set of guiding principles or rules. In fact, most statements about methodology are no more than models themselves – that is, mental constructs that attempt to represent a degree of understanding of the process whereby knowledge is acquired, and judged as being adequate for some purpose. It is therefore important to acknowledge the range of circumstances and conditions within which a particular methodology is both developed and applied. For example, the experimental method is shown by Harré (1981) to be employed for a wide range of reasons and purposes (Table 7.1), and in a wide variety of forms. This reflects the fact, among others, that as scientific knowledge about a phenomenon increases, so the methods required to extend that knowledge further are likely to be adapted.

In reality, both the specific explanation of particular geomorphological events (cases), and general explanations of geomorphological phenomena, commonly demand a methodology in which a complex migration occurs between the poles that are represented in

Table 7.1　The uses of experiment as identified by Harré (1981)

A.　As formal aspects of method
　　1. To explore the characteristics of a naturally occurring process
　　2. To decide between rival hypotheses
　　3. To find the form of a law inductively
　　4. As models to simulate an otherwise unresearchable process
　　5. To exploit an accidental occurrence
　　6. To provide null or negative results
B.　In the development of the content of a theory
　　7. Through finding the hidden mechanism of a known effect
　　8. By providing existence proofs
　　9. Through the decomposition of an apparently simple phenomenon
　　10. Through demonstration of underlying unity within apparent variety
C.　In the development of technique
　　11. By developing accuracy and care in manipulation
　　12. Demonstrating the power and versatility of apparatus

typical dichotomies. These dichotomies can therefore be seen not to be truly dichotomous, but merely devices to simplify, summarise and misrepresent what are, in fact, continua. This chapter considers, and seeks to deconstruct, some related, apparent dichotomies, and assesses their continually changing roles in the methodology and practice of geomorphology in particular, and the environmental sciences in general (of which geomorphology is a case).

DICHOTOMIES AND EXPERIMENTS

The linked 'dichotomies' defined in Table 7.2 form the basis for discussing the thesis that, as research into a phenomenon continues, a continual bidirectional, spiralling migration occurs between end members, represented by the left-hand and right-hand columns of this table. This has implications for the conduct of field research in geomorphology, affecting the research methods adopted by both individuals and research communities. Different individuals researching aspects of a particular problem may be simultaneously at opposite poles of the dichotomy, but as research questions evolve, communities may shift position in directions conditioned by the emerging research needs and paradigms. The left-hand column in Table 7.2 summarises a broadly empirical and 'positivist' approach to research, typically characterised as concerned with observational and experimental evidence. For example, a relationship between drainage basin morphometric variables and mean annual flood might be considered a typical outcome of such a research method, being a statistical generalisation of the relationship between operationalised and measurable variables. The research method employed in the construction of this relationship is extensive – it requires the sampling of a large number of drainage basins for which the independent ('causal') and the dependent ('response') variables are measured and related (an example of experimental method A.3 in Table 7.1). The variables employed are representative of both 'form' and 'product'. A 'realist' approach, on the other hand, employs methods appropriate to a world-view or ontology in which a distinction is drawn between three levels of a phenomenon (Bhaskar 1989; Richards 1994). These are the underlying mechanisms and the intellectual structures that represent them, events caused by those mechanisms in particular circumstances, and observations of those events. It recognises that observation is contingent on both the occurrence of observable events and the presence of capable observers (with appropriate technology), and that events are also contingent, in this case on an appropriate conjunction of mechanisms and the necessary space–time context to allow them to operate and create events. This construction of the world and our interpretation of it implies that empirical observation alone cannot reveal causal behaviour, and that theoretical analysis underpins realist research (Sayer 1992). Since identification of an association between observed and measured variables is itself no basis for the explanation sought by a realist methodology, this commonly requires intensive research of individual cases. A typical example is an investigation of the mechanisms of hillslope hydrology which seeks to uncover them by detailed study at a single site. The throughflow pit at a single point on the slope yields data, but the hydrological processes are interpreted for the upslope hillside on the basis of theoretical consideration of the behaviour of water flowing through the particular soils observed on the slope.

Table 7.2 Some linked apparent dichotomies in the scientific methodologies employed by geomorphologists

Ontology	**'Positivism'**	**'Realism'**
Epistemology	Empirical	Theoretical
	Concrete	Abstract
	Extensive research	Intensive research
	Large-N	Small-N
	Samples	**Cases**
Subject	Form and product	Process and mechanism

Note: The intention is to suggest that, in practice, research moves back and forwards between these poles, and that the labels attached to the methodologies are often semantic devices rather than rigid definitions; hence, the inverted commas.

These methodological distinctions are of importance in environmental sciences like geomorphology because they influence the way in which investigations in such sciences are undertaken in practice. If the 'model' for scientific activity is a conventional view of positivist experimentation, when applied to field-based sciences this leads to an emphasis on sampling theory, extensive 'large-N' studies, statistical methods and empirical generalisation. Laboratory experimentation involves physical isolation of the system being studied; this experimental closure allows manipulation of causal relationships so that the regular behaviour observed (constant conjunction) permits law-like statements to be made. However, this is subject to the charge that such law-like statements, derived from this form of experimental activity, only reflect the behaviour 'created' by the act of experimental closure, and are laws made in the laboratory rather than 'laws of nature' (Bhaskar 1989). Multiple causes operate together in 'open' systems in the natural environment, and their effects may be self-cancelling in certain contexts, so that no observable events occur, or so that events relate to causal processes inconsistently (as in the case of the variable storm-related slope and channel responses described in the upper Severn catchment by Newson 1980). Thus there can be no simple link from a field observation to identification of a causal mechanism; extensive, large-N studies are therefore necessary in order that 'closure' can be statistically created, by methods such as partial correlation.

While extensive research reveals patterns through statistical manipulation, intensive research may involve a detailed study of a single, or a small number, of case(s); the objective is then to provide an explanation of the mechanisms generating the observed patterns in an extensive investigation (Yatsu 1992). The small-N case study (Ragin and Becker 1992) provides 'detailed examination of an event (or series of related events) which the analyst believes exhibits (or exhibit) the operation of some identified general theoretical principle' (Mitchell 1983, p. 192). Generalisation from a case study to other cases is not through empirical extrapolation using statistical inference; rather, 'the validity of extrapolation depends not on the typicality or representativeness of the case but upon the cogency of the theoretical reasoning' (Mitchell 1983, p. 207). However, this theoretical reasoning identifies intellectual structures that seek to represent natural mechanisms, and the degree of generality of these structures may vary. For example, the laws of conservation of mass and momentum represent highly general natural constraints within which fluid dynamic mechanisms operate, but when embodied in intellectual structures such as versions of the Navier–Stokes equations, certain assumptions are made that limit their applicability (constant density, steady flow, hydrostatic pressure distribution, depth averaging).

Experimental investigations of salt weathering provide a useful illustration of the differences between the approaches in the columns of Table 7.2. Climatic cabinets can be used to explore the different rates of rock breakdown under controlled conditions (rock sample size and shape, and temperature and humidity cycles). Appropriately controlled experiments (Goudie 1974) allow differential rates of rock sample breakdown to be measured, and permit ranking of the susceptibility of different rocks to salt weathering (by a given salt), and ranking of the efficacy of different salts (acting on a given rock). However, simple observation of differential rates of breakdown identifies neither the precise rock properties responsible for susceptibility to weathering, nor the actual mechanisms of destruction. These mechanisms can be identified by theoretical consideration of the stresses imposed by crystal growth, and stress–strain behaviour leading to crack-tip propagation (Whalley et al. 1982). They are then necessarily observed and measured by different techniques (and experiments).

The dichotomies in Table 7.2 are, however, somewhat artificial, and the implied characterisation of positivism and realism should not be treated as defining different sets of rules for scientific activity. In the first place, it is not that positivism is a scientific method that eschews theory, but that positivist or empiricist models of scientific activity have erroneously implied that observation and measurement can take place without theory. The researcher who relates morphometric variables to mean annual flood in fact selects those variables on the basis of theory, implying that there is no such thing as pure empiricism. The dichotomies do, however, become useful initial guides in relation to certain practical aspects of research, particularly when distinguishing between 'samples' and 'cases' as both the objectives and the outcomes of field research. In summary, an initial dichotomy can be identified between large-N, extensive field studies whose outcome is often empirical generalisation about forms or products; and small-N, intensive field-based case studies whose outcome is often a theoretical understanding of process–form relationships. However, these two styles of research, and the generalisations and explanations that they generate, are inextricably linked, and in any area of geomorphological enquiry there is a continual spiralling between them. This reflects the movement of the study from outside the case(s) in more extensive investigations, to inside a case in an intensive investigation (when new questions may be posed that require the embedding of additional extensive enquiries within the intensive case study).

In the case of extensive large-N studies, there are many familiar rules to guide the selection of the cases about which generalisation may subsequently be made – this is traditionally a matter of sampling theory (Son 1973). The theoretical basis for this allows errors to be accounted for and both estimated as uncertainties in prediction confidence, and protected against by the collection of a suitably large sample size. However, there is much less theoretical clarity about the process in which concepts, which are often 'chaotic conceptions' (Sayer 1992, p. 202), are converted into measurable variables. In intensive, small-N case studies the rules for case selection are even less clearly defined and well known, although the site-specific boundary conditions are critical for understanding of the case (because they determine whether the mechanisms being investigated will produce particular kinds of events). The selection of a field area for a case study therefore demands appraisals (i) of what it is considered to be a case of; and (ii) of those characteristics that may allow observable events, that are interpretable in terms of the mechanisms or processes about which understanding is sought. Otherwise, the choice of a particular case may predetermine the mechanisms the researcher can investigate.

The relationships and trends between extensive and intensive, large-N and small-N studies, and the problems of experimental design in the latter, are illustrated in this chapter in three main ways. Firstly, the general historical development of approaches to the explanation of a particular geomorphological phenomenon, based on a commonly occurring shift from large-N to small-N studies, is highlighted by the example of river meandering. Secondly, some detailed characteristics of intensive research are illustrated through a review of a research design for the investigation of the seasonal evolution of the character of subglacial drainage (Richards et al. 1996), an example of a geomorphological research project in which the role of direct observation is severely circumscribed and innovative methodology is essential. What emerges from analysis of this case study is that intensive research frequently demands study of interaction and coincidence, and therefore is crucially dependent on simultaneous study of several related phenomena and processes. Finally, the importance of establishing the boundary conditions for the investigation of earth surface processes is considered, in terms of both the need to evaluate carefully the choice of field location, and the need to develop innovative methods of observation and measurement.

FROM LARGE-N TO SMALL-N: INCREASED
KNOWLEDGE AND CHANGING METHOD

One context for a move from the left- to the right-hand column in Table 7.2 is historical; extensive research is commonly necessary in the early stages of an investigation, but may give way to intensive research later, as observation and empirical generalisation about form and product necessitate theoretical consideration about mechanism. This is demonstrated clearly by the changing modes of investigation of channel pattern, especially river meandering.

Rivers are diverse and spatially variable, but our appreciation of this seems to have diminished over the years. Indeed, the tendency has been to emphasise the similarity of river meanders over a broad scale range, particularly on the basis of bivariate plots of morphometric variables displaying statistically linear relationships across several orders of magnitude (Figure 7.1(a)). One of the consequences both of these simple quantitative relationships derived by extensive, empirical research employing large-N sampling designs in fluvial geomorphology, and of traditional river engineering methods such as channelisation for flood control, has been to emphasise, and even create, similarity among rivers. The empirical generalisations which reinforce this are those such as the channel width–meander wavelength relationship (Leopold and Wolman 1960), which seems to imply that all rivers have bends of similar shape. These relationships are supported by early notions which account for meander development from initially straight river reaches (e.g. Dury 1969). In reality, rivers rarely evolve thus except when straightened artificially; rather there is continual adjustment from variable and arbitrary states by varying combinations of erosional and depositional processes as discharge regime and sediment supply

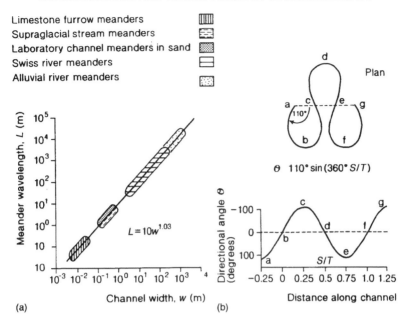

Figure 7.1 Examples of morphometric models of river meanders. (a) The classic regression of meander wavelength on channel width (after Leopold and Wolman 1960); (b) the sine-generated curve model of meander shape (after Langbein and Leopold 1966)

alter. In contrast to these geomorphological and engineering traditions, any focus on aesthetics and ecology in river management is liable to emphasise the conservation value of uniqueness. This is illustrated by the failures – and expensive restorations – of river engineering schemes, such as the straightening and subsequent renaturalisation of the Kissimmee River in Florida (Boon 1992). There appears to be a contradiction in these different emphases on similarity and difference, but one which is being rapidly eroded by new approaches to the study of river morphology which are more sensitive to the local contexts within which generally occurring processes operate, and which recognise that similarity lies in the fundamentals of process (hydrodynamic and sediment transport) and not in the morphologies of meander bends.

Bivariate empirical relationships among wavelength, width and radius of curvature are static descriptions of morphology involving simple, discrete, quantitative parameters. These simple parametric descriptions of meander bends developed into attempts to represent meander morphology more continuously, for example by employing a single-parameter sine-generated curve model of the bend shape which implies that bends are symmetrical about their axes (Figure 7.1(b): Langbein and Leopold 1966). Processes (for example, of energy dissipation) were then inferred from the bend geometry, which is, however, still represented in a static manner. Since maximum bank erosion tends to be displaced downstream from the apex of a bend, symmetrical forms are improbable, and a variety of evidence suggests that bends are more generally asymmetric in plan shape. This focus on bank erosion has in turn demanded greater emphasis on bends as dynamic, migrating forms, and a wide range of modes of bend migration has been identified.

Understanding of bend development requires an examination of processes such as secondary circulation, bank erosion, sediment transport from the base of eroding banks and across point bars, and deposition of point bar sediments. Study of these phenomena emphasises meanders as dynamic features having a diversity of behaviour. For example, many become asymmetric in shape as bank erosion occurs downstream from the apex, although there are many other styles of migration. Such study also links meander migration with the structuring of floodplain sedimentology, and implies that the classic model of inward-directed flow at the bed because of secondary flow, leading to fining upwards of lateral accretion deposits, is only one of several such relationships.

Bends therefore display a wide range of migration styles which result in shapes that are far from uniform and symmetrical (Figure 7.2). There are delayed inflection bends (Figure 7.2(a)) in high-power rivers with rapid bank erosion and strong secondary circulation, and with sandy bedload. In these, erosion occurs on the outer bank, the flow hugs the bank, evacuates products of bank erosion rapidly, scours along the base of the bank, and a delayed switch of the flow across to the opposite bank occurs. The bend is therefore strongly asymmetric, and may become a gooseneck bend which bends back up-valley (Lapointe and Carson 1986). However, there are also premature inflection bends (Figure 7.2(b)) in steep rivers with very high stream power and coarse gravelly bedload. These rivers develop over-widened bends, in which the point bar is deposited as the flow spreads across it without displaying any inward-directed cross-stream flow, and a deeply scoured pool is formed against a resistant bank out of which the flow 'squirts' across against the opposite bank again (Carson 1986). There is also the bend characterised by concave-bank bench deposition (Figure 7.2(c)), in which confined meanders on narrow floodplains in rivers carrying heavy suspended sediment load but little bedload develop a sharp bend in which erosion occurs on the inner, convex bank, causing over-widening at the apex and a dead zone against the concave bank. Silt deposition occurs on the outer, concave bank, and the inner convex experiences erosion, and the floodplain accretes by silt deposition (Page and Nanson 1982). This richness in the behaviour of meander bends is also reflected in an increasing awareness that bends in different environments – especially different sedimentological environments – may have shapes that by traditional standards are extremely irregular, but which in terms of the flow and sediment transport processes that both create the bend shape and are modified by it, are perfectly explicable.

This development of a more diversified view of the nature of meander bends reflects a change in the method of investigation, to small-N case studies involving meticulous observation of sediment transport rates and paths, three-dimensional flow dynamics, and their interdependence with channel form (a typical example is the work of Dietrich and Smith 1983). Additionally, it is evident that small-N studies involve comparative assessment of the effects of mechanisms at different times and places. Carson's (1986) study of premature inflection bends on the Canterbury Plains in New Zealand is typical of the combination of detailed observation of site characteristics and in-depth understanding of general processes that case studies require. Indeed, Carson criticises reliance on mathematical approaches, implying that they have helped to preserve the myth of uniform circular motion. However, increasing use of computational fluid dynamics numerical modelling programs based on two-and three-dimensional solutions of the shallow-water (Saint Venant) equations now allows numerical simulation of flow in channels with arbitrary plan geometries and bed topographies, ranging from what have been called 'non-

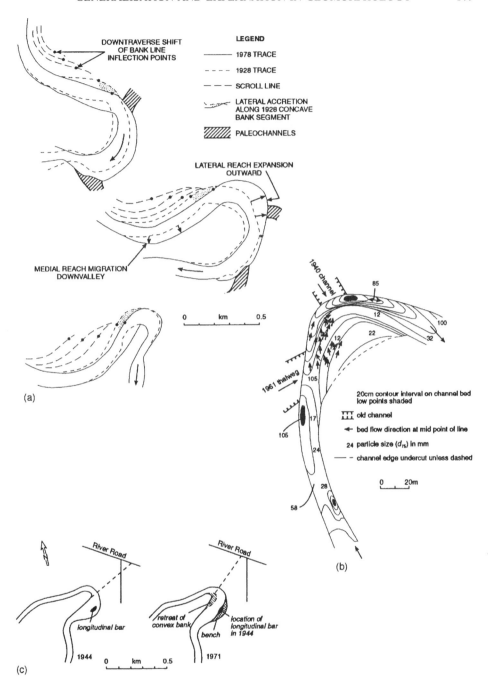

Figure 7.2   Different kinds of meander bend in different environments. (a) Delayed-inflection bends on the Rouge River, Quebec (after Lapointe and Carson 1986); (b) a premature-inflection bend on the Waireka Stream, Canterbury Plains, New Zealand (after Carson 1986); (c) a bend on the Murrumbidgee River, New South Wales, Australia, characterised by outer-bank deposition (after Page and Nanson 1982)

classical meander bends' (Hodskinson 1995) to even more complex braided channels (Lane et al. 1994, 1995; see below). These approaches demand intensive, and generally very detailed field measurement of the channel bed topography, depending on the channel size and sedimentology, to form the boundary condition for the application of the numerical models. This necessitates the shift from large-N studies in which the many sampled cases are represented by simple parameters, to small-N studies in which the single (or few) cases are represented by complex sets of measurements, representing three- and even four-dimensional boundary conditions. Thus, a wide variety of bend shapes, migration styles and floodplain sedimentologies exists depending on local conditions of stream power, floodplain width, floodplain sediment and sediment load. This diversity has critical implications for river management, demanding greater attention to the local conditions of an individual site. What proves to be a suitable management strategy in one place will not, necessarily, be equally successful elsewhere. Channel changes, both natural and designed, must be interpreted in relation to the spatial interaction between flow and sediment transport processes and the bend morphology.

A CASE OF A CASE STUDY: EXPLANATION CONFIRMED BY COINCIDENCE

In a large-N study, measurements are external to the individual sampled 'cases'; cases are sampled according to some variant of a random sampling strategy, and a simple measurement is performed on each one. In a small-N study, the case is selected, and is then the subject of a large number of observations and measurements internal to it. However, this case may also become a variate in a large-N study. The distinction between extensive and intensive research in Table 7.2 is fuzzy, since an intensive case study can also involve extensive monitoring. The distinction therefore rests less on the quantity of observational evidence than on the procedures involved in generalising the results of the study; as noted above, it is not the empirical evidence of form or product that is extrapolated, but the theoretical understanding of processes that allows explanation of the behaviour of the systems of which the case study is representative.

Traditionally, great stress is placed on random sampling in order to justify statistical inference, or on replication in order to validate conclusions based on sampled cases. Both have the effect of reducing the level of detailed observation with which the individual cases can be treated. In an in-depth single case study, the loss of inferential power arising from the absence of multiple cases may be compensated by considering the inter-dependency of different but related phenomena observable within the case. This basis of an explanatory investigation in interdependency can be seen particularly in research fields where conventional observation is difficult, and field glaciology is the classic example considered below. Much scientific activity aims to facilitate observation, and the visual sense appears to have such primacy that 'seeing is believing'. Accordingly, science is often employed to develop technologies to enable visual observation. As Hacking (1983) has argued, the existence of a coincidence among the images obtained by, for example, optical and electron microscopes and the forms predicted by the scientific theory of the thing observed is enough to lead to belief in the reality of that thing. An alternative form of coincidence that strengthens belief in a theory arises when multiple properties observed in a case study coincidentally lead to similar conclusions about the behaviour of the system of which the case is a representative. Thus, the replication of similar measurement of supposedly similar samples (cases) that characterises large-N investigation is replaced in small-N case-study research by a research design which employs a multiplicity of different observations of related phenomena. Replication *per se* is replaced by forms of comparative assessment, of the conclusions developed from one case study with those derived in other case studies where the boundary conditions are different (at different times or in different places).

Field observation of glacial processes is fraught with difficulty, and interpretation of subglacial processes, for example, has always employed indirect indicators, such as the quality of outflow meltwater (Collins 1979). However, recent research has begun to employ the methodology of intensive case studies in which several phenomena are monitored together (for example, Hooke and Pohjola 1994; Richards et al. 1996; Lawler et al. 1996), with their interdependence being used to assist interpretation, and with numerical modelling providing both a predictive and testing role. Thus instead of monitoring electrical conductivity alone, as a surrogate for total dissolved solids, several water quality properties (cation, anion and suspended sediment concentrations, pH, pCO_2 and stable isotopes) are all monitored, together with dye-tracing experiments and water balance studies. Interpretation of aspects of the temporal covariation of these variables permits explanation of the seasonal evolution of subglacial drainage from a distributed form (e.g. linked cavities) to a channelised system, in terms of the up-glacier retreat of the transient snowline. When the surface snow cover has melted and glacier ice is exposed, the reduced albedo (coupled with higher energy inputs later in the summer) increases the amplitude of diurnal melt cycles, which destabilises the subglacial drainage. Suspended sediment and cation concentrations reveal when water chemistry is non-conservative because of rapid chemical interaction of meltwater with freshly abraded and finely divided sediment. The open-or closed-system nature of the subglacial weathering environment in the subglacial drainage system can therefore be interpreted. Dye-tracing experiments provide another check, distinguishing slow routing in distributed drainage from fast routing in subglacial channels, and water pressures monitored in moulins (or in boreholes drilled by hot-water drills) distinguish pressurised from free-surface flow in conduits.

Such a multivariate research design demands highly labour- and capital-intensive monitoring of the several indicators that allow testing of hypotheses suggested by one set of data against another, but it results in rigorous, detailed understanding of processes at a glacier bed. An example of such research is the study of glacial hydrology undertaken at the Haut Glacier d'Arolla, in Valais, Switzerland, outlined by Richards et al. (1996). This began by establishing the surface and bed topographies of the glacier by conventional survey and radio-echo sounding. Digital elevation models (DEMs) were then used to estimate the subglacial drainage network structure from the maps of the contributing area draining over the subglacial potential surface (Sharp et al. 1993). Dye-tracing experiments (over 500 in this case) permitted reconstruction of the subglacial catchments of the outlet streams, and therefore provided a check on the drainage system structure. In addition, the dye travel time from the moulins into which the injections were made to the outlet stream varies seasonally, and the shapes of the dye return curves distinguish rapid throughput in a conduit system from delayed flow in distributed drainage. Multivariate hydrochemical data (particularly the cation sum, pH and pCO_2) discriminate between open-and closed-system waters, interpreted in terms of both access to atmospheric CO_2 and chemical kinetics during the mixing of quickflow and delayed flow, and confirmed by water balance data which distinguish periods of net water storage and drainage. Finally, the data were integrated by a physically based numerical model. This simulates spatially distributed melt over the glacier surface using a surface energy balance submodel with hourly meteorological data inputs, and which accounts for surface albedo variations and topographic shading. A coupled conduit flow submodel then routes the simulated hourly meltwater input to moulins and crevasses, through the subglacial drainage network defined by the DEM and confirmed by dye-tracing data. Conduit dimensions in this submodel are varied by simulation of the seasonally changing balance between closure under overburden pressure and wall melting by the energy dissipated by the flowing water (Figure 7.3), and water pressure data provide a check on the validity of the model's predictions of the occurrence of surcharging in the conduits.

This example illustrates how understanding of subglacial hydrology is revealed in a detailed case study, involving an integrated, multivariate, multi-process programme of data collection and analysis. This provides what might be considered to be a realist interpretation of glacier hydrology, through a mutually reinforcing explanatory system, although it is clear that it contains elements of both columns in Table 7.2. Nevertheless, i. illustrates an important methodological trend in physical geographical research – towards in-depth case-study research involving multiple, simultaneous investigation of a wide range of interdependent phenomena in order to improve process understanding. Returning to Harré's (1981) classification of the uses experiment in Table 7.2, this kind of investigation appears to be closest to B.10, although not identical to this type. Open-system, field-based experiments in realist environmental science appear to need a new category in this typology, emphasising interaction.

BOUNDARY CONDITIONS: CONTEXTS FOR MECHANISMS

Small-N case study research is invariably set in a particular location – at a field site, for example. The principles that underpin selection of the site – the object of the case study – are much less clearly formulated than those that guide sampling design in large-N investigations. Often, selection is based on convenience, and on logistic rather than scientific grounds. Furthermore, if small-N research seems logically 'weaker' than large-N research, this may also reflect the fact that the phenomena of which the selected case is at first considered to be representative may change as the case study progresses, and accumulated knowledge acquired during the study reveals aspects of the case that were unknown at the outset. For example, in the study of glacial hydrology the substrate conditions may not be known initially, but may become apparent during the research programme. Interpretation of data may then have to be adjusted in the light of this information.

The philosophy of realism implies a structuring of the world in which observable events are contingent on appropriate circumstances for mechanisms to act. Methodologically, this means that interpretation of the nature of mechanisms from observation of events demands an understanding of the role played by local conditions in time and space. In laboratory science, the experiment is designed such that an interpretable outcome arises:

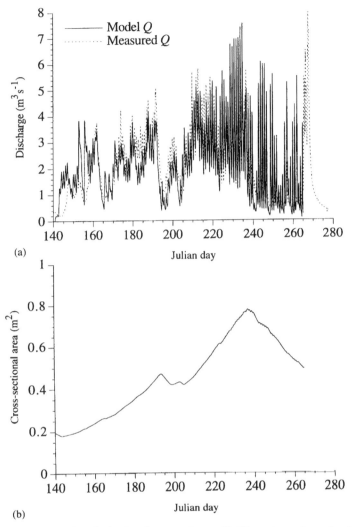

Figure 7.3 (a) Observed and simulated seasonal runoff (Q) patterns from the Haut Glacier d'Arolla, Valais, Switzerland. Simulated patterns are based on an energy-balance melt model, a drainage network structure reconstructed from the map of cumulated drainage area over the subglacial hydraulic potential surface, and connecting the mapped moulin locations to the melt-stream portal, and a hydraulic sewer-flow routing model; (b) the simulated evolution of the width of the main subglacial conduit at about 1000 m upstream from the portal, based on the calculated balance of wall melting and closure by the ice overburden pressure

in field science where the experiment is conducted in the 'naughty world' (Kennedy 1979), the outcome is only interpretable if the experimental conditions are fully understood. This suggests that the manner in which a field location is chosen, and the characteristics of that location, need to be given much more detailed attention than is commonly the case. It does not suffice to provide a bland statement of the broad climatic and geological characteristics of the field location; what is important is identification of the properties of the site that have led to its selection in order that the mechanisms under investigation might be expected to produce observable events consistent with the hypotheses being evaluated. For example, the hypothesis that bed degradation occurs in a glacial meltwater stream on the rising limb of the diurnal hydrograph may not be supportable if the upstream boundary conditions are affected by changes in sediment supply (Lane et al. 1996).

Furthermore, debates in the literature may be more productive if it is recognised that process interpretations based on research in one location may be appropriate for that site, but not for a different one. Thus, different degrees of emphasis on back-to-back secondary circulation cells (Ashmore et al. 1992) and shear layers (Biron et al. 1993) as key aspects of turbulent motion and bed scour at tributary junctions may reflect not the fundamentals of the processes, but rather the different ways in which they are manifested in different environments. There is no reason to suppose that hydrodynamic processes and associated sediment transport will necessarily be comparable for a case where shallow, rapid flows of high Froude number combine over a rough gravel-bed boundary with equal depths in the two tributaries, and a case where deeper flows of low Froude number combine over a sand bed of low relative roughness and unequal depths. However, the flows in both types of confluence obey the same fundamental physical laws, and the same constraints of conservation of mass and momentum.

These issues imply that meticulous description of local boundary conditions is now as critical to successful experimentation and interpretation as the development of fundamental theory, and that increased attention must be focused on innovative means of describing and measuring these boundary conditions. Two brief examples illustrate this point. The first concerns the application of computational fluid dynamics (CFD) to the examination of two-and three-dimensional spatial patterns of flow velocity in braided river reaches (Lane et al. 1994). Engineering applications of CFD often involve regular, uniform channels, and of major concern is the turbulence model used in the closure of the shallow-water equations employed in numerical simulation. In fluvial geomorphological studies of natural channels with complex three-dimensional bed topographies, DEMs must be used to represent the boundary condition for flow modelling (Richards et al. 1995). This requires efficient data acquisition, and terrestrial analytical photogrammetry and stereo-matching procedures offer great potential (Lane et al. 1993). Geomorphological studies of flows over such complex, irregular surfaces (Figure 7.4) may be more concerned with description of this (external) boundary condition, rather than with the choice of (internal) turbulence model. The second example of innovative definition of boundary conditions is in the simulation of flow in soils with macro-pores. The physics of soil water movement, whether as Darcian flow or pipe-flow, are generally understood. However, description of the boundary conditions is extremely difficult because of the destructive nature of most methods for describing the internal structure of soils. Magnetic resonance imaging (MRI) does provide a non-destructive means of resolving the macro-pore structure (Figure 7.5), and therefore a basis for defining the geometry of the irregular conduits within which the flow must be modelled as a turbulent rather than a laminar process (Amin et al. 1993).

These two examples of attempts to represent the boundary characteristics for complex natural flows involve detailed case studies. In terms of Table 7.2, they might be considered 'realist' case studies – their focus is on the precise characteristics of a particular 'place'

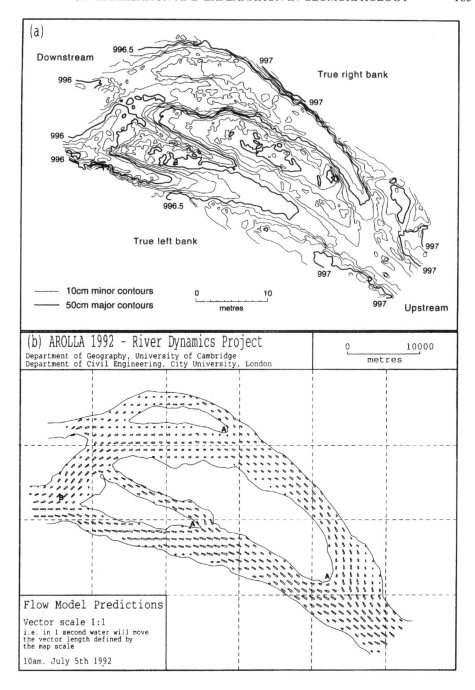

Figure 7.4 (a) A contour map of the terrain model of the bed topography of a study reach on the meltwater stream of the Haut Glacier d'Arolla on the 5 July 1992 at 10 a.m.; (b) the simulated spatial pattern of velocity vectors of flow through this reach at the same time, obtained using a depth-average solution of the Navier–Stokes equations

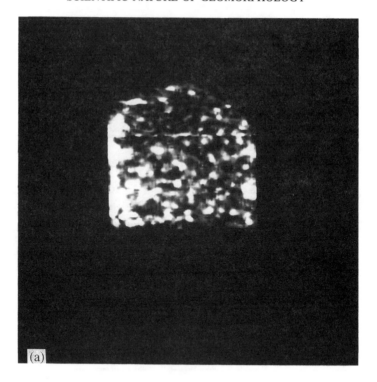

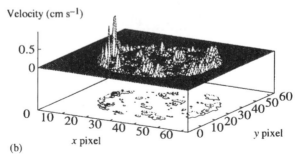

Figure 7.5 (a) A magnetic resonance image of a cylindrical sample of saturated soil, showing the distribution of water held in the larger pores; (b) a velocity distribution for flow through a horizontal surface within a soil sample, again obtained by magnetic resonance imaging

that determine the way general processes produce unique patterns. They demonstrate that the primacy accorded to analysis of the physics of a problem has given way to the necessity for sophistication in the description of local characteristics. In general, in geomorphology and hydrology, the fundamental theories relating to mechanisms operating at relevant scales – for example, those of fluid dynamics and sediment transport – are now relatively well understood. Intellectual effort in the future is therefore more likely to have to focus on the means of acquiring suitable data on the specific local environment (in time and space) in which those mechanisms are acting.

CONCLUSIONS

Geomorphologists – and more broadly, physical geographers and environmental scientists – have often employed experimental methods, although perhaps with a narrower conception of their role than that implied by Table 7.1. They have particularly employed physical laboratory experiments (Schumm et al. 1987) and statistical experiments (e.g. experimental design; Chorley 1966). Although some physical experiments essentially provide qualitative interpretation of possible natural processes, traditionally a laboratory experiment has embodied elements of both physical and statistical design, as it is necessary to consider the number of 'factors', 'treatments' and 'replicates', in order to assess the significance of the influence of factors and treatments on the variance of a response variable. However, emphasis on these quantitative aspects rarely expands the horizons of the experiment markedly. By contrast, modern field experiments may involve a combination of physical and numerical methods, in which the latter have a mathematical–physical basis rather than a statistical basis, and provide a key bridge between the results of the laboratory experiment and the field environment it seeks to simulate. In addition, these experiments are more likely to develop laterally, through the introduction of new measurements which seek to improve the physical basis for understanding the role of the control variables.

These new elements of experimentation suggest two important conclusions. Firstly, the classical 'statistical experimental design' structure may render itself redundant as the knowledge it creates undermines the utility of formally structured experimentation. Later experiments can involve unique natural combinations of factors, being case studies in which the results of each (unique) treatment are explained by a variety of ancillary measurements and by physical understanding and associated numerical modelling. Secondly, the answer to a classical empiricist who criticises a study for lacking 'replication' is that this may be unnecessary if alternative scientific devices exist to allow interpretation, explanation, extrapolation and prediction (of which modelling is perhaps the most important). This conclusion is vitally important for the practice of generalisation from representative experimental catchments, for example, where the expense of replication renders a traditional experimental design impracticable.

Scientific explanation and theoretical understanding may therefore occur particularly through small-N case studies, in which multiple investigations are undertaken simultaneously, guided by 'lateral thinking' which emphasises horizontal connections among related factors. The best experiments are eclectic, and involve 'horizontal' linkage of disparate approaches rather than 'vertical' extension, in terms of increasing N and emphasising replication. However, there are experiments within experiments, and case studies within case studies – comparative studies of river confluences or soil types, for example. Conventional 'extensive' monitoring and experimentation are embedded within 'intensive' case study, as in the glacial hydrology project outlined above, which includes dye-tracing experiments, the monitoring of discharge and solute and suspended sediment concentrations, echo sounding, surveying, meteorological and flow modelling and borehole experiments. Hence the conclusion that the dichotomies in Table 7.2 are more semantic than real.

Case studies, however, may develop at a later stage in a research project than large-N studies, and may be introduced to test ideas about mechanisms. It is necessary to understand mechanisms and processes at a reasonable level of sophistication before a case study can be undertaken. Then, a key issue is description of the local boundary conditions within which those processes operate. This shifts the balance of the science away from developing general laws, towards accounting for the uniqueness of the events produced by their operation in different places. The example of the study of braided stream processes outlined above illustrates this; detailed spatial data on river-bed topography, sedimentology and bed roughness are required as a boundary condition for physically based flow modelling, and this allows analysis of the topographic control of flow direction, and ultimately allows consideration of bedload routing through the study reach, and of spatially distributed patterns of erosion and deposition that allow the dynamic behaviour of the channel to be understood and simulated. Such a specific objective in the case of this case study must surely represent a desirable general goal for geomorphology, providing this framework within which to explore the dynamic, time-dependent behaviour of specific landforms.

ACKNOWLEDGEMENTS

The author acknowledges helpful discussions with Stuart Lane, Neil Arnold and Gao Amin, and their assistance in the preparation of Figures 7.3, 7.4, and 7.5; Jenny Wyatt for drawing the figures; and the National Environment Research Council for Research Grants GR9/547, GR3/7004, GR3/8114 and GR3/9715.

REFERENCES

Amin, M.H.G., Chorley, R.J., Richards, K.S., Bache, B.W., Hall, L.D. and Carpenter, T.A. 1993. Spatial and temporal mapping of water in soil by magnetic resonance imaging. *Hydrological Processes*, **7**, 279–286.

Ashmore, P.E., Ferguson, R.I., Prestegaard, K., Ashworth, P.J. and Paola, C. 1992. Secondary flow in coarse-grained braided river confluences, *Earth Surface Processes and Landforms*, **17**, 299–312.

Bhaskar, R. 1989. *Reclaiming Reality: A Critical Introduction to Contemporary Philosophy*, Verso Press, London, 218 pp.

Biron, P., de Serres, B., Roy, A.G. and Best, J.L. 1993. Shear layer turbulence at an unequal depth channel confluence, in *Turbulence: Perspectives on Flow and Sediment Transport*, edited by N.J. Clifford, J.R. French and J. Hardisty, Wiley, Chichester, pp. 197–213.

Boon, P.J. 1992. Essential elements in the case for river conservation, in *River Conservation and Management*, edited by P.J. Boon, P. Calow and G.E. Petts, Wiley, Chichester, pp. 11–33.

Carson, M.A. 1986. Characteristics of high energy 'meandering' rivers: the Canterbury Plains, New Zealand, *Geological Society of America Bulletin*, **97**, 886–895.

Chorley, R.J. 1966. The application of statistical methods to geomorphology, in *Essays in Geomorphology*, edited by G.H. Dury, Heinemann, London, pp. 275–387.

Collins, D.N. 1979. Hydrochemistry of meltwaters draining from an alpine glacier, *Arctic and Alpine Research*, **11**, 307–324.

Dietrich, W.E. and Smith, J.D. 1983. Influence of the point bar on flow through curved channels. *Water Resources Research*, **19**, 1173–1192.

Dury, G.H. 1969. Relation of morphometry to runoff frequency, in *Water, Earth and Man*, edited by R.J. Chorley, Methuen, London, pp. 419–430.

Goudie, A.S. 1974. Further experimental investigation of rock weathering by salt and other mechanical processes, *Zeitschrift für Geomorphologie*, Supplement Band, **21**, 1–12.

Hacking, I. 1983. *Representing and Intervening*, Cambridge University Press, Cambridge, 287 pp.

Harré, R. 1981. *Great Scientific Experiments*, Oxford University Press, Oxford, 216 pp.

Hodskinson, A. 1995. Flow structures in river bends with chute cut-off and large scale flow separation, in *Coherent Flow Structures in Open Channels: Origins, Scales and Interactions with Sediment Transport and Bed Morphology*, Conference, University of Leeds, 10–12 April 1995, Abstracts volume, p. 27.

Hooke, R. LeB. and Pohjola, V. 1994. Hydrology of a segment of a glacier located in an overdeepening, *Journal of Glaciology*, **40**, 140–148.

Kennedy, B.A. 1979. A naughty world, *Transactions, Institute of British Geographers* NS, **4**, 550–558.

Lane, S.N., Richards, K.S. and Chandler, J.H. 1993. Developments in photogrammetry; the geomorphological potential, *Progress in Physical Geography*, **17**, 306–328.

Lane, S.N., Richards, K.S. and Chandler, J.H. 1994. Application of distributed sensitivity analysis to a model of turbulent open channel flow in a natural river channel, *Proceedings of the Royal Society of London*, Series A, **447**, 49–63.

Lane, S.N., Richards, K.S. and Chandler, J.H. 1995. Within reach spatial pattern of process and channel adjustment, in *River Geomorphology*, edited by E.J. Hickin, Wiley, Chichester, pp. 105–130.

Lane, S.N., Richards, K.S. and Chandler, J.H. 1996. Discharge and sediment supply controls on erosion and deposition in a dynamic alluvial channel, *Geomorphology*, **15**, 1–15.

Langbein, W.B. and Leopold, L.B. 1966. River meanders – theory of minimum variance, *United States Geological Survey, Professional Paper*, 422-H.

Lapointe, M.F. and Carson, M.A. 1986. Migration patterns of an asymmetric meandering river: the Rouge River, Quebec, *Water Resources Research*, **22**, 731–743.

Lawler, D.M., Björnsson, H. and Dolan, M. 1966. Impact of subglacial geothermal activity on meltwater quality in the Jökulsá á Sólheimasandi system, southern Iceland, *Hydrological Processes*, **10**, 557–578.

Leopold, L.B. and Wolman, M.G. 1960. River meanders, *Geological Society of America Bulletin*, **71**, 769–794.

Mitchell, J.C. 1983. Case and situation analysis, *Sociological Review*, **31**, 187–211.

Newson, M. 1980. The geomorphological effectiveness of floods – a contribution stimulated by two recent events in mid-Wales, *Earth Surface Processes*, **5**, 1–16.

Outhwaite, W. 1987. *New Philosophies of Social Science: Realism, Hermeneutics and Critical Theory*, Macmillan, London, 137 pp.

Page, K.J. and Nanson, G.C. 1982. Concave bank benches and associated floodplain formation, *Earth Surface Processes and Landforms*, 7, 529–543.

Phillips, D.C. 1992. *The Social Scientists Bestiary: A Guide to Fabled Threats to, and Defences of, Naturalistic Social Science*, Pergamon, Oxford, 166 pp.

Ragin, C.C. and Becker, H. (Eds) 1992. *What is a case? Exploring the Foundation of Social Inquiry*, Cambridge University Press, Cambridge, 242 pp.

Richards, K. 1994. 'Real' geomorphology revisited, *Earth Surface Processes and Landforms*, **19**, 277–281.

Richards, K., Arnold, N., Lane, S., Chandra, S., El-hames, A., Mattikalli, N. and Chandler, J.H. 1995. Numerical landscapes: static, kinematic and dynamic, *Zeitschrift für Geomorphologie*, Supplement Band, **101**, 201–220.

Richards, K., Sharp, M., Arnold, N., Gurnell, A., Clark, M., Tranter, M., Nienow, P., Brown, G., Willis, I. and Lawson, W. 1996. An integrated approach to modelling hydrology and water quality of glacierised catchments. *Hydrological Processes*, **10**, 479–508.

Sack, D. 1992. New wine in old bottles: the historiography of a paradigm change, in *Geomorphic Systems*, edited by J.D. Phillips and W.H. Renwick, Elsevier, Amsterdam, pp. 251–263.

Sayer, A. 1992. *Method in Social Science: A Realist Approach*, 2nd edn, Routledge, London, 313 pp.

Schumm, S.A., Mosley, M.P. and Weaver, W.E. 1987. *Experimental Fluvial Geomorphology*, Wiley, Chichester, 413 pp.

Sharp, M., Richards, K., Willis, I., Arnold, N., Nienow, P., Lawson, W. and Tison, J.-L. 1993. Geometry, bed topography and drainage system structure of the Haut Glacier d'Arolla, *Earth Surface Processes and Landforms*, **18**, 557–571.
Son, R.K. 1973. *A Manual of Sampling Techniques*, Heinemann, London, 384 pp.
Whalley, W.B., Douglas, G.R. and McGreevy, J.P. 1982. Crack tip propagation and associated weathering in igneous rocks, *Zeitschrift für Geomorphologie*, **26**, 33–54.
Yatsu, E. 1992. To make geomorphology more scientific, *Transactions, Japanese Geomorphological Union*, **13-2**, 87–124.

8 Climatic Hypotheses of Alluvial-fan Evolution in Death Valley Are Not Testable

Ronald I. Dorn

Department of Geography, Arizona State University

ABSTRACT

For the last two decades, climatically based interpretations have been the major focus in evolutionary studies of dryland alluvial fans. Climatic hypotheses to explain the development of alluvial fans in southern Death Valley, however, do not fare well when they are assessed with criteria used by philosophers of science. First and foremost, dating techniques do not have the chronometric precision or accuracy to correlate Pleistocene fan aggradation (or hiatuses in deposition) to Pleistocene climatic changes. Second, there is only one clear correlation of fan aggradation and a climatic interval: in the hyperarid Holocene in Death Valley. A single temporal correlation, no matter how many data points go into it, does not comprise abundant support for a climatic hypothesis. Third, climatic hypotheses are difficult to separate from nonclimatic explanations of Death Valley fan evolution. Fourth, it is very difficult to successfully predict or model Death Valley fan behavior from climatic hypotheses. Lastly, it is not possible to falsify the competing hypothesis that aggradation on Death Valley alluvial fans is entirely from high-magnitude meteorological storms that are not necessarily tied to any climatic regime. This analysis indicates that climatic hypotheses are not testable for Pleistocene-age fans in Death Valley, and perhaps for alluvial fans in other drylands.

INTRODUCTION

Dryland alluvial fans have been viewed through the spectacles of a variety of geomorphic paradigms (Table 8.1). In the last two decades, however, the dominant perspective has

The Scientific Nature of Geomorphology: Proceedings of the 27th Binghamton Symposium in Geomorphology held 27–29 September 1996. Edited by Bruce L. Rhoads and Colin E. Thorn. © 1996 John Wiley & Sons Ltd.

Table 8.1 Twentieth-century geomorphological models used in alluvial-fan research

Model	Summary
Evolutionary	Alluvial fans occur in a youthful stage in the arid lands cycle of erosion (Davis 1905)
Climatic	Climatic changes influence the weathering, stream flow, mass movement and sediment supply in the drainage basin above the fan. The climatic changes influence the base-level of closed basins, gullying, weathering and soil development on fan deposits (Bull 1991; Dorn 1994; Lustig 1965; Melton 1965; Tuan 1962; Wells et al. 1990)
Dynamic equilibrium	Alluvial fans represent a dynamic equilibrium in the transportation of coarse debris from range to basin (Denny 1967)
Steady state	The relationship of fan area and drainage basin area tends toward a steady state, that can shift as forcing relationships change (Hooke 1968; Jansson et al. 1993)
Tectonic	Faulting influences the entrenchment and location of deposition on a fan, the preservation of older fan deposits, and morphometric parameters (Bull and McFadden 1977; Clarke 1989; Hooke 1972; Rockwell et al. 1984)
Intrinsic factors	Fan-head trenching and movement of the intersection point downfan can be explained by intrinsic threshold responses, for example by oversteepening of the fan-head slope, or by drainage basin ruggedness influencing fan incision (Hawley and Wilson 1965; Hooke and Rohrer 1979; Humphrey and Heller 1995; Schumm et al. 1987; Viseras and Fernández 1995; White 1991)
Allometry	Alluvial fans are not in a steady state. Boundary conditions of drainage basin, climate, and tectonism change over time (Bull 1975)
Combination	Alluvial fans aggrade in response to a combination of forcing factors (Bull 1977; Germanoskiy and Miller 1995; Hooke and Dorn 1992; Ritter et al. 1995)

been to relate morphogenetic events on alluvial fans to climatic changes – supported in part by the corollary pillars of climatic interpretations of weathering (Pedro and Sieffermann 1979), soil development (Wright 1992), hillslope erosion (Gerson 1982), and a *perspective* that 'a fundamental goal of earth science is to develop a more complete understanding of mechanisms and rates of climate change so that credible estimates of past global conditions can be constructed' (Drummond et al. 1995, p. 1031). The purpose of this chapter is to assess whether it is possible to test climatic hypotheses of alluvial-fan evolution, at least for the fans in Death Valley, eastern California.

The first section of this chapter introduces how geomorphologists link climatic changes to the evolution of dryland fans. By drylands, I mean semiarid, arid, and hyperarid climates – as defined by Meigs (1953). A comprehensive review of the different climatic hypotheses is beyond the scope of this chapter. My purpose, instead, is to present the general categories of climatic models under consideration as explanations for dryland-fan evolution. Although I focus on Death Valley, I draw analogs from fans in other drylands.

The second section provides an introduction into the nature of climatic changes experienced in the last 100 000 years. This time period covers the last glacial/postglacial cycle, and it is the best dated glacial cycle in terms of dryland-fan research. There are many scales of climatic change in this period (Gates and Mintz 1975), with higher-frequency fluctuations nested within longer-term oscillations. In this chapter I deal with three time scales: decades, where meteorological records are applicable; centuries to millennia, the focus of high-resolution paleoclimatic datasets; and tens of thousands of years, appropriate for analysis by orbital forcing mechanisms. Within the last three years, there has been a shift in the paradigm of paleoclimatology. Before, climatic changes were thought to occur gradually and over time scales of 10^4–10^5 years. The new paradigm, introduced in the second section, stresses the importance of sudden and dramatic climatic fluctuations on millennial and century time scales.

The key issue of this chapter revolves around whether these sudden climatic changes can be linked to alluvial-fan evolution. Linkages are clear where fanglomerate is in physical contact with glacial moraines or with lacustrine sediment. Outside of such contexts, making the climate–fan connection is more difficult. It cannot be accomplished through sedimentological analyses, because all types of fan sediment occur in all climates. Instead, most researchers have been forced into making temporal correlations with the aid of dating techniques. The third section of this chapter, therefore, explores whether it is possible to use dating techniques to correlate aggradation (or hiatuses in deposition) with millennial-or century-scale climatic changes.

I had previously advocated a position that climatic changes exerted a major control on fan morphogenesis in southern Death Valley (Dorn 1988, 1994; Dorn et al. 1987). In the fourth and fifth sections of this chapter, I now argue that climatic hypotheses of alluvial-fan evolution are not testable in Death Valley – even with the application of a new, higher-resolution chronometric technique. I conclude with the position that climatic hypotheses of alluvial-fan evolution in Death Valley (and probably in other drylands), while still possibly correct, are not testable at the present time.

CLIMATIC HYPOTHESES OF DRYLAND ALLUVIAL-FAN EVOLUTION

Climatic changes have been related to alluvial-fan morphogenesis through four different process-based explanations. These hypotheses are not necessarily mutually exclusive, even for the same fan. Often, however, different hypotheses are used in different climatic settings. In drainage systems that have been glaciated, the *paraglacial* hypothesis holds that debris generated by glaciers overwhelms the fluvial system, producing alluvial fans. In the *periglacial* hypothesis, cryogenic processes weather and transport enough debris to build fans. The most popular perspective today, *transition to a drier climate*, invokes a reduction in vegetative resistance to particle erosion from slopes. Other authors contend that *humid-period aggradation* did occur when deserts had a more moisture-effective, semiarid climate.

Paraglacial Hypothesis

The importance of glaciers on alluvial-fan development in drylands was recognized early in the twentieth century (Trowbridge 1911, p. 739):

> Glaciation has played a large part in the deposition of the [eastern] Sierra bajada [in the Owens Valley of California]. Glaciers prepared immense amounts of material in the mountain canyons for transportation by streams. At the same time they furnished great volumes of water to act as the transporting agent during the melting-season.

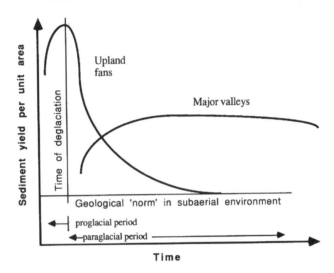

Figure 8.1 The paraglacial model of alluvial-fan aggradation in response to an abundance of sediment generated by glaciers, adapted from Church and Slaymaker (1989)

Six decades later the term 'paraglaciation' was applied to uplands in British Columbia (Church and Ryder 1972; Ryder 1971) to mean:

> ... nonglacial processes that are directly conditioned by glaciation. It refers both to proglacial processes, and to those occurring around and within the margins of a former glacier that are the direct result of the earlier presence of the ice. It is specifically contrasted with the term 'periglacial', which does not imply the necessity of glacial events occurring ... (Church and Ryder 1972, p. 3059).

The paraglacial model (Figure 8.1) has also been used in lower latitudes (e.g., Dorn 1994; Dorn et al. 1991; Meyer et al. 1995; Ritter et al. 1995).

The problem of whether climatic hypotheses of fan aggradation are testable is not at issue where deposits are traceable to glacial moraines (e.g. Birkeland 1965, p. 56; Coleman and Pierce 1981). In these circumstances, there is a direct-spatial linkage between the climatically driven forcing function of glaciation and fan aggradation. The testability of the other three climatic hypotheses in Death Valley is the focus of this chapter.

Periglacial Hypothesis

A classic hypothesis for dryland-fan aggradation in unglaciated drainages is that Pleistocene frost weathering and solifluction on upland slopes generated an abundant load that overwhelmed fluvial systems and led to Pleistocene aggradation on alluvial fans (Zeuner 1959). During colder periods, periglacial activity dominated many western US upland elevations that flank drylands (Dohrenwend 1984; Péwé 1983). Periglacial processes are capable of weathering and transporting (Clark 1987) enough sediment to build alluvial fans in cold regions (Blikra and Longva 1995) and drylands (Catto 1993; Dorn 1988, 1994; Melton 1965; Wasson 1977; Williams 1973).

Transition to a Drier Climate

The importance of a reduction of vegetation cover in enhancing hillslope erosion and fan aggradation has long been recognized (Eckis 1928; Huntington 1907; Zeuner 1959). Bull and Schick (1979) and later Bull (1991) refined the basic model (Figure 8.2), for example, by explaining that the response of fans can be time-transgressive, and depends upon the direction and magnitude of the climatic change:

> Replenishment of the hillslope sediment reservoir is as important as erosion in the production of an aggradation event. Conditions that favor rapid and progressive increases in hillslope plant and soil cover may be infrequent or may require long time spans ... Aggradation of desert valleys occured because of rapid stripping of a thin hillslope sediment reservoirs after a change to markedly less vegetation cover or an increase in intense summer-type precipitation events, or both (Bull 1991, pp. 281, 284).

This general hypothesis has become the most popular explanation for alluvial-fan evolution in drylands (Blair et al. 1990; Dorn 1988, 1994; Gile et al. 1981; Harvey 1990; Iriondo 1993; Kale and Rajaguru 1987; Meyer et al. 1992; Peterson et al. 1995; Ritz et al. 1995; Slate 1991; Throckmorton and Reheis 1993; Wells et al. 1987, 1990).

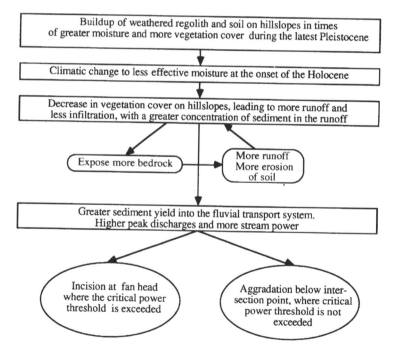

Figure 8.2 Alluvial-fan aggradation during times of transition from more humid to more arid climates, where sediment is generated in a drainage not influenced by periglacial or glacial activity, adapted from Bull (1991)

Aggradation During Humid Periods

A persistent thread in studies of dryland fans is that streams were actively depositing sediment during moisture-effective periods. This theme repeats in the literature for drylands in Australia (Williams 1973), Africa and the Middle East (Dardis et al. 1988; Goldberg 1984; Maizels 1990; Talbot and Williams 1979; Wilson 1990), and North America (Barsch and Royse 1972; Bull 1991, p. 55; Dorn 1994; Harvey and Wells 1994; Huckleberry 1996; Lustig 1965; Mills 1982; Mulhern 1982; Peterson et al. 1995; Ponti 1985; Tuan 1962). Lustig (1965, p. 185) argued that stream flows with high water-to-sediment ratios in wetter periods would deposit material widely over a fan surface. In contradistinction, others argue for the effectiveness of debris flows (Harvey and Wells 1994), or the start of a more humid period being able to transport sediment more effectively (Melton 1965; Thomas and Thorp 1995, p. 203). I have argued that even with a slight increase in biomass, from desert scrub to dwarf conifer woodland, erosion of hillslopes would continue (Dorn 1988, 1994) and might increase since erosion maxima occur in semiarid regions today (Knox 1983; Langbein and Schumm 1958).

The aforementioned models link periods of aggradation or hiatuses in sediment transport to climatic changes. Hence, the next section turns to a review of current advances in paleoclimatology – focusing on the last glacial/postglacial cycle.

PARADIGM SHIFT IN QUATERNARY RESEARCH

During the last glaciation (Wisconsinian, marine oxygen isotope stages 2, 3, 4), periodic 'armadas' of icebergs were released from the Laurentide ice sheet into the northern Atlantic. Iceberg releases produced layers in marine cores that are poor in foraminifera shells and rich in ice-rafted debris derived from Canada. These layers, first recognized by Hartmut Heinrich (1988), have been found in cores throughout the North Atlantic (Bond et al. 1993; Dowdeswell 1995; Mayewskiy et al. 1994). The most recent 'Heinrich layer' corresponds to the Younger Dryas (Andrews et al. 1995), and there were six others from 10 000 to 70 000 years ago (Broecker 1994).

Steadily accumulating evidence reveals that the climate of the globe changed to the metronome of whatever controlled Heinrich events, perhaps changes in oceanic circulation (Birchfield et al. 1994; Broecker 1994) or maybe fluctuations in tropical water vapor (Lowell et al. 1995). The short and sudden return to a glacial world during the Younger Dryas, the latest Heinrich event (Andrews et al. 1995), was globally synchronous (Denton and Hendy 1994; Gosse et al. 1995; Islebe et al. 1995; Kudrass et al. 1991; Wright 1989).

One indication that there has been a paradigm shift in Quaternary research is the wide variety of paleoclimatic records that have been correlated with Heinrich events, including: Greenland (Bond et al. 1993) and Antarctic (Bender et al. 1994) ice cores; sea-surface temperatures (Maslin et al. 1995); pollen in Florida (Grimm et al. 1993); rock magnetism in Europe (Thouveny et al. 1994); foraminifera off the coast of California (Thunell and Mortyn 1995); iceberg rafting in the North Pacific (Kotilainen and Shackleton 1995); spring deposits in southern Nevada (Quade 1994); glacial advances and paleolakes in western North America (Benson et al. 1995; Clark and Bartlein 1995; Gosse et al. 1995;

Phillips et al. 1994); glacial advances in South America (Lowell et al. 1995); monsoons (Sirocko *et al.* 1996); and even loess in China (Porter and Zhisheng 1995).

Heinrich events only record the major Wisconsinian iceberg armadas. There is growing evidence globally and regionally for higher-frequency millennial- and century-scale climate instability in the Pleistocene (e.g. Bond and Lotti 1995; Chappellaz et al. 1993; Fronval et al. 1995; Keigwin and Jones 1994; Kotilainen and Shackleton 1995) and Holocene (e.g. Blunier et al. 1995; Meese et al. 1994; Roberts et al. 1994; Scuderi 1994; Weisse et al. 1994). There may have been iceberg-related climate instabilities in the period from 70 000 to 13 000 years ago (Keigwin et al. 1994).

The aforementioned references represent a small fraction of the 'snowball' of publications in the last three years. Sudden and dramatic climatic changes have replaced orbital forcing as the main focus of Quaternary climate change research. Climatic changes coincident with millennial- and submillennial events hold far-reaching implications for our understanding of global climate change. Conventional *perceptions* of Quaternary climatic change are usually driven by, and correlated with, stages assigned to relatively gentle $\delta^{18}O$ curves of 'global' ice volume change (Imbrie et al. 1984, 1993; Martinson et al. 1987). There is comparatively little uncertainty that global ice-volume curves record the slow buildup and decay of continental ice, but ice-sheet volume fluctuates much more slowly than the sudden and dramatic climatic changes that appear to be characteristic of the Quaternary around the globe.

There are relatively few high-resolution records of climatic change in terrestrial settings, especially in the Death Valley region. Within the last few years, however, there has been a growing dataset indicating that century- and millennial-scale climatic changes did strongly influence hydrologic and geomorphic systems in the western USA (e.g. Allen and Anderson 1993; Benson et al. 1995; Clark and Bartlein 1995; Gosse et al. 1995; Phillips et al. 1994; Quade 1994; Smith and Bishoff 1993; Thunell and Mortyn 1995). I am convinced by the burgeoning database that ocean–atmosphere–terrestrial processes were strongly coupled in the Pleistocene – felt in the western USA by changes in atmospheric circulation (cf. Clark and Bartlein 1995). However, if the reader believes that there is still 'insufficient evidence' to conclude that sudden and dramatic, century- and millennial-scale climatic fluctuations influenced Death Valley and other terrestrial drylands, then the rest of the chapter becomes an exercise in the subjunctive.

The core issue in this chapter is the link between these sudden and dramatic climatic changes and fan evolution – through temporal correlations. The next section addresses whether techniques used to date alluvial-fan deposits (or hiatuses in deposition) are up to the task.

IMPRECISION IN CORRELATION TECHNIQUES

In order to relate dryland fans to climate, alluvial-fan researchers have been forced to turn to indirect chronometric correlations, because sedimentological characteristics cannot be tied to any particular climatic interval. Water-laid, debris-flow, and sieve-flow deposits all exist in a variety of climates. While climatic inferences are readily made for paraglacial fans that are physically tied to glacial moraines or fanglomerate that physically interdigitates with lacustrine sediment, in other circumstances chronometric correlations have been an important methodology.

At an extremely simplistic level, the following general approach is employed: a deposit X is dated to fall within climatic period Y. If there is a regional temporal pattern, and local lithotectonic or intrinsic factors are ruled out, a climatic signal is discerned. In this section, I evaluate different chronometric methods used to make correlations. If I appear too critical, however, it is because the only relevant issue in this section is whether available age-determination methods have the precision and accuracy to make a correlation between events on dryland fans and the century- to millennial-scale climatic changes that dominated the late Pleistocene.

A major problem in dryland-fan research is the paucity of age control in stratigraphic contexts. While accelerator mass spectrometry (AMS) ^{14}C (Linick et al. 1989) and uranium-series mass spectrometry (Edwards et al. 1986) do have sufficient precision to test correlations with millennial-scale climatic change, suitable materials for these techniques are extremely rare in dryland fanglomerate. For example, there are only two published Pleistocene ^{14}C measurements from within fan sediment in Death Valley (Hooke and Dorn 1992). Multiple ^{14}C ages do exist for fanglomerate elsewhere (e.g. Kale and Rajaguru 1987; Pohl 1995), for example in Holocene fans exiting ranges with conifers (e.g. Meyer et al. 1995; Slate 1991; Throckmorton and Reheis 1993). The issue, however, is the extreme paucity of stratigraphic age control for dryland fans before the Holocene.

Volcanic tephras diagnostic of a particular eruption have been used as isochronous units (Beaty 1970; Throckmorton and Reheis 1993). In effect, tephras provide upper and lower age limits. Volcanic ashes by themselves can be used to disprove a climatic correlation, for example if an investigator found a Holocene-age ash in a unit thought to be late Pleistocene. Tephras are particularly valuable when they can be directly linked to a climatic event, for example finding the same tephra in lacustrine sediment. Unfortunately, tephras have a limited spatial and temporal distribution in fanglomerate.

In light of the paucity of datable material in stratigraphic contexts, alluvial-fan researchers have turned to surface-exposure dating methods (cf. Dorn and Phillips 1991). Most surface chronometric methods produce a relative sequence. Morphostratigraphic relationships establish whether a fan segment is inset into or overlaps over another segment (Hooke 1972; Hunt and Mabey 1966). Soil development (Gile et al. 1981), changes in the degree of varnish or desert pavement development (Swadley and Hoover 1989), and changes in remotely sensed characteristics (White 1993) have been used to establish an ordering among deposits. The problem is simple: relative dating methods only provide information on order and cannot be used to correlate a fan unit with any particular time interval, let alone a climatic period. Correlations with discrete climatic intervals must rely on calibrated-, correlative-, and numerical-dating methods.

Calibrated dating methods regress a relative age signal against independently established numerical ages. For example, different soil properties are tabulated into a soil development index (Harden 1982) that is used to assign calibrated ages (Reheis et al. 1989; Switzer et al. 1988). Even if calibration points are valid, the uncertainties inherent in the method (Switzer et al. 1988) yield errors that are much larger than the length of millennial-scale climatic events. A similar problem in inadequate precision exists for cation-ratio dating of rock varnish (Dorn 1994). Soils and cation-ratio dating only have the precision to establish that dryland fan deposition occurred during the drier Holocene. Correlations with Pleistocene climatic changes are beyond the chronometric resolution of these techniques.

The inherent limitations of correlative-dating methods have not inhibited climatic interpretations. For example, visual differences in varnish appearance have been used to assign correlated ages to deposits, based on varnish characteristics at chronometrically constrained sites (McFadden et al. 1989). Varnish appearance has also been used to make climatic correlations (Harvey and Wells 1994). This is all despite serious uncertainties in using varnish appearance to estimate age, such as tremendous surface-to-surface variability in rates of varnish development (Bednarik 1979; Colman and Pierce 1981, p. 2; Dorn 1983; Dorn and Oberlander 1982; Dragovich 1984; Friedman et al. 1994; Grote and Krumbein 1992; Haberland 1975; Linck 1928; Lucas 1905; Rivard et al. 1992; Viereck 1964; Whitley et al. 1984) – issues that have been ignored by those who attempt to estimate exposure age in this manner.

Morphostratigraphic relationships can be used to establish correlative ages (Dorn 1988; Wells et al. 1987), where some units are older or younger than a given numerical age. In the case of dryland fans that spatially intersect paleolake shorelines, deposits resting over a terminal-Pleistocene shoreline would be Holocene (Gilbert 1890; Russell 1885), but fans cut by terminal Pleistocene shorelines could be correlated with any earlier Pleistocene climatic period (Hawley and Wilson 1965).

Much of the numerically dated material provides only minimum ages for sediment deposition. AMS ^{14}C ages on weathering rinds (Dorn 1994) tell when organic matter stopped exchanging CO_2 with the atmosphere – essentially when rock varnish encapsulated the weathering rinds. Although these ^{14}C ages postdate surface exposure by approximately 10% (Dorn et al. 1992b), even this uncertainty makes definitive correlations with millennial-scale climatic events impossible. Similarly, ^{36}Cl, uranium-series, and radiocarbon ages on pedogenic carbonate (Hooke and Dorn 1992; Liu et al. 1994; Peterson et al. 1995) must postdate fan deposition by an uncalibrated amount of time that it took the carbonate to form. In addition, pedogenic carbonates do not appear to be a closed system (Stadelman 1994).

Numerical ages have been assigned to fan units with the *in situ* buildup of cosmogenic ^{10}Be/^{26}Al (Bierman et al. 1995; Nishiizumi et al. 1993; Ritz et al. 1995) and ^{36}Cl (Liu et al. 1996). Claims of high-precision fan dating with cosmogenic nuclides being able to 'exploit these terrestrial archives of climate change' (Bierman et al. 1995, p. 449) ignore fundamental methodological limitations.

1. The 'tightest' datasets have a 1σ precision for 'apparent' exposure ages of 25–30%; this error alone invalidates Pleistocene climatic correlations.
2. Fire spalling is a serious problem for cosmogenic nuclides such as ^{10}Be and ^{26}Al that are only produced by spallation (Bierman and Gillespie 1991), but less so for ^{36}Cl that is also produced by neutron activation (Zreda et al. 1994); the only noncircular solution assesses boulder erosion with varnish microlaminations (Liu 1994).
3. Cosmogenic nuclides have uncertainties associated with 'inheritance' of nuclide buildup prior to clast emplacement in a fan (Dorn and Phillips 1991). Using measurements of cobbles in the most recent deposits in order to address issues of signal

inheritance (Bierman et al. 1995) assumes that late Holocene fluvial 'storage' of alluvium was similar to Pleistocene fluvial 'storage' – a very uncertain assumption (Church and Slaymaker 1989; Leece 1991; White 1991).

4. Boulder and cobble geometry can change over time, especially when sampling occurs on debris-flow deposits that erode or on desert pavements that are mobile (Mabbutt 1979).

5. The uncertainties associated with production rates are difficult to quantify at present, but these errors add at least another 20% to the uncertainty for $^{10}Be/^{26}Al$ ages. Cosmogenic nuclides have great potential to inform on rates of geomorphic processes, but these methods do not yield precise enough or accurate enough ages to make a definitive correlation between fan aggradation and Pleistocene climatic changes.

This entire discussion has assumed that dating techniques are employed flawlessly, and that no errors in accuracy are introduced. However, there are quite a number of technical issues that could affect accuracy. Consider a method whose results are often accepted uncritically, ^{14}C dating. There are serious uncertainties associated with sample pretreatment. Young organic molecules move with water and can adsorb to organics and clay-sized minerals in samples (Gu et al. 1995; Hedges et al. 1993; Heron et al. 1991; Österberg et al. 1993). Inaccurate ages may result when organics are not pretreated, or when conventional pretreatment does not remove these young organics (Gillespie 1991). My point is that even conventional dating methods such as radiocarbon are experimental, especially when they are used to date dryland alluvial fans. There are many uncertainties surrounding the history of the carbon atoms that are actually measured.

There is a more general concern, related to a systematic bias in the way that samples are collected for age determination. Traditionally, the first step is the genesis of morpho-stratigraphic maps – based on relative dating methods of characterizing fan surfaces. Then, samples are collected on these different morphologic units – with the assumption that the entire fan segment is temporally equivalent. If this assertion is erased, significant morphogenetic events may have occurred, but may not have been sampled. In other words, what is now recognized as a single fan unit may truly be composed of many time-transgressive elements – each of which occurred in response to a different forcing. This uncertainty amplifies concerns over the accuracy and precision of the dating results by an unknown amount.

In summary, I am not advocating a position that these new chronometric insights are unimportant. On the contrary, the aforementioned techniques provide valuable insight into rates of geomorphic processes and rates of landscape evolution in drylands. My only point in this section is that available chronometric methods do not have sufficient temporal resolution to correlate dryland Pleistocene alluvial-fan evolution with century- or millennial-scale climatic records. In the next section, I reassess research on the Death Valley fans in light of the new paradigm of Quaternary climatic change and in light of the above limitations in dating methods.

DEATH VALLEY FANS

Reevaluation of Prior Data

I have advocated three different climatic hypotheses to help explain the evolution of alluvial fans debouching from the Panamint Range (Figure 8.3) into Death Valley (Dorn

1994). Periglacial activity, dated in the upper Panamint Range to the last glacial period, could have supplied sediment in colder periods. Erosion of hillslopes likely generated sediment in the wetter period of the late Wisconsinian when Lake Manly occupied the floor of the valley and *Yucca* scrub and dwarf conifers grew on the lower hillslopes. Lastly, when the climate changed from the semiarid latest Pleistocene to the hyperarid Holocene, the vegetation could no longer hold the sediment in place – leading to fan aggradation. These hypotheses were supported by morphostratigraphic relationships, conventional ^{14}C measurements, weathering rind ^{14}C ages, ^{14}C ages on organics in pedogenic carbonate rinds, uranium-series ages, [10]Be/[26]Al ages, calibrated- and correlative-varnish methods, and soil development (Dorn 1988, 1994; Dorn et al. 1987; Hooke and Dorn 1992; Nishiizumi et al. 1993; Stadelman 1994).

Climatic hypotheses to explain fan evolution in Death Valley may be true, but I now contend that it is not possible to test them at the present time. Consider the dataset for Hanaupah Canyon alluvial fan (Figure 8.4). The age range for the eroding Q1 unit is too long to correlate with even oxygen-isotope stages (Martinson et al. 1987). The age ranges for the Q2 and Q3 units are similarly too long to correlate with any global (Broecker 1994; Keigwin et al. 1994; Kotilainen and Shackleton 1994), regional (Phillips et al. 1990, 1994; Smith and Bischoff 1993; Thunell and Mortyn 1995), or locally derived (Ku et al. 1994; Lowenstein et al. 1995; Szabo et al. 1994; Winograd et al. 1992) climatic signal. A climatic correlation is not possible, even for the best constrained Pleistocene aggradational unit, an orange-colored tributary fan on the northwest side of Hanaupah Canyon fan – younger than 16 000 years, but older than 12 000 years (Dorn 1994). This period straddles Heinrich event 1, a wet event ∼ 14 000 years ago that was felt throughout the western

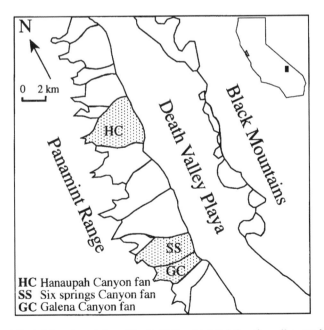

Figure 8.3   Alluvial fans in southern Death Valley, highlighting fans discussed in this chapter

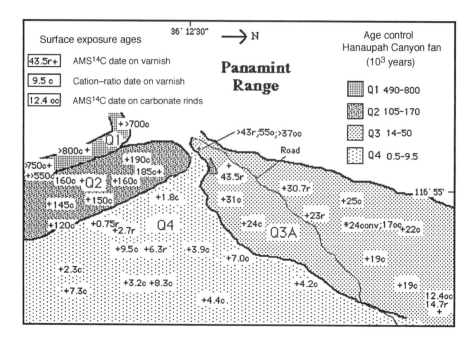

Figure 8.4 Oblique aerial photograph and corresponding map of Hanaupah Canyon alluvial fan, Death Valley. The symbol '+' identifies sampling sites for radiocarbon (r), cation-ratio (c), ^{14}C ages on organic matter from pedogenic carbonate rinds (oc), and a conventional ^{14}C measurement on a charcoal sample (conv). All ages are in 10^3 years. Fan segments correspond to those used by Hooke and Dorn (1992)

USA (Benson et al. 1995; Broecker 1994). It is not possible to correlate aggradation of the tributary fan to Heinrich event 1, drier periods on either side, or the climatic transitions between.

The only alluvial fan unit that correlates with a climate period is Q4. Although transgressive in space and time, deposition of Q4 occurred throughout the Holocene, a largely hyperarid climate interval (Lowenstein et al. 1995; Wells and Woodcock 1985). Climatic changes did occur in the Holocene (Blunier et al. 1995; Bryson, 1992; Meese et al. 1994; Roberts et al. 1994; Scuderi 1994; Weisse et al. 1994), but in Death Valley (Lowenstein et al. 1995; Wells and Woodcock 1985) and globally (Bryson 1992), the magnitude of Holocene climatic changes was far less than within the Pleistocene and between the Pleistocene and the Holocene. I think it reasonable, therefore, to acknowledge at least this correlation. As discussed in a later section, a singular temporal correlation is insufficient evidence to support a climatic interpretation. I note that the inability of these chronometric data to test climatic hypotheses, however, does not detract from or conflict with nonclimatic interpretations of fan evolution in Death Valley (e.g. Denny 1965; Hooke 1968, 1972; Hooke and Dorn 1992; Hunt and Mabey 1966; Jansson et al. 1993).

There is an entirely new way of linking alluvial fans to climatic changes in drylands. The next section explores, in the context of Death Valley, whether this higher-resolution approach can be used to test climatic hypotheses of fan evolution.

Evaluating Varnish Microlaminae as a Tool to Test Climatic Hypotheses

Optical varnish microlaminae (VML) exist in millimeter-scale depressions on rock surfaces. VML are analogous to lake and ocean sediment. They accumulate over time and yield climatic information. Orange (manganese-poor) and black (manganese-rich) varnish corresponds to dry and wet climates (Cremaschi, 1996; Jones 1991; Liu and Dorn 1996). In a study of some 2900 rock-surface depressions in 420 ultrathin sections from 360 rocks in Death Valley and the surrounding region, Liu (1994) determined that VML are organized into distinct layering units. Figure 8.5 illustrates the ^{14}C, uranium-series, and ^{10}Be/^{26}Al calibrated sequence for optical microlaminations for Death Valley, California.

VML have been applied to the study of alluvial-fan evolution in Death Valley (Liu 1994; Liu and Dorn 1996), where the basal layer of rock varnish provides a minimum age for the subaerial exposure of the underlying boulder. Details on sampling density, sampling procedures, and sample preparation are presented elsewhere (Liu 1994; Liu and Dorn 1996). However, I must clarify that the surface appearance of rock varnish in the field can be a misleading indicator of age. Time information is obtained by the oldest (bottom) microlamination of rock varnish, as seen with a light microscope in ultrathin sections obtained from different rock-surface depressions on a boulder. This oldest (bottom) layer provides a minimum age for the exposure of that particular boulder through correlation with the calibrated sequence (Figure 8.5).

A mappable pattern emerges when multiple boulders are sampled, side by side, and over a fan (Figure 8.6(a) and (b)). In these maps, the name of the alluvial-fan segment corresponds to the basal layer of the VML sequence in replicate samples. For example, those fan segments mapped as LU-3 all have varnishes where the oldest VML is LU-3. By mapping the basal varnish layers of the preceding and subsequent fan deposits, Liu (1994) and Liu and Dorn (1996) have determined that it is possible to obtain experimental,

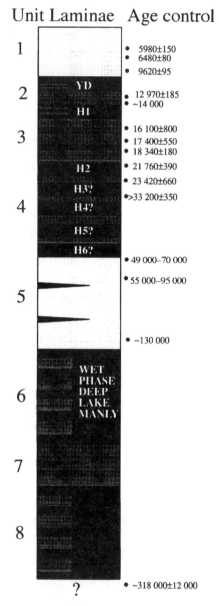

Figure 8.5 An idealized sequence of varnish layering units in Death Valley. Each varnish layering unit is on the order of tens of micrometers in thickness. Shading corresponds with what would be seen in an ultrathin section under a light microscope: light grey is yellow-orange and Mn-poor; dark grey is orange and Mn-intermediate; and black is dark and Mn-rich. YD and H1–H6 indicate possible correspondences between black layers and the Younger Dryas (YD) and Heinrich events H1 through H6. Age control, identified on the side of the sedimentary sequence, comes from radiocarbon ages (<35 000 years), uranium-series (49 000 to 130 000 years), and ^{10}Be/^{26}Al (318 000 years); the ages are tabulated in Liu and Dorn (1996). Each chronometric measurement, however, only provides a maximum age for the varnish stratigraphy on that boulder. Hence, the validity of the time scale exists within the limitations of the dating methods discussed in the text

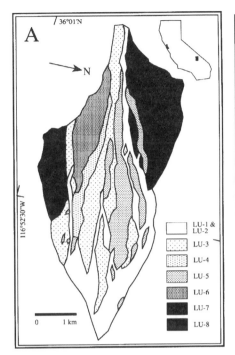 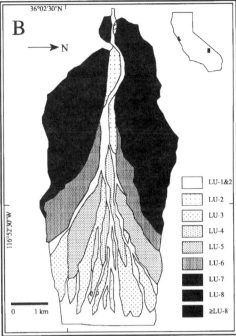

Figure 8.6 Maps of Galena fan (A) and Six Springs fan (B) (adapted from Liu and Dorn 1996). Geomorphic units on each fan deposit are correlated with the basal layering unit in replicate varnish samples (see Liu 1994; Liu and Dorn 1996)

minimum ages for fan segments in Death Valley for the past $\sim 300\,000$ years (Figure 8.5). The calibration is firm, however, only for the last $\sim 24\,000$ years (Liu and Dorn 1996). VML also make it possible to correlate fan aggradation with wet and dry intervals, because these intervals are recorded directly in sedimentary strata (rock varnish layers) that rest directly on top of the fan deposit.

Although I believe that the VML method, developed by Liu (1994), is the easiest to use and the most accurate varnish technique yet developed, and that VML provide millennial-scale correlations of spatially disjunct geomorphic surfaces for the last $\sim 24\,000$ years, this technique cannot yet be used to test climatic interpretations of alluvial-fan development, for a couple of reasons. First, the rate of varnish formation is so slow that submillennial climatic changes are not necessarily recorded; it may be possible, however, in the future to systematically examine the fastest-growing varnishes. Second, available VML data do not indicate a clear correlation between fan aggradation and climatic change. Consider Figure 8.6, maps of two adjacent alluvial fans in southern Death Valley. Fan deposition occurred in all of the climatic intervals that are recognized in Death Valley varnishes, and in other paleoclimatic records (Lowenstein et al. 1995; Phillips et al. 1990, 1994; Smith and Bischoff 1993; Thunell and Mortyn 1995; Winograd et al. 1992). In other words, fan aggradation appears to have been continuous within the period of record, at least to the limit of the chronometric resolution of the VML method. Since deposition has been essentially 'nonstop' for the last $\sim 300\,000$ years on time scales of 10^3–10^5 years, and since information on the volume of this sediment is lacking, it is not possible to define the role of climate on fan formation in Death Valley through chronometric means.

EVALUATING CLIMATIC HYPOTHESES FOR DEATH VALLEY FANS

I contend here that all climatic models of alluvial-fan evolution that have been applied to Death Valley (periglacial, humid-period, transition to drier climate) fall short on key criteria developed by philosophers of science (Copi 1982; Farr 1983; Hempel 1966; Newton-Smith 1981; Popper 1966): (1) quantity of data explained; (2) ability to test the hypothesis; (3) consistency with established theoretical frameworks and accepted theories; (4) predictive capabilities; and (5) inability to falsify a competing hypothesis.

Quantity of Data Explained

The only clear match between a climatic interval (wet, transition, or dry) and dryland fan aggradation (or a hiatus in aggradation) in Death Valley is during the arid (Lowenstein et al. 1995; Wells and Woodcock 1985) Holocene (Dorn 1988, 1994; Hooke and Dorn 1992). Holocene aggradation also occurred in the Mojave Desert to the south (Dorn 1994; Wells et al. 1990), in southern Nevada to the east (Peterson et al. 1995), and western Nevada to the north (Slate 1992; Throckmorton and Reheis 1993).

 The temporal correlation in Death Valley of fan aggradation during the drier Holocene, however, does not comprise an abundance of data in support of any of the climatic hypotheses. A basketball player who is able to make one shot may be a good shooter, but it is only one shot. (I do not mean to infer that the funds and labor spent on the corpus of 84 Holocene age determinations on Death Valley fans have been a waste, but the only issue here is the testability of climatic models.) More problematic is the realization that these data only falsify a model, not even proposed in the literature, that deposition only occurs in more humid periods.

Inability to Test Climatic Hypotheses

The inability of present-day techniques to match fan-segment age (or hiatuses in aggradation) with a climatic interval (with the exception of the Holocene) implies that it is not possible to test any of the three extant climatic hypotheses for Death Valley. It is impossible to test whether a dryland fan aggraded or stabilized in response to any of the century- to millennial-scale climatic changes in the Pleistocene. Even if a chronometric match occurred, the time-transgressive nature of dryland geomorphic systems would make temporal correlations ambiguous. In the context of Death Valley, different elevations in a single drainage basin in the Panamint Range would respond to climatic change differently and in a time-transgressive fashion (Melton 1965; Bull 1991). Certainly, chronometric information can be used to make sound and logical deductions. But the poor temporal resolution inherent in the time-transgressive nature of the hillslope–fan system, combined with uncertainties in precision and accuracy of dating methods, combined with increasingly precise paleoclimatic information, mean that available surface and subsurface chronometric data are not precise enough or accurate enough to test climatic hypotheses for Death Valley.

There is also a fundamental issue over correlation and causation. The correlation of alluvial-fan units with climatic periods can only suggest a climatic cause, not prove one. To illustrate this concern, consider Montgomery and Dietrich's (1992) model of the importance of the position of the channel head. In their discussion, a decrease in vegetation cover from either climatic- or land-use change moves the channel head up the slope – entraining hillslope debris into the fluvial system. A 50-year-long dry phase could cause an upslope movement in channel heads and excavate abundant hillslope debris in the midst of a 1000-year period of more effective moisture. The cause of any correlation could, therefore, be a temporal illusion based on an unstated assumption of climate stability during a given period. There is simply no way to test this complication in Death Valley with available chronometric techniques.

Consistency of Hypotheses with Established Theory

There are other complications that make it very difficult to test climatic hypotheses for fan evolution in Death Valley. A few of the more prominent issues, well recognized in the alluvial-fan literature, are sketched in this section.

Intrinsic geomorphic factors affecting fan evolution are difficult to separate from climatic factors (Field 1994). Intrinsic variables may force loci of deposition to switch (Beaty 1974; Hooke 1987), promote fan-head entrenchment (Germanoskiy and Miller 1995; Schumm et al. 1987; Weaver 1984), or redirect deposition through drainage piracy (Clarke 1989; Denny 1965). Some of the internal feedbacks include the role of tributary streams that empty at the fan head (Hawley and Wilson 1965, p. 22; Dorn 1994), in-basin storage (Lecce 1991; MacArthur et al. 1990; White 1991), and drainage-basin size (Melton 1965; Bull 1991; Wilcox et al. 1995).

Tectonic factors can be isolated from climate (Bull 1991; Ritter et al. 1995), but long-term rates of tectonic activity, as well as the timing of specific events, must be understood for each fan. In addition, tectonically altered spatial variability in stream power can influence the location of incision, aggradation, transportation, and depositional settings (Bull 1991; Bull and McFadden 1977; Hooke 1972; Jansson et al. 1993; Rockwell et al. 1984).

Volumes of aggradational units are an important missing link for climatic interpretations in Death Valley, because larger volumes imply more erosion. Volumes, however, are largely unknown. What evidence does exist suggests considerable spatial and temporal variability. For example, a <16 000 years to >12 000 years (Dorn 1994) orange-colored tributary fan, at the north side at the head of Hanaupah Canyon, is over 8 m thick against the hillslope; but it thins out completely a few hundred metres in the distal direction. The LU-4 unit (Figure 8.6) on Galena Canyon fan, resting on top of a petrocalcic paleosol, ranges from a thickness of ≥ 8 m to 2 m. There is also evidence, reconstructed from the partial erosion of pre-LU-4 fan units, that pedimentation occured before or during the time that LU-4 deposits aggraded; this can be seen, for example, at the wave-eroded outlier at the northeast corner of Hanaupah Canyon alluvial fan. Although the results of future, volumetrically based studies may indeed be fully consistent with climatic models, it is very difficult to test hypotheses of climatically driven sediment transfer when the volumes of sediment transfer are unknown.

Predictive Capabilities

A valid criterion for assessing scientific hypotheses is prediction, both success in prediction and ability to predict. Consider the failure of a climatic model in predicting fault hazards in Mongolia. Offset fan segments were formerly correlated with the last glacial/interglacial transition – leading to estimates of fault movement ~ 20 mm yr^{-1}. Yet, new ^{10}Be ages reveal that faulting rates have been >16 times slower (<1.2 mm yr^{-1}) than predicted using climatic models (Ritz et al. 1995).

In the case of Death Valley, ambiguous links between climatic changes and sediment erosion in drainage basins make predictions extremely difficult. The basic assumption of the transition-to-drier-climate model is that more vegetation holds sediment in place. This assumption is supported by regional analyses (cf. Knox 1983) and case studies of deforestation (Kesel and Lowe 1987; Meyer et al. 1995). Erosion rates are now highest in semiarid (cf. Bailey 1979) climates (Clayton 1983; Knox 1983; Langbein and Schumm 1958; Walling and Kleo 1979). Key spatial issues for Death Valley drainage basins are the starting position of erosion maxima and the direction of the climatic change. Yet, spatial positions of erosion maxima on hillslopes are not known in Death Valley at present, and certainly not for different times in the past.

Eventually, it may be possible to extrapolate erosion information from fossil plant assemblages (cf. Spaulding 1990; Wells and Woodcock 1985), but not at present. There are no quantitative links between vegetation-specific data and resistance of sediment to hillslope erosion (see discussion in Eybergen and Imeson 1989). Furthermore, vegetation changes can lag behind climatic changes (Bull 1991). Information is lacking in Death Valley to answer such basic site-specific questions as: was resistance to erosion greater on the lower slopes in the Pleistocene under a semiarid cover of *Yucca* scrub and dwarf conifers, or during the Holocene under a hyperarid regime of sparse *Larrea* coverage? This sort of ambiguity is heightened by clear warnings of the geomorphic impact of climatic change in other dryland contexts:

> ... when regional or local scales are considered, the relationship between climate and environmental conditions become problematic. Local factors such as topography, lithology and soils play a decisive role in the spatial redistribution of water resources ... The relationship between climate and environment in arid and semi-arid areas is even more problematic when climate change is considered ... The data presented show that the effect of a climate change in such areas is highly controlled by the surface conditions prevailing in the area prior to climatic change ... (Yair 1994, p. 223–224).

Inability to Falsify Competing Hypothesis

It is not possible in Death Valley to falsify the competing hypothesis that sediment transfer (hillslope erosion to fan deposition) in dryland alluvial fans is from high-magnitude (low-frequency) meteorological events – unrelated to any particular climatic condition. Extreme meteorological events (or short-term climatic fluctuations) within a longer climatic state are often responsible for geomorphic changes (Graf 1988; Kochel and Ritter 1990; Macklin et al. 1992; Meyer et al. 1995; Roberts et al. 1994; Shick 1974; Thomas and Thorp 1995). Furthermore, the 'impacts of extreme events within present-day climate regimes may mimic those of palaeoclimates ...' (Thomas and Thorp 1995, p. 195). Sediment mobilization may have even been tied to extreme drought followed by extreme rainfall, or in response to a fire followed by extreme rainfall (Germanoskiy and Miller 1995). There is abundant historic and prehistoric evidence to suggest the importance of extreme meteorological events in aggradation (Allen and Anderson 1993; Beaty 1974; Bowman 1988; Brookes et al. 1982; Dorn et al. 1992a; Eybergen and Imeson 1989; Field 1994; Grossman and Gerson 1987; Hooke 1987; Kesel and Lowe 1987; Williams and Guy 1973). In the case of Death Valley, it is simply not possible to falsify this relevant, plausible, and simple competing hypothesis to climatic models.

In summary, extant climatic hypotheses of alluvial-fan evolution in Death Valley do not fare well when they are evaluated with criteria suggested by philosophers of science. This does not mean that climatic explanations are incorrect, only that they are not testable at this time.

CONCLUDING REMARKS

Climatic hypotheses of alluvial-fan evolution in Death Valley have serious deficiencies, the most significant of which is they cannot be tested. Many of the difficulties in testing climatic models that I have isolated for Death Valley may also apply to research on other dryland fans. The fundamental question of temporal correlation is certainly applicable elsewhere: is it even possible to correlate, in time, climatic changes with fan aggradation or with hiatuses in deposition? The answer varies with the time scale of climate change. My answer would be 'probably' for the twentieth century where meteorological records are available; it would probably be possible to compile and map many historic aggradational events – and then to correlate them with meteorological data. My answer is 'possibly' for the orbital time scale of 10^4–10^5 years; correlations between alluvial-fan segments and broad climatic intervals are problematic under the old paradigm of gradual climatic changes (Figure 8.7), but still possible.

Correlations between sudden and dramatic, century- to millennial-scale Pleistocene climatic changes and dryland alluvial-fan events (aggradational or hiatus in deposition) are not possible within, and probably outside of, Death Valley. The best available chronometry in the southwestern USA, for example, places fan aggradation within any number of different climatic intervals (Figure 8.8). The very difficult problem is that the Pleistocene experienced high-frequency and high-magnitude oscillations in climate. The issue is not the worth of available chronometric information; the value of age determinations in providing insights into the rates of dryland geomorphic processes should be decoupled from their utility in testing climatic hypotheses. Chronometric methods available to measure the ages of Pleistocene dryland-fan deposits are simply not up to the task of correlation with century- or even millennial-scale climatic changes. To mix metaphors, the target has moved so far back that the light at the end of the tunnel is no longer visible.

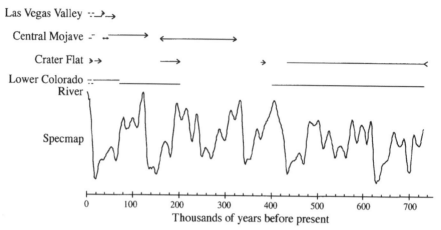

Figure 8.7   A visual comparison between the aggradation of alluvial-fan units in the southwestern USA and broad climatic intervals in the Pleistocene as portrayed by the Specmap record of global ice-volume change (Imbrie et al. 1993; Martinson et al. 1987). The Pleistocene alluvial-fan chronologies presented are those with the highest available chronometric resolution from: the lower Colorado River (Bull 1991); Crater Flat in southern Nevada (Peterson et al. 1995); the central Mojave Desert (Wells et al. 1990); and Las Vegas Valley (Quade 1986; Quade and Pratt 1989). Arrows indicate that ranges may be maximums or minimums. Aggradation units were placed lower and higher in any given record, giving the appearance of dashed lines, in order to clearly delineate truly separate deposits that would have otherwise 'run together'.

As in Death Valley, only one clear correlation can now be made in the southwestern USA between a climatic event and a fan unit: during the Holocene. Calibrated, correlative, and numerical ages are precise enough to constrain the ages of certain fan deposits to this drier-warmer climatic period (e.g. Bull 1991; Dorn 1994; Peterson et al. 1995; Reheis et al. 1989; Slate 1991; Throckmorton and Reheis 1993; Wells et al. 1987, 1990). I argue that, as in Death Valley, a singular correlation does not constitute an abundance of data in favor of a hypothesis. The issue here is not the value of these numerous Holocene datasets in answering important geomorphic questions, or their value for building process–response models, but in their ability to test climatic hypotheses. There are complications even in this single temporal correlation, because the response of fans to climatic change has been time-transgressive in the Holocene (Bull 1991), because smaller magnitude climatic changes did occur during the Holocene (Slate 1991; Bryson 1992), and because Holocene fan aggradation has occurred in a variety of climatic regions (e.g. Bull 1991; Church and Slaymaker 1989; Dorn 1994; Meyer et al. 1995; Peterson et al. 1995).

If my analysis for Death Valley can indeed be extended to other dryland alluvial-fan sites, the implication would be clear. Climatic hypotheses for fan evolution would have to be reevaluated: by geomorphologists developing theory; by the global change research community; by those assessing tectonic hazards; by remote sensing specialists who map fans with the assumption that certain units are diagnostic of a particular climate; by policy-makers deciding issues of flood insurance; by those looking for widespread proxy data in arid continental settings to test climate models; or by cognate disciplines such as biogeography or pedology interested in relating spatial data to landscape evolution in drylands.

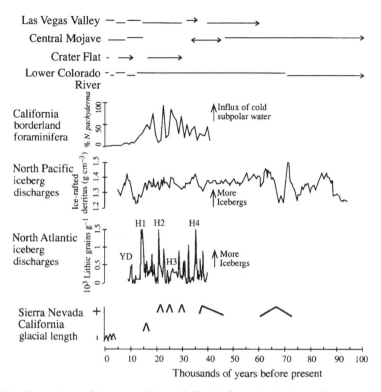

Figure 8.8 Comparison of the aggradation of alluvial-fan units in the southwestern USA (from Figure 8.7) with records of millennial-scale climatic change. This diagram focuses on the last 100 000 years, because this period has a higher chronometric resolution. The millennial-scale climate records are from: foraminifera in the Tanner Basin in the southern California borderlands (Thunell and Mortyn 1995); iceberg rafting events in the North Atlantic, with the Younger Dryas (YD) and Heinrich events (H1–H4) indicated (Bond and Lotti 1995); glacial advances in the Sierra Nevada (Bach and Elliott-Fisk 1996; Dorn 1996; Zreda and Phillips 1994; Zreda et al. 1994); and ice rafting in the North Pacific (Kotilainen and Shackleton 1995)

The discussion here has been limited to dryland fans, but similar issues may exist in the interpretation of evolutionary changes as a function of discrete climatic intervals in other morphoclimatic settings (Butzer 1980; Thomas and Thorp 1995). The paradigm shift toward sudden and dramatic climatic changes throws down a gauntlet for geomorphologists who attempt to relate past climatic events to landforms. There may be troubled times ahead for those who attempt to use chronometric techniques to correlate hillslope and fluvial systems with late Pleistocene climatic changes.

ACKNOWLEDGEMENTS

Supported by NSF PYI Award, National Geographic Society Grant 84-2961, and ASU for sabbatical support. Thanks to D. Dorn for field assistance, S. Campbell for assistance on graphics, T. Liu for developing and sharing a method that allows a direct correlation between fan development and climate, J.W. Bell, W. Graf, R. LeB. Hooke, N. Meek, and W.B. Bull and many students for discussions on different components of this material, and comments on the manuscript by J. Ritter, S. Wells and the editors. However, the opinions stated here are my own.

REFERENCES

Allen, B.D. and Anderson, R.Y. 1993. Evidence from western North America for rapid shifts in climate during the last glacial maximum, *Science*, **260**, 1920–1923.

Andrews, J.T., Jennings, A.E., Kerwin, M., Kirby, M., Manley, W., Miller, G.H., Bond, G. and MacLean, B. 1995. A Heinrich-like event, H-O (DC-0): source(s) for detrital carbonate in the North Atlantic during the Younger Dryas chronozone, *Paleoceanography*, **10**, 943–952.

Bach, A. and Elliott-Fisk, E.L. 1995. Soil development on late-Pleistocene moraines at Pine Creek, east-central Sierra Nevada, California, *Physical Geography*, **17**, 1–28.

Bailey, H.P. 1979. Semi-arid climate: their definition and distribution, in *Agriculture in Semi-arid Environments*, edited by A.E. Hall and G.H. Cannell, Springer-Verlag, Berlin, pp. 73–97.

Barsch, D. and Royse, C.F.J. 1972. A model for development of Quaternary terraces and pediment-terraces in the Southwestern United States of America, *Zeitschrift für Geomorphologie*, NF, **16**, 54.

Beaty, C.B. 1970. Age and estimated rate of accumulation of an alluvial fan, White Mountains, California, USA, *American Journal of Science*, **268**, 50–70.

Beaty, C.B. 1974. Debris flow, alluvial fans and a revitalized catastrophism, *Zeitschrift für Geomorphologie*, Supplement, **21**, 39–51.

Bednarik, R.G. 1979. The potential of rock patination analysis in Australian archaeology – Part 1, *The Artefact*, **4**, 14–38.

Bender, M., Sowers, T., Dickson, M.-L., Orchardo, J., Grootes, P., Mayewskiy, P.A. and Meese, D.A. 1994. Climate correlations between Greenland and Antarctica during the past 100 000 years, *Nature*, **372**, 663–666.

Benson, L., Kashgarin, M. and Rubin, M. 1995. Carbonate deposition, Pyramid Lake subbasin, Nevada: 2. Lake levels and polar jet stream positions reconstructed from radiocarbon ages and elevations of carbonates (tufas) deposited in the Lahontan basin, *Palaeogeography, Palaeoclimatology, Palaeocology*, **117**, 1–30.

Bierman, P. and Gillespie, A. 1991. Range fires: a significant factor in exposure-age determination and geomorphic surface evolution, *Geology*, **19**, 641–644.

Bierman, P., Gillespie, A. and Caffee, M. 1995. Cosmogenic ages for earthquake recurrence intervals and debris flow fan deposition, Owens Valley, California, *Science*, **270**, 447–450.

Birchfield, E.G., Wang, H. and Rich, J. 1994. Century/millennium internal climate oscillations in an ocean–atmosphere–continental ice sheet model, *Journal of Geophysical Research*, **99** (C6), 12 459–12 470.

Birkeland, P.W. 1965. Reno to Mount Rose. Tahoe City, Truckee and return, in *INQUA VII Congress Guide Book for Field Conference I. Northern Great Basin and California*, edited by C. Warhaftig, R.B. Morrison and P.W. Birkeland, Nebraska Academy of Sciences, Lincoln, pp. 48–59.

Blair, T.C., Clark, J.S. and Wells, S.G. 1990. Quaternary continental stratigraphy, landscape evolution, and application to archeology: Jarilla piedmont and Tularosa graben floor, White Sands Missile Range, New Mexico, *Geological Society of America Bulletin*, **102**, 749–759.

Blikra, L.H. and Longva, O. 1995. Frost-shattered debris facies of Younger Dryas age in the coastal sedimentary successions in western Norway: palaeoenvironmental implications, *Palaeogeography, Palaeoclimatology, Palaeocology*, **118**, 89–110.

Blunier, T., Chappellaz, J., Schwander, J., Stauffer, B. and Raynaud, D. 1995. Variations in atmospheric methane concentration during the Holocene epoch, *Nature*, **374**, 46–49.

Bond, G., Broecker, W., Johnson, S., McManus, J., Labeyrie, L., Jouzel, J. and Bonani, G. 1993. Correlations between climate records from North Atlantic sediments and Greenland ice, *Nature*, **365**, 143–147.

Bond, G.C. and Lotti, R. 1995. Iceberg discharges into the North Atlantic on millennial time scales during the last glaciation, *Science*, **267**, 1005–1010.

Bowman, D. 1988. The declining but non-rejuvenating base level – Lisan Lake, the Dead Sea area, Israel, *Earth Surface Processes and Landforms*, **13**, 239–249.

Broecker, W.S. 1994. Massive iceberg discharges as triggers for global climate change, *Nature*, **372**, 421–424.

Brookes, I.A., Levine, L.D. and Dennell, R.W. 1982. Alluvial sequence in central west Iran and implications for archaeological survey, *Journal Field Archaeology*, **9**, 285–299.

Bryson, R.A. 1992. Simulating past and forecasting future climates, *Environmental Conservation*, **20**, 339–346.

Bull, W.B. 1975. Allometric change of landforms, *Geological Society of America Bulletin*, **86**, 1489–1498.

Bull, W.B. 1991. *Geomorphic Responses to Climatic Change*, Oxford University Press, Oxford, 326 pp.

Bull, W.B. and McFadden, L.D. 1977. Tectonic geomorphology north and south of the Garlock Fault, California, in *Geomorphology in Arid Regions*, edited by D.O. Doehring, Proceedings volume of the Eighth Annual Geomorphology Symposium held at the State University of New York, Binghamton, pp. 115–138.

Bull, W.B. and Schick, A.P. 1979. Impact of climatic change on an arid watershed: Nahel Yael, southern Israel, *Quaternary Research*, **11**, 153–171.

Butzer, K. 1980. Holocene alluvial sequences: problems of dating and correlation, in *Timescales in Geomorphology*, edited by R.A. Cullingford, J. Lewin and D.A. Davidson, Wiley, Chichester, pp. 333–354.

Catto, N.R. 1993. Morphology and development of an alluvial fan in a permafrost region, Aklavik Range, Canada, *Geografiska Annaler*, **75A**, 83–92.

Chappellaz, J., Blunier, T., Raynaud, D., Barnola, J., Schwander, J. and Stauffer, B. 1993. Synchronous changes in atmospheric CH4 and Greenland climate between 40 and 8 kyr BP, *Nature*, **366**, 443–445.

Church, M. and Ryder, J.M. 1972. Paraglacial sedimentation, a consideration of fluvial processes conditioned by glaciation, *Geological Society of America Bulletin*, **83**, 3059–3072.

Church, M. and Slaymaker, O. 1989. Disequilibrium of Holocene sediment yield in glaciated British Columbia, *Nature*, **337**, 452–454.

Clark, M.J. 1987. Geocryological inputs to the alpine sediment system, in *Glacio-fluvial Sediment Transfer*, edited by A.M. Gurnell and M.J. Clark, Wiley, New York, pp. 33–58.

Clark, P.U. and Bartlein, P.J. 1995. Correlation of the late Pleistocene glaciation in the western United States with North Atlantic Heinrich events, *Geology*, **23**, 483–486.

Clarke, A.O. 1989. Neotectonics and stream piracy on the Lytle Creek Alluvial Fan, Southern California, *California Geographer*, **29**, 21–42.

Clayton, K.M. 1983. Climate, climatic change and rates of denudation, in *Studies in Quaternary Geomorphology. Proceedings of the VIth British–Polish Seminar*, edited by D.J. Briggs and R.S. Waters, Institute of British Geographers, London, pp. 157–167.

Colman, S.M. and Pierce K. 1981. Weathering rinds on andesite and basaltic stones as a Quaternary age indicator, western United States, *US Geological Survey Professional Paper*, **1210**, 1–56.

Copi, I.M. 1982. *Introduction to Logic*, 6th edn, Macmillan, New York, 604 pp.

Cremaschi, M. 1996. The desert varnish in the Messak Settafet (Fezzon, Libyan Sahara), age, archaeological context, and paleo-environmental implication, *Geoarchaeology Journal*, **11**, in press.

Dardis, G., Beckedahl, H., Bowyer-Bower, T. and Harvey, P. 1988. Soil erosion forms in southern Africa, in *Geomorphological Studies in Southern Africa*, edited by G.F. Dardis and B. Moon, Balkema, Rotterdam, pp. 187–213.

Davis, W.M. 1905. The geographical cycle in arid climate, *Journal of Geology*, **13**, 381–407.

Denny, C.S. 1965. Alluvial fans in the Death Valley region of California and Nevada, *US Geological Survey Professional Paper*, **466**, 62 pp.

Denny, C.S. 1967. Fans and pediments, *American Journal of Science*, **265**, 81–105.

Denton, G.H. and Hendy, C.H. 1994. Younger dryas age advance of Franz Josef Glacier in the Southern Alps of New Zealand, *Science*, **264**, 1434–1437.

Dohrenwend, J.C. 1984. Nivation landforms in the western Great Basin and their paleoclimatic significance, *Quaternary Research*, **22**, 275–288.

Dorn, R.I. 1983. Cation-ratio dating: a new rock varnish age determination technique, *Quaternary Research*, **20**, 49–73.

Dorn, R.I. 1988. A rock varnish interpretation of alluvial-fan development in Death Valley, California, *National Geographic Research*, **4**, 56–73.

Dorn, R.I. 1994. Alluvial fans an an indicator of climatic change, in *Geomorphology of Desert Environments*, edited by A.D. Abrahams and A.J. Parsons, Chapman & Hall, London, pp. 593–615.

Dorn, R.I. 1996. Radiocarbon dating glacial moraines with the aeolian biome: test results at Bishop Creek, Sierra Nevada, California, *Physical Geography*, **17**, in press.

Dorn, R.I., Clarkson, P.B., Nobbs, M.F., Loendorf, L.L. and Whitley, D.S. 1992a. New approach to the radiocarbon dating of rock varnish, with examples from drylands, *Annals of the Association of American Geographers*, **82**, 136–151.

Dorn, R.I., DeNiro, M.J. and Ajie, H.O. 1987. Isotopic evidence for climatic influence on alluvial-fan development in Death Valley, California, *Geology*, **15**, 108–110.

Dorn, R.I., Jull, A.J.T., Donahue, D.J., Linick, T.W., Toolin, L.J., Moore, R.B., Rubin, M., Gill, T.E. and Cahill, T.A. 1992b. Rock varnish on Hualalai and Mauna Kea Volcanoes, Hawaii, *Pacific Science*, **46**, 11–34.

Dorn, R.I. and Oberlander, T.M. 1982. Rock varnish, *Progress in Physical Geography*, **6**, 317–367.

Dorn, R.I. and Phillips, F.M. 1991. Surface exposure dating: review and critical evaluation, *Physical Geography*, **12**, 303–333.

Dorn, R.I., Phillips, F.M., Zreda, M.G., Wolfe, E.W., Jull, A.J.T., Kubik, P.W. and Sharma, P. 1991. Glacial chronology of Mauna Kea, Hawaii, as constrained by surface-exposure dating, *National Geographic Research and Exploration*, **7**, 456–471.

Dowdeswell, J.A. 1995. Iceberg production, debris rafting, and the event and thickness of Heinrich layers (H-1, H-2) in North Atlantic sediments, *Geology*, **23**, 301–304.

Dragovich, D. 1984. Desert varnish as an age indicator for Aboriginal rock engravings: a review of problems and prospects, *Archaeology in Oceania*, **19** (2), 48–56.

Drummond, C.N., Patterson, W.P. and Walker, J.C.G. 1995. Climatic forcing of carbon–oxygen isotopic covariance in temperature-region marl lakes, *Geology*, **23**, 1031–1034.

Eckis, R. 1928. Alluvial fans in the Cucamonga district, southern California, *Journal of Geology*, **36**, 111–141.

Edwards, R.L., Chen, J.H. and Wasserburg, G.J. 1986. U-238, U-234, Th-230, Th-232 systematics and precise measurement of time over the past 500 000 years, *Earth and Planetary Science Letters*, **81**, 175–192.

Eybergen, F.A. and Imeson, A. 1989. Geomorphological processes and climate change, *Catena*, **16**, 307–319.

Farr, J. 1983. Popper's hermeneutics, *Philosophy of the Social Sciences*, **13**, 157–176.

Field, J.J. 1994. Surficial processes on two fluvially dominated alluvial fans in Arizona, *Arizona Geological Survey Open-File Report*, **94-12**, 1–31.

Friedman, E., Goren-Inbar, N., Rosenfeld, A., Marder, O. and Burian, F. 1994. Hafting during Mousterian times – further indication, *Journal of the Israel Prehistoric Society*, **26**, 8–31.

Fronval, T., Jansen, E., Bloemendal, J. and Johnsen, S. 1995. Oceanic evidence for coherent fluctuations in Fennoscandian and Laurentide ice sheets on millennium timescales, *Nature*, **374**, 443–446.

Gates, W.L. and Mintz, Y. 1975. *Understanding Climatic Change. A Program for Action*, National Academy of Sciences, Washington DC.

Germanoskiy, D. and Miller, J.R. 1995. Geomorphic response to wildfire in an arid watershed, Crow Canyon, Nevada, *Physical Geography*, **16**, 243–256.

Gerson, R. 1982. Talus relics in deserts: a key to major climatic fluctuations, *Israel Journal of Earth Sciences*, **31**, 123–132.

Gilbert, G.K. 1890. Lake Bonneville, *US Geological Survey Monograph*, **1**, 1–438.

Gile, L.H., Hawley, J.W. and Grossman, R.B. 1981. Soils and geomorphology in the Basin and Range area of southern New Mexico. Guidebook to the Desert Project, *New Mexico Bureau of Mines and Mineral Resources Memoir*, **39**, 1–222.

Gillespie, R. 1991. Charcoal dating – oxidation is necessary for complete humic removal, *Radiocarbon*, **33** (2), 199.

Goldberg, P. 1984. Late Quaternary history of Quadesh Barnea, northeastern Sinai, *Zeitschrift für Geomorphologie*, NF, **28**, 193–217.

Gosse, J.C., Evenson, E.B., Klein, J., Lawn, B. and Middleton, R. 1995. Precise cosmogenic ^{10}Be measurements in western North America: support for a global Younger Dryas cooling event, *Geology*, **23**, 877–880.

Graf, W.L. 1988. *Fluvial Processes in Dryland Rivers*, Springer-Verlag, Berlin, 346 pp.

Grimm, E.C., Jacobson, G.L., Watts, W., Hansen, B. and Maasch, K. 1993. A 50 000-year record of climate oscillations from Florida and its temporal correlation with the Heinrich events, *Science*, **261**, 198–200.

Grossman, S. and Gerson, R. 1987. Fluviatile deposits and morphology of alluvial surfaces as indicators of Quaternary environmental changes in southern Negev, Israel, in *Desert Sediments: Ancient and Modern. Geological Society Special Publications No. 37*, edited by L. Frostick and I. Reid, Geological Society, London, pp. 17–29.

Grote, G. and Krumbein, W.E. 1992. Microbial precipitation of manganese by bacteria and fungi from desert rock and rock varnish, *Geomicrobiology*, **10**, 49–57.

Gu, B., Schmitt, J., Chen, Z., Liang, L. and McCarthy, J. 1995. Adsorption and desorption of different organic matter fractions in iron oxide, *Geochimica et Cosmochimica Acta*, **59**, 219–220.

Haberland, W. 1975. Untersuchungen an Krusten, Wustenlacken und Polituren auf Gesteinsoberflachen der nordlichen und mittlerent Saharan (Libyen und Tchad), *Berlinger Geographische Abhandlungen*, **21**, 1–77.

Harden, J.W. 1982. A quantitative index of soil development from field descriptions: example from a chronosequence in central California, *Geoderma*, **18**, 1–28.

Harvey, A.M. 1990. Factors influencing Quaternary alluvial fan development in southeast Spain, in *Alluvial Fans: A Field Approach*, edited by A.H. Rachocki and M. Church, Wiley, New York, pp. 247–269.

Harvey, A.M. and Wells, S.G. 1994. Late Pleistocene and Holocene changes in hillslope sediment supply to alluvial fan systems: Zzyzx, California, in *Environmental Change in Drylands: Biogeographical and Geomorphological Perspectives*, edited by A.C. Millington and K. Pye, Wiley, London, pp. 67–84.

Hawley, J.W. and Wilson, W.E. 1965. Quaternary geology of the Winnemucca area, Nevada, *Desert Research Institute Technical Report*, **5**.

Hedges, J.I., Keil, R.G. and Cowie, G.L. 1993. Sedimentary diagenesis: organic perspectives with inorganic overlays, *Chemical Geology*, **107**, 487–492.

Heinrich, H. 1988. Origin and consequences of cyclic ice rafting in the Northeast Atlantic Ocean during the past 130 000 years, *Quaternary Research*, **29**, 143–152.

Hempel, C.G. 1966. *Philosophy of Natural Science*, Prentice-Hall, Englewood Cliffs, NJ, 116 pp.

Heron, C., Evershed, R.P. and Goad, L.J. 1991. Effects of migration of soil lipids on organic residues associated with buried potsherds, *Journal of Archaeological Science*, **18**, 641–659.

Hooke, R.L. 1968. Steady-state relationships on arid-region alluvial fans in closed basins, *American Journal of Science*, **266**, 609–629.

Hooke, R.L. 1972. Geomorphic evidence for late Wisconsin and Holocene tectonic deformation, Death Valley, California, *Geological Society of America Bulletin*, **83**, 2073–2098.

Hooke, R.L. 1987. Mass movement in semi-arid environments and the morphology of alluvial fans, in *Slope Stability*, edited by M.G. Anderson and K.S. Richards, Wiley, New York, pp. 505–529.

Hooke, R.L. and Dorn, R.I. 1992. Segmentation of alluvial fans in Death Valley, California: new insights from surface-exposure dating and laboratory modelling, *Earth Surface Processes and Landforms*, **17**, 557–574.

Hooke, R.L. and Rohrer, W.L. 1979. Geometry of alluvial fans: effect of discharge and sediment size, *Earth Surface Processes*, **4**, 147–166.

Huckleberry, G. 1996. Geomorphology and surficial geology of Garden Canyon, Huachulca Mountains, Arizona, *Arizona Geological Survey Open-File Report* **96–5**, 1–19.

Humphrey, N.F. and Heller, P.L. 1995. Natural oscillations in coupled geomorphic systems: an alternative origin for cyclic sedimentation, *Geology*, **23**, 499–502.

Hunt, C.B. and Mabey, D.R. 1966. Stratigraphy and structure, Death Valley, California, *US Geological Survey Professional Paper*, **494A**.

Huntington, E. 1907. Some characteristics of the glacial period in non-glaciated regions, *Geological Society of America Bulletin*, **18**, 351–388.

Imbrie, J., Boyle, E., Clemens, S., Duffy, A., Howard, W.K., Kukla, G., Kutzbach, J., Martinson, D., McIntyre, A., Mix, A.C., Molfino, B., Morley, J., Peterson, L., Pisias, N., Prell, W., Raymo, M., Shackleton, N. and Toggweiler, J. 1993. On the structure and origin of major glacial cycles: 1. Linear responses to Milankovitch forcing, *Paleooceanography*, **7**, 701–738.

Imbrie, J., Hays, J.D., Martinson, D.G., McIntyre, A., Mix, A.C., Morley, J., Pisias, N., Proll, W. and Shackleton, N.J. 1984. The orbital theory of Pleistocene climate: support from a revised chronology of the marine O-18 record, in *Milankovitch and Climate*, edited by A. Berger, Reidel, Dordrecht, pp. 269–305.

Iriondo, M. 1993. Geomorphology and late Quaternary of the Chaco (South America), *Geomorphology*, **7**, 289–303.

Islebe, G.A., Hooghiemstra, H. and van der Borg, K. 1995. A cooling event during the Younger Dryas Chron in Costa Rica, *Palaeogeography, Palaeoclimatology, Palaeoecology*, **117**, 73–80.

Jansson, P., Jacobson, D. and Hooke, R.L. 1993. Fan and playa areas in southern California and adjacent parts of Nevada, *Earth Surface Processes and Landforms*, **18**, 108–119.

Jones, C.E. 1991. Characteristics and origin of rock varnish from the hyperarid coastal deserts of northern Peru, *Quaternary Research*, **35**, 116–129.

Kale, V.S. and Rajaguru, S.N. 1987. Late Quaternary alluvial history of the northwestern Deccan upland region, *Nature*, **325**, 612–614.

Keigwin, L.D., Curry, W.B., Lehman, S.J. and Johnson, S. 1994. The role of the deep ocean in North Atlantic climate change between 70 and 130 kyr ago, *Nature*, **371**, 323–326.

Keigwin, L.D. and Jones, G.A. 1994. Western North Atlantic evidence for millennial-scale changes in ocean circulation and climate, *Journal of Geophysical Research*, **99** (C6), 12 397–12 410.

Kesel, R.H. and Lowe, D.R. 1987. Geomorphology and sedimentology of the Toro Amarillo alluvial fan in a humid tropical environment, Costa Rica, *Geografiska Annaler*, **69A**, 85–99.

Knox, J.C. 1983. Responses of river systems to Holocene climates, in *Late Quaternary Environments of the United States*, Vol. 2, *The Holocene*, edited by H.E. Wright, Jr, University Minnesota Press, Minneapolis, pp. 26–41.

Kochel, R.C. and Ritter, D.F. 1990. Complex geomorphic response to minor climatic changes, San Diego County, CA, *Hydraulics/Hydrology of Arid Lands. Proceedings of International ASCE Symposium, San Diego, July 30–August 2*, pp. 148–153.

Kotilainen, A.T. and Shackleton, N.J. 1995. Rapid climate variability in the North Pacific Ocean during the past 95 000 years, *Nature*, **377**, 323–326.

Ku, T.L., Luo, S., Lowenstein, T.K., Li, J. and Spencer, R.J. 1994. U-series chronology for lacustrine deposits of Death Valley, California: implications for late Pleistocene climate changes, *Geological Society of America Abstracts with Programs*, **26** (7), A-169.

Kudrass, H.R., Erlenkeuser, H., Vollbrecht, R. and Weiss, W. 1991. Global nature of the Younger Dryas cooling event inferred from oxygen isotope data from Sulu Sea cores, *Nature*, **349**, 406–409.

Langbein, W.B. and Schumm, S.A. 1958. Yield of sediment in relation to mean annual precipitation, *Transactions American Geophysical Union*, **39**, 1076–1084.

Lecce, S.A. 1991. Influence of lithologic erodibility on alluvial fan area, western White Mountains, California and Nevada, *Earth Surface Processes and Landforms*, **16**, 11–18.

Linck, G. 1928. Über Schutzrinden, *Chemie die Erde*, **4**, 67–79.

Linick, T.W., Damon, P.E., Donahue, D.J. and Jull, A.J.T. 1989. Accelerator mass spectrometry: the new revolution in radiocarbon dating, *Quaternary International*, **1**, 1–6.

Liu, B., Phillips, F.M., Elmore, D. and Sharma, P. 1994. Depth dependence of soil carbonate accumulation based on cosmogenic ^{36}Cl dating, *Geology*, **22**, 1071–1074.

Liu, B., Phillips, F.M., Pohl, M. and Sharma, P. 1996. An alluvial surface chronology based on cosmogenic ^{36}Cl dating, Ajo Mountains, southern Arizona, *Quaternary Research*, **45**, 30–37.

Liu, T. 1994. Visual microlaminations in rock varnish: a new paleoenvironmental and geomorphic tool in drylands, Ph.D. thesis, Department of Geography, Arizona State University, Tempe, 173 pp.

Liu, T. and Dorn, R.I. 1996. Understanding spatial variability in environmental changes in drylands with rock varnish microlaminations, *Annals of the Association of American Geographers*, **86**, 187–212.

Lowell, T.V., Heusser, C.J., Andersen, B., Moreno, P., Hauser, A., Heusser, L., Schlüchter, C., Marchant, D. and Denton, G.H. 1995. Interhemispheric correlation of late Pleistocene glacial events, *Science*, **269**, 1541–1549.

Lowenstein, T.K., Spencer, R.J., Roberts, S.M., Yang, W., Ku, T.-L. and Forester, R.M. 1995. Death Valley salt cores: 200 000 year paleoclimate record from mineralogy, fluid inclusions, sedimentary structures and ostracodes, *Geological Society of America Abstracts with Program*, **27** (6), 321.

Lucas, A. 1905. *The Blackened Rocks of the Nile Cataracts and of the Egyptian Deserts*, National Printing Department, Cairo, 58 pp.

Lustig, L. K. 1965. Clastic sedimentation in Deep Springs Valley, California, *US Geological Survey Professional Paper*, **352-F**, 131–190.

Mabbutt, J.A. 1979. Pavements and patterned ground in the Australian stony deserts, *Stuttgarter Geographische Studien*, **93**, 107–123.

MacArthur, R.C., Harvey, M.D. and Sing, E.F. 1990. Estimating sediment delivery and yield on alluvial fans, *Hydraulics/Hydrology of Arid Lands. Proceedings of International ASCE Symposium, San Diego, July 30–August 2*, pp. 700–705.

McFadden, L.D., Ritter, J.B. and Wells, S.G. 1989. Use of multiparameter relative-age methods for age estimation and correlation of alluvial fan surfaces on a desert piedmont, eastern Mojave Desert, *Quaternary Research*, **32**, 276–290.

Macklin, M.G., Rumsby, B.T. and Heap, T. 1992. Flood alluviation and entrenchment: Holocene valley-floor development and transformation in the British uplands, *Geological Society of America Bulletin*, **104**, 631–643.

Maizels, J. 1990. Raised channel systems as indicators of palaeohydrologic change: a case study from Oman, *Palaeogeography, Palaeoclimatology, Palaeoecology*, **76**, 241–277.

Martinson, D.G., Pisias, N.G., Hays, J.D., Imbrie, J., Moore, T.C.J. and Shackleton, N.J. 1987. Age dating and the orbital theory of the ice ages: development of a high-resolution 0 to 300 000 year chronostratigraphy, *Quaternary Research*, **27**, 1–29.

Maslin, M.A., Shackleton, N.J. and Plaumann, U. 1995. Surface temperature, salinity, and density changes in the northeast Atlantic during the last 45 000 years: Heinrich events, deep water formation, and climatic rebounds. *Palaeoceanography*, **10**, 527–544.

Mayewskiy, P., Meeker, L., Whitlow, S., Twickler, M., Morrison, M., Bloomfield, P., Bond, G., Alley, R., Gow, A., Grootes, P., Meese, D., Ram, M., Taylor, K. and Wumkes, W. 1994. Changes in atmospheric circulation and ocean ice cover over the North Atlantic during the last 41 000 years, *Science*, **263**, 1747–1751.

Meese, D.A., Gow, A.J., Grootes, P., Mayewskiy, P.A., Ram, M., Stuiver, M., Taylor, K.C., Waddington, E.D. and Zielinski, G.A. 1994. The accumulation record from the GISP2 core as an indicator of climate change throughout the Holocene, *Science*, **266**, 1680–1682.

Meigs, P. 1953. World distribution of arid and semi-arid homoclimates, *UNESCO Arid Zone Research Series*, **1**, 203–209.

Melton, M.A. 1965. The geomorphic and paleoclimatic significance of alluvial deposits in southern Arizona, *Journal of Geology*, **73**, 1–38.

Meyer, G., Wells, S.G., Balling, R.C.J. and Jull, A.J.T. 1992. Response of alluvial systems to fire and climatic change in Yellowstone National Park, *Nature*, **357**, 147–150.

Meyer, G.A., Wells, S.G. and Jull, A.J.T. 1995. Fire and alluvial chronology in Yellowstone National Park: climatic and intrinsic controls on Holocene geomorphic processes, *Geological Society of America Bulletin*, **107**, 1211–1230.

Mills, H.H. 1982. Piedmont-cove deposits of the Dellwood quadrangle, Great Smoky Mountains, North Carolina, USA: morphometry, *Zeitschrift für Geomorphologie*, **26**, 163–178.

Montgomery, D.R. and Dietrich, W.E. 1992. Channel initiation and the problem of landscape scale, *Science*, **255**, 826–830.

Mulhern, M.E. 1982. Lacustrine, fluvial and fan sedimentation: a record of Quaternary climatic change and tectonism, Pine Valley, Nevada, *Zeitschrift für Geomorphologie*, Supplementband, **42**, 117–133.

Newton-Smith, W. 1981. *The Rationality of Science*, Routledge and Kegan Paul, London, 287 pp.

Nishiizumi, K., Kohl, C., Arnold, J., Dorn, R., Klein, J., Fink, D., Middleton, R. and Lal, D. 1993. Role of in situ cosmogenic nuclides ^{10}Be and ^{26}Al in the study of diverse geomorphic processes, *Earth Surface Processes and Landforms*, **18**, 407–425.

Österberg, R., Lindqvist, I. and Mortensen, K. 1993. Particle size of humic acid, *Soil Science Society of America Journal*, **57**, 283–285.

Pedro, G. and Sieffermann, G. 1979. Weathering of rocks and formation of soils, in *Review of Research on Modern Problems in Geochemistry*, edited by F.R. Siegel, UNESCO, Paris, pp. 39–55.

Peterson, F.F., Bell, J.W., Dorn, R.I., Ramelli, A.R. and Ku, T.L. 1995. Late Quaternary geomorphology and soils in Crater Flat, Yucca Mountain area, southern Nevada, *Geological Society of America Bulletin*, **107**, 379–395.

Péwé, T.L. 1983. The periglacial environment in North America during Wisconsin time, in *Late Quaternary Environment of the United States*, Vol. 1, *The Late Pleistocene*, edited by S.C. Porter, University of Minnesota Press, Minneapolis, pp. 157–189.

Phillips, F.M., Campbell, A.R., Smith, G.I. and Bischoff, J.L. 1994. Interstadial climatic cycles: a link between western North America and Greenland? *Geology*, **22**, 1115–1118.

Phillips, F.M., Zreda, M.G., Smith, S.S., Elmore, D., Kubik, P.W. and Sharma, P. 1990. A cosmogenic chlorine-36 chronology for glacial deposits at Bloody Canyon, eastern Sierra Nevada, California, *Science*, **248**, 1529–1532.

Pohl, M. 1995. First radiocarbon ages on organics from piedmont alluvium, Ajo Mountains, Arizona, *Physical Geography*, **16**, 339–353.

Ponti, D.J. 1985. The Quaternary alluvial sequence of the Antelope Valley, California, *Geological Society America Special Paper*, **203**, 79–96.

Popper, K. 1966. *The Open Society and Its Enemies*, Vol. II, Princeton University Press, Princeton, NJ, 367 pp.

Porter, S.C. and Zhisheng, A. 1995. Correlation between climate events in the North Atlantic and China during the last glaciation, *Nature*, **375**, 305–308.

Quade, J. 1986. Late Quaternary environmental changes in the upper Las Vegas Valley, Nevada, *Quaternary Research*, **26**, 340–357.

Quade, J. 1994. Black mats and Younger Dryas (?) recharge of valley aquifers near Yucca Mountain, southern Nevada, *Geological Society of America Abstracts with Programs*, **26** (7), A-62.

Quade, J. and Pratt, W.L. 1989. Late Wisconsin groundwater discharge environments of the southwestern Indian Springs Valley, southern Nevada, *Quaternary Research*, **31**, 351–370.

Reheis, M.C., Harden, J.W., McFadden, L.D. and Shroba, R.R. 1989. Development rates of late Quaternary soils, Silver Lake Playa, California, *Soil Science Society of America Journal*, **53**, 1127–1140.

Ritter, J.B., Miller, J.R., Enzel, Y. and Wells, S.G. 1995. Reconciling the roles of tectonism and climate in Quaternary alluvial fan evolution, *Geology*, **23**, 245–248.

Ritz, J.F., Brown, E.T., Boules, D.L., Philip, H., Schlupp, A., Raisbeck, G.M., Yiou, F. and Enkhtuvshin, B. 1995. Slip rates along active faults estimated with cosmic-ray exposure dates: application to the Bogd fault, Gobi-Altaï, Mongolia, *Geology*, **23**, 1019–1022.

Rivard, B., Arvidson, R.E., Duncan, I.J., Sultan, M. and Kaliouby, B. 1992. Varnish sediment, and rock controls on spectral reflectance of outcrops in arid regions, *Geology*, **20**, 295–298.

Roberts, N., Lamb, H.F., El-Hamouti, N. and Barker, P. 1994. Abrupt Holocene hydroclimatic events: palaeolimnological evidence from north-west Africa, in *Environmental Change in Drylands: Biogeographical and Geomorphological Perspectives*, edited by A.C. Millington and K. Pye, Wiley, London, pp. 163–175.

Rockwell, T.K., Keller, E.A. and Johnson, D.L. 1984. Tectonic geomorphology of alluvial fans and mountain fronts near Ventura, California, in *Tectonic Geomorphology*, edited by M. Morisawa and J.T. Hack, Proceedings of the 15th Annual Binghamton Geomorphology Symposium, Binghamton, pp. 183–207.

Russel, I.C. 1885. Geologic history of Lake Lahontan, a Quaternary lake of northwestern Nevada, *US Geological Survey Monograph*, **11**, 1–287.

Ryder, J.M. 1971. Some aspects of the morphology of paraglacial alluvial fans in south-central British Columbia, *Canadian Journal of Earth Sciences*, **8**, 1252–1264.

Schick, A.P. 1974. Formation and obliteration of desert stream terraces – a conceptual analysis, *Zeitschrift für Geomorphologie*, Supplementband, **21**, 88–105.

Schumm, S.A., Mosley, M.P. and Weaver, W.E. 1987. *Experimental Fluvial Geomorphology*, Wiley, New York, 413 pp.

Scuderi, L.A. 1994. Solar influences on Holocene treelines altitude variability in the Sierra Nevada, *Physical Geography*, **15**, 146–165.

Sirocko, F, Garbe, O., McIntyre, A. and Molfino, B. 1996. Teleconnections between the subtropical monsoons and high-latitude climates during the last deglaciation, *Science*, **272**, 526–529.

Slate, J.L. 1991. Quaternary stratigraphy, geomorphology, and geochronology of alluvial fans, Fish Lake Valley, Nevada and California, in *Guidebook for Field Trip to Fish Lake Valley, California–Nevada*, edited by M.C. Reheis, Friends of the Pleistocene, Pacific Cell, pp. 94–113.

Smith, G.I. and Bischoff, J.L. 1993. Core OL-92 from Owens Lake, southeast California, *US Geological Survey Open File Report*, **93-683**, 1–398.

Spaulding, W.G. 1990. Vegetation and climatic development of the Mojave Desert: the last glacial maximum to the present, in *Packrat Middens. The Last 40 000 Years of Biotic Change*, edited by J.L. Betancourt, T.R. Van Devender and P.S. Martin, University of Arizona Press, Tucson, pp. 166–169.

Stadelman, S. 1994. Genesis and post-formational systematics of carbonate accumulations in Quaternary soils of the Southwestern United States, Ph.D. thesis, Department of Agronomy, Texas Tech University, Lubbock, 124 pp.

Swadley, W.C. and Hoover, D.L. 1989. Geologic map of the surficial deposits of the Topopah Spring Quadrangle, Nye County, Nevada, *US Geological Survey Miscellaneous Investigation Series Map*, **I-2018**.

Switzer, P., Harden, J.W. and Mark, R.K. 1988. A statistical method for estimating rates of soil development and ages of geological deposits: a design for soil chronosequence studies, *Mathematical Geology*, **20**, 49–61.

Szabo, B.J., Kolesar, P., Riggs, A., Winograd, I. and Ludwig, K.R. 1994. Paleoclimatic inferences from a 120 000-yr calcite record of water-table fluctuation in Browns Room of Devils Hole, Nevada, *Quaternary Research*, **41**, 59–69.

Talbot, M.R. and Williams, M.A.J. 1979. Cyclic alluvial fan sedimentation on the flanks of fixed dunes, Janjari, central Niger, *Catena*, **6**, 43–62.

Thomas, M.F. and Thorp, M.B. 1995. Geomorphic response to rapid climatic and hydrologic change during the late Pleistocene and early Holocene in the humid and sub-humid tropics, *Quaternary Science Reviews*, **14**, 193–207.

Thouveny, N., de-Beaulieu, J.-L., Bonifay, E., Cree, K., Guiot, J., Icole, M., Johnsen, S., Jouzel, J., Reille, M., Williams, T. and Williamson, D. 1994. Climate variations in Europe over the past 140 kyr deduced from rock magnetism, *Nature*, **371**, 503–506.

Throckmorton, C.K. and Reheis, M.C. 1993. Late Pleistocene and Holocene environmental changes in Fish Lake Valley, Nevada–California: geomorphic response of alluvial fans to climate change, *US Geological Survey Open File Report*, **93-620**, 1–82.

Thunell, R.C. and Mortyn, P.G. 1995. Glacial climate instability in the northeast Pacific Ocean, *Nature*, **376**, 504–506.

Trowbridge, A.C. 1911. The terrestrial deposits of Owens Valley, California, *Journal of Geology*, **19**, 706–747.

Tuan, Y.-F. 1962. Structure, climate and basin landforms in Arizona and New Mexico, *Annals Association American Geographers*, **52**, 51–68.

Viereck, A. 1964. An examination of patina on stone tools with regard to possible dating, *South African Archaeological Bulletin*, **19**, 276–279.

Viseras, C. and Fernández, J. 1995. The role of erosion and deposition in the construction of alluvial fan sequences in the Guadix Formation (SE Spain), *Geologie en Mijnbouw*, **74**, 21–33.

Walling, D.A. and Kleo, A.H.A. 1979. Sediment yields of rivers in areas of low precipitation – a global view, *International Association of Scientific Hydrologists Publication*, **128**, 479–492.

Wasson, R.J. 1977. Late-glacial alluvial fan sedimentation in the Lower Derwent Valley, Tasmania, *Sedimentology*, **24**, 781–799.

Weaver, W.E. 1984. Geomorphic thresholds and the evolution of alluvial fans, *Geological Society America Abstracts with Program*, **16**, 688.

Weisse, R., Mikolajewicz, U. and Maier-Reimer, E. 1994. Decadal variability of the North Atlantic in an ocean general circulation model, *Journal of Geophysical Research*, **99** (C6), 12 411–12 421.

Wells, P.V. and Woodcock, D. 1985. Full-glacial vegetation of Death Valley, California. Juniper woodland opening to Yucca semidesert, *Madrono*, **32**, 11–23.

Wells, S.G., McFadden, L.D. and Dohrenwend, J.C. 1987. Influence of late Quaternary climatic changes on geomorphic and pedogenic processes on a desert piedmont, eastern Mojave Desert, California, *Quaternary Research*, **27**, 130–146.

Wells, S.G., McFadden, L.D. and Harden, J. 1990. Preliminary results of age estimations and regional correlations of Quaternary alluvial fans within the Mojave Desert of Southern California, in *At the End of the Mojave: Quaternary Studies in the Eastern Mojave Desert*, edited by R.E. Reynolds, S.G. Wells and R.J.I. Brady, San Bernardino County Museum Association, Redlands, pp. 45–53.

White, K. 1991. Geomorphological analysis of piedmont landforms in the Tunisian Southern Atlas using ground data and satellite imagery, *The Geographical Journal*, **157**, 279–294.

White, K. 1993. Image processing of thematic mapper data for discriminating piedmont surficial materials in the Tunisian Southern Atlas, *International Journal of Remote Sensing*, **14**, 961–977.

Whitley, D.S., Baird, J., Bennett, J. and Tuck, R.G. 1984. The use of relative repatination in the chronological ordering of petroglyph assemblages, *Journal of New World Archaeology*, **6**, 19–25.

Wilcox, B.P., Reid, K., Pitlick, J. and Allen, C. 1995. Scale relationships for runoff and sediment from semiarid woodland hillslopes, New Mexico, *EOS*, **76**, 242.

Williams, G.E. 1973. Late Quaternary piedmont sedimentation, soil formation and paleoclimates in arid south Australia, *Zeitschrift für Geomorphologie*, NF, **17**, 102–125.

Williams, G.P. and Guy, H.P. 1973. Erosional and depositional aspects of Hurricane Camille in Virginia, 1969, *US Geological Survey Professional Paper*, **804**, 47–54.

Wilson, M.C. 1990. Dating late Quaternary humid paleoclimatic episodes in the Sahel of West Africa from paleosol carbonate nodules, *Programme and Abstracts, First Joint Meeting, AMQUA and CANQUA, Waterloo, Canada*, 34 pp.

Winograd, I.J., Coplen, T.B., Landwehr, J.M., Riggs, A.C., Ludwig, K.R., Szabo, B.J., Kolesar, P.T. and Revesz, K.R. 1992. Continuous 500 000-year climate record from vein calcite in Devils Hole, Nevada, *Science*, **258**, 255–260.

Wright, H.E. Jr. 1989. The amphi-Atlantic distribution of the Younger Dryas paleoclimatic oscillation, *Quaternary Science Reviews*, **8**, 295–306.

Wright, V.P. 1992. Paleopedology: stratigraphic relationships and empirical models, in *Weathering, Soils, and Paleosols*, edited by I.P. Martini and W. Chesworth, Elsevier, Amsterdam, pp. 475–499.

Yair, A. 1994. The ambiguous impact of climate change at a desert fringe: northern Negev, Israel, in *Environmental Change in Drylands: Biogeographical and Geomorphological Perspectives*, edited by A.C. Millington and K. Pye, Wiley, London, pp. 199–227.

Zeuner, F.E. 1959. *The Pleistocene Period; Its Climate, Chronology, and Faunal Successions*, Hutchinson Scientific and Technical, London, 447 pp.

Zreda, M.G. and Phillips, F.M. 1994. Surface exposure dating by cosmogenic chlorine-36 accumulation, in *Dating in Exposed and Surface Contexts*, edited by C. Beck, University of New Mexico Press, Albuquerque, pp. 161–183.

Zreda, M.G., Phillips, F.M. and Elmore, D. 1994. Cosmogenic ^{36}Cl accumulation in unstable landforms. 2. Simulations and measurements on eroding moraines, *Water Resources Research*, **30**, 3127–3136.

9 Physical Modelling in Fluvial Geomorphology: Principles, Applications and Unresolved Issues

Jeff Peakall

Department of Earth Sciences and School of Geography, University of Leeds, UK

Phil Ashworth

School of Geography, University of Leeds, UK

Jim Best

Department of Earth Sciences, University of Leeds, UK

ABSTRACT

The relationships between fluvial process and form are often extremely difficult to quantify using conventional field and numerical computational techniques. Physical modelling offers a complementary technique to these methods and may be used to simulate complex processes and feedbacks in many geomorphic phenomena. Depending on the temporal/spatial scale of a particular research problem, physical models may be either 1:1 replicas of the field prototype, scale with Froude number only, have distorted scales or serve as unscaled experimental analogues that attempt to reproduce some properties of a prototype. This chapter presents a critique of the underlying principles that determine the degree to which physical models accurately replicate the form and dynamics of natural alluvial systems. Examples are presented of each modelling technique to illustrate both the advantages and inherent limitations of these different approaches and highlight the contribution of physical modelling in the study of fluvial geomorphology and sedimentology.

Three issues are identified for achieving significant progress in the scale modelling of fluvial systems: (i) incorporation into models of variables such as multiple time scales, flood hydrographs, fine-grained sediment, cohesion, surface tension and floodplain

The Scientific Nature of Geomorphology: Proceedings of the 27th Binghamton Symposium in Geomorphology held 27–29 September 1996. Edited by Bruce L. Rhoads and Colin E. Thorn. © 1996 John Wiley & Sons Ltd.

vegetation which will increase the degree of model realism; (ii) continued development and implementation of a range of measurement techniques; and (iii) detailed model:prototype verification across a range of scales. Whilst these steps will increase significantly the power and attractiveness of scale modelling in the earth sciences, simple analogue models will continue to enable testing of new concepts across the full range of spatial and temporal scales.

INTRODUCTION

Many problems in fluvial geomorphology involve complex, multivariate situations, often at large spatial and temporal scales (see Kirkby, Chapter 10 this volume). These topics have traditionally been addressed through detailed fieldwork combined with theoretical and numerical modelling. Whilst mathematical models have promoted major advances in our understanding of the complex interrelationships involved in sediment production, transfer and deposition in dynamic fluvial environments (cf. Pickup 1988; Ikeda and Parker 1989; Kirkby 1994), they necessarily involve simplifications and use of empirical coefficients derived from limited input data. A complementary technique that has developed in parallel with these computational simulations is physical modelling, which has two principal advantages. First, the formative processes can be observed, usually in a reduced time-frame, within a controlled and manageable laboratory environment. Second, physical models may allow incorporation of variables which are not known *a priori* and which may have markedly non-linear effects on the resultant dynamics or morphology. However, these advantages are counterbalanced by prototype to model scaling difficulties which result in increasing simplification and abstraction from reality as spatial and temporal scales increase. Additionally, it is clearly important to establish and quantify the influences of processes which may be non-linear in their scaling between model and prototype (e.g. particle settling velocity, see p. 233) and their consequent effect on morphology.

Physical modelling techniques can be classified both by their specificity (degree to which the model replicates a prototype) and the temporal/spatial scale at which they are most applicable (see Figure 9.1). For the smallest spatial and temporal scales, a 1:1 replica of flow and sediment dynamics can be re-created in the laboratory with little or no difference from the natural prototype. These models have, for example, been instrumental in investigating the morphology and controlling variables of bedform generation both in sands and gravels (e.g. Guy et al. 1966; Allen 1982; Southard and Boguchwal 1990a). However, even in these 1:1 models, care must be taken in considering temperature/viscosity influences (Southard and Boguchwal 1990b), applying such flume results to much deeper natural flows (Williams 1970; Southard 1971), and accounting for the influence of sidewalls on the experimental results (Crickmore 1970; Williams 1970). At large spatio-temporal scales, prototypes must be scaled down both to compress the time scale and allow the model to be accommodated within the constraints of available laboratory space. For the largest prototypes, a true scaled modelling approach becomes untenable and purely 'analogue' models must be employed. However, when viewed in the light of imposed modelling constraints, these large-scale models (e.g. studies of base level controls on fluvial incision) can provide invaluable insights into the behaviour of complex natural phenomena.

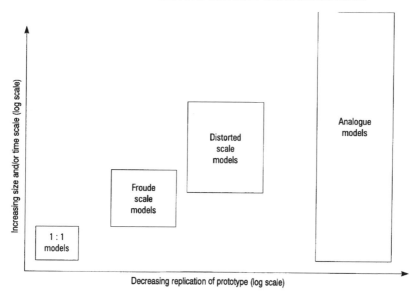

Figure 9.1   Schematic view of the balance between model specificity and spatial/temporal scales for different modelling techniques. There is an overall decrease in the replication of prototype characteristics from 1:1 models through Froude scale models (FSMs), distorted scale models, to analogue models. It should, however, be noted that the modelling of a single parameter (e.g. sediment transport) within a distorted scale model can be more accurate than in an FSM. The spacing of the boxes is schematic, but illustrates two key points. First, there is a significant decrease in replicability when moving from 1:1 to scaled models, and from scaled to analogue models. In contrast, the transition from FSM to distorted scale models is associated with a smaller loss of model replicability. Secondly, the chosen spatial and temporal scales for the 1:1, FSM and distorted models illustrate the relative size at which the modelling techniques are generally used

Early attempts to model fluvial and coastal processes include the pioneering work of Fargue (reported in Zwamborn 1967), Thomson (1879), Reynolds (1887) and Gilbert (1914, 1917). Although Fargue and Reynolds scaled some key controlling variables (e.g. the horizontal and vertical distance and tidal period by Reynolds), it was not until the development of dimensional analysis by Buckingham (1915) that scale modelling techniques in engineering were widely adopted (e.g. ASCE 1942; Murphy 1950). Since these two benchmark publications, numerous physical modelling texts have been published including the influential works of Yalin (1971) on scaling theory, Franco (1978) and Shen (1991) on movable-bed modelling, and Schumm et al. (1987) on a range of analogue modelling techniques.

It is now widely accepted across a range of disciplines that physical modelling offers a number of advantages to the scientist interested in a number of landscape evolution processes (see Hooke 1968; Mosley and Zimpfer 1978; Ashmore 1982; Warburton and Davies, in press). Physical models have been used successfully to investigate various issues in fluvial geomorphology over a range of scales, including:

1. Confluence morphology (e.g. Mosley 1976; Ashmore and Parker 1983; Best 1988; Ashmore 1993);
2. Fluvial sediment transport (e.g. Ashmore 1988, 1991a; Ashworth et al. 1992a; Hoey and Sutherland 1991; Young and Davies 1991; Warburton and Davies 1994a);

3. Bar deposition and migration (e.g. Ashmore 1982; Southard et al. 1984; Lisle et al. 1991; Ashworth 1996);
4. Channel change (e.g. Davies and Lee 1988; Leddy et al. 1993; Ashmore 1991b);
5. Channel pattern development (e.g. Leopold and Wolman 1957; Schumm and Khan 1972; Ashmore 1991b);
6. River response to changing extrinsic variables such as tectonics (e.g. Ouchi 1985; Jin and Schumm 1987), aggradation (e.g. Ashworth et al. 1994; Peakall 1995) and base level (e.g. Wood et al. 1993, 1994; Koss et al. 1994).

With such a great variety of physical modelling applications and their increasing use both in geomorphology and sedimentology, it is now appropriate to evaluate the underlying principles that ensure accurate prototype–model scaling and present an assessment of the degree to which replication of the prototype is achieved in such models.

This chapter briefly reviews the theoretical basis of scale modelling to provide a context for identifying the key issues that must be addressed before scale modelling can achieve its full potential. Examples from several models are used to illustrate both the widespread appeal of physical modelling and the progressive decrease in model replicability with increasing ratio of prototype:model scales (Figure 9.1). Traditional 1:1 hydraulic flume models, which have been used widely at the smallest scales of geomorphological interest (Figure 9.1), will not be reviewed here. Examples of such modelling approaches and their application within fluvial environments are contained in Allen (1982), Southard and Boguchwal (1990a) and Best (1996). This chapter instead concentrates on the scales of geomorphological interest ranging from the river channel to the drainage basin. Unresolved issues in the scaling of such systems are discussed, together with an appraisal of future developments which may greatly increase the use and application of scaled physical models within the earth sciences.

CLASSIFICATION OF PHYSICAL MODELS

At the simplest level, physical models of rivers have been traditionally classified on the grounds of specificity, i.e, how closely they replicate a prototype, and by the controlling boundary conditions (Chorley 1967; Schumm et al. 1987). Two types of boundary condition are recognised in the engineering literature: fixed-bed studies, which have non-erodible boundaries and no sediment transport, and movable-bed experiments where the substrate is free to move within a constrained or unconstrained channel. The majority of geomorphological models have movable beds, with either constrained channels (e.g. most bedform studies) or unconstrained channels where the edges of the experimental apparatus serve as the ultimate constraint (e.g. most river/channel network models).

Towards one end of the specificity continuum (Figure 9.1), scale models attempt to represent exactly some, or all, of the key parameters of the system, either from a specific prototype or from general values. Scale models are based on similarity theory, which produces a series of dimensionless parameters that fully characterise the flow. In an idealised situation every variable should be perfectly scaled in the model; however, in the majority of experiments it is not possible to fulfil this requirement. Consequently, the flow Reynolds number is relaxed (see discussion on p. 227) while remaining in the fully turbulent flow regime, but the Froude number is scaled correctly. Relaxation of the flow

Reynolds number allows more flexibility in the model scaling than variation in the Froude number which must be far more tightly constrained. This technique is known as Froude scale modelling (FSM) and has been used successfully in movable-bed modelling of river anabranches and in fixed-bed modelling of flow interaction with artificial structures such as spillways, conduits and breakwaters (e.g. French 1985; Owen 1985). A perfect FSM must achieve geometric, kinematic (motion) and dynamic (force) similarity between model and prototype. FSM studies may model either specific prototypes or scaled versions of a general geomorphic feature. The latter class has been referred to as 'generic Froude scale models' (Church quoted in Ashmore 1991a, b; Ashworth et al. 1994; Warburton and Davies, in press).

For the study of large-scale geomorphological features, such as estuaries and major rivers (e.g. Novak and Cábelka 1981; Klaassen 1991), model geometry may be distorted by increasing the vertical to horizontal scaling ratio, which enables small models to be built or large prototypes to be studied. To achieve precise modelling of sediment transport, a supplementary slope is often added and changes are made to the velocity and discharge scales (e.g. Franco 1978). It is not possible to fully predict the magnitudes of these various adjustments and therefore a process of verification is used, whereby variables are systematically altered until the model reproduces changes observed in the prototype. In this chapter, these models are referred to as 'distorted' models, but it should be noted that similarity of Froude number is also generally achieved in these experiments.

At the other end of the specificity continuum, models can be considered as small landforms in their own right and have been referred to as 'similarity of process' models (Hooke 1968). These models must obey gross scaling relationships and reproduce certain features of the prototype, but since the model is not scaled from either a specific prototype or from generic data, none of the model processes which can be quantified may be applied directly to specific field examples. In this chapter, unscaled, similarity of process studies are referred to as 'analogue' models. The term 'analogue' (Chorley 1967) has been used previously to refer to models that reproduce certain features of a natural system even though the driving forces, processes, materials and geometries may differ from the original. Analogue models are most applicable to the largest prototype:model scales where even distorted scale models cannot be applied successfully (Figure 9.1).

PRINCIPLES OF SCALE MODELLING

An overview of the underlying principles of scale modelling and a definition of key variables is necessary in order to discuss some of the different approaches that have been adopted. Since these scaling laws and their derivation are comprehensively reviewed by Yalin (1971), Langhaar (1980) and covered in detail in many other texts (e.g. Henderson 1966; French 1985; Chadwick and Morfett 1986), only a brief review is given here.

Basic Scaling Laws

Two examples of open-channel flow are used to derive, by dimensional analysis, the most important scaling parameters for physical models. The first and simplest example is that of fixed-bed modelling since only the flow and boundary parameters need to be considered.

The second example of movable-bed modelling is complicated by the additional consideration of sediment transport.

Fixed-bed modelling (case without mobile sediment)

In order to use dimensional analysis, the quantities that control a given system must first
be selected and expressed in terms of their fundamental units. For open channel flow with
a fixed bed, these controlling variables are usually taken as (Yalin 1971):
- properties of the fluid – the dynamic viscosity (μ) and density (ρ) .
- boundary conditions of the channel, normally hydraulic radius (R) and surface
  roughness (k_s)
- bed slope (S)
- average downstream velocity (U) and
- gravitational constant (g)

Three governing variables must be chosen to obtain a solution from these seven variables (μ, ρ, R, k_s, S, U and g) using dimensional analysis. These principal variables are
generally taken as ρ, R and U since they generate two key flow parameters, the Froude and
flow Reynolds numbers. The method generates $n - 3$ dimensionless terms where n is the
number of variables and 3 is the number of governing variables. These are referred to as
'pi' (Π) terms. In the example considered here, four terms are produced:

$$\Pi_1 = \frac{\rho R U}{\mu} = Re \tag{1}$$

$$\Pi_2 = \frac{U}{\sqrt{gR}} = Fr \tag{2}$$

$$\Pi_3 = \frac{k_s}{R} \tag{3}$$

$$\Pi_4 = S \tag{4}$$

The four Π terms represent the flow Reynolds number (Π_1), the Froude number (Π_2),
the relative roughness (Π_3) and the channel bed slope (Π_4). If the ratio between prototype
and model is kept identical for all four of these terms, the model would be an exact
representation of the prototype. However, this situation is rarely attainable in hydraulic
modelling as illustrated by considering the Froude and Reynolds numbers. Since water is
used in most experimental studies, the density and viscosity of the fluid are the same in
the model ($_m$) and prototype ($_p$), assuming a constant temperature, and therefore the
Reynolds number can be rearranged to give

$$U_p R_p = U_m R_m \tag{5}$$

and therefore

$$\lambda_u = \frac{U_m}{U_p} = \frac{R_p}{R_m} \tag{6}$$

where λ_u is the scaling ratio of velocity.

In the case of the Froude number, since acceleration due to gravity, g, remains constant for both model and prototype, then

$$\frac{U_m^2}{U_p^2} = \frac{R_m}{R_p} \tag{7}$$

and therefore

$$\lambda_u = \frac{U_m}{U_p} = \left(\frac{R_m}{R_p}\right)^{0.5} \tag{8}$$

Equations (6) and (8) can only be resolved if R_m is equal to R_p. Thus, the flow Reynolds number is commonly relaxed, with the proviso that in the case of open channel flows it remains within the fully turbulent flow regime ($Re > 500$). As noted previously, this form of modelling is referred to as Froude scale modelling (FSM).

Movable-bed modelling (case with mobile sediment)

If a movable rather than fixed bed is considered, the flow can be considered as a two-phase flow with both fluid and particles. The following set of parameters is used to describe these flows: μ, ρ, S, R, g and two parameters which describe the sediment, ρ_s (the sediment density) and D (the characteristic grain size of the sediment). Some of these variables can be replaced by other dependent parameters. For example, the shear velocity $U_* = \sqrt{gRS}$ can replace S and the immersed specific weight of grains in the fluid $\gamma_s = g(\rho_s - \rho)$ can replace g giving μ, ρ, R, D, ρ_s, U_* and γ_s. These variables also produce $n - 3$ or 4Π terms

$$\Pi_1 = \frac{R}{D} \tag{9}$$

$$\Pi_2 = \frac{\rho_s}{\rho} \tag{10}$$

$$\Pi_3 = \frac{\rho U_* D}{\mu} = Re_* \tag{11}$$

$$\Pi_4 = \frac{\rho U_*^2}{\gamma_s D} \tag{12}$$

The Π_1 and Π_2 terms represent relative roughness of the sediment and relative density respectively, while the term Π_3 is the grain Reynolds number (Re_*), which is a measure of the roughness of the bed relative to the thickness of the viscous sub-layer. Equation (12) expresses the Shields relationship which is normally rearranged as

$$\tau^* = \frac{\tau}{(\rho_s - \rho)gD} \tag{13}$$

where τ is the bed shear stress responsible for initiating sediment transport for a particular grain size, D, and τ^* is the dimensionless shear stress. Together, τ^* and Re_* form the axes of the Shields entrainment diagram (Figure 9.2). The scatter of points on the Shields diagram may be used to define an entrainment threshold known as the critical shear stress, τ_c^*, above which a flow is capable of transporting sediment. The Shields diagram also shows τ_c^* becoming constant (approximately 0.056) at high values of Re_*. Recent debate on the critical threshold value of Re_*, where the flow may be deemed to be fully rough and turbulent with the minimal effect of viscous forces (Re_{*crit}), is discussed on p. 228, but it should be noted that this value occurs in the range $Re_* = 5$ to > 70. It has been proposed that Re_{*crit} may be used to define the optimal length scale, λ_L, for modelling (Yalin 1971):

$$\lambda_L = \left(\frac{Re_{*crit}}{Re_{*p}} \right)^{2/3} \tag{14}$$

where Re_{*p} refers to the prototype. When modelling a two-phase flow, the specific dimensionless properties of the fluid, Re and Fr (equations (1) and (2)), must also be satisfied.

UNRESOLVED ISSUES IN FROUDE SCALE MODELLING

Flow and Hydraulic Constraints

Grain Reynolds number, Re_*

The principal criterion for choosing the optimal length scale for an FSM study is a function of the ratio between the prototype Re_* and the minimum Re_* for a fully rough flow field (equation (14)). There is, however, still debate on the interpolation and interpretation of the original Shields diagram (Kennedy 1995). Rouse (1939) added a threshold line to a form of the Shields diagram and later established a critical value of $Re_* = 400$ (Rouse 1950). Subsequent workers have redrawn the line so that it is an asymptote with constant τ_c^* at $Re_* > 350$ (Richards 1982) and 1000 (Henderson 1966; Novak and

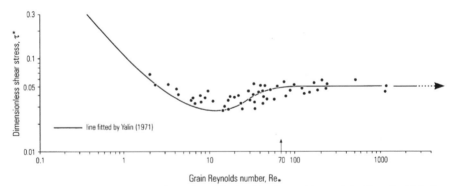

Figure 9.2   The Shields curve as plotted by Yalin (1971, Figure 6.2, p. 154). Note the high degree of scatter amongst the data points which makes the interpolation of the trend line highly subjective. The flow becomes fully rough (i.e. the Shields line becomes horizontal) at a value of 70 using this interpolation. After Ashworth et al. (1994) and reproduced by permission of John Wiley and Sons Ltd, from *Process Models and Theoretical Geomorphology*, edited by M.J. Kirkby, copyright 1994, John Wiley and Sons Ltd

Cábelka 1981). Earlier work by Nikuradse (1933) on pipe-flow boundary conditions produced a similar, but much better constrained plot to that of Shields, with the fully rough category being defined as $Re_* > 70$ (Schlichting 1968). Yalin's (1971) version of the Shields diagram (Figure 9.2) is also an asymptote with τ_c^* constant at $Re_* > 70$ (see Yalin 1971, Figure 3.3, p. 58) and this value has been adopted in the majority of recent FSM studies (e.g. Ashmore 1982; Davies and Lee 1988; Young and Davies 1990; Ashmore 1991a, b; Hoey and Sutherland 1991; Warburton and Davies 1994a).

Parker (1979) divorced the concepts of boundary roughness and constant τ_c^* by arguing that the flow becomes hydraulically rough if $Re_* > 15$. Ashworth et al. (1994) also note that the scatter in the Shields diagram allows for either a 'dip' or a horizontal line to be plotted from values greater than $Re_* = 5$ and argue that the minimum model grain Reynolds number should be 15 (Ashworth et al. 1994, p. 119). Jaeggi (1986) advocates that the minimum value of Re_* that can be used in models is 5 since ripple formation occurs with lower grain Reynolds number. However, Jaeggi (1986) also notes that Re_* values in the transitional field (i.e. before a constant τ_c^* is reached) incorrectly reproduce the initial sediment entrainment conditions.

Prototype verifications of FSMs rarely include Re_* data from anabranches or parts of channels that contain fine-grained sediment or shallow water depths (e.g. backwaters, lateral bars, bank and bartops, and reactivated abandoned channels) although such locations may be expected to contain smooth or transitional Re_* (see calculations in Ashworth and Best, in press). The influence of transitional roughness on the morphological and sediment transport characteristics of many models (Jaeggi 1986), as well as more careful consideration of the range of Re_* found in the field, is clearly a central issue which must be addressed in future FSM studies.

Transitional and supercritical flow

Many braided river FSM studies produce large trains of standing waves indicating areas of supercritical flow (see Figure 9.3; Table 9.1). Field studies also indicate that braided channels may possess Froude numbers greater than unity (e.g. Fahnestock 1963; Williams and Rust 1969; Boothroyd and Ashley 1975; Bristow and Best 1993), but these conditions appear to be less temporally and spatially extensive in the field than in many models. Transitional ($500 > Re > 2000$), supercritical ($Fr > 1$) flow conditions are also evident in many FSM studies (e.g. Ashmore 1982, 1993; Ashworth et al. 1994; Peakall 1995) as highlighted by the presence of oblique rhomboidal standing waves and associated low-relief bedforms (see Figure 9.4, after Karcz and Kersey 1980). This situation suggests that the present scaling ratios based on Re_* may need revision, since model Froude numbers may be too high and flow Reynolds numbers too low. There are at least two possible explanations for this discrepancy.

1. The high Froude numbers in the model may be attributable to incorrect scaling of the bed roughness which may considerably influence the velocity distribution. A predominance of supercritical flow in the model may be explained by velocities being too high because there is less skin roughness when compared to the field prototype. The absence of the very coarsest fractions of the grain size distribution in the model ($> D_{95}$) may be a contributing factor as may the use of rounded/subrounded sand as representative of the field sediment which will lead to a marked drop in flow

Figure 9.3 View of a train of standing waves along the thalweg of the main 'active' channel in a 1 : 20 braided gravel-bed river FSM study (Ashworth et al. 1992b). Flow is from top to bottom and the braidplain is approximately 1 m wide. Avulsion of the main channel has left an abandoned channel complex along the right bank and flow is concentrated into one dominant channel which is eroding the outer left bank at the beginning of full braidplain development

Table 9.1 Compilation of key hydraulic variables from recent FSM studies

Paper	We	Fr	Re	Re_*
Ashmore (1991a)	ND	0.56–0.93	920–4760	36–103
Ashmore (1991b)	11–65[a]	0.91–1.30	1893–6870	89–140
Ashworth et al. (1994)	ND	0.43–0.61[b]	1885–2632[b]	18–56[b]
Ashworth (1996)	8–132[a]	0.64–1.89	3580–14 500	178–586
Hoey and Sutherland (1991)	5[a,c]	0.44–1.54	ND	32–141
Warburton and Davies (1994a, b)	1–16[d]	0.36–1.20[d]	2230–3400[c]	42–106[d]
Young and Davies (1990)	ND	0.30–0.60	ND	76–115[e]

[a] For papers that do not quote water temperature, variables are calculated assuming a value of 15 °C.
[b] Values calculated using average maximum depths and average velocities.
[c] Initial condition only.
[d] Range is calculated based on two standard deviations from the mean.
[e] From Young (1989).
For Re_* calculations, the D_{50} is used by Ashmore (1991a); the D_{90} by Young and Davies (1990), Ashmore (1991b), Ashworth et al. (1994) and Warburton and Davies (1994a, b); and the $D_{100}/1.8$ by Hoey and Sutherland (1991).
ND = No data quoted.
We = Weber number (see equation 18).

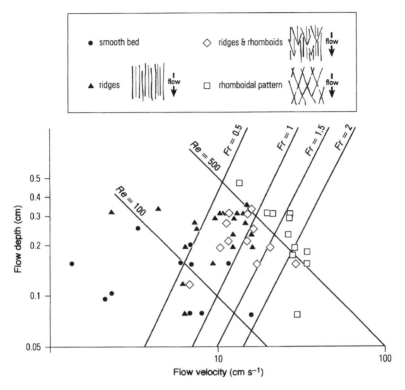

Figure 9.4 Relationship between flow Reynolds number (*Re*), Froude number (*Fr*) and the characteristic low-relief bedforms often found in physical models of braided rivers. Note the predominance of ridges and rhomboidal bedforms at transitional flow Reynolds number (500 < *Re* > 2000) and supercritical (*Fr* > 1) flow. Plot is redrawn from Karcz and Kersey (1980)

resistance and particle interlocking (see Church et al. 1991 and discussion below concerning sediment transport). Clearly, other factors such as discharge and slope may also be involved in this *Re*, *Fr* and *Re*$_*$ relationship.

2. The choice of a critical grain Reynolds number may be inappropriate since both Nikuradse's (1933) work on relative roughness in pipes and the Shields curve use sediment with a uniform grain size distribution. Their plots therefore have *Re*$_*$ numbers calculated using a unimodal grain size distribution whilst the majority of FSM experiments calculate *Re*$_*$ using the D_{90} of a heterogeneous grain size distribution (see Table 9.1).

The presence of low particle Reynolds numbers (*Re*$_*$ < 5) in an FSM may lead to the formation of ripples which may not have scaled equivalents in the field prototype (Jaeggi 1986). Changes in morphology of these small, smooth-boundary bedforms in fluctuating local flow depths may also lead to the presence of standing waves and supercritical flow bedforms in the FSM. Jaeggi (1986) suggests the need to coarsen the bed material in an FSM to avoid formation of these bedforms which, although possible, may be contradictory to other model study objectives if the fine tail of the grain size distribution is of interest.

The supercritical to subcritical flow transition may also produce hydraulic jumps which have been observed to cause headward erosion in alluvial fan models (Parker 1996) and produce abrupt gravel to sand size sorting in downstream fining models (Paola et al. 1992). Significant headward erosion of anabranches has also been observed within braided river FSM studies (e.g. Ashmore 1982, p. 218), but it is unknown how common hydraulic jumps are in the field, particularly in braided channels with high width:depth ratios.

Particle Settling

The particle fall velocity in a stationary fluid can be considered using two different equations. Stokes' law considers only viscous resistance forces and is of the form

$$U_f \propto D^2 \tag{15}$$

where U_f is the fall velocity of a particle and D is the grain size. For particles smaller than 0.1 mm in water this relationship holds very well, but Stokes' law does not account for boundary layer separation behind a falling particle and a consequent increase in the fluid drag. For large particles, Newton derived an expression, known as the impact law, that incorporates the effects of boundary layer separation

$$U_f \propto D^{0.5} \tag{16}$$

The impact law is not a particularly good approximation of experimental results, even for particles larger than 1 mm (see Figure 9.5), and has many inadequacies when the particles are non-spherical. The combined experimental curve does, however, break into two distinct linear segments, one characterised by Stokes' law and the other broadly delineated by the impact law. When both model and prototype grain sizes fall exclusively within either of these two areas, a linear scaling ratio can be applied. However, if the prototype grain size is in the 'impact region' and the model is in the field of Stokes' law, then the function is nonlinear and consequently cannot be perfectly scaled. The main result of this nonlinearity is that the fall velocities of the particles relative to the downstream velocity are much slower in the model than the prototype, although the particle time constant (the ratio of the particle response time to the characteristic eddy turnover time, see Elghobashi, 1994) and relationship between particle size and turbulence in the model and prototype must also be taken into account. Saltating particles in the model may consequently have larger hop lengths and heights than the geometric scale ratio would suggest. This limitation could be overcome by altering the grain-size distribution as suggested by Jaeggi (1986) for initiation of sediment movement, but this solution would also affect the mode of sediment transport.

Bedload Transport and Deposition

A scale ratio for sediment transport rate in models with fully turbulent flow, $(\lambda_t)_s$, can be derived by dimensional analysis (Yalin 1963, 1971),

$$(\lambda_t)_s = (\lambda_L)^{1.5} \tag{17}$$

However, Yalin (1971) has demonstrated that if Re_* is below the critical threshold for a fully turbulent boundary and the fluid and temperature are the same in the prototype and

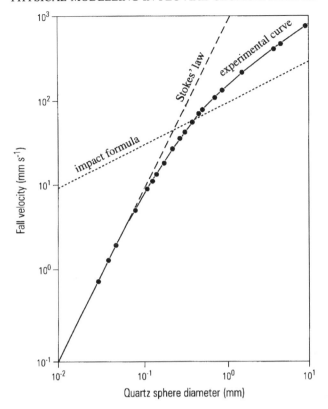

Figure 9.5   Graph of fall velocity as a function of grain diameter, for water at 20 °C, plotted against the predictions of Stokes' law and the impact formula. From Leeder (1982) with the permission of the author using data from Gibbs et al. (1971)

model, then it is impossible to achieve dynamic similarity of sediment transport. The majority of FSM studies fall within the range of suggested critical grain Reynolds numbers (see above; Table 9.1) and therefore may compromise sediment transport similarity.

Several studies have compared the observed bedload sediment transport from an FSM with established transport equations (e.g. Ashmore 1988; Hoey and Sutherland 1989; Young and Davies 1990, 1991; Warburton and Davies 1994a). Young and Davies (1990) compared the empirically based equations of Schoklitsch (1962) and Bagnold (1980) with their flume data and found a very strong agreement (see Figure 9.6). The Bagnold (1980) equation had the strongest correlation with an average under-prediction of 18% for steady flows and just 1% for unsteady flows. Ashmore (1988) and Hoey and Sutherland (1989) also demonstrated that the Bagnold (1980) formula was in good agreement with model transport rates. The formation of bedload pulses or waves in flumes has also been studied by several authors (e.g. Ashmore 1988; Kuhnle and Southard 1988; Young and Davies 1990, 1991; Hoey and Sutherland 1991; Warburton and Davies 1994a) and the associated short-term variations in sediment transport rates may account for much of the scatter in

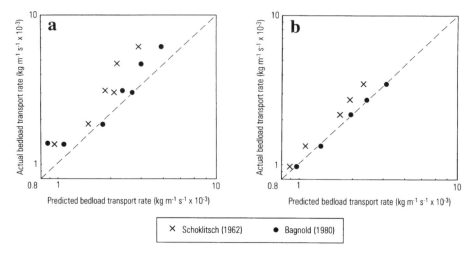

Figure 9.6   Bedload transport rate predictions from Young and Davies (1990) for (a) steady flows, and (b) unsteady flows. Dotted line represents a perfect 1 : 1 relationship. After Young and Davies (1990), and reproduced with permission of the authors and the New Zealand Journal of Hydrology

correlations of time or channel-averaged bedload transport rate with discharge (e.g. Figure 9.6).

The possible influence of sediment shape and a limited size gradation on bedload transport rates within some FSM studies has been noted by Church and Jones (1982) and Church et al. (1991). Most gravel bed rivers are composed of angular clasts with a very large size range whilst many models use subangular to well-rounded sand grains and do not model the very largest grains. These differences may explain the preferential mobility in many models of the largest grains which tend to roll rapidly across finer-grained substrates ('overpassing') and are preferentially deposited as accreting avalanche faces at bartails and on bartops (e.g. Leopold and Wolman 1957; Ashmore 1982).

Water Surface Tension

Surface tension is the tensile force per unit length (N m^{-1}) acting at the fluid surface. The tensile force results from the difference between the internal molecular forces of a liquid and the forces between liquid molecules and an adjacent surface, and varies as a function of temperature. In rivers, the effects of surface tension are largely insignificant (Dingman 1984, p. 85) although biofilms may help stabilise the sediment surface, but, if a model has too large a vertical scale and consequently too small a flow depth, surface tension can be important. The addition of a surface tension term, σ, into the previous dimensional analysis gives a Π term known as the Weber number (*We*)

$$We = \frac{\rho U^2 h}{\sigma} \tag{18}$$

where h is the average flow depth.

The Weber number represents the ratio between the inertial and surface tension forces. The velocity and time scales for perfect scaling of surface tension effects (ASCE 1942) are

$$\lambda_u = \left(\frac{\lambda_\sigma}{\lambda_h \lambda_\rho}\right)^{0.5} \tag{19}$$

$$\lambda_T = \left(\frac{\lambda_h^3 \lambda_e}{\lambda_\sigma}\right)^{0.5} \tag{20}$$

In the case of an FSM, where the fluid density, ρ, is kept the same in both model and prototype and where λ_u is controlled by correctly scaling the Froude number and relaxing the flow Reynolds number, it is not possible to scale the Weber number correctly. However, a similar argument can be made for the Weber number to that used for the relaxation of the flow Reynolds number, namely that so long as the surface tension effects are insignificant then exact scaling is unnecessary. Unfortunately, there is no current consensus on the critical *We* value where surface tension begins to strongly influence sediment transport and deposition, although suggested values range from 10 to 120 (see discussion in Peakall and Warburton, in press). Most small river experiments have Weber numbers that fall within or marginally below the suggested range of critical values (Table 9.1) which suggests that a degree of surface tension induced distortion may have been added to the models. There is therefore a clear need for FSM studies to calculate and publish Weber numbers for a range of channel geometries.

Cohesion and Vegetation

Cohesion

Clay minerals are cohesive due to a combination of two forces, the weak van der Waals' forces which all matter is subject to, and ionic bonds which form through the process of cation exchange between clay minerals. These intermolecular forces act as a major constraint on scale modelling, because coarse sand and gravel in the prototype can only be scaled down as far as silt sizes within the model, without adding cohesion. The cohesive forces of clay are also scale independent so that inclusion of a proportion of clay in the model will lead to unrealistic rates of erosion and channel change. This was demonstrated in the analogue meander model of Schumm and Khan (1972) where the addition of a low concentration of clay caused channel erosion to cease. Consequently, the difficulty in modelling fine-grained sediment limits the length, or geometric scale, of the model and can lead to a truncation of the grain-size distribution. However, it has been shown that it is possible to use inert silica flour as fine as 1 μm as a substitute for prototype silt/fine sand grain sizes (e.g. Parker et al. 1987; García 1993; Leddy et al. 1993; Ashworth et al. 1994). Some of this very fine material may travel solely in suspension or be repeatedly deposited and re-entrained. The use of fine-grained silica flour also leads to significant capillary forces which helps simulate prototype cohesion. This property is desirable in many fluvial models because several geometric variables, such as sinuosity and hydraulic geometry, may change with the degree of cohesion (Schumm 1960).

Vegetation

Although vegetation plays an important role in strengthening the banks and floodplain (e.g. Zimmerman et al. 1967; Smith 1976), particularly in coarse-grained rivers where inter-particle cohesion is largely unimportant, there are very few examples of physical models that incorporate the effects of vegetation. Recent experimental work has documented the interaction between within-channel vegetation and flow structure (e.g. Ikeda and Kanazawa 1995; Tsujimoto 1996) but the significance and magnitude of these effects have yet to be considered within physical scale models. Marsden (1981) experimented with different densities of toothpicks and planted wheat, rape, cress, lawn and budgie seed, before successfully growing mustard on the floodplain of an analogue braided model. The results demonstrate an optimal planting density for maximum floodplain accretion, and Marsden (1981) recommended that the approach be extended to larger models that could incorporate scale effects.

Scaling of Time

One of the primary objectives of physical modelling is to change the rate of the formative processes, thus permitting study of landform evolution over long prototype time periods. Two different approaches to the modelling of time are possible, one based on dimensional analysis and the other on magnitude-frequency analysis.

Dimensional analysis of time scales

The time scale for mean flow velocity $(\lambda_t)_u$ is given by dimensional analysis as $(\lambda_t)_u = (\lambda_L)^{0.5}$. This scale differs from the previously derived time scale for sediment transport, $(\lambda_t)_s$, which has a scale ratio of $(\lambda_L)^{1.5}$ (equation (17)). Similarly, the fall velocity of a particle as characterised by Stokes' law, has a time scale of $(\lambda_t)_{u_g} = (\lambda_L)^{-1}$. Yalin (1971) also notes a series of other time scales relevant to scale modelling of river channels

$$(\lambda_t)_y = (\lambda_L)^2 \tag{21}$$

$$(\lambda_t)_x = (\lambda_L)^{0.5} \tag{22}$$

$$(\lambda_t)_m = (\lambda_L)^{-1} \tag{23}$$

where $(\lambda_t)_y$ is vertical erosion/accretion; $(\lambda_t)_x$ is the downstream displacement of individual sediment grains; and $(\lambda_t)_m$ is the grain motion during saltation in either the horizontal or vertical dimensions. Vertical bed surface change is therefore the fastest time scale operating in the model relative to the prototype (see Figure 9.7), followed by sediment transport rate, the displacement of sediment or fluid in the downstream direction, particle fall velocity and the individual motion of grains during saltation:

$$(\lambda_t)_y < (\lambda_t)_s < (\lambda_t)_x \ (\lambda_t)_u < (\lambda_t)_m \ (\lambda_t)_{u_g} \tag{24}$$

These different time scales may cause confusion when trying to interpret experimental results. For example, there has been debate as to whether short-term fluctuations in bedload rates should be scaled in terms of either total sediment transport rate, $(\lambda_L)^{1.5}$, or downstream displacement of grains, $(\lambda_L)^{0.5}$ (Ashmore 1988; Young and Davies 1991).

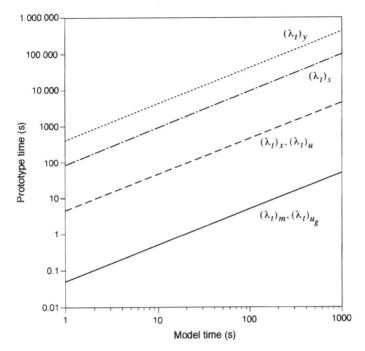

Figure 9.7   Model:prototype time scales for different processes in a 1:20 FSM. $(\lambda_t)_y$, $(\lambda_t)_s$, $(\lambda_t)_u$, $(\lambda_t)_m$ and $(\lambda_t)_{u_g}$ are the time scales for vertical erosion/accretion, sediment transport, downstream displacement of individual sediment grains, flow velocity, grain motion during saltation and particle fall velocity respectively (see text for more details). Some processes such as vertical erosion are much faster in the model than in the prototype whilst others, such as fall velocity and grain motion during saltation, are slower

This issue may have important ramifications for the modelling of alluvial architecture. For example, since vertical erosion and deposition are much faster than horizontal accretion, there may be distortion in the size of scours and overbank splays preserved in aggrading river models. Initial FSM work on braided river aggradation (Ashworth et al. 1994; Peakall 1995) suggests that the influence of multiple time scales is limited, but detailed experiments are still required to resolve this issue. The question of time scaling within hydraulic models and the period required for 'equilibrium' to be reached in the initial stages of experiments is clearly a subject that warrants further attention, especially if such models are to be used to investigate long-term alluvial channel behaviour.

Hydrograph scaling

A complementary approach to modelling time within physical models is through use of the geomorphological concept of event magnitude–frequency. Wolman and Miller (1960) suggested that there is a good correlation between the 'dominant' discharge (i.e. that which does most 'geomorphological work' in terms of sediment transport) and the bankfull discharge. Most physical models use a constant discharge which approximates to bankfull (e.g. Leopold and Wolman 1957; Ashmore 1982). The sequential simulation of a number of medium to high magnitude (low recurrence interval) flood hydrographs (e.g. 50 or 500 year flood), enables additional time-scale compression by increasing the 'geomorphological work' completed in a period of model time. Most rivers have a relatively short 'memory effect' from large flood events, but care must be taken to avoid exceeding a magnitude threshold where channel recovery to its previous state is imperceptibly slow (Carling 1988). A limited number of modelling studies have used hydrographs but have not specifically examined magnitude–frequency effects (e.g. Anastasi 1984; Davies and Lee 1988; Young and Davies 1990, 1991; Leddy et al. 1993; Ashworth et al. 1994; Peakall 1995). The modelling of hydrographs has a number of other advantages in addition to enhanced time-scale compression. For example, many important fluvial processes such as overbank sedimentation, avulsion, bend cutoff and bar dissection usually occur both at peak discharges and on the waning limb of the hydrograph. Additionally, several field studies have shown the clear differences in sediment transport between the rising and falling limbs of the hydrograph (Reid and Frostick 1984; Reid et al. 1984) and the marked impact of flow variability on bedform and bar formation (Hein and Walker 1977; Church and Jones 1982; Welford 1994; Julien and Klaassen 1995).

Two aspects must be considered when scale modelling hydrographs: scaling of the water discharge and time. The scale ratio for discharge is given as

$$\lambda_Q = (\lambda_L)^{2.5} \tag{25}$$

In contrast, the time scale for the fluid is $(\lambda_L)^{0.5}$ and model hydrographs are therefore flatter than their prototype equivalents (Leddy 1993).

Aggradation and Alluvial Architecture

Most scale modelling studies have concentrated on describing the two-dimensional planform and surface geomorphology of alluvial channels. However, there is an increasing demand for extending this work into three dimensions by modelling subsidence or aggradation and therefore preserving the alluvial architecture through time. This issue is even more pressing with the recent development of three-dimensional 'process/geometry-based' alluvial architecture models (e.g. Webb 1994, 1995; Mackey and Bridge 1995) which require calibration and testing.

Logistical constraints prevent the use of most flumes for subsidence/aggradation experiments but recent work by Ashworth and Best (1994), Ashworth et al. (1994) and Peakall (1995) show that it is possible to promote aggradation and basin-wide sedimentary fill of scaled braided channels. Using a constant aggradation rate, repeated flood hydrographs and an FSM of a gravel-braided river, Ashworth and Best (1994) and Peakall (1995) showed that the preserved alluvial architecture closely resembles that seen in field outcrop and core. Figure 9.8 shows a cross-stream section through such a preserved deposit, which clearly delineates a number of key sedimentary niches of different grain size. Amalgamation of geometric and spatial information for each niche class in successive sections at closely spaced intervals, permits reconstruction of the subsurface sedimentology and, when combined with surface morphological data, the full three-dimensional alluvial architecture. Recent experiments (Ashworth and Best 1994; Peakall 1995) have started to quantify the impact of allocyclic controls (e.g. a change in

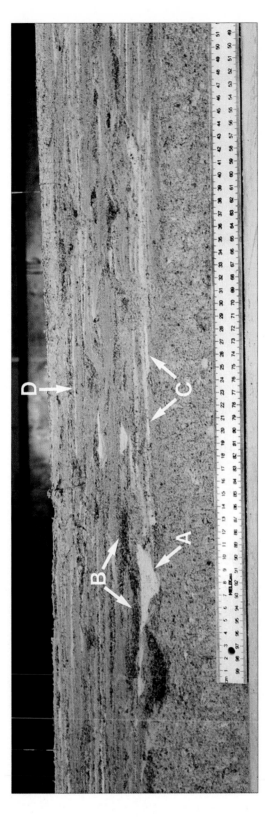

Figure 9.8 Preserved alluvial architecture from a 1:20 braided river FSM (Peakall 1995). The section is perpendicular to the mean flow direction and the bottom 8 cm of sediment is the original unsorted channel bed sediment. Areas of white silica flour deposition represent fine sand/silt in the prototype whilst the darker coarse sand is equivalent to gravel. Note the clear differentiation of sedimentary niches with sharp erosion surfaces – common depositional units include abandoned channel plugs (A), bar core (B) and lee deposits (C), and thin overbank and bartop splays (D). More details of the modelling rationale and niche classification are given in Ashworth et al. (1994) and Peakall (1995)

aggradation rate and the imposition of different magnitudes of lateral tectonic tilt) on fluvial deposition. Future work will consider changing hydrograph type and base level control.

Although the methodology for scale modelling of aggradation is still being developed, there is tremendous potential for answering the 'what if' scenario by systematically changing an auto or allocyclic control on alluvial deposition. The main drawback with experiments concerning aggradation is that, even with a compression of time, it is impossible to reconstruct basin sedimentation rates over geological time periods (e.g. 10^3–10^6 years). However, it may be the case that short-term, 'instantaneous' erosion/deposition events dominate the preserved alluvial record so that the gradual, long-term, basin-wide subsidence rate is less important for the preservation of *individual* depositional niches (Ashworth and Best 1994). Clearly, there is an immediate need for more experiments that employ a range of aggradation rates.

EXAMPLE OF A FROUDE SCALE MODEL

Braided gravel bed rivers are some of the most difficult environments to study in the field since the majority of planform change occurs during flood when the flow is highly turbulent and turbid, thereby limiting observation of near-bed processes. The pioneering work of Ashmore (1982, 1988, 1991a, b, 1993) was the first to highlight the potential of the FSM approach for understanding and quantifying complex and dynamic braided fluvial environments. Two aspects of this work are described here: channel confluence kinetics and the development of braiding. Both illustrate the power and potential of a Froude scale modelling approach.

Channel junctions are key nodes within braided networks and form the critical areas of flow convergence/divergence that are instrumental in the process of braiding. Past scaled models using both fixed (Mosley 1976; Best 1987, 1988) and mobile banks (Mosley 1976), have considered the details of the confluence zone in terms of the flow dynamics and sediment transport pathways. Additionally, fieldwork has provided invaluable insights into the processes operative at these sites (e.g. Roy and Roy 1988; Ashmore et al. 1992; Biron et al. 1993). However, the importance of channel junctions at larger spatial and temporal scales could not be addressed by these studies. Ashmore (1993), however, considers the kinetics of channel junctions in an FSM and presents models for the migration of channel junctions and their influence on downstream sedimentation. An example from this work demonstrates destruction of a post-confluence medial bar through downstream migration of the upstream junction (Figure 9.9). Such qualitative and semi-quantitative studies may be used to propose generalised models of confluence zone migration (Figure 9.10).

Ashmore's work on confluence dynamics links to the broader issue of bar and channel pattern development. In a sequence of papers, Ashmore (1982, 1991b, 1993) has successfully used an FSM to classify the main mechanisms of braiding, explain the processes controlling bar formation and down-bar fining, and relate the internal generation of bedload pulses to channel change and bar dissection/creation. One of the most influential papers (Ashmore 1991b) unambiguously defined the causes of braiding both at the initial stage of a single channel (e.g. Ashmore 1991b, Figure 3, p. 330) and when a fully braided

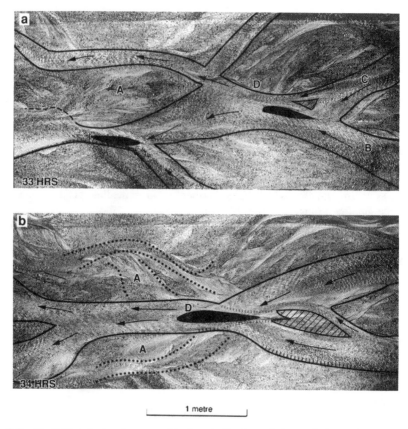

Figure 9.9   Medial bar destruction caused by longitudinal translation and change in total discharge of an upstream confluence. (a) The medial bar complex (A) formed by rapid progradation of sediment in the left confluent channel (B) and subsequent channel bifurcation. Waning of the flow in the right confluent channel (C) led to the formation of a new confluence (D) further downstream. (b) Later, lateral migration of the confluent channels caused the confluence zone (D) to migrate downstream, eventually triggering an avulsion across the centre of the medial bar, leaving two isolated remnants (A). Figure and interpretation reproduced from Ashmore (1993) with permission of the author and the Geological Society of London

network has developed (see Figure 9.11). The main braiding mechanisms identified were through deposition of a central bar, chute cut-off of point bars, conversion of a single transverse unit bar to a mid-channel bar and dissection of multiple bars. The chute cut-off mechanism was the most common process of braiding in Ashmore's experiments (Figure 9.11). This process may be very common in single low-sinuosity gravel-bed streams (cf. Lewin 1976; Carson 1986) and is the dominant transformation process from an initial single channel with alternate bars to a braided network that occurs at the beginning of most scale modelling experiments of braided rivers (Leddy 1993). By relating the different mechanisms of braiding to the local flow conditions (excess shear stress), channel cross-sectional geometry and bedform regime, Ashmore's (1991b) work provided valuable insights into the critical conditions necessary for channel bifurcation and mid-channel bar growth. Ashmore's experiments used a constant water discharge, truncated grain-size

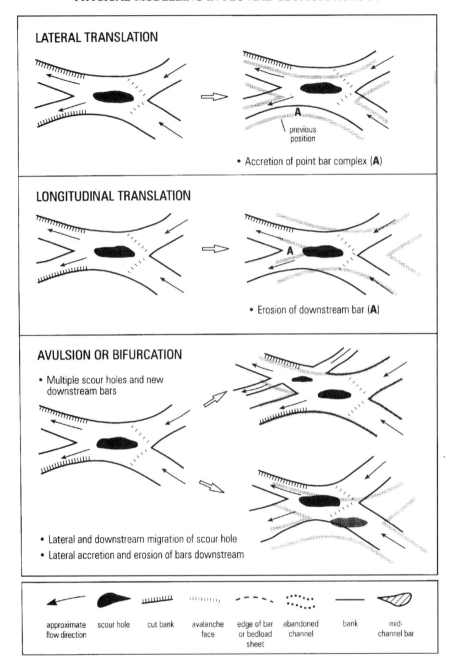

Figure 9.10   Schematic summary of observed modes of confluence movement and sedimentation in response to the migration of confluent anabranches. After Ashmore (1993) and reproduced by permission of the author and the Geological Society of London

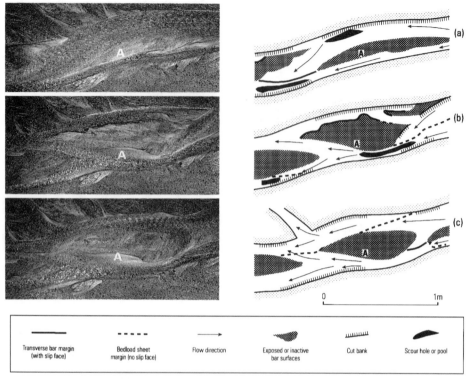

Figure 9.11 Chute cut-off in an established braided channel. (a) Point bar A begins to develop. (b) After a period of growth and increasing sinuosity, point bar A is cut off leaving portions of the original channel abandoned and developing a new point bar on the opposite bank. (c) Point bar at A is converted to a medial bar by a second chute cut-off. The elapsed time between successive photographs is 1 hour. Flow is from right to left. Interpretation and diagram after Ashmore (1991b, p. 331), reproduced with the permission of the author and the NRC Research Press

distribution and had flow Reynolds and Froude numbers that were often transitional and supercritical, respectively (see Table 9.1 and the presence of oblique diagonal standing waves in many of Ashmore's photographs). However, these experiments clearly illustrate the full power and potential of an FSM and suggest that it is not always necessary to scale all parameters strictly to produce realistic predictive models.

DISTORTED SCALE MODELLING (MOVABLE-BED MODELLING)

For the study of large geomorphological scales, fine prototype sediments or precise modelling of sediment transport and deposition, models may have to be distorted. Geometrically distorted models have a small vertical:horizontal scale ratio in order to model large prototypes, whilst maintaining adequate model flow depth. Perhaps the most spectacular example of a distorted geometric model is that of the entire Mississippi basin by the US Corps of Engineers at a horizontal scale of 1:2000 and vertical scale of 1:100, on a 100 ha site (reported in Novak and Cábelka 1981, p. 163). Distortion of the geometric scale is also a prerequisite in many cases where the precise modelling of sediment movement is attempted since this increases model shear stresses. A number of other scales are also commonly adjusted when modelling sediment movement in rivers, including valley slope, grain size, sediment density, flow velocity and discharge. The combination of these scale adjustments is usually referred to as 'movable-bed modelling' in the engineering literature and refers to alteration of more than just the boundary conditions.

One example of movable-bed modelling is McCollum's (1988) study of the Apalachicola River in the southern United States. The model reproduced a 7 km section of the Apalachicola River which suffered from persistent sedimentation, requiring annual dredging to maintain a navigation channel. Horizontal and vertical scales were 1:120 and 1:80 respectively and crushed coal with a specific gravity of 1.3 g cm^{-3} was used to simulate the sandy bed of the Apalachicola, thereby avoiding introduction of cohesion into the model. Model verification consisted of replicating a 12-month prototype discharge record and comparing the observed model changes with hydrographic surveys. The channel slope and discharge ratio were also adjusted to achieve a good model:prototype agreement. After initial verification, a number of channel improvement schemes were tested using a wide range of discharges. Based on these tests, a series of 'L'-shaped dikes were proposed as the most effective method for maintaining the navigation channel.

McCollum's (1988) study illustrates the potential of movable-bed modelling for studying large prototypes, fine-sediment and specific sediment transport problems. There appears to be great potential for utilising the same techniques within fluvial geomorphology, as illustrated by Klaassen (1991) for the Brahmaputra River (Bangladesh) and Davies and Griffiths (in press) for flow–sediment transport relationships in the Waimakariri River (New Zealand).

ANALOGUE MODELLING

Analogue models have been used to study a wide range of fluvial scales from small channels to entire drainage networks. Although Schumm et al. (1987) illustrate the many conceptual and practical advantages this approach offers for geomorphology, some of the small geomorphological scales investigated in the past using analogue models can now be modelled in far more detail using FSM and movable-bed techniques (see Figure 9.1 and earlier discussion). The main advantages of analogue models are speed and simplicity in setting up experiments and the reduced space and budget costs which constrain other modelling approaches. These advantages are illustrated using recent studies of base level change and alluvial fan aggradation.

Base Level Control and Sequence Stratigraphy

The continental shelf/slope system is of such a large size that a scale model would be prohibitively expensive, if not impossible, to construct. In addition, important attributes such as cohesion and vegetation are difficult to incorporate into such models. However, the effect of base level change on these systems has recently been studied using analogue models of: (i) a fan forming in a drainage ditch (Posamentier et al. 1992), (ii) a single channel in a stream table (Wood et al. 1993) and (iii) a drainage network developed in a stream table using a rainfall simulator (Koss et al. 1994). These analogue models have been used to test sequence stratigraphic concepts that are difficult to examine using conventional computer modelling techniques. The analogue models illustrate several important points:

• Depositional systems 'tracts' and the bounding surfaces between them are scale independent (Posamentier et al. 1992; Koss et al. 1994).
• There is a significant lag time between base level fall and coarse-grained sediment reaching the lowstand fan (Wood et al. 1993; Koss et al. 1994).
• A large number of incised valleys form at the shelf/slope interface, only one of which connects to the main river system (Wood et al. 1993; Koss et al. 1994).
• Base level rise is frequently accompanied by significant slumping on the walls of incised valleys (Wood et al. 1993).
• The rate of base level fall and the shelf angle are important controls on sediment deposition and preservation (Wood et al. 1993, 1994).

Whilst these results cannot be directly quantified, they demonstate one of the major advantages of analogue models in that they can be used to test some of the latest hypotheses concerning large-scale dynamics of sedimentary basins which are extremely difficult to verify by any other technique.

Alluvial Fan Aggradation

Analogue modelling can provide extremely valuable information on general alluvial fan and channel dynamics as shown by Schumm et al. (1987), Bryant et al. (1995) and Whipple et al. (1995). Bryant et al. (1995) tested the hypothesis that the frequency of channel avulsions is linked to the sedimentation rate by forming a simple alluvial fan (sediment cone) and varying the sediment input through time. Whilst the alluvial fan cannot be directly compared with field examples, the experiments clearly demonstrate the general principle that avulsion frequency is directly related to sedimentation rate at all but the highest rates. This conclusion has significant implications for models of alluvial architecture, many of which assume that the avulsion frequency is invariant (e.g. Bridge and Leeder 1979; Mackey and Bridge 1992). More recently, analogue modelling of alluvial fan dynamics has been used in an applied context to simulate the redistribution and aggradation of mine waste tailings from a single point source (Parker 1996).

ADVANCES AND TRENDS IN PHYSICAL MODELLING

The application of physical modelling within fluvial geomorphology has produced major advances in the last two decades, many of which have resulted from, or been produced by, improvements in the methodology by which experiments are conducted. Apart from the solution of scaling issues, future advances in scale modelling will undoubtedly be accompanied by the development of appropriate measurement technology for these models. A clear parallel to this situation is shown by work over the past 30 years concerning bedform generation in scaled hydraulic flume models. This work has progressed from the qualitative description of bedform morphology and their gross fluid dynamic controls (e.g. Simons et al. 1961; Guy et al. 1966), to quantification of sediment transport rates and investigation of bedform response to changing flow conditions (Guy et al. 1966; see Allen 1982) through to studies of the detailed flow, turbulence and sediment dynamics associated with a range of bedforms (e.g. Raudkivi 1966; McLean et al. 1994; Bennett and Best 1995; Nelson et al. 1995). Increases in the quantitative assessment of the processes governing bedform stability have been brought about by changes in instrumentation during this period, which have been partly responsible for the increasing levels of resolution incorporated within numerical models of bedform generation. Perhaps the greatest potential for increasing both the applicability of scale models to larger scales (of the order of the channel width or braidplain scale) and confidence in the degree of prototype agreement, lies in developing new methods of flume experimentation. Several areas appear ripe for development at present.

Quantification of Flow

Most past FSM studies have recorded only the basic attributes of water flow through model fluvial channels. These velocity measurements have usually been obtained by surface float tracing (e.g. Ashworth 1996) or use of pitot tubes (Ashmore 1982, 1996). However, these measurements are often very difficult, if not impossible, to perform in shallow flow depths, or across the width of the channel, and therefore many model studies present flow/hydraulic data predominantly from large channels. Instrumentation in these model channels is extremely difficult yet clearly central to verifying the range of flow conditions within an FSM. The solution to these problems may lie in the application of several available methods of instrumentation for quantifying flow structure:

1. Hot-film or laser Doppler anemometry (e.g. Durst et al. 1987; Tritton 1988) where flow depths and experimental configuration permit. Although hot-film probes may possess calibration difficulties, especially in flows with sediment transport, the sensors often are small enough to be used in shallow depths. Laser Doppler anemometry may provide a non-intrusive methodology for recording flow velocities/turbulence by focusing the beams through the water surface, but will only work in clear liquids with little sediment transport.
2. Ultrasonic Doppler anemometry (e.g. Takeda 1991, 1995) offers the potential for obtaining high-resolution, often multi-point, measurements in opaque fluids although current systems may only be of use in channels with flow depths of the order of several centimetres or greater.
3. Particle tracking and particle image velocimetry (PIV; Linden et al. 1995; Seal et al. 1995) offers great potential to quantify particle velocities (of perhaps fluid and sediment) within the flow and may provide the best method for yielding channel-wide estimates of flow velocity both on the water surface and within the flow through use of neutrally buoyant particles.

Quantification of Topography

Many FSM studies have quantified channel change through continuous recording using video cameras in conjunction with limited topographic surveying/point gauging. One area which would immediately yield valuable information on scale model topography would be the rapid, and automated, quantification of bed heights within the scale model.

The appropriate methodology to accomplish this goal may consist of the application of ultrasonic bed profilers (e.g. Kuhnle 1993; Best and Ashworth 1994) which can resolve heights down to 0.1 mm. The development of photogrammetric methods (Ashmore, personal communication 1995) or use of laser light sheets (Rice et al. 1988; Römkens et al. 1988) may also yield suitable technology.

Quantification of Sediment Transport

Several studies have documented sediment transport rates within model studies and used these to discuss phenomena such as the presence and importance of bedload pulses within braided channels (e.g. Ashmore 1988). However, prediction of more local channel change (avulsion for example) and development of braid networks requires more detailed quantification of the rates of sediment transport *within* individual model channels. Few studies have sought to address this topic and the introduction of samplers into these flows is fraught with difficulties, not just in the disturbance to the flow field, but in the design and efficiency of the sediment samplers themselves. Apart from use of high-resolution and continuous ultrasonic bed profilers, which may be used to quantify bed height change, the use of PIV techniques to track different size (i.e. colour) grains may yield valuable estimates of transport rates and pathways, possibly of individual size fractions within the sediment load. Refinement of acoustic devices which have been developed to monitor bedload noise (e.g. Thorne et al. 1989; Hardisty 1993; Rouse 1994) could yield another methodology for estimating transport rates, whilst individual particle trajectories in clear flows may be tracked using high-speed video (1000 frames per second, cf. García et al. 1996). Other tracer techniques, perhaps based on thermal imaging of grain paths within the flow, may provide a tool for providing the much-needed quantification of the links between flow, sediment transport and channel change.

Quantification of Sedimentary Architecture

If FSMs can be used to examine the subsurface geometry and internal bedding characteristics of fluvial deposits, great potential exists for obtaining true three-dimensional descriptions of such deposits. Apart from detailed trenching and description of these sediments, use of miniaturised geophysical techniques, such as seismic imaging or resistivity techniques, may provide invaluable tools for quantifying subsurface sedimentary structure and connectivity between key depositional elements.

SUMMARY

The use and application of physical modelling within fluvial geomorphology lies at a crossroad. Work over the past 20 years has yielded considerable advances in our knowledge of many complex fluvial processes and forms and has progressed to incorporate realistic scaling assumptions from the scale of the sediment grain to that of the river channel. Scale models are now powerful tools for testing mathematical models because they can closely approximate the idealised assumptions that underpin many numerical models. Further progress in this field may only be possible if three issues are addressed:

1. Incorporation of more realistic model parameters into FSM studies, such as flood hydrographs, fine-grained sediment and cohesion;
2. Development and implementation of methodology to better quantify flow, sediment transport and morphological change;
3. Additional and more complete testing/verification of FSMs against their prototype conditions.

Although unscaled or 'analogue' models can shed much light on large temporal and spatial scale processes and products, these studies must always be considered in terms of their underlying simplifications and drawbacks and must not be interpreted as true scale models. Such 'analogue' models may increasingly be used to investigate the role of allocyclic factors on sedimentation, such as local tectonic and base level controls, but their departure from true scaling perhaps demands more complete field verification than more rigidly scaled Froude scale models.

ACKNOWLEDGEMENTS

Many of the ideas expressed in this chapter have developed from work sponsored over the past six years by BP Exploration. We are grateful to BP for award of a Ph.D. studentship to Jeff Peakall and grants to establish the scale modelling/aggradation facility at Leeds. This modelling has also been supported by a grant from the Royal Society and more recently funding from NERC (GR9/01640) and ARCO Oil (USA) to Ashworth and Best. Marcelo García and Bruce Rhoads provided helpful suggestions to improve the clarity of this contribution. Peter Ashmore kindly supplied original photographs of his flume experiments for Figures 9.9 and 9.11.

REFERENCES

Allen, J.R.L. 1982. *Sedimentary Structures: Their Character and Physical Basis*, Elsevier, Amsterdam, 539 pp..

Anastasi, G. 1984. Simulazinoe di regime torrentizio su modello fisico a fondo mobile mediante micro-computer, in *Memorie XIX convegno di idraulica e construczioni idrauliche*, Pavia, Italy, 6–8 September 1984, Paper A9, 10 pp.

ASCE 1942. *Hydraulic Models, The American Society of Civil Engineers Manual of Practice*, **25**, American Society of Civil Engineers, New York, 110 pp.

Ashmore, P.E. 1982. Laboratory modelling of gravel braided stream morphology, *Earth Surface Processes and Landforms*, **7**, 201–225.

Ashmore, P.E. 1988. Bedload transport in braided gravel-bed stream models, *Earth Surface Processes and Landforms*, **13**, 677–695.

Ashmore, P.E. 1991a. Channel morphology and bed load pulses in braided, gravel-bed streams, *Geografiska Annaler*, **68**, 361–371.

Ashmore, P.E. 1991b. How do gravel-bed rivers braid? *Canadian Journal of Earth Sciences*, **28**, 326–341.

Ashmore, P.E. 1993. Anabranch confluence kinetics and sedimentation processes in gravel-braided streams, in *Braided Rivers*, edited by J.L. Best and C.S. Bristow, Geological Society Special Publications, 75, pp. 129–146.

Ashmore, P.E., Ferguson, R.I., Prestegaard, K.L., Ashworth, P.J. and Paola, C. 1992. Secondary flow in anabranch confluences of a braided, gravel-bed stream, *Earth Surface Processes and Landforms*, **17**, 299–311.

Ashmore, P.E. and Parker, G. 1983. Confluence scour in coarse braided streams, *Water Resources Research*, **19**, 392–402.

Ashworth, P.J. 1996. Mid-channel bar growth and its relationship to local flow strength and direction, *Earth Surface Processes and Landforms*, **21**, 103–123.

Ashworth, P.J. and Best, J.L. 1994. The scale modelling of braided rivers of the Ivishak Formation, Prudhoe Bay II: shale geometries and response to differential aggradation rates, *Final BP Project Report*, August 1994, 247 pp.

Ashworth, P.J. and Best, J.L. in press. Discussion of 'the use of hydraulic models in the management of braided gravel-bed rivers' by Warburton, J. and Davies, T.R.H., in *Gravel-bed Rivers in the Environment*, edited by P.C. Klingeman, R.L. Beschta, P.D. Komar and J.B. Bradley, Wiley, New York.

Ashworth, P.J., Best, J.L. and Leddy, J.O. 1992b. The scale modelling of braided rivers of the Ivishak Formation, Prudhoe Bay, *Final BP Project Report Phase 2*, September 1992, 76 pp.

Ashworth, P.J., Best, J.L., Leddy, J.O. and Geehan, G.W. 1994. The physical modelling of braided rivers and deposition of fine-grained sediment, in *Process Models and Theoretical Geomorphology*, edited by M.J. Kirkby, Wiley, Chichester, pp. 115–139.

Ashworth, P.J., Ferguson, R.I. and Powell, M.D. 1992a. Bedload transport and sorting in braided channels, in *Dynamics of Gravel-bed Rivers*, edited by P. Billi, R.D. Hey, C.R. Thorne and P. Tacconi, Wiley, Chichester, pp. 497–513.

Bagnold, R.A. 1980. An empirical correlation of bedload transport rates in natural rivers, *Proceedings of the Royal Society of London*, **372A**, 453–473.

Bennett, S.J. and Best, J.L. 1995. Mean flow and turbulence structure over fixed, two-dimensional dunes: implications for sediment transport and bedform stability, *Sedimentology*, **42**, 491–513.

Best, J.L. 1987. Flow dynamics at river channel confluences: implications for sediment transport and bed morphology, in *Recent Developments in Fluvial Sedimentology*, edited by F.G. Ethridge, R.M. Flores and M.D. Harvey, Special Publication of the Society of Economic Palaeontologists and Mineralogists 39, pp. 27–35.

Best, J.L. 1988. Sediment transport and bed morphology at river channel confluences, *Sedimentology*, **35**, 481–498.

Best, J.L. 1996. The fluid dynamics of small-scale alluvial bedforms, in *Advances in Fluvial Dynamics and Stratigraphy*, edited by P.A. Carling and M. Dawson, Wiley, Chichester, 67–125.

Best, J.L. and Ashworth, P.J. 1994. A high-resolution ultrasonic bed profiler for use in laboratory flumes, *Journal of Sedimentary Research*, **A64**, 674–675.

Biron, P., de Serres, B., Roy, A.G. and Best, J.L. 1993. Shear layer turbulence at an unequal depth channel confluence, in *Turbulence: Perspectives on Flow and Sediment Transport*, edited by N.J. Clifford, J.R. French and J. Hardisty, Wiley, Chichester, pp. 197–213.

Boothroyd, J.C. and Ashley, G.M. 1975. Process, bar morphology and sedimentary structures on braided outwash fans, northeastern Gulf of Alaska, in *Glaciofluvial and Glaciolacustrine Sedimentation*, edited by A.V. Jopling and B.C. McDonald, Society of Economic Paleontologists and Mineralogists Special Publication 23, pp. 193–222.

Bridge, J.S. and Leeder, M.R. 1979. A simulation model of alluvial stratigraphy, *Sedimentology*, **26**, 617–644.

Bristow, C.S and Best, J.L. 1993. Braided rivers: perspectives and problems, in *Braided Rivers*, edited by J.L. Best and C.S. Bristow, Geological Society Special Publication 75, pp. 1–11.

Bryant, M., Falk, P. and Paola, C. 1995. Experimental study of avulsion frequency and rate of deposition, *Geology*, **23**, 365–368.

Buckingham, E. 1915. Model experiments and the forms of empirical equations, *Transactions of the American Society of Mechanical Engineers*, **37**, 263–292.

Carling, P. 1988. The concept of dominant discharge applied to two gravel-bed streams in relation to channel stability thresholds, *Earth Surface Processes and Landforms*, **13**, 355–367.

Carson, M.A. 1986. Characteristics of high-energy 'meandering' rivers: the Canterbury Plains, New Zealand, *Geological Society of America Bulletin*, **97**, 886–895.

Chadwick, A.J. and Morfett, J.C. 1986. *Hydraulics in Civil Engineering*, Harper Collins, London, 492 pp.

Chorley, R.J. 1967. Models in geomorphology, in *Models in Geography*, edited by R.J. Chorley and P. Haggett, Methuen, London, pp. 59–96.

Church, M. and Jones, D. 1982. Channel bars in gravel-bed rivers, in *Gravel-bed Rivers*, edited by R.D. Hey, J.C. Bathurst and C.R. Thorne, Wiley, Chichester, pp. 291–338.

Church, M., Wolcott, J.F. and Fletcher, W.K. 1991. A test of equal mobility in fluvial sediment transport: behaviour of the sand fraction, *Water Resources Research*, **27**, 2941–2951.

Crickmore, M.J. 1970. Effect of flume width on bedform characteristics, *Journal of the Hydraulics Division, Proceedings of the American Society of Civil Engineers*, **96**, 473–496.

Davies, T.R.H. and Griffiths, G.A. in press. Physical model study of stage–discharge relationships in a braided river gorge, *Journal of Hydrology (New Zealand)*. **35**, 2.

Davies, T.R.H. and Lee, A.L. 1988. Physical hydraulic modelling of width reduction and bed level change in braided rivers, *Journal of Hydrology (New Zealand)*, **27**, 113–127.

Dingman, S.L. 1984. *Fluvial Hydrology*, W.H. Freeman, New York, 383 pp.

Durst, F., Melling, A. and Whitelaw, J.H. 1987. *Principles and Practice of Laser-Doppler Anemometry*, 2nd edn, G. Braun, Karlsruhe, 405 pp.

Elghobashi, S. 1994. On predicting particle-laden turbulent flows, *Applied Scientific Research*, **52**, 309–329.

Fahnestock, R.K. 1963. Morphology and hydrology of a glacial stream – White River, Mount Rainier, Washington, *US Geological Survey Professional Paper 422A*, 70 pp.

Franco, J.J. 1978. Guidelines for the design, adjustment and operation of models of the study of river sedimentation problems, Instruction Report H-78-1, *US Waterways Experimental Station*, Vicksburg, Miss., 57 pp.

French, R.H. 1985. *Open-channel Hydraulics*, McGraw-Hill, New York, 739 pp.

García, M.H. 1993. Hydraulic jumps in sediment-driven bottom currents, *Journal of Hydraulic Engineering*, **119**, 1094–1117.

García, M.H., Niño, Y. and López, F. 1996. Laboratory observations of particle entrainment into suspension by turbulent bursting, in *Coherent Flow Structures in Open Channels*, edited by P.J. Ashworth, S.J. Bennett, J.L. Best and S.J. McLelland, Wiley, Chichester, 63–68.

Gibbs, R.J., Mathews, M.D. and Link, D.A. 1971. The relationship between sphere size and settling velocity, *Journal of Sedimentary Petrology*, **41**, 7–18.

Gilbert, G.K. 1914. Transportation of debris by running water, *US Geological Survey Professional Paper 86*, 263 pp.

Gilbert, G.K. 1917. Hydraulic mining debris in the Sierra Nevada, *US Geological Survey Professional Paper 105*, 154 pp.

Guy, H.P., Simons, D.B. and Richardson, E.V. 1966. Summary of alluvial channel data from flume experiments, 1956–1961, *US Geological Survey Professional Paper 462-I*, 96 pp.

Hardisty, J. 1993. Monitoring and modelling sediment transport at turbulent frequencies, in *Turbulence: Perspectives on Flow and Sediment Transport*, edited by N.J. Clifford, J.R. French and J. Hardisty, Wiley, Chichester, pp. 35–59.

Hein, F.J. and Walker, R.G. 1977. Bar evolution and development of stratification in the gravelly, braided Kicking Horse River, British Columbia, *Canadian Journal of Earth Sciences*, **14**, 562–570.

Henderson, F.M. 1966. *Open Channel Flow*, Macmillan, New York, 552 pp.

Hoey, T.B. and Sutherland, A.J. 1989. Self formed channels in a laboratory sand tray, *Proceedings, 23rd Congress, International Association for Hydraulic Research*, Ottawa, Canada, pp. 41–48.

Hoey, T.B. and Sutherland, A.J. 1991. Channel morphology and bedload pulses in braided rivers: a laboratory study, *Earth Surface Processes and Landforms*, **16**, 447–462.

Hooke, R.L. 1968. Model geology: prototype and laboratory streams: discussion, *Geological Society of America Bulletin*, **79**, 391–394.

Ikeda, S. and Kanazawa, M. 1995. Organised vortex structures in turbulent flows with flexible water plants, in *Coherent Flow Structures in Open Channels*, Leeds, England, 10–12th April 1995, *Abstract Volume*, 29 pp.

Ikeda, S. and Parker G. (eds). 1989. *River Meandering*, American Geophysical Union, Water Resources Monograph 12, 485 pp.

Jaeggi, M.N.R. 1986. Non distorted models for research on river morphology, *Proceedings of the Symposium on Scale Effects in Modelling Sediment Transport Phenomena*, August 1986, International Association of Hydrological Sciences, pp. 70–84.

Jin, D. and Schumm, S.A. 1987. A new technique for modelling river morphology, in *International Geomorphology, 1986 Part I*, edited by V. Gardiner, Wiley, Chichester, pp. 681–690.

Julien, P.Y. and Klaassen, G.J. 1995. Sand-dune geometry of large rivers during floods, *Journal of Hydraulic Engineering*, **121**, 657–663.

Karcz, I. and Kersey, D. 1980. Experimental study of free-surface flow instability and bedforms in shallow flows, *Sedimentary Geology*, **27**, 263–300.

Kennedy, J.F. 1995. The Albert Shields story, *Journal of Hydraulic Engineering*, **121**, 766–772.

Kirkby, M.J. (ed.). 1994. *Process Models and Theoretical Geomorphology*, Wiley, Chichester, 417 pp.

Klaassen, G.J. 1991. On the scaling of braided sand-bed rivers, in *Movable Bed Physical Models*, edited by H.W. Shen, Kluwer Academic, New York, pp. 59–72.

Koss, J.E., Ethridge, F.G. and Schumm, S.A. 1994. An experimental study of the effects of base-level change on fluvial, coastal plain and shelf systems, *Journal of Sedimentary Research*, **B64**, 90–98.

Kuhnle, R.A. 1993. Incipient motion of sand–gravel sediment mixtures, *Journal of Hydraulic Engineering*, **119**, 1400–1415.

Kuhnle, R.A. and Southard, J.B. 1988. Bed load transport fluctuations in a gravel bed laboratory channel, *Water Resources Research*, **24**, 247–260.

Langhaar, H.L. 1980. *Dimensional Analysis and Theory of Hydraulic Models*, Robert E. Krieger, Florida, 178 pp.

Leddy, J.O. 1993. Physical scale modelling of braided rivers: avulsion and channel pattern change, M.Phil.thesis, University of Leeds, 130 pp.

Leddy, J.O., Ashworth, P.J. and Best, J.L. 1993. Mechanisms of anabranch avulsion within gravel-bed braided rivers: observations from a scaled physical model, in *Braided Rivers*, edited by J.L. Best and C.S. Bristow, Geological Society Special Publication, **75**, pp. 119–127.

Leeder, M.R. 1982. *Sedimentology: Process and Product*, Unwin Hyman, London, 344 pp.

Leopold, L.B. and Wolman, M.G. 1957. River channel patterns: braided, meandering and straight, *US Geological Survey Professional Paper 282-B*, pp. 39–85.

Lewin, J. 1976. Initiation of bed forms and meanders in coarse-grained sediment, *Geological Society of America Bulletin*, **87**, 281–285.

Linden, P.F., Boubnov, B.M. and Dalziel, S.B. 1995. Source-sink turbulence in a rotating stratified fluid, *Journal of Fluid Mechanics*, **298**, 81–112.

Lisle, T.E., Ikeda, H. and Iseya, F. 1991. Formation of stationary alternate bars in a steep channel with mixed-size sediment: a flume experiment, *Earth Surface Processes and Landforms*, **16**, 463–469.

McCollum, R.A. 1988. Blountstown Reach, Apalachicola River; movable-bed model study, Technical Report HL-88-17, *US Waterways Experimental Station*, Vicksburg, 39 pp.

Mackey, S.D. and Bridge, J.S. 1992. A revised FORTRAN program to simulate alluvial stratigraphy, *Computers and Geosciences*, **18**, 119–181.

Mackey, S.D. and Bridge, J.S. 1995. Three-dimensional model of alluvial stratigraphy: theory and application, *Journal of Sedimentary Research*, **B65**, 7–31.

McLean, S.R., Nelson, J.M. and Wolfe, S.R. 1994. Turbulence structure over two-dimensional bedforms: implications for sediment transport, *Journal of Geophysical Research*, **99**, 12 729–12 747.

Marsden, N. 1981. Model simulation of effect of vegetation on braided rivers, *Project Report*, Department of Agricultural Engineering, Lincoln College, Canterbury, New Zealand, 68 pp.

Mosley, M.P. 1976. An experimental study of channel confluences, *Journal of Geology*, **84**, 535–562.

Mosley, M.P. and Zimpfer, G.L. 1978. Hardware models in geomorphology, *Progress in Physical Geography*, **2**, 438–461.

Murphy, G. 1950. *Similitude in Engineering*, Ronald Press, New York, 302 pp.

Nelson, J.M., Shreve, R.L., McLean, S.R. and Drake, T.G. 1995. Role of near-bed turbulence structure in bed load transport and bed form mechanics, *Water Resources Research*, **31**, 2071–2086.

Nikuradse, J. 1933. Strömungsgesetze in rauhen Rohren, VDI-Forschungsheft, 361. English translations: *NACA Technical Memo. 1292* and Petroleum Engineer (1940) March, 164–166; May, 75, 78, 80, 82; June, 124, 127, 128, 130; July, 38, 40, 42; August, 83, 84 and 87.

Novak, P. and Cábelka, J. 1981. *Models in Hydraulic Engineering*, Pitman, London, 459 pp.

Ouchi, S. 1985. Response of alluvial rivers to slow active tectonic movement, *Geological Society of America Bulletin*, **96**, 504–515.

Owen, M.W. 1985. Ports and harbours, in *Developments in Hydraulic Engineering – 3*, edited by P. Novak, Elsevier, London, pp. 263–311.

Paola, C., Parker, G., Seal, R., Sinha, S.K., Southard, J.B. and Wilcock, P.R. 1992. Downstream fining by selective deposition in a laboratory flume, *Science*, **258**, 1757–1760.

Parker, G. 1979. Hydraulic geometry of active gravel rivers, *Journal of the Hydraulics Division, Proceedings of the American Society of Civil Engineers*, **105**, 1185–1201.

Parker, G. 1996. Some speculations on the relation between channel morphology and channel-scale flow structures, in *Coherent Flow Structures in Open Channels*, edited by P.J. Ashworth, S.J. Bennett, J.L. Best and S.J. McLelland, Wiley, Chichester, pp. 423–458.

Parker, G., García, M., Fukushima, Y. and Yu, W. 1987. Experiments on turbidity currents over an erodible bed, *Journal of Hydraulic Research*, **25**, 123–147.

Peakall, J. 1995. The influences of lateral ground-tilting on channel morphology and alluvial architecture, Ph.D. thesis, University of Leeds, 333 pp.

Peakall, J. and Warburton, J. in press. Surface tension in small hydraulic river models – the significance of the Weber number, *Journal of Hydrology (New Zealand)*, **35**, 2.

Pickup, G. 1988. Hydrology and sediment models, in *Modelling Geomorphological Systems*, edited by M.G. Anderson, Wiley, Chichester, pp. 153–215.

Posamentier, H.W., Allen, G.P. and James, D.P. 1992. High resolution sequence stratigraphy – the East Coulee Delta, Alberta, *Journal of Sedimentary Petrology*, **62**, 310–317.

Raudkivi, A.J. 1966. Bed forms in alluvial channels, *Journal of Fluid Mechanics*, **26**, 507–514.

Reid, I., Brayshaw, A.C. and Frostick, L.E. 1984. An electromagnetic device for automatic detection of bedload motion and its field applications, *Sedimentology*, **31**, 269–276.

Reid, I. and Frostick, L.E. 1984. Particle interaction and its effect on the thresholds of initial and final bedload motion in coarse alluvial channels, in *Sedimentology of Gravels and Conglomerates*, edited by E.H. Koster and R.H. Steel, Canadian Society for Petroleum Geology Memoir 10, pp. 61–68.

Reynolds, O. 1887. On certain laws relating to the régime of rivers and estuaries, and on the possibility of experiments on a small scale. A Report of the British Association, in Reynolds, O., *Papers on Mechanical and Physical Subjects*, Vol. II, 1881–1900, Cambridge University Press, 1901, pp. 326–335.

Rice, C.T., Wilson, B.N. and Appleman, M. 1988. Soil topography measurements using image processing techniques, *Computers and Electronics in Agriculture*, **3**, 97–107.

Richards, K.S. 1982. *Rivers, Form and Process in Alluvial Channels*, Methuen, London, 361 pp.

Römkens, M.J.M., Wang, J.Y. and Darden, R.W. 1988. A laser microrelief-meter, *Transactions American Society of Agricultural Engineers*, **31**, 408–413.

Rouse, H. 1939. An analysis of sediment transportation in light of fluid turbulence, SCS-TP-25, Sediment Division, *US Department of Agriculture, Soil Conservation Service*, Washington, DC.

Rouse, H. (ed.). 1950. *Engineering Hydraulics*, Wiley, New York, 1039 pp.

Rouse, H.L. 1994. Measurement of bedload gravel transport: the calibration of a self-generated noise system, *Earth Surface Processes and Landforms*, **19**, 789–800.

Roy, A.G. and Roy, R. 1988. Changes in channel size at river confluences with coarse bed material, *Earth Surface Processes and Landforms*, **13**, 77–84.

Schlichting, H. 1968. *Boundary Layer Theory*, 6th edn, McGraw-Hill, New York, 748 pp.

Schoklitsch, A. 1962. *Handbuch des Wasserbaues*, 3rd edn, Springer-Verlag, Vienna, 475 pp.

Schumm, S.A. 1960. The shape of alluvial channels in relation to sediment type, *US Geological Survey Professional Paper 352-B*, 30 pp.

Schumm, S.A. and Khan, H.R. 1972. Experimental study of channel patterns, *Geological Society of America Bulletin*, **83**, 1755–1770.

Schumm, S.A., Mosley, M.P. and Weaver, W.E. 1987. *Experimental Fluvial Geomorphology*, Wiley, New York, 413 pp.

Seal, C.V., Smith, C.R., Akin, O. and Rockwell, D. 1995. Quantitative characteristics of a laminar unsteady necklace vortex system at a rectangular block–flat plate juncture, *Journal of Fluid Mechanics*, **286**, 117–135.

Shen, H.W. (ed.). 1991. *Movable Bed Physical Models*, Kluwer Academic, Boston, 171 pp.

Simons, D.B., Richardson, E.V. and Albertson, M.L. 1961. Flume studies using medium sand (0.45 mm), *US Geological Survey Water Supply Paper 1498-A*, 76 pp.

Smith, D.G. 1976. Effect of vegetation on lateral migration of anastomosed channels of a glacial meltwater river, *Geological Society of America Bulletin*, **87**, 857–860.

Southard, J.B. 1971. Representation of bed configurations in depth–velocity–size diagrams, *Journal of Sedimentary Research*, **41**, 903–915.

Southard, J.B. and Boguchwal, L.A. 1990a. Bed configurations in steady unidirectional water flows, part 2, synthesis of flume data, *Journal of Sedimentary Petrology*, **60**, 658–679.

Southard, J.B. and Boguchwal, L.A. 1990b. Bed configurations in steady unidirectional water flows, part 3, effects of temperature and gravity, *Journal of Sedimentary Petrology*, **60**, 680–686.

Southard, J.B., Smith, N.D. and Kuhnle, R.A. 1984. Chutes and lobes: newly identified elements in braiding in shallow gravelly streams, in *Sedimentology of Gravels and Conglomerates*, edited by E.H. Koster and R.J. Steel, Canadian Society of Petroleum Geology Memoir 10, pp. 51–59.

Takeda, Y. 1991. Development of an ultrasound velocity profile monitor, *Nuclear Engineering and Design*, **126**, 277–284.

Takeda, Y. 1995. Instantaneous velocity profile measurement by ultrasonic doppler method, *International Journal of the Japanese Society of Mechanical Engineers*, **38B**, 8–16.

Thomson, J. 1879. Flow round river bends, *Proceedings of the Institute of Mechanical Engineers*, pp. 456–460.

Thorne, P.D., Williams J.J. and Heathershaw, A.D. 1989. In situ acoustic measurements of marine gravel threshold and transport, *Sedimentology*, **36**, 61–74.

Tritton, D.J. 1988. *Physical Fluid Dynamics*, 2nd edn, Clarendon Press, Oxford, 519 pp.

Tsujimoto, T. 1996. Coherent fluctuations in a vegetated zone of open-channel flow: causes of bedload lateral transport and sorting, in *Coherent Flow Structures in Open Channels*, edited by P.J. Ashworth, S.J. Bennett, J.L. Best and S.J. McLelland, Wiley, Chichester. pp. 375–396.

Warburton, J. and Davies, T.R.H. 1994a. Variability of bedload transport and channel morphology in a braided river hydraulic model, *Earth Surface Processes and Landforms*, **19**, 403–421.

Warburton, J. and Davies, T.R.H. 1994b. Variability of bedload transport and channel morphology in a braided river hydraulic model, errata, *Earth Surface Processes and Landforms*, **19**, Issue 8, ii.

Warburton, J. and Davies, T.R.H. in press. The use of hydraulic models in the management of braided gravel-bed rivers, in *Gravel-bed Rivers in the Environment*, edited by P.C. Klingeman, R.L. Beschta, P.D. Komar and J.B. Bradley, Wiley, New York.

Webb, E.K. 1994. Simulating the three-dimensional distribution of sediment units in braided-stream deposits, *Journal of Sedimentary Research*, **B64**, 219–231.

Webb, E.K. 1995. Simulation of braided-channel topology and topography, *Water Resources Research*, **31**, 2603–2611.

Welford, M.R. 1994. A field test of Tubino's model of alternate bar formation, *Earth Surface Processes and Landforms*, **19**, 287–297.

Whipple, K.X., Parker, G. and Paola, C. 1995. Experimental study of alluvial fans, in *Proceedings of the International Joint Seminar on Reduction of Natural and Environmental Disasters in Water Environment*, edited by J.H. Sonu, K.S. Lee, I.W. Seo and N.G. Bhowmik, Seoul National University, 18–21 July, pp. 282–295.

Williams, G.P. 1970. Flume width and water depth effects in sediment transport experiments, *US Geological Survey Professional Paper 562-H*, 37 pp.

Williams, P.F. and Rust, B.R. 1969. The sedimentology of a braided river, *Journal of Sedimentary Petrology*, **39**, 649–679.

Wolman, M.G. and Miller, J.P. 1960. Magnitude and frequency of forces in geomorphic processes, *Journal of Geology*, **68**, 54–74.

Wood, L.J., Ethridge, F.G. and Schumm, S.A. 1993. The effects of rate of base-level fluctuation on coastal-plain, shelf and slope depositional systems: an experimental approach, in *Sequence Stratigraphy and Facies Associations*, edited by H.W. Posamentier, C.P. Summerhayes, B.U. Haq and G.P. Allen, Special Publication of the International Association of Sedimentologists 18, pp. 43–53.

Wood, L.J., Ethridge, F.G. and Schumm, S.A. 1994. An experimental study of the influence of subaqueous shelf angles on coastal plain and shelf deposits, in *Siliciclastic Sequence Stratigraphy: Recent Developments and Applications*, edited by P. Weimer and H.W. Posamentier, Association of American Petroleum Geologists Memoir 58, pp. 381–391.

Yalin, M.S. 1963. An expression for bed-load transportation, *Journal of the Hydraulics Division, Proceedings of the American Society of Civil Engineers*, **89**, 221–250.

Yalin, M.S. 1971. *Theory of Hydraulic Models*, Macmillan, London, 266 pp.

Young, W.J. 1989. Bedload transport in braided gravel-bed rivers, Ph.D. thesis, University of Canterbury, New Zealand, 187 pp.

Young, W.J. and Davies, T.R.H. 1990. Prediction of bedload transport rates in braided rivers: a hydraulic model study, *Journal of Hydrology (New Zealand)*, **29**, 75–92.

Young, W.J. and Davies, T.R.H. 1991. Bedload transport processes in a braided gravel-bed river model, *Earth Surface Processes and Landforms*, **16**, 499–511.

Zimmerman, R.C., Goodlett, J.C. and Corner, G.H. 1967. The influence of vegetation on channel form of small streams, *Publication of the International Association of Scientific Hydrology*, **75**, 255–275.

Zwamborn, J.A. 1967. Solution of river problems with movable bed hydraulic models, Symposium paper S.24, Council for Scientific and Industrial Research, Pretoria, South Africa, 40 pp.

MODELING: PROSPECTS AND PROBLEMS

Contemporary geomorphology is increasingly embracing the notion that theoretical concepts from the basic sciences, such as physics and chemistry, provide foundational principles for guiding the development of theories about landscape dynamics. Models, especially mathematical models, provide a structural framework for articulating these theories in a formalized, precise manner. Where the link between landscape processes and the underlying physics or chemistry is strong, we have the classic case of using a model to try to reduce landscape dynamics to underlying physical and chemical mechanisms. In other cases, aggregated variables are employed within more loosely formulated theoretical frameworks in an effort either to predict landform behavior or to determine landscape stability.

Modeling obviously is a specific example of a methodology or technique in geomorphology. Perhaps no other technique has as much mystique or is the center of as much intradisciplinary enthusiasm and skepticism as modeling. Although the mathematical and scientific literacy of geomorphologists has increased enormously over the past 45 years, the discipline still includes a minority that is highly proficient in mathematical modeling, and a majority that either is mathematically illiterate or has only a basic proficiency in mathematics. Geomorphology will benefit greatly from widespread understanding of the strengths and weaknesses of modeling. Such understanding will serve to demystify modeling, allowing its true utility to be objectively evaluated by the community of geomorphologists at large.

Michael Kirkby was invited to provide a context for, and evaluation of, the role of models in geomorphology. In undertaking this task, he emphasizes their importance as thought experiments, the need to maintain simplicity, and the constraint that models can only provide 'possible' explanations. Deborah Lawrence examines the scales at which application of continuum flux models is appropriate and highlights the still underexploited roles of dimensional analysis and scaling in geomorphological modeling. She emphasizes model construction, but also briefly examines some issues related to solution procedures.

Keith Beven uses his extensive experience with hydrological models to explore in detail the notion of models providing 'possible,' rather than 'certain' explanations. He points out that alternative models using a wide variety of parameter sets frequently all produce good fits– a trait that he labels 'model equifinality'. He then suggests ways that this seeming limitation may be exploited.

Sensitivity to initial conditions and unpredictable behavior are patterns in system and model behavior that have long frustrated the generalization that is commonly sought in scientific modeling. Jonathan Phillips discusses the potential role and limitations of nonlinear dynamical systems concepts in explaining such behavioral traits. Of course, the fundamental issue is whether or not such sophisticated descriptions of behavioral patterns can be successfully welded to a body of explanatory theory. Peter Haff provides a detailed consideration of the limitations of models as predictive devices using sediment-transport models as the vehicle for his discussion. He draws particular attention to the difficulties associated with scaling up results and the discovery, and role, of emergent variables as scale is increased.

10 A Role for Theoretical Models in Geomorphology?

Michael J. Kirkby

School of Geography, University of Leeds

ABSTRACT

Simulation models provide one of the crucial links between the study of process and the study of landforms, the two traditional activities of geomorphology. Only in exceptional cases can significant changes in landforms be observed directly, so that models provide a means of extrapolating from short-term process measurements to the long-term evolution of macroscopic landforms. The role of models is discussed, mainly in the context of hillslope profile models.

Models provide both improved understanding and forecasting capability. Preferred models are physically based, generally starting from the continuity of mass equations which also provide the formal link between space and time rates of change. Beyond this, many models for landform evolution are still based on gross simplifications of the detailed process mechanics. Effective models should generally be simple, subject to sufficient generality to allow transfer between areas, and of sufficient richness to link to cognate work. Furthermore, there is an important duty on geomorphologists to reconcile models at different spatial and temporal scales.

Although most simulation models may be used in a forecasting mode, perhaps their more important role is as a qualitative thought experiment, testing whether we have a sufficient and consistent theoretical explanation of landscape processes. The best model can only provide a possible explanation which is more consistent with known data than its current rivals. Every field observation, and especially the more qualitative or anecdotal ones, provides an opportunity to refute, or in some cases overturn, existing models and the theories which lie behind them.

The Scientific Nature of Geomorphology: Proceedings of the 27th Binghamton Symposium in Geomorphology held 27–29 September 1996. Edited by Bruce L. Rhoads and Colin E. Thorn. © 1996 John Wiley & Sons Ltd.

INTRODUCTION

The science of geomorphology is severely constrained by the normally slow rates of landform change. Our data typically come from cross-sectional studies, either over time in stratigraphic sequences or over space as the distribution of current landforms. In addition we can make relatively short-term process studies in the field or the laboratory. Our preferred interpretations, however, are generally for time and space scales which increase together, from event-based erosion plot models to regional or global models with geological time scales (Figure 10.1). Other combinations can be found in the literature, but are generally less satisfactory. For example, there are many hillslope evolution models which are able to run for a million years or more, but they are deficient in providing the regional setting, particularly in terms of tectonics and basal boundary conditions. Similarly many remote sensing studies provide global snapshots of landscapes, but the geomorphology can only properly be interpreted in terms of regional geological history.

Models, seen as simplifying abstractions of reality, provide the basis for aggregating from the scales of the observations to the scales of interest. In principle a physically based understanding, obtained from process studies, can be applied to explain both current spatial distributions and at-a-point stratigraphic sections, and these cross-sectional studies can be used to calibrate and validate the physical model for more general application at all time and space scales.

We can implement this approach most readily if we are able to make some equilibrium assumptions: preferably equilibrium with respect to either time or space. Note, however, that conditions for equilibrium are themselves scale dependent. We may choose to assume that forms have reached equilibrium over time, because we believe that the duration of uniform conditions is long relative to the system response time. We may then interpret the spatial distribution of forms as a set of equilibrium responses to different climatic, lithological or tectonic environments. Still assuming equilibrium over time, we are able to interpret the stratigraphic record in terms of the palaeogeography of the site relative to

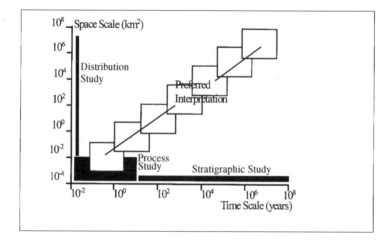

Figure 10.1   The relationship between field studies (shaded) and geomorphological interpretations

shorelines, plate margins, etc. Alternatively, we may choose to assume that equilibrium has been reached over space, because we believe that the lateral extent of uniform conditions lies within a uniform erosional or depositional environment. We then interpret the spatial distribution of landforms in terms of, say, Davisian stage; and the stratigraphic record as responses to environmental change.

In the simplest possible view, we adopt narrow uniformitarianism which allows us to argue directly from analogy. Using this method, we may, for example, infer the impact of global climate change in an area by looking at the landforms and processes acting in an analogue area, which matches the lithology and expected climate or land-use scenario for our area of interest. However, we quickly run into difficulties both because there are no exact analogues (the well-known failure of naïve uniformitarianism) and because of the wide range of relevant relaxation times which coexist in the landscape.

A CONTEXT: HILLSLOPE PROFILE MODELS

To provide a more concrete basis for discussing the properties of models in geomorphology or physical geography, we will focus on one-dimensional models for slope profiles, catenas or flow strips. These typically refer to spatial scales of 0.01–1 km^2, and have been used at time scales ranging from seconds to millions of years (Figure 10.2). As suggested above, some of these time scales are more appropriate and fruitful than others for the single profile, and the slope profile models may also be considered as components within catchments or other larger areal units. The models have been categorised for convenience into three ranges of time scales, which might be loosely linked to Schumm and Lichty's (1965) steady state, graded and cyclic spans, although it is recognised that these categories were originally formulated for rivers rather than hillslopes. For each time span, the lower limit might correspond to the fundamental computational iteration, and the upper limit to

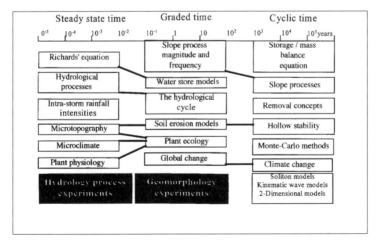

Figure 10.2   Relevant models for slope profiles at a range of time scales. Only the links between time spans are shown

the duration of 1000 such iterations, loosely related to the relevant time span for which a model may sensibly run.

As interpreted here, the time spans refer to hillslope sediment transport rather than to the hillslope form, and the periods involved then relate more closely to those for the states of the fluvial system. Steady-state time is interpreted as a period in which the slope form is essentially static, but with considerable dynamism in soil and surface hydrology. Graded time refers to a period over which hillslope sediment transport can be assigned a mean value and a distribution of magnitudes and frequencies. Cyclic time refers to a period over which sediment transport begins to have a cumulative impact on the form of the hillslope, and over which the parts of the hillslope profile interact with one another via sediment transport. In this sense cyclic time, as used here, includes periods in which the hillslope is both 'graded' and over which it shows net evolution.

Historically, slope models have most commonly combined a mass balance or continuity equation with a set of process 'laws' which express the variation of sediment transport rate in terms of topographic variables, typically as power laws in slope gradient and distance from the divide. Gradient is clearly a direct driver for sediment transport, and distance represents the collecting area for flow, which is important for wash transport. In this form, removal is generally transport (or flux) limited, the processes are effectively limited to diffusive and wash transport, and models may readily (if not realistically) run for iterations of 1000 years, and total periods which represent millions of years. Greater richness and realism may be introduced at the expense of some increased complexity.

The range of slope processes may be widened to include solution and mass movements, as well as taking on the distinctions between rainsplash, rainflow and rillwash. The addition of solution rates also allows changes in regolith depth to be simulated. By introducing the concept of travel distance, the mass balance framework can accommodate both transport (flux) and detachment (supply) limited removal, allowing better representation of mass movement processes especially, and allowing grain-size selective wash. Although one-dimensional models cannot deal with the nonlinearities associated with channel and valley extension, the criteria for stability may be evaluated in one dimension. There is also scope, little realised as yet, for Monte Carlo simulation in the parameter space in order to provide realistic model validation. Perhaps the most important extension is the use of realistic scenarios for changing process rates over time, indirectly reflecting the major climate changes of the Quaternary.

Using time steps representing a month, or a storm event, it is possible to build up to periods of up to 100 years, which are relevant to long-term planning horizons and to global change scenarios for climate, land-use and CO_2. This corresponds to the graded time span over which discussions of process magnitude and frequency are relevant. It is also the time frame relevant to an understanding of plant ecology and to seasonal hydrological changes. This medium time scale is the span of most geomorphological process measurements and most hydrological and meteorological records. An important class of geomorphological studies is related to soil erosion from runoff plots, in which the pattern of rilling, microtopography and vegetation is one significant set of controls (e.g. Kirkby et al. in press).

Five-minute time steps, building up to a period of a few days, represent the time scale appropriate for most detailed hydrological, meteorological and plant eco-physiological process studies. At this scale we may be concerned with representing microtopography, infiltration patterns and overland flow paths in an explicit way; and it is at this level that we generally feel most secure in claiming a physical understanding of the processes. Examples of this level of theory include the Richards equation for movement of water in soils, Reynolds and Darcy–Weissbach roughness equations built into a kinematic cascade for overland flow relationships and the Penman–Monteith equation for evapotranspiration.

These examples will be used to explore our levels of understanding and to assess how far we are from the desired objective of creating a theory, or a family of models, which can be validated against field data, and used to link our knowledge of process and form. In the context of hillslope profiles, it is also worth noting that the stratigraphic dimension in Figure 10.1 is very poorly developed because hillslopes are essentially erosional forms, with minimal preservation potential, so that the need to rely on models is particulary strong.

WHAT SHOULD A MODEL PROVIDE?

Models ideally provide both an insight into the functioning of the natural environment, and a means of forecasting the range of likely outcomes, either in a forecasting sense or to assess the impact of alternative policies. The priority given to these objectives differs according to needs, but in the best practice there should be no conflict. A useful model will be able to make reliable forecasts for environments other than those for which they were originally constructed. As we attempt to combine our best field data to provide estimates for large areas and over long times (Figure 10.1), we will inevitably need to work well outside our original data set, so that we urgently need 'good' models!

Better Understanding

One essential features of any numerical model is that it must provide a logically sufficient and consistent explanation of the process or form it represents. Thus any model offers a view, usually simplified, of our understanding of the system of interest. The insights associated with a useful model, particularly with a conceptually simple model, generally have a much greater impact than any specific forecasts, because they provide components for work in related subfields, and allow new progress to be made, building on the understanding gained. In principle, the same benefits may accrue to non-numerical models which have a formal logical basis, but in practice the logical basis of many qualitative models is less exact, and therefore less effective at revealing any lack of consistency or incompleteness in the explanation.

Once a level of understanding has been achieved, there is generally some scope to assess the criticality and quantitative importance of each process and each state or storage to the overall explanation. Ideally a process of distillation can lead to an essential core of theory, eliminating secondary factors.

These processes of explanation and distillation have always been at the heart of theory development. Numerical modelling may be used to help this process, but may also obscure it with an overemphasis on the quality of the numerical forecasts, and on deriving parameter values rather than meaningful and consistent relationships. Intellectual insight into the working of natural processes is the crucial tool which allows us to go beyond the inductive content of even the best-designed experiment, and structure our world with a network of scientific theories. This search for deeper understanding must lie at the heart of the most significant modelling activity, in geomorphology as in all science, and underpins the critical dialogue between the development of theory and the design of critical experiments.

An important aspect of our search for understanding is to reconcile theories at different scales. Self-consistent theories at each scale typically make generalisations which are not immediately seen to be compatible with those on the next scale, and an important stage in mature theory development is to reconcile theories across scale differences.

Forecasting Potential

Forecasting is both a useful activity and a means of testing the validity and range of our understanding. While the best models may be those with an elegant simplicity, the best forecasts have to apply the understanding of principles to real examples, often in combination with other less complete theories and with strong empirical components.

A model or theory needs to be validated against field data, by comparing its forecasts with observations. In many cases, validation can only be achieved indirectly, since we cannot, for example, observe landscape change over geological time periods. In any case, there is no absolute criterion for acceptance of a theory, and Popper's (1972) view of theories as open to rejection but not acceptance, seems to provide a practicable programme for research. Validation also requires some statistical assessment of goodness of fit, which can be obtained from a distribution of acceptable model outcomes, from a distribution of acceptable real-world outcomes, or some combination of both. However, although forecasts strive towards empirical accuracy, it is clear that most geomorphological models are far from achieving it.

WHAT MAKES A GOOD MODEL?

If a model can be formally constructed from a body of existing theory with the addition of any necessary new development, if the theory takes into account the dominant processes operating, and if the theories are at the scale of interest, then the new theory should be as fundamental as that from which it is derived. Because many of our existing models have only an informal link to an accepted body of theory, development of new models has generally been slow. However, there are a number of factors which characterise a good model; an explicit physical basis, simplicity, generality, richness and the potential for scaling up.

Physical Basis

Models range from totally empirical black box models, often based on regression or neural net methods, to those where all parameters are independently determined physical constants, like $E = mc^2$. Even black box models usually have some physical basis, in the selection of variables and in choosing the form of the regression, arithmetic or logarithmic for example, which expresses some preconceptions about the form of the dependence. For example, the universal soil loss equation estimates soil loss as the product (rather than say the sum) of a series of factors.

In geomorphology, few models rise far above empiricism, and most 'physically based' models are simply pushing the level of empiricism one level further down. For example, we may have a sound physical basis for relating wash sediment transport to water discharge, but we are still forced to assign one or more empirical soil erodibility parameters. As the physical basis improves, identification of model parameters becomes more consistent and allows closer links to be made to other theories, loosely along this scale:

Parameters
    calibrated by model optimisation
    can be consistently obtained from measured values
    directly identifiable from field measurements
    'universal' across a wide range of theory

Calibrated parameters can, in general, only be obtained by optimising forecast outcomes against real data. At the next level, the process of parameter estimation becomes the major part of model development, so that many of the factors in the Universal Soil Loss Equation can best be obtained from a set of look-up tables, based on crop types, soil series, conservation practices, etc. At the next higher physical level, we may hope to measure directly soil parameters related to, say, infiltration capacity which have a physical meaning in both our infiltration model and in the field. It is an advantage if these parameters relate to easily identified characteristics such as surface form or vegetation, as there is then much better knowledge of inherent variability and its spatial structure. Finally we might hope to develop models in which the parameters had a wider physical meaning, beyond the confines of the particular model. A few parameters, such as the gravitational acceleration, g, have this wider context, but they are generally in a minority.

Where models have a strong physical basis, this usually provides consistency with other theories. Such consistency helps to support both their validity, by providing additional theoretical support, and their acceptability, by providing more users within the scientific community. Consistency also allows theories to be woven together to explain other related phenomena, and may help to demonstrate consistency across scales, and/or relevant ways to aggregate or disaggregate.

For the hillslope models outlined in Figure 10.2, an important physical basis is provided by the mass balance or storage equation, which already constrains the overall behaviour of any model. The equation guarantees continuity of mass in a sense which is central to the models, since the transfer and storage of water and/or soil masses are at their heart. For the long-term models, the remaining physical content is contained in the slope process rate 'laws', which are generally highly empirical, at least in the simplest versions of the models. Nevertheless, even loosely specified physical principles can have a powerful effect in constraining forecasts within reasonable bounds. Consider the following rather non-specific statements about rates of sediment transport on slopes:

> Net sediment transport is in a downslope direction.
> Overland flow is zero at the divide, and increases downslope.
> Wash transport increases more rapidly than overland flow discharge.
> Not all sediment transport depends on overland flow.

These can most easily be interpreted as suggesting that sediment transport takes the form

$$Q_s = K\Lambda^p + Bx^m\Lambda^n$$

where Λ is slope gradient, x is distance from the divide, and p, m, n, K and B are empirical constants with $m > 1$. The only additional (empirical) assumption implicit in this expression is that the relationships are power laws. The first term represents the non-flow-dependent or 'diffusive' processes, and the second the flow-dependent 'wash' processes. This form of expression was first proposed by Musgrave (1947). One commonly used simple form (Kirkby 1976) is a particular case in which the slope exponents p, n are equal, $m = 2$ and B has been replaced by K/u^2:

$$Q_s = K\left[1 + \left(\frac{x}{u}\right)^2\right]\Lambda$$

Simplicity

Many models suffer from an excess of complexity, few from being too simple. Although some complex detail may be needed to apply a model in a specific context, the central concept of a good model must be simple. This may be argued pragmatically, both by analogy with existing successful models, and by experience of trying to create models. One part of the need for simplicity comes from the need for the model to be understood and communicated; the other part from the need for the modeller to understand how the model works. As a rule of thumb, it is difficult to construct the core of a model in which more than three (usually more than two) dominant processes interact at a time. For example, an important part of fluvial hydraulics is built on the Reynolds and Froude numbers, which are concerned with deciding which two sets of forces need to be considered in any situation. Similarly, it may be argued that landscape form is controlled primarily by the processes which are dominant in landscape denudation. We may consider the ratio of mass movement rates to wash rates; or of wash rates to solution rates as similar dimensionless ratios which determine the hillslope 'regime'. These ratios are controlled by a number of factors, including climate, topographic situation and lithology.

Figure 10.3 sketches a simple realisation of the process domains for wash, mass movement processes and solution, following the concepts of Langbein and Schumm (1958), and Langbein and Dawdy (1964). Comparing wash and mass movements, it shows rapid mass movements dominant on steep slopes and wash dominant below the landslide threshold gradient, both around the semi-arid peak, and in the humid tropics. For the temperate wash minimum and for very arid areas, wash processes are slow enough for

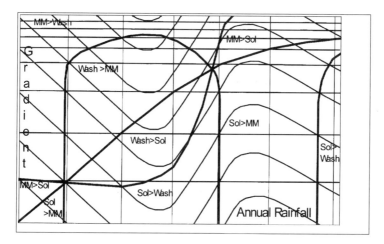

Figure 10.3 Schematic dominance zones for wash, mass movement and solution processes. Axes represent logarithmic scales of annual rainfall and slope gradient. Horizontal lines represent isolines of denudation by mass movement, which is represented as showing a low initial rate of increase for soil creep, and a more rapid increase beyond a threshold for landslides. Vertical lines represent isolines for solutional denudation, increasing linearly with rainfall, at a high rate where rainfall is less than potential evapotranspiration and at a lower rate thereafter. The parallel curves for wash indicate a semi-arid peak, a temperate minimum and a tropical increase, all at rates proportional to gradient. The heavy curves represent the thresholds of equal rates defined for each pair of processes, dividing the field into dominance fields. The captions indicate the two dominant processes in each field

slow mass movements to dominate. Taking account of solution, we may schematically define fields dominated by a pair of dominant processes. Thus the mass movement (MM) > solution (Sol) field may be characterised by periodic stripping of the soil by mass movement, at a rate determined by bedrock weathering; whereas the Sol > MM field is characterised by a creeping saprolite regolith. MM > Wash gives boulder veneered rock slopes, while Wash > MM gives size-sorted pediment and fan surfaces, the two types often linked at a semi-arid break in slope.

Since the majority of geomorphic processes are significantly nonlinear, there is great scope for the development of nonlinearity and chaotic unpredictability in model beha- viour. In practice this is often constrained by the strongly diffusive nature of many landscape processes, which allows landscapes, and models representing them, to run forward over time in a stable manner. Diffusive models, or models with a strong diffusive component, tend to converge as a negative exponential towards stable forms. This con- vergence creates strong equifinality in the evolution, ultimately towards peneplains for a stable tectonic regime, which creates difficulties in assigning a unique model to an observed landform. It also means that models cannot be run backwards in time, since arbitrarily small initial perturbations then grow exponentially over the reversed time scale.

Large numbers of parameters also tend to provide many opportunities for equifinality in the output, usually including cases where it is qualitatively plain that the 'right' answer is being produced by the 'wrong' set of processes. Furthermore, the process of parameter estimation, either by optimisation or by measurement, becomes increasingly laborious, expensive and indeterminate as the number of relevant parameters increases.

The use of minimal parameter sets also presents some practical difficulties. The quality of fit is rarely as perfect as with larger sets, and it is frequently tempting to tweak the model with a few extra parameters, sometimes at the expense of retaining the underlying physical understanding. There is always a trade-off between parameter number and 'objective' measures of fitting efficiency, which can be evaluated through the loss in degrees of freedom as extra parameters are added.

In principle, real systems may not be simple enough to justify a simple model. Past successes and failures suggest that complex explanations can be built up in small steps from secure foundations. Where this route is not available, or not yet available, then we can only make progress by seeking simple generalisations at the whole-system scale, and rarely if ever by building on a dubious infrastructure. This message is being amply documented as we are urged to scale up to an understanding of global issues.

Generality and Richness

There is always a danger that models for particular areas have no validity outside the field area. A model rises above pure numerical description only when it has some transferability to other areas, or may be applied in other contexts. Clearly any improvement in the physical basis helps to enhance transferability, as it improves the consistency of model parameters. These are important components of the concept of model generality. Models also differ in richness: in how much they help to explain. A rich slope profile model may be able to give some information about soil and vegetation conditions, in addition to the bare form of the profile. Such a model provides greater opportunities for cognate understanding and reduces the risk of equifinal outcomes and model misidentification.

Even a well-specified physical model may, however, face difficulties in being transferred to a new area. One of the main problems is that different processes may be dominant, so that transferability may only be possible over a limited range. For example, a hillslope model may only be valid provided that the pair of dominant processes remains the same. Models may adopt different strategies towards greater transferability, or greater relevance to the conditions of a specific site. If transferability is a high priority, then we may prefer to seek generic rather than specific physical understanding. For example, if a geomorphological model includes explicit response to vegetation cover, the optimal model for a local scale may reflect the particular dominant species, and their response to small differences in environment. For transferring to other areas, or other conditions, it is necessary to include individual curves for each species or functional group, or directly represent the envelope relationship across the whole range of species which might replace the current dominant species, based perhaps on considerations of energy conversion. This is shown schematically in Figure 10.4. In the same way, a broadly relevant hillslope model might either include all relevant processes, or use an envelope relationship. In this case, a series of individual process models may be preferable, because the variety of hillslope processes responds to rather different variables and because of the relatively small number of functional process groups. Even the simplistic view illustrated in Figure 10.3 may be too complex to lend itself to the envelope curve approach, so that geomorphologists may prefer to consider each group of processes separately, and strive to include all processes which might be dominant within the planned range of model transferability.

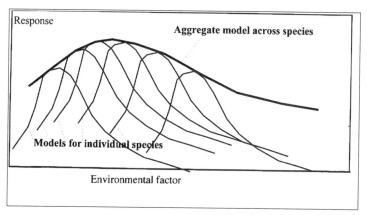

Figure 10.4   Schematic relationship between local and global optima in a model, for the example of vegetation species responses

An important component of richness lies in the notion of the net information gain of the model, defined conceptually as the net change in information content or entropy, comparing output to input values. Highly distributed hydrological models have very large input data requirements, and in many cases are used only to forecast the output hydrograph, so that their net information gain is strongly negative. Lumped models, such as TOPMODEL (Beven and Kirkby 1979), have a more favourable net gain, using a small number of parameters to forecast both outflow hydrographs and, in this example, distributions of saturated area or soil moisture deficits for many points in the catchment.

The information gain is significantly affected by uncertainties attached both to input and output data. Uncertainties in inputs are generally transmitted to the outputs, with gain or attenuation according to the sensitivity of the model to each input. The quality of the output, however, generally also responds to the uncertainties and simplifications built into the model as a representation of reality. Some of this uncertainty is linked to our qualitative categorisation of the model, loosely along this scale:

Principle	e.g. continuity of mass
Law	e.g. Newtonian gravitation
Theory	e.g. plate tectonics
Hypothesis	e.g. the geomorphological unit hydrograph (Rodriguez-Iturbe and Valdéz 1979)
Conjecture	e.g. landscape entropy (Leopold and Langbein 1962)

It is clear that few models or theories in geomorphology are far from the bottom of this scale, and that few would agree about which models belong in which categories.

Hillslope models can clearly benefit from attempts to increase their richness. The slope profile is a simple, but a relatively spare description of the landscape. The description, and the associated models, can be greatly enriched by adding the information on soils, vegetation and generalised moisture conditions which are more commonly found in a soil survey than in a geomorphological model. At the very least, this argues for incorporating solute processes, vegetation and hydrology into slope models in an appropriate way.

Potential for Scaling Up and Down

In many cases the physical mechanisms which drive process rates depend strongly on the scale of interest, so that a physical understanding must include a knowledge of how those drivers, and therefore the dominant processes and parameters, change with scale. For example, sediment transport is commonly driven by distance or areas from the divide in a long-term slope evolution model, but this is known to be a surrogate for discharge and the distribution of event discharges for a model with a shorter time span. For a whole-slope model, wash processes may be considered to be transport limited but, at the scale of a runoff plot, travel distances become important and removal is more controlled by detachment factors.

If, and generally only if, a model has an explicit and well-understood physical basis, there is, in principle, the scope to apply it at a range of different scales. Fine-scale models should be capable of aggregation up to coarser scales, although the reverse process of disaggregation is not generally possible without additional insights. In principle, aggregation may be achieved by integrating over relevant frequency distributions, provided that these have a well-behaved structure. In practice this condition is usually met where the distributions have well-behaved means and variances, and that they remain well behaved when combined with the nonlinearities of the system. These conditions are met if, for example, the underlying distributions are normal, exponential or gamma in form, at least for extreme values, and the nonlinearities in, say, sediment transport are in the form of power laws. The extremes of the distribution will then take forms like $x^n \exp(-x)$ or $x^n \exp(-x^2)$, which are themselves of gamma or normal form, and so still have finite means and variances. Two examples of this kind of aggregation are the conversion of event-based wash erosion to an integrated average in long-term slope evolution models (Kirkby and Cox 1995) and the integration of flow depths over microtopography (Kirkby et al. 1995).

In the first of these simplified examples, the overland flow runoff production, j, from a single storm of rainfall r, assuming a fixed soil-water storage threshold, h, is

$$j = (r - h)$$

The frequency density of days with rainfall r is, at the simplest, approximated by the exponential distribution (or as a sum of exponential and gamma terms):

$$N(r) = N_0 \exp(-r/r_0)$$

where N_0, r_0 are empirical parameters fitted to the distribution of daily rainfalls. Summing over this distribution, we obtain the total overland flow production:

$$J = \int_h^\infty (r - h)N(r)\mathrm{d}r = N_0 r_0 \exp(-h/r_0)$$

The sediment yield, assumed proportional to discharge squared, for a single rainfall event of r is

$$t \propto (r - h)^2$$

Again summing over the frequency distribution, the total sediment yield is

$$T \propto \int_h^\infty (r - h)^2 N_0 \exp(-r/r_0)\mathrm{d}r = 2N_0 r_0^2 \exp(-h/r_0)$$

In this case the aggregation is a simple summation process, but it should be observed that the main parameter of the distribution, r_0 appears explicitly in the aggregated form for the climatic erodibility as a strong control on the long-term value.

The second example is for aggregation of overland flows and sediment transport across a rough surface, which is here envisaged as a series of grooves running up- and downslope rather than as terraces or furrows along the contour. For a flow over such a micro-topography, stage h may be defined relative to the mean elevation of the surface. The distribution of elevations on the surface can be well approximated empirically by a normal distribution, also referred to the mean elevation, and characterised by a standard deviation, h_0. Points on this surface at elevation z occur with probability density:

$$p(z) = \frac{1}{h_0\sqrt{2\pi}} \exp\left(\frac{-h^2}{2h_0^2}\right)$$

Using this probability density as a weighting, we have, for the mean flow depth, z_0, total water (q) and sediment (S) discharge:

$$z_0 = \frac{1}{h_0\sqrt{2\pi}} \int_{-\infty}^h (h - z) \exp\left(\frac{-z^2}{2h_0^2}\right)\mathrm{d}z$$

$$q = \frac{c}{h_0\sqrt{2\pi}} \int_{-\infty}^h (h - z) \exp\left(\frac{-z^2}{2h_0^2}\right)\mathrm{d}z$$

$$S = \frac{Kc^2}{h_0\sqrt{2\pi}} \int_{-\infty}^h (h - z)^2 \exp\left(\frac{-z^2}{2h_0^2}\right)\mathrm{d}z$$

Again it may be seen that the integrals over these distributions remain well behaved, and that the roughness, h_0, appears in the integrated forms as an important determinant of the aggregate rate. These expressions, which can be integrated numerically, show that, for a given overland flow discharge, rougher surfaces create greater concentration of the flow into depressions, and consequently greater sediment transport. The influence of the roughness term is greater at low flows, whereas at flows high enough to inundate the entire surface, the effect is weaker.

It may been seen that, in both of these examples, the effect of processes at the finer scales modifies the rates at the coarser scales. Thus both storm distribution and micro-topography may influence long-term rates of slope evolution. In the latter case, it is probable that microtopography not only influences the mean rate of wash transport, but that its variation downslope may also affect the rate of change of sediment transport downslope. Furthermore, the erosional history of the slope is, in turn, likely to control the evolution of the microtopography, in a feedback loop which is largely ignored in current slope models.

An important unifying concept is the span of relevance for each process, in transferring between scales. For example, discussions of landform change in semi-arid environments are concerned primarily with the impacts of climate and imposed land-use. Within discussions of climate impact, there has long (e.g. Leopold and Miller 1954; Cooke and Reeves 1977) been a discussion of the relative importance of changes in total precipitation and in its frequency distribution. The analysis above clearly shows explicit dependence on the frequency distribution of daily rainfalls (through r_0), as well as on total rainfall (roughly equal to $N_0 r_0$). It is clearly legitimate to ask also about what scales are most relevant, even within the frequency distribution.

There are many scaling issues to be addressed. Within the topic of wash erosion, even, we are still not clear exactly how to model the influences of variations in rainfall intensity within storms, although we know it to be important (Yair et al. 1978). The influence of surface stoniness is also clearly important. On soil-covered humid slopes, wash shows little size selectivity, whereas on some stony semi-arid slopes there is strong size sorting which can largely counteract the effect of gradient between about 10° and 30°, leading to sharp breaks in slope.

In many cases, a theoretical understanding of how aggregation is achieved can also provide important insights into the magnitude and frequency distribution of the process. For example, the aggregation of event sediment yields into long-term averages, set out in simplified form above, also gives estimates of return periods for dominant events, with clear implications for, among other things, the design of field experiments. Where dominant return periods are long (> 100 years, say) for example, there is little point in carrying out monitoring experiments for a few years, and measurements should be based primarily on extensive surveys.

MODEL VALIDATION?

For most hillslope models, serious attempts at formal model validation are at a very early stage. Validation is more advanced for hydrological models and substantive work has also been done on assessing uncertainties (e.g. Beven and Binley 1992) for some simpler hydrological models. Only a relatively crude approach to uncertainty in forecasts can, however, be applied to most distributed models.

Validation must, in many cases, be preceded by extensive calibration for the less physically based parameters, usually based on optimisation methods, although there are practical difficulties in exhaustive optimisation where there are many parameters. Further problems arise from uncertainties in the calibration data sets, with which forecasts are compared. For many geomorphological models, including those for hillslope evolution, there are major uncertainties about initial landscape forms and dates, about boundary conditions, particularly for basal removal, and about the variations in long-term process rates, taking account of climatic and anthropogenic changes. Thus the best that can generally be achieved is to strive for consistency with both what is known about plausible process rates and their variations, and what is known about the external conditions and constraints. This conclusion may be taken in two ways, which may both be fruitful. On one hand, it argues for treating most models as thought experiments at least as much as for practical forecasting. Alternatively, it requires us to improve our measures of goodness of fit, so that we may evaluate our progress in providing better parameter values or more satisfactory models.

If we construct a landscape evolution model for a slope profile or a catchment, we need to begin from one or a set of logically selected initial forms at a given date; with the changing rates of slope processes in relation to regional knowledge of climate, sea level, land-use and other relevant conditions; and with the surrounding area as reflected in the behaviour of the model boundary conditions, at the outlet(s) and/or divide(s). Goodness of fit to an observed topographic form can be assessed directly as a least-squares departure or similar measure.

How do we combine this measure with other numerical measures, such as that for the water and sediment yield from the profile/catchment outlet? Immediately we have to make a subjective judgement about relative weightings. We may also have forecasts of vegetation cover, soil depth and distributions of soil moisture, among others. It is likely that the quality of these data will be highly variable, particularly in a large area. Some data will be in the form of quantitative surveys, but much will be qualitative, for example in maps of soil capability or erosion sensitivity. Our assessment of changing process rates over time relies on other knowledge (or models) which link rates with climate and land-use, so that this aspect of the landscape model performance is also based on our confidence in other theoretical relationships.

A probabilistic of fuzzy logic scheme may be one way to assess what constitutes acceptable goodness of fit for each distinct criterion. Formally we seek to maximise the probability that a model, defined by a parameter set which includes an error distribution, will provide an acceptable forecast, taking all criteria into consideration. It is important to proceed along some such path of formal optimisation and validation in a framework which allows criteria of different types to be evaluated together, and thus to formalise and confirm our progress in making better forcasts. It is, however, clear that there are many subjective judgements within this process. In other words, there is now, and is likely to remain for the foreseeable future, a key role, and I would argue the key role, for models as essentially qualitative thought experiments. The quantitative nature of most computational and mathematical models obscures this role, by appearing to offer a precision which is generally spurious. Nevertheless, we are still in the position where only a few of our variables are known within an accuracy of $\pm 50\%$, and where one good field observation can overturn, or at least radically modify, what had previously seemed to be a well-founded model.

CONCLUSION

Models in geomorphology form an essential bridge between the scales at which we can observe and the generally larger scales at which we seek explanations and forecasts. At present our models are most effective as thought experiments which help to refine our understanding of the dominant processes acting. This theoretical understanding is best achieved though models which have a strong physical basis, a high degree of generality, and a richness and power which link them to other areas of environmental and earth sciences. One major challenge lies in the search for improved methods for testing the quality of model forecasts, taking full account of the richness in the forecasts, and combining qualitative and quantitative criteria for success. The other major challenge for modelling is to address problems of up-scaling, for application in global change and geophysical contexts. This not only requires new model concepts, but an explicit understanding of compatibility between scales of interest.

REFERENCES

Beven, K. and Binley, A.M. 1992. The future of distributed models: calibration and predictive uncertainty, *Hydrological Processes*, **6**, 279–298.

Beven, K.J. and Kirkby, M.J. 1979. A physically based, contributing area model of basin hydrology, *Hydrological Sciences Bulletin*, **24**, 43–69.

Cooke, R.U. and Reeves, R.W. 1977. *Arroyos and Environmental Change in the American Southwest*, Clarendon Press, Oxford, 213 pp.

Kirkby, M.J. 1976. Tests of the random network model and its application to basin hydrology, *Earth Surface Processes*, **1**, 197–212.

Kirkby, M.J., Abrahart, R., McMahon, M.D., Shao, J. and Thornes, J.B. in press. MEDALUS soil erosion models for global change, paper presented at the 3rd International Geomorphology Conference, Hamilton, Ontario (for publication in an edited volume).

Kirkby, M.J. and Cox, N.J. 1995. A climatic index for soil erosion potential (CSEP), including seasonal factors, *Catena*, **25**, 333–352.

Kirkby, M.J., McMahon, M.L. and Abrahart, R.J. 1995. *The MEDRUSH model, MEDALUS II Final Report*, 1/1/93–30/9/1995.

Langbein, W.B. and Dawdy, D.R. 1964. Occurrence of dissolved solids in surface waters of the United States, *USGS Professional Paper 501-D*, 115–117.

Langbein, W.B. and Schumm, S.A. 1958. Yield of sediment in relation to mean annual precipitation, *Transactions American Geophysical Union*, **39**, 1076–1084.

Leopold, L.B. and Langbein, W.B. 1962. The concept of entropy in landscape evolution, *USGS Professional Paper 500-A*, 20 pp.

Leopold, L.B. and Miller, J.P. 1954. A post-glacial chronology for some alluvial valleys in Wyoming, *USGS Water Supply Paper 1261*, 90 pp.

Musgrave, G.W. 1947. The quantitative evaluation of factors in water erosion, a first approximation, *Journal of Soil and Water Conservation*, **2**, 133–138.

Popper, K.R. 1972. *The Logic of Scientific Discovery*, 6th revised impression, Hutchinson, London.

Rodriguez-Iturbe, I. and Valdéz, J.B. 1979. The geomorphological structure of hydrological response, *Water Resources Research*, **15**, 1490–1520.

Schumm, S.A. and Lichty, R.W. 1965. Time, space and causality in geomorphology, *American Journal of Science*, **263**, 110–119.

Yair, A., Sharon, D. and Lavee, H. 1978. An instrumented watershed study of partial area contribution runoff in the arid zone, *Zeitschrift für Geomorphologie*, Supplementband, **29**, 71–82.

11 Physically Based Modelling and the Analysis of Landscape Development

D. S. L. Lawrence

Postgraduate Research Institute for Sedimentology, The University, Reading

ABSTRACT

Recent interest in the numerical simulation of drainage basin and land surface evolution raises several issues concerning the status of physically based modelling in the analysis of landscape development. Most models use a continuum description for characterizing material transport within and across system boundaries. The scale at which continuum flux models break down is often not considered in the specification of mathematical models of surface development, despite its importance in the modelling of hillslope processes contributing to channel initiation and in the field testing of physically based models. Dimensional analysis and scaling are important modelling techniques, which are still relatively underused in descriptions of landscape development, and can simplify an analysis by eliminating redundant or extraneous variables and terms in governing equations as appropriate for the scales of interest. They can also be used to identify key dimensionless quantities which must have similar values if a laboratory or other quantitative study is to be considered representative of a selected field environment. Many of the constitutive functions for material transport used in contemporary models are very similar, at least in general form, to those found in earlier studies. Current work in this area is concerned with the relative contributions of advective versus diffusive transport and with the inclusion of threshold versus non-threshold processes for channel initiation in quantitative models. The results of numerical simulations that rely on the calculation of constitutive quantities on a point-by-point basis according to a set of prescribed rules are not necessarily comparable to numerical solutions based on a discretization of the fully coupled model system. Although geomorphologists have tended to prefer the first of these approaches for its ease of implementation, it has yet to be established that this choice is a sound one and that such simulation techniques do not produce misleading or erroneous results.

The Scientific Nature of Geomorphology: Proceedings of the 27th Binghamton Symposium in Geomorphology held 27–29 September 1996. Edited by Bruce L. Rhoads and Colin E. Thorn. © 1996 John Wiley & Sons Ltd.

INTRODUCTION

Physically based modelling and the mathematical formalism associated with model derivation and solution represents a fundamental research methodology used pervasively throughout the earth sciences. Its introduction in the study of landscape development is often associated with the models presented by Culling (1960, 1965), Scheidegger (1961), Ahnert (1967, 1973, 1976) and Kirkby (1971, 1976), although as pointed out by Cox (1977) credit for the first mathematical continuum model of slope development should perhaps be given to Jeffreys (1918). This chapter reviews the current status of this general methodology (which is often referred to simply as 'mathematical modelling' but which strictly speaking is more appropriately described as 'quantitative constitutive modelling') and includes discussions of the continuum hypothesis, the use of dimensional analysis and dimensionless groups to simplify models, the development of constitutive functions to characterize material transfer, and the coupling of these concepts into formal models. Rather than consider the wide array of topics within geomorphology in which this type of modelling has been applied, the emphasis herein will be on its use in describing land surface development at the hillslope to drainage basin scale. This area of work not only represents a contemporary, albeit much more quantitative, counterpart to the models of early geomorphologists such as Gilbert (1877, 1909) and Davis (1892, 1899), but is also of considerable recent interest due to a need for coupled hydrologic/geomorphic models which can be used to assess the response of surfaces to land-use or climatic changes. This latter demand motivates the development of quantitative geomorphic models which are compatible with, and contribute appropriate variables to, the somewhat more refined physical descriptions of transport processes employed by surface and subsurface hydrologists (and in some cases fluvial geomorphologists) but which also account for ongoing changes in the configuration of a surface.

The development of a mathematical model is often presented as a sequence of procedures (e.g. Thomas and Huggett 1980) in which one generates a hypothesis, constructs a mathematical model based on identified variables, simplifying assumptions and boundary conditions, develops a model solution which may be more or less quantified depending on the degree of specificity of the relationships between model variables, and finally compares model predictions with observations. If a lack of adequate correspondence between the observed and predicted variables is found, then the model is either partially or entirely redesigned to minimize the discrepancies and thus provide better predictions. Alternatively, a new set of observations which more closely reflects the variables, simplifying assumptions and boundary conditions used in the model, may be collected or compiled for the purposes of model testing. The sequence of procedures is then repeated until the objectives underlying model development are achieved. Assuming this larger framework for mathematical model development, this chapter focuses primarily on the process of model construction, although issues pertaining to solution methods or field testing are pointed out where they illustrate how physically based modelling can be more fully exploited in the study of landscape development than it currently is. The compilation and analysis of data for the field testing of hypotheses derived either from the modelling process or otherwise are not discussed herein as they are considered elsewhere in this volume.

THE CONTINUUM HYPOTHESIS

The most fundamental physical concept invoked in the development of mathematical models of landscape evolution is the principle that mass is conserved and can therefore be accounted for with a mass balance. This idea underlies not only quantitative models, but is also central to the systems approach to modelling popularized by Chorley (1962) and even to Davis' (1899), Gilbert's (1877) and Hack's (1960) discussions of land surface change with time. The 'mass' of primary interest is that portion of earth materials that extends above a reference datum and is of varying compositions, volumes and densities. One accordingly defines system boundaries which allow one to characterize the influx and outflux of mass, but which also provide for a system large enough so that the phenomenon of interest can be investigated. Given these general requirements, it is not surprising that early quantitative models of surface development focused on hillslopes (e.g. Scheidegger 1961; Ahnert 1967; Kirkby 1971) as the system boundaries can be clearly distinguished, often taking the form of a divide at the top of the slope and a stream at the bottom, with the slope laterally extending to infinity. The length of a single hillslope also seems to pose a scale that is potentially tractable in the field for model testing. It should not be overlooked, however, that these early mathematical models of slope development were not constructed in isolation from the qualitative discussions of landscape change that preceded them, and that they provided a tool for evaluating many of the concepts posed earlier in which the scale of interest had already been defined (e.g. Figure 3 of Hack 1960).

The quantification of a physical mass balance is generally based on a continuity equation which provides a foundation for the formal structure of the model. This equation usually takes a differential form, although other forms are possible, and characterizes both the influx and outflux of mass across system boundaries and the internal redistribution of material with time. This theoretical structure is thus very consistent with the scientific questions posed by geomorphologists. Problems may arise, however, in the interpretation and testing of models if the proposed continuum scale used to model the flux accounted for by the mass balance is not carefully considered in the modelling process.

The continuum hypothesis prescribes the existence of a length scale at which the properties of the system can be described by some representative average so that constitutive relations describing material flux and redistribution can be defined. Additionally, for all physical systems, an upper and lower bound may exist on the magnitude of the length scale at which a particular continuum model can be applied, and these will vary significantly between systems. A classical application of this hypothesis is in the study of fluid flow for which one distinguishes a length scale that is significantly larger than the scale at which one would 'see' the behaviour of individual molecules and which must be smaller than the scale at which large variations in physical properties would undermine the approach (Figure 11.1). Clearly, if the fluid of interest is water, then the lower bound on the appropriate length scale for defining system properties, such as fluid viscosity, will be much smaller than it is for, say, flowing granular materials. In both cases, the upper bound on the continuum length scale is often the physical boundaries of the vessel in which the fluid is contained, but it may also be more subtly associated with variations in physical properties that occur over relatively large length scales and therefore cannot readily be described by a single average value. Similarly, the continuum hypothesis underlies the description of porous media flow based on Darcy's law

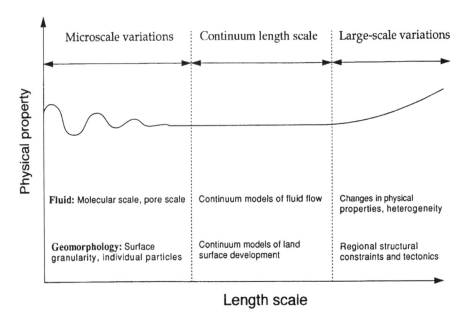

Figure 11.1 Variations in a typical physical property such as viscosity, hydraulic conductivity or surface roughness as a function of increasing length scale

such that the lower bound at which this description is applicable is given by the length scale associated with the size of individual pores and the upper bound is often influenced by regional heterogeneity in hydraulic conductivity and may be difficult to distinguish.

The application of a continuum model for land surface development presumes that one can identify a length scale at which denudational processes can be described as a material flux that is not predetermined by a superimposed structure or spatial pattern external to the model. As has long been recognized by geomorphologists, there exists a spatial scale at which regional structural features and tectonic processes significantly influence patterns of surface deformation. These features and processes effectively impose an upper bound on the length scale of applicability of many continuum models for surface evolution, as discussed by Scheidegger (1992) in somewhat different terms. The incorporation of this upper bound into quantitative denudation models is often as an initial surface configuration or as an external forcing function, but the coupling of denudation models with regional tectonic models (e.g. Koons 1989) can serve a similar role. Thus, as is the case with both the fluid and porous media examples described in the previous paragraph, upper bounds on the length scale at which a particular continuum flux model can be applied often simply represent either the presence of distinct physical boundaries or limitations on the range of variables incorporated into a particular continuum model for material flux.

The lower bound on a continuum length scale for modelling surface denudation is critical in many geomorphic studies and may be imposed by a variety of physical factors. As is true for the upper bound, the magnitude of this length scale depends upon the specificity of the model, especially the transport functions used to describe the flux, and moreover on the field environment of interest. In many cases, the granularity of the surface itself, the surface structure imparted by the presence of vegetation, or perhaps the mean path length of an individual granular particle while in transit, effectively impose a lower limit on the length scale at which a continuum transport model can be applied. This is not to say that the net or average effects of these physical factors on the overall flux of sediment cannot be incorporated into a model of landscape evolution; however, they rarely explicitly appear in constitutive model formulations for surface sediment flux. In most cases, the scale of applicability for continuum sediment transport models may be at least an order of magnitude larger than the surface detail that frequently attracts the eye in the field.

Early quantitative modellers could largely neglect the issue of identifying an intermediate length scale associated with continuum model applicability as they were primarily interested in the general relationship between process and form observed in hillslope profiles (e.g. Smith and Bretherton 1972; Kirkby 1971) and on surfaces in which a preexisting slope–valley configuration was superimposed (e.g. Ahnert 1976). However, contemporary models of simultaneous land surface and drainage basin evolution that seek to couple hillslope and channel network development must tackle this question of scale directly. The initiation and development of a drainage network itself introduces a series of multiple length scales (e.g. Tarboton et al. 1988; Rigon et al. 1993) and these may not necessarily correspond to the upper and lower bounds on a continuum length scale for modelling the denudation of unchannelized slopes. Of paramount importance is the problem of channel initiation and maintenance and specifically the length scale at which this occurs in the landscape (e.g. Dietrich et al. 1992; Montgomery and Dietrich 1992), the extent to which this length scale represents physical factors which are internal to the slope system or externally imposed (e.g. Loewenherz 1991a, 1994a; Izumi and Parker 1995), and the feasibility of incorporating a physical understanding of this scalar process into coupled models of drainage basin evolution (e.g. Willgoose et al. 1990, 1991a; Howard 1994). These questions are central to the development of physically based models for land surface development that can be applied both to further our understanding of the natural landscape and to assess the potential impacts associated with disturbances to the system. The following two sections of this chapter will consider the formulation of transport functions for use in continuum flux models that are consistent with the physical scales of interest.

DIMENSIONLESS GROUPS AND PHYSICAL SCALINGS

The development of a theoretical model, whether pursued using qualitative or quantitative methods, relies on the identification of a set of dependent variables or configurations that one seeks to explain and a set of independent variables or events that appear to contribute to that explanation. The identification of the appropriate variables for inclusion in a mathematical model and the relationships between those variables is to a certain degree an inductive process, such that the opportunity for spontaneity, serendipity and the exercise of 'common sense' is as vital to the success of this methodology as it is to more qualitative techniques (cf. Baker and Twidale 1991). Beyond the application of the conservation of mass as discussed in the previous section (and perhaps also the conservation of momentum in the case of modelling very rapid surface processes), geomorphologists are often left with relatively little guidance from physics or other basic

sciences as to a quantitative form of surface weathering and transport processes. This is
due largely to the differing spatial and temporal scales of interest, but also reflects the
complexity of the phenomena they seek to explain. In practice, transport models are often
formulated based on a combination of qualitative field or laboratory observations, the
results of statistical analyses of quantified data, and physical reasoning. This section will
discuss how dimensional analysis, a physical modelling technique that is relatively
underused by geomorphologists, can assist in simplifying the array of variables that are
incorporated into formal mathematical statements and also ensure that the resulting formal
model focuses on relevant spatial and temporal scales.

The use of dimensional analysis is often thought of in its most elementary form,
dimensional homogeneity, in which each group of terms in a quantitative statement must
have the same dimensional representation in order for the statement to be physically
meaningful. Accordingly, the physical dimensions associated with empirical coefficients
in transport functions are often predetermined by the other variables used in an expression. For example, a very simple one-dimensional model for a steady-state hillslope
profile in which the rate of surface transport is locally dependent on the slope is given by

$$\frac{\partial h}{\partial t} + \frac{\partial q_s}{\partial x} = 0 \tag{1}$$

$$q_s = kS = -k\left(\frac{dh}{dx}\right) \tag{2}$$

$$\frac{\partial h}{\partial t} = k\frac{\partial^2 h}{\partial x^2} \tag{3}$$

where h is the elevation at a point on the surface, q_s is the sediment flux across the surface,
S is the local surface slope, x and t are space and time coordinates respectively, and k is an
empirical constant. In order for this set of equations to be dimensionally homogeneous,
both the surface sediment flux, q_s, and the empirical constant k must have the physical
dimensions $[L^2/T]$. This latter constant is referred to as a diffusion coefficient for surface
transport due to its role in equation (3) and its associated physical dimensions.

An equally important use of dimensional analysis is in the development of dimensionless groups based on the variables contributing to a physical phenomenon. These
dimensionless groups serve two major functions in physical modelling. Firstly, they
provide a basis for comparing models or data independent of the underlying physical (e.g.
spatial) scales. Secondly, they can be used to identify terms in a set of governing equations
which may be neglected under a given set of conditions, so that the resulting formal model
is specific to the physical scales of interest, and does not necessarily apply more generally
to the entire range of possible values for all of the mathematical variables. According to
Buckingham's Π theorem, the number of independent dimensionless groups that are
necessary to describe fully a phenomenon known to involve n variables is equal to the
number $n - r$, where r is usually the number of physical dimensions (such as mass, length
or time) underlying the dimensional form of the variables. (For further discussion of this
theorem and its application in physically based modelling see Langhaar 1951.) Thus, the
total number of those groups will always be less than the number of original variables, so
that this technique can be used to simplify problem statements, both those which are used
in the development of formal models and those which underlie the design of field or

laboratory experiments. However, the actual form that the dimensionless groups take (i.e. how the variables are arranged) is not unique, so that the choice of groupings must be guided by sound physical judgment and constant reference to the physical problem of interest.

Although process geomorphologists frequently encounter dimensionless groups developed in other scientific disciplines, such as the Reynolds number and the Froude number for describing fluid flow regimes and the Shields criterion for the incipient motion of uniform sediments, relatively little explicit attention has been given to the fundamental dimensionless groups that characterize the processes contributing to landscape evolution. They have been used for many years, however, and both 'Horton's laws' for describing drainage networks (Horton 1932, 1945) and the hypsometric integral for presenting the distribution of surface elevations within a basin (Strahler 1952) utilize dimensionless groups allowing the attributes of different systems to be compared independent of spatial scale. Similarly, the results of early quantitative models of hillslope development (e.g. Scheidegger 1961; Kirkby 1971; Ahnert 1976) are often presented in a scaled or dimensionless form. All of these examples represent geometric scalings in which variables having fundamental physical dimensions of length or area, such as the surface elevation, the distance downslope, or the basin area above a given elevation, are scaled by quantities which are constants for a given field environment and have the same fundamental physical dimension. Typical scaling factors which are used to normalize the above variables are the maximum elevation, the total slope length, or the total area of a basin, respectively. The resulting dimensionless variables usually take on values from 0 to 1, and this facilitates the comparison of systems representing very different absolute spatial scales.

As models of land surface evolution incorporate increasingly detailed and coupled physical processes, an explicit recognition of the role of dimensional analysis and scaling in model development beyond simple geometric normalizations, such as those presented above, becomes essential. Although this is beginning to emerge in recent work (e.g. Willgoose et al. 1991a; Loewenherz 1991b, 1994a), this technique has yet to be fully exploited in landscape modelling studies. Willgoose et al. (1991b) discuss the use of dimensional scaling in some detail and present a set of dimensionless groups that characterize both transient states and conditions of dynamic equilibrium for their own landscape evolution model (i.e. Willgoose et al. 1991a). Many of the fundamental physical scales underlying the derivation of those groups can be readily interpreted, such as the vertical and horizontal magnitude of the drainage contributing area; however, others, such as the runoff and channel initiation scales, require further explanation and are quite specific to the constitutive functions used in this model. Nevertheless, there are two distinct advantages in their use of a dimensionless model over one in dimensional form. First of all, the number of variables necessary for describing the system is reduced (in their case, the 30 original dimensional variables are effectively replaced by approximately 15 dimensionless groupings), and this provides for a much more parsimonious use of numerical simulations or other techniques for obtaining model results. Additionally, the dimensionless groups establish a basis for comparing model results with field or laboratory experimental studies in that they specify the parameters that must be of similar magnitude in order for two physical systems to be considered similar. This latter advantage is very aptly illustrated by the authors in their analysis of the catchment modelling laboratory experiments reported by Parker (1977) and Schumm et al. (1987).

Their analysis demonstrates that there is not necessarily a similarity of process between those experiments and typical field scale catchments due to a dominance of diffusive transport in the experimental configuration as compared with the field environment. The dimensionless group, which should be of a similar value, describes the ratio of diffusive transport to fluvial sediment transport on the surface and is approximately 100 times greater for the experimental catchment than would generally be associated with a field scale catchment.

A related use of dimensional analysis and scaling in physical modelling is in simplifying the equations governing a physical system by assigning relative magnitudes to the multiple physical scales appearing in a problem statement. This technique facilitates the mathematical analysis and interpretation of a physical problem, as demonstrated in the stability analyses for land surface development and channel initiation considered by Smith and Bretherton (1972) and Loewenherz (1991a, b, 1994a). In those models, the governing equations that are actually evaluated to assess the conditions leading to channel initiation are considerably simplified from the more general form of the governing equations. This simplification is achieved by identifying the relevant physical variables specific to the problem of interest (e.g. the hillslope length, the average depth of the surface water flux, the rate of tectonic uplift), establishing the relative magnitude of key variables (e.g. the average depth of the surface water relative to the length of the hillsope, which for most cases of land surface evolution one would assume to be very small), using the associated scalings to generate dimensionless forms of the variables, and finally distinguishing those terms in the dimensionless equations that dominate the behavior of the system from those which contribute negligibly. The associated dimensionless groups pose constraints on the range of conditions represented by the model solutions. Some of these constraints are quite simple, such as the dimensionless 'erosion time scale' specified by Smith and Bretherton (1972), which takes the form

$$t^* \sim \frac{\varepsilon \alpha t}{D_0} \tag{4}$$

where t is the time scale associated with measurable changes in the surface, α is the average effective rainfall rate on the surface, D_0 scales both the vertical and horizontal lengths of the surface, ε is a dimensionless scaling factor and is posed as being very small (i.e. $\ll 1$). Thus, equation (4) directly implies that the time-dependent solutions for their model apply to surfaces for which the nominal surface lowering rate D_0/t is very small relative to the rainfall rate, α, driving the surface water flux. This constraint is clearly appropriate for many, but perhaps not all, problems of long-term surface denudation.

The physical scalings derived from a dimensional analysis can also provide a basis for expanding and coupling physical models to represent system behaviour at multiple length scales. This strategy is illustrated in the physically based model for channel initiation presented by Loewenherz (1994a) which couples a kinematic model for sediment transport at the hillslope scale with a hydrodynamic model for local sediment flux at the scale of channel initiation. The principal distinctions between this work and that of Smith and Bretherton (1972) lie in the differentiation of downslope and lateral length scales associated with surface erosional processes, and in the explicit characterization of the effects of surface water hydrodynamics on the advective transport of sediment. Two key dimensionless parameters which govern the coupling of the hydrodynamic with the kinematic model are given by

$$\delta = \frac{D}{B} = \frac{B}{L} = \left(\frac{D}{L}\right)^{1/2} \tag{5}$$

$$\Sigma = \frac{L}{UT} \tag{6}$$

where D, L and B are length scales corresponding to the depth of the surface water and the downslope and transverse (cross-slope) dimensions of the surface respectively, such that δ is very small under the physical conditions associated with channel initiation by surface flow. The timescale parameter, Σ, incorporates the mean downslope velocity of surface water, U, and the amount of time associated with large-scale effects on the evolution of the surface, as represented by T, in addition to the downslope length of the surface, L. This parameter is also expected to be very small for time scales at which hydrodynamic effects in the surface water flux can be expected to contribute to the local evolution of the surface. Although it is very difficult to provide useful solutions, either analytical or numerical, for the surface evolution problem if the hydrodynamic model is applied at the scale of the entire drainage surface, the use of this physical scaling facilitates the analysis by identifying the length scale at which surface water hydrodynamics will begin to contribute significantly, rather than negligibly, to the behaviour of the system.

CONSTITUTIVE FUNCTIONS FOR MATERIAL TRANSPORT

Following the identification of relevant independent and dependent variables and the clarification of the physical scales of interest, the relationship between variables must be specified if one seeks to generate model solutions. In developing models of denudational processes, one is primarily concerned with identifying constitutive functions for weathering and surface transport processes. Although the earliest mathematical models of slope profile development neglect physical discussions of the constitutive functions used in solving those models, both Ahnert (1967, 1976, 1977) and Kirkby (1971, 1976) pay careful attention to this issue in the development of their models and incorporate the findings of contemporaneous research in the specification of transport functions. The constitutive functions for sediment transport that are currently used to simulate landscape evolution (e.g. Willgoose 1991; Howard 1994) are very similar in their general form to those Ahnert and Kirkby originally proposed. There has nevertheless been considerable recent attention (Dietrich and Dunne 1993; Kirkby 1994) given to the different classes of transport functions that need to be quantified in order for a model of landscape evolution to be representative of the range of conditions occurring in the natural environment. This section will highlight some of the issues arising out of those discussions, particularly the differentiation of advective and diffusive transport processes, the use of general versus specific functional forms in analysis, and threshold versus non-threshold processes and their relationship to the more general issues of the continuum length scale underlying the physically based model.

Many of the constitutive models for material transport which are used in modelling land surface development can be reduced to the general set of variables given by

$$q_s = f(S, q_w) \qquad (7)$$

where q_s is the sediment flux, S is the local surface slope and q_w is the surface water flux. This basic functional relationship is often further simplified by using the contributing area per unit contour width as a surrogate for the amount of surface water that will accumulate at a point, so that material transport is given entirely as a function of surface geometry. The variables presented in equation (7) are often used in the somewhat more specific form given by

$$q_s = F(S^n, q_w^m) \qquad (8)$$

where n and m are empirical constants usually assumed to be > 0, and F is a linear or bilinear function of the variables S^n and q_w^m. As discussed by Kirkby (1971) and Smith and Bretherton (1972), this form of the equation for sediment flux reflects reported field and laboratory observations of a range of sediment transporting processes.

Two very different system behaviours can be distinguished based on the relative sensitivity of the sediment transport to the surface slope versus the surface water flux, as measured by the magnitude of the partial derivatives of the transport function F, with respect to S versus q_w (Smith and Bretherton 1972; Kirkby 1980). The first class of physical processes is generally diffusive in character in that material transport occurs in response to a gradient, which in this case is given by the slope of the surface elevation. Diffusive processes tend to stabilize physical systems as they eliminate strong gradients by locally redistributing material from zones of high concentration to those of lower concentration, thus reducing the magnitude of the gradient. In models of land surface development, diffusive processes contribute to the general smoothing of the surface in that they preferentially erode the steepest elements of the landscape. The second class of processes, representing those transport phenomena which rely on the aid of a transporting medium, such as the surface water, q_w, to move material through the system, are advective or concentrative in character. In contrast to diffusive processes, these processes contribute to the local incision and dissection of the surface, both by providing a mechanism for the removal of material from the system which may be independent of the local surface slope and by enhancing the tendency for local surface water accumulation which then increases the capacity for material entrainment and removal. This differentiation between diffusive and advective sediment transport is broadly analogous with functional forms of transport models used in other areas of physical science, such as heat transfer and porous media flow. It is of both physical and mathematical significance as the coupling of a strictly diffusive transport function with a continuity equation results in a parabolic partial differential equation, while a purely advective system is hyperbolic in character. This distinction in turn has implications for the solution techniques, either analytical or numerical, which are most appropriate for evaluating system behaviour. (For more thorough discussions of this topic, see e.g. Pinder and Gray 1977 or Ames 1992 for the case of numerical techniques; Hassani 1991 for the case of analytical methods.)

For two-dimensional models of surface development, the material transport function must specify both the magnitude and local direction of the sediment flux and is often of the form given by

$$\mathbf{q}_s = F(S^n, q_w^m)\mathbf{m} \tag{9}$$

where $\mathbf{m}$ is the direction of the local slope and thus determines the pathways for sediment across the surface. However, if the local pathways associated with advective sediment flux are to be explicitly distinguished from those associated with the diffusive flux, as is necessary for evaluating channel initiation (Loewenherz 1991b, 1994a), then the constitutive function must include this distinction and thus be written as

$$\mathbf{q}_s = F_A(S^{n_1}, q_w^{m_1})\mathbf{r} + F_D(S^{n_2}, q_w^{m_2})\mathbf{m} \tag{10}$$

where $\mathbf{r}$ is the local direction of the surface water, F_A is the magnitude of the advective component of sediment transport, F_D is the magnitude of the diffusive component, and the subscripts 1 and 2 are used to distinguish the empirical constants associated with advective transport and diffusive transport respectively. It is important to note that equations (9) and (10) are still rather general constitutive functions as they do not indicate specific values for n and m, nor do they specify multiplicative or additive constants for the variables. However, they can be used in conjunction with constraints on the behaviour of the derivatives of the functions to provide some very useful qualitative insights into the processes of landscapes development and especially into the relative roles of diffusive versus advective transport in channel initiation (e.g. Smith and Bretherton 1972; Kirkby 1980; Loewenherz 1991a, 1994a). Furthermore, and of particular relevance in light of the 'detachment-limited' simulation model recently presented by Howard (1994), the general formulations given by equations (9) and (10) neither neglect nor preclude the inclusion of constant factors or coefficients into a fully quantified problem statement. They are thus not strictly limited to the analysis of transport-limited conditions, although much previous work has relied on this assumption. Accordingly, the addition of a critical shear stress as a specified constant in the sediment transport function does not in itself establish a new class of solutions that are distinct from those encompassed by the analysis of equation (9). The magnitude of the physical scales implied by these additional constants (i.e. just how large the transporting capacity must be before sediment transport will occur, the time scale over which one can anticipate that significant transport will occur, and the size of the surface area likely to be affected once transport is initiated) may, however, have implications for the continuum flux model used in the analysis.

Recent discussions of hillslope transport processes, especially those contributing to channel initiation, point towards a fundamental distinction between threshold and non-threshold processes in landscape development (Dietrich and Dunne 1993; Kirkby 1994; Montgomery and Dietrich 1989). Processes characterized by distinct thresholds contribute to the initiation of surface channels with well-defined banks, and non-threshold, more 'gradual', processes are associated with valley formation. In other types of physical problems, thresholds often represent a discontinuity in the equations or conditions governing transport, such as those which are associated with a shock wave in a compressible flow. An incipient surface channel with a well-defined channel head and banks often represents a morphological discontinuity (e.g. see illustrations in Montgomery and Dietrich 1989; Dietrich and Dunne 1993), and the equations governing the flux of water and sediment across the adjacent unchannelized surface will be different from those associated with transport in the channel. Therefore, the concept of a morphological threshold may be quite appropriate in these circumstances, particularly at this scale of resolution. However, for the purposes of modelling, the more fundamental question is to what extent can certain types of hillslope processes that ultimately result in the initiation of surface channels (e.g. surface wash and landsliding) be differentiated as threshold versus non-threshold processes. The resolution of this issue seems to be directly linked to the scale of observation of sediment-transporting events as compared with the continuum length scale that underlies the physically based model. In other words, although sheetwash erosion is characterized as a continuum process in most models of land surface development, the microscale process consists of a series of discrete, 'threshold-based' events. Similarly, in environments where landsliding is the dominant hillslope-forming process, land surface morphology is often the product of numerous discrete events, such that in principle, there may be a larger length scale at which this ongoing series of events could also be modelled as a continuum process. The distinction between these two categories is thus directly related to the magnitude of the length scale of interest relative to the typical length scale that characterizes surface deformation during a single event. If the magnitude of these scales is similar, then an appropriate continuum scale cannot be defined. However, if the length scale that is being modelled is much larger than that which is affected by microscale processes (which may or may not include local thresholds), then the continuum hypothesis is valid and the aggregate effects of this process on surface development can be defined using a continuum model.

MATHEMATICAL MODELS OF LAND SURFACE EVOLUTION

Formal mathematical models of land surface development are usually derived by applying the continuity equation to account for the transfer of mass within the system in the general form

$$\frac{\partial h}{\partial t} = -(\nabla \cdot q_s)\mathbf{m} + T \tag{11}$$

where h is the surface elevation, q_s is the sediment flux that is given by a constitutive function similar to those discussed in the previous section, $\mathbf{m}$ is a vector describing the local surface gradient, and T represents any external forcing, such as tectonic uplift, which is contributing to time-dependent changes in the configuration of the land surface. The divergence operator, $(\nabla \cdot)$, constrains the redistribution of mass, and takes a variety of specific forms, depending on the spatial coordinate system which is used in the problem. An additional continuity equation is often also employed to account for the flux of water across the surface, so that it can be described as a local variable for use in the constitutive function. The dependent variable, h, is the actual elevation of the surface at a point in space and time so that this equation describes the relationship between process (the sediment flux) and form (the elevation) for evolving surfaces. When the constitutive function consists only of diffusive (i.e. slope-dependent) processes, equation (11) reduces to a diffusion equation, and the time-dependent behaviour of the landscape is analogous to other diffusive physical processes, such as the molecular diffusion of solutes. The conditions of dynamic equilibrium are described by simply setting the time rate of change of the surface, i.e. $\partial h/\partial t$, equal to zero, so that the system represents a balance between tectonic uplift, T, and the flux of sediment through the system.

Beyond the simple case of a strictly diffusive system in dynamic equlibrium, it can be surprisingly difficult to provide actual mathematical solutions, either analytical or numerical, for the time-dependent surface evolution problem described by equation (11) for cases which are of interest to geomorphologists. This is due to several factors, including:

1. The multidimensional nature of the problem;
2. The relatively nonlinear character of the transport functions for the surface sediment flux which arises from the exponents on S and q_w in equations (9) and (10);
3. The presence of advective transport which precludes the modelling of the system as strictly diffusive and changes the governing equation from one that is parabolic to one that may exhibit both parabolic and hyperbolic behaviours;
4. The spatially non-uniform behaviour of transport processes, particularly as associated with the general downslope increase in the importance of advective transport as surface water progressively accumulates from upslope contributions;
5. The physical discontinuity which may be associated with the presence of surface channels.

Full mathematical solutions for the governing equations, which consider multidimensional surfaces, encompass both diffusive and advective transport, and incorporate the tendency for surface channel initiation and development, are thus virtually non-existent in the geomorphic literature.

Two alternative approaches are, however, used to effectively 'solve' and thereby obtain results for landscape evolution models. The first of these is represented by the general technique of numerical simulation, in which a set of rules which describe the processes of surface erosion on a local basis and are derived from the constitutive functions and a mass balance accounting (e.g. Ahnert 1976; Kirkby 1986; Howard 1994) is applied to a lattice of points representing the surface and then reapplied at subsequent time steps, so that the problem is marched forward in time. The simulation technique is very attractive due to its intuitive character, its low demand on computational capacity (as the large matrices used in the simulation are never inverted as they would be under many numerical solution schemes), and the ease with which subsystems can be coupled to generate a larger model system. However, although rule-based simulation results are often considered by geomorphologists to be analogous to an analytical or at least an approximate numerical solution for the fully coupled differential equations governing the system, we have yet to establish the range of conditions for which this is in fact true. Of particular concern are the general nonlinear character of the transport functions and the associated potential for instability in the physical system, which in many other areas of physically based modelling (e.g. fluid dynamics) virtually precludes the use of simple simulation techniques. As models for surface evolution become increasingly sophisticated, and particularly as mechanisms for channel initiation are identified and incorporated into constitutive functions, the reliability of simulation results relative to those provided by full numerical solutions based on a discretization of the fully coupled model system deserves attention.

Analytical techniques, such as asymptotic and stability analyses, which evaluate various aspects of the system behaviour within well-defined limits (e.g. Smith and Bretherton 1972; Loewenherz 1991a, b; 1992, 1994a, b) have also been used to obtain results from land surface evolution models. These methods, while often unable to provide complete solutions, can nevertheless establish important constraints on the behaviour of the physical system (e.g. conditions associated with tendency for surface channel initiation) and can also be used to check the validity of numerical results. Although they remain relatively inaccessible to many geomorphologists, their role may become more prominent as other cognate disciplines (e.g. geophysics, civil engineering) continue to take an increasing interest in landscape evolution and as the availability and use of reliable computer software capable of algebraic and other symbolic manipulation (e.g. Mathematica by Wolfram Research, Inc.) become more widespread.

REFERENCES

Ahnert, F. 1967. The role of the equilibrium concept in the interpretation of landforms of fluvial erosion and deposition, in *L'evolution des versants*, edited by P. Macar, *Les Cong, et Colloques de l'Univ. de Liege*, **40**, 23–41.

Ahnert, F. 1973. COSLOP 2 – a comprehensive model program for simulating slope profile development, *Geocom Bulletin*, **6**, 99–122.

Ahnert, F. 1976. Brief description of a comprehensive three-dimensional process–response model of landform development, *Zeitschrift für Geomorphologie*, NF Supplementband, **25**, 29–49.

Ahnert, F. 1977. Some comments on the quantitative formulation of geomorphological processes in a theoretical model, *Earth Surface Processes*, **2**, 191–201.

Ames, W.F. 1992. *Numerical Methods for Partial Differential Equations*, 3rd edn, Academic Press, London, 451 pp.

Baker, V.R. and Twidale, C.R. 1991. The reenchantment of geomorphology, *Geomorphology*, **4**, 73–100.

Chorley, R.J. 1962. Geomorphology and general systems theory, *US Geological Survey Professional Paper*, 500-B.

Cox, N.J. 1977. A note on a neglected early model of hillslope development, *Zeitschrift für Geomorphologie*, NF Supplementband, **21**, 354–356.

Culling, W.E.H. 1960. Analytical theory of erosion, *Journal of Geology*, **68**, 336–344.

Culling, W.E.H. 1965. Theory of erosion on soil-covered slopes, *Journal of Geology*, **71**, 127–161.

Davis, W.M. 1892. The convex profile of bad-land divides, *Science*, **20**, 245.

Davis, W.M. 1899. The geographical cycle, *Geographical Journal*, **14**, 481–504.

Dietrich, W.E. and Dunne, T. 1993. The channel head, in *Channel Network Hydrology*, edited by K. Beven and M.J. Kirkby, Wiley, Chichester, pp. 175–219.

Dietrich, W.E., Wilson, C.J., Montgomery, D.R., McKean, J. and Bauer, R. 1992. Erosion thresholds and land surface morphology, *Geology*, **20**, 675–679.

Gilbert, G.K. 1877. *Report on the Geology of the Henry Mountains*, US Geological Survey, Washington, DC 160 pp.

Gilbert, G.K. 1909. The convexity of hilltops, *Journal of Geology*, **17**, 344–350.

Hack, J.T. 1960. Interpretation of erosional topography in humid temperate regions, *American Journal of Science*, **258A**, 80–97.

Hassani, S. 1991. *Foundations of Mathematical Physics*, Allyn and Bacon, Boston, 918 pp.

Horton, R.E. 1932. Drainage basin characteristics, *Transactions of the American Geophysical Union*, **13**, 350–361.

Horton, R.E. 1945. Erosional development of streams and their drainage basins; hydrophysical approach to quantitative morphology, *Bulletin of the Geological Society of America*, **56**, 275–370.

Howard, A.D. 1994. A detachment-limited model of drainage basin evolution, *Water Resources Research*, **30**, 2261–2285.

Izumi, N. and Parker, G. 1995. Inception of channelization and drainage-basin formation–upstream-driven theory, *Journal of Fluid Mechanics*, **283**, 341–363.

Jeffreys, H. 1918. Problems of denudation, *Philosophical Magazine*, **36**, 179–190.

Kirkby, M.J. 1971. Hillslope process–response models based on the continuity equation, in *Slopes: Form and Process*, edited by D. Brunsden, Institute of British Geographers, Special Publication, **3**, 15–30.

Kirkby, M.J. 1976. Deterministic continuous slope models, *Zeitschrift für Geomorphologie*, NF Supplementband, **25**, 1–19.

Kirkby, M.J. 1980. The stream head as a significant geomorphic threshold, in *Thresholds in Geomorphology*, edited by D.R. Coates and J.D. Vitek, Allen and Unwin, London, pp. 53–73.

Kirkby, M.J. 1986. A two-dimensional simulation model for slope and stream evolution, in *Hillslope Processes*, edited by A.D. Abrahams, Allen and Unwin, London, pp. 203–222.

Kirkby, M.J. 1994. Thresholds and instability in stream head hollows: a model of magnitude and frequency for wash processes, in *Process Models and Theoretical Geomorphology*, edited by M.J. Kirkby, Wiley, Chichester, pp. 294–314.

Koons, P.O. 1989. The topographic evolution of collisional mountain belts: a numerical look at the Southern Alps, New Zealand, *American Journal of Science*, **289**, 1041–1069.

Langha, H.L. 1951. *Dimensional Analysis and Theory of Models*, Wiley, New York, 166 pp.

Loewenherz, D.S. 1991a. Stability and the initiation of channelized surface drainage: a reassessment of the short wavelength limit, *Journal of Geophysical Research* **96(B)**, 8453–8474.

Loewenherz, D.S. 1991b. Stability and the role of length scale dependent material transport in the initiation and development of surface drainage networks, Ph.D. thesis, University of California, Berkeley, 206 pp.

Loewenherz, D.S. 1992. Finite amplitude development of surface incisions derived from sheetwash erosion, *EOS, Transactions of the American Geophysical Union*, 73, Fall Meeting Supplement, 212.

Loewenherz, D.S. 1994a. Hydrodynamic description for advective sediment transport and rill initiation, *Water Resources Research*, **30**, 3203–3212.

Loewenherz, D.S. 1994b. Theoretical constraints on the development of surface rills: mode shapes, amplitude limitations and implications for non-linear evolution, in *Process Models and Theoretical Geomorphology*, edited by M.J. Kirkby, Wiley, Chichester, pp. 315–333.

Montgomery, D.R. and Dietrich, W.E. 1989. Source areas, drainage density and channel initiation, *Water Resources Research*, **25**, 1907–1918.

Montgomery, D.R. and Dietrich, W.E. 1992. Channel initiation and the problem of landscape scale, *Science*, **255**, 828–830.

Parker, R.S., 1977. Experimental study of drainage basin evolution and its hydrologic implications, Ph.D. thesis, Colorado State University, Fort Collins.

Pinder, G.F. and Gray, G. 1977. *Finite Element Simulation in Surface and Subsurface Hydrology*, Academic Press, New York, 295 pp.

Rigon, R., Rinaldo, A., Rodriguez-Iturbe, I., Bras, R.L. and Ijjaszvasques, E. 1993. Optimal channel networks–a framework for the study of river basin morphology, *Water Resources Research*, **29**, 1635–1646.

Scheidegger, A.E. 1961. Mathematical models of slope development, *Bulletin of the Geological Society of America*, **72**, 37–59.

Scheidegger, A.E. 1992. Limitations of the system approach in geomorphology, *Geomorphology*, **5**, 213–217.

Schumm, S.A., Mosley, M.P. and Weaver, W.E., 1987. *Experimental Fluvial Geomorphology*, Wiley, New York, 413 pp.

Smith, T.R. and Bretherton, F.P. 1972. Stability and the conservation of mass in drainage basin evolution, *Water Resources Research*, **8**, 1506–1529.

Strahler, A.N. 1952. Hypsometric (area-altitude) analysis of erosional topography, *Bulletin of the Geological Society of America*, **63**, 1117–1142.

Tarboton, D.G., Bras, R.L. and Rodriguez-Iturbe, I. 1988. The fractal nature of river networks, *Water Resources Research*, **24**, 1317–1322.

Thomas, P.W. and Huggett, R.J. 1980. *Modelling in Geography: A Mathematical Approach*, Harper and Row, London, 338 pp.

Willgoose, G., Bras, R.L. and Rodriguez-Iturbe, I. 1990. A model of river basin evolution. *EOS, Transactions of the American Geophysical Union*, **71**, 1806–1807.

Willgoose, G., Bras, R.L. and Rodriguez-Iturbe, I. 1991a. A coupled channel network growth and hillslope evolution model. 1. Theory, *Water Resources Research*, **27**, 1671–1684.

Willgoose, G., Bras, R.L. and Rodriguez-Iturbe, I. 1991b. A coupled channel network growth and hillslope evolution model. 2. Nondimensionalization and applications, *Water Resources Research*, **27**, 1685–1696.

12 Equifinality and Uncertainty in Geomorphological Modelling

Keith Beven

Centre for Research on Environmental Systems and Statistics, Lancaster
University

ABSTRACT

Recent experience in modelling hydrological systems has revealed that good fits to the
available data can be obtained with a wide variety of parameter sets that usually are
dispersed throughout the parameter space. This problem of *equifinality* of different model
structures or parameter sets is discussed in the context of previous uses of the term in
geomorphology. The consequences of equifinality are uncertainty in inference and prediction. Recognition of such uncertainties, however, may suggest ideas for hypothesis
formulation and testing by creative experiment and monitoring that will lead to the
elimination of some of the possible model scenarios.

INTRODUCTION

The availability of increasingly powerful computers has made possible the study of
complex environmental systems by numerical experiment. Prime examples are the
advances made in numerical weather forecasting and the scenario modelling of the
atmosphere and oceans for the prediction of the effects of possible climate change
resulting from anthropogenic pollution. There has also been a recent flurry of research
publications on models of geomorphological development (see p. 301).

In most environments, the geomorphological development of the landscape and processes of erosion, deposition and weathering, are dependent on the flow of water.
Consequently the modelling of geomorphological processes must necessarily depend on
the modelling of hydrological processes with all its complications of dynamic surface and

*The Scientific Nature of Geomorphology: Proceedings of the 27th Binghamton Symposium in Geomorphology held 27–29
September 1996.* Edited by Bruce L. Rhoads and Colin E. Thorn. © 1996 John Wiley & Sons Ltd.

subsurface contributing areas forced by an unpredictable sequence of events of different magnitude. In turn, the modelling of hydrological processes must necessarily depend on the form of the landscape, with its control over convergent and divergent flow paths, soil and vegetation development. This interaction of hydrological and geomorphological processes will shape the development of the landscape over long periods of time within the context of climate change and tectonic change.

In what follows we will first consider the implications for geomorphology of recent studies in hydrological modelling. This has a long history, driven by the needs of prediction for water resources management. Over the last 10 years, the increases in computer power have been used in two main ways. The first has been to create ever more complex models of both hillslopes and river flows, with the aim of introducing as much physical understanding of the processes as possible (see for example Bates and Anderson 1993; Bathurst et al. 1995; Refsgaard and Storm 1995). In this way the hydrologist is attempting to emulate the atmospheric modeller but in a system that is less amenable to study in such ways because of the lack of knowledge of the subsurface part of the hydrological cycle. The second approach has been to use the computer power to make many thousands of runs of simpler models to explore the different predictions in the 'parameter space' and, in particular, how well different parameter sets fit the observed data. The results have been revealing. It has been widely found for both hydrological and geochemical models that many different models are *behavioural*, i.e. fit the data to an acceptable level, with the behavioural parameter values dispersed widely through the parameter space.

This chapter is primarily concerned with the implications of this *model* (rather than system) equifinality. It will be shown that geomorphological models should be expected to exhibit similar model equifinality, resulting in uncertainty and limitations on geomorphological predictability. Equifinality also leads to problems of inference where parameter values are determined by calibration, since such values will be conditional on the values of the other parameters in the model. Discussion of the problem of equifinality leads to the conclusion that progress in geomorphological modelling will depend on creative hypothesis formulation and testing by experiment and monitoring that will lead to the elimination of some of the feasible models.

MODELLING WITH DATA – THE HYDROLOGICAL EXPERIENCE

Only recently have coupled models of hillslope hydrology and sediment production and transport and of channel form, discharge and sediment transport started to appear (e.g. Bathurst et al. 1995). In virtually all hydrological analysis and models that take some account of catchment topography, the 'landscape' element (catchment characteristics) are considered to be fixed (e.g. Beven et al. 1995; Refsgaard and Storm 1995). No feedbacks between hydrology and geomorphology are generally considered (despite the continuing requirement for the hydrologist to re-evaluate rating curves for the conversion of stage to discharge, particularly after extreme events). This has been partly due to a lack of computer power in the past, partly because of a lack of measurement techniques and data, but is primarily attributable to a lack of interest of hydrologists in sediments. The problem has been that the prediction of stream discharges and modelling of water flow pathways assuming everything else constant has been sufficiently challenging. In fact, only relatively recently have the topographic controls on flow pathways been reflected in the model structures used by hydrologists (see for example Stephenson and Freeze 1974; Beven and Kirkby 1979; O'Loughlin 1981; Abbott et al. 1986; Beven et al. 1995; Ambroise et al. 1996).

The hydrologist has, however, had some major advantages in modelling flow processes over the geomorphologist wanting to model sediment transport processes. One is simply that the time scales of interest to the hydrologist are more compatible with dissertation, research grant and research career time scales, rather than the generally longer scales needed to integrate the effects of processes to a level of significant geomorphic change. The time scales of hydrological process theories are also short, and are not, in fact, amenable to application over long periods of time. In addition, because of the importance of water resources management, considerably more effort has been expended by both researchers and government agencies in hydrological data collection. The data are still limited but there are many sites for which rainfall, discharge and evapotranspiration data are available; fewer sites where limited internal state data such as soil moisture or water table information are available; and just a few sites where detailed measurements of the spatial patterns of flows have been undertaken using tracers and detailed sampling of state variables.

At well-gauged sites, therefore, it has been possible to examine the magnitude–frequency characteristics of hydrological events directly and to calibrate models of flood and drought frequencies, monthly water balances, and lumped and distributed models of groundwater and river flows, and lumped and distributed models of catchment hydrographs both for individual events and continuous (discrete time step) simulation. Hydrological models are now used routinely for flood forecasting, sometimes with real-time updating, surface and groundwater reservoir management and design, predicting the effects of land-use and climate change on runoff and flow extremes, pollution incident prediction and a range of other purposes. There is a vast literature on model structures (see Wheater et al. 1993 and Singh 1995, for recent reviews of available models), ranging from the purely functional unit hydrograph, still used to advantage in modern transfer function form (see Duband et al. 1993; Jakeman and Hornberger 1993; Young and Beven 1994), to the solution of stochastic differential equations for flow in a heterogeneous soil or groundwater system (e.g. Jensen and Mantoglou 1992).

Hydrological models are, in fact, a particularly interesting class of environmental models. At the small scale, the theory of water flows as embodied in the Navier–Stokes equations is relatively well understood. These equations are, however, very difficult to solve: in general because of the nonlinearity of the equations and the problems of closure associated with the velocity fluctuations in turbulent flows; and in specific applications because of the poor knowledge of boundary conditions, both locally and for the flow domain as a whole. Thus, it has been normal in developing model structures to resort to semi-empirical physical theory of surface and subsurface flows (Darcy's law, Richards' equation, the St. Venant equations, or more functional representations at larger scales). Some processes are quite well represented (at least locally) by the resulting descriptions; for others (such as evapotranspiration from a vegetated surface) the descriptive equations may be only poorly developed. For each process, however, this introduces parameters of the model that must be calibrated for individual applications, either by direct measurement, some indirect relationship with some characteristics of the catchment, or by adjustment to fit model predictions to observed responses. Where direct or indirect measurement of such parameter values is possible, it has often revealed that the parameter of such models cannot necessarily be assumed constant in space or time, leading to further difficulties of calibration for general applications.

A recent review of techniques for the calibration of parameter values has been presented in the context of hydrological models by Sorooshian and Gupta (1995). The availability of observed output discharges in hydrology for comparison with model predictions makes such calibration possible. It is, indeed, necessary since it has proven very difficult either to measure or estimate the parameter values of hydrological models a priori, even for the most 'physically based' hydrological models (see discussion in Beven 1989). This is in part because the scales of measurement of parameter values tend to be very different from the scale at which the model requires 'effective' parameter values to be specified. It has been known for a considerable time that calibration of hydrological models by comparison of observed and predicted variables is fraught with difficulties, because of model non-linearities (particularly those associated with threshold parameters), interaction between parameter values, insensitive parameter values and the effects of error in the observations. Where a quantitative measure of goodness of fit is used to assess model performance, these effects can result in very complex 'response surfaces' in the parameter space, with flat areas and multiple local optima that creates considerable difficulties for automatic optimisation techniques (see Blackie and Eeles 1985; Duan et al. 1992; Beven 1993).

It might be expected that improving the theoretical basis of model structures would help in this respect. It is now recognised, however, that this is not necessarily the case (see Beven 1989; 1993; Grayson et al. 1992; Jakeman and Hornberger 1993). There are a number of reasons for this. Increasing the physical basis of a model will usually increase the number of parameter values that must be supplied to a model while the data available for calibration may not increase commensurately. Even the simplest hydrological models tend to have more parameters than can be justified by the data available for calibration; they are *overparameterised* in a systems identification sense (see for example Kirkby 1975).

In addition, even the most physically based theory available has been developed at small scales for 'homogeneous' systems. Some processes, such as flow through structured, macroporous soil and extraction of water by root systems, are not adequately described at application scales by the available equations; heterogeneities and time variability within such a nonlinear system may mean that it may not be possible to relate local measured values to the effective parameter values required at the model grid scale; while many of the boundary conditions required may be essentially unknowable (Beven 1995a, b).

As a result, it has been suggested that all hydrological models can easily be invalidated as descriptions of reality and that even the most 'physically based' models must be considered as merely conceptual descriptions as used in practice (see Beven 1989), and not very good descriptions at that. The process of modelling is then saved by the process of calibration; the models normally have sufficient degrees of freedom in their parameters to be able to fit the observed data with an acceptable degree of accuracy, at least provided our standards of acceptability are not too high.

Confirmation of such models by prediction of another period of data (a split-record test) is a relatively weak test. A much stronger confirmation test would be independent check of the predicted *internal* states of the system. This is also problematic, however, since most internal variables have to be measured at scales much smaller than the grid or catchment scales of the model predictions. Predictions and measurements will then refer to different *incommensurate* quantities, making validation difficult. Concepts of validation, verification and confirmation of models have recently been much discussed in the hydrological literature (see Konikow and Bredehoeft 1992; Oreskes et al. 1994).

One interesting feature of hydrological models that has been revealed recently by computationally intensive explorations of parameter response surfaces is that, for most models, there may be many combinations of parameter values that will provide almost equally good fits to the observed data (see for example Duan et al. 1992; Beven 1993). For any given calibration period and chosen goodness-of-fit measure there will be one set of parameter values that gives the global optimum. There will, however, be many other parameter sets, in many cases from very different parts of the parameter space, that give almost as good fits. A little thought will suggest that this should not be unexpected, due to the problems of parameter calibration outlined above, together with the effects of error in the model structure, in the input and boundary data that drive it, and error in the observed variables themselves. Changing the calibration period or the goodness-of-fit index will give a different ranking of parameter sets in fitting the observations. In short, there is no single parameter set (or model structure) that can be taken as characteristic in simulating the system of interest; there is consequently a degree of model equifinality in reproducing the observations with model predictions.

This problem is, in fact, worse since one result of the lack of an adequate hydrological theory is that there may also be competing model descriptions of a catchment system, as well as competing parameter sets within a given model structure. They may differ in conception in one or more elements, the details of approximate solution techniques (such as different base functions for finite element solutions) or have totally different bases and parameter definitions. Even a cursory examination of the literature will reveal a plethora of models in hydrology with no clear basis for making a scientifically reasoned choice between them. Choice is more normally made for *ad hoc* reasons: the model is already on the computer; it is in the public domain; it is not too expensive to run; I have experience of previous applications; it can make use of the soil and topographic data already loaded on the geographic information system (GIS); the model has been used in this type of environment/for this type of problem before; it is the model I developed. The latter reason normally takes precedence over other considerations.

The problem can be compounded if the interest is not in the hydrology alone but in variables and processes that depend on the hydrology (weathering, solute, sediment and pollutant transport). This introduces additional model components with additional parameter values, all with the same problems of measurement scales, spatial and temporal heterogeneity, dependence on the model structure, together with the possibility of interactions between the hydrological parameters. This will generally increase the possibility of many different parameter combinations being able to fit the available measured data, especially when the available measurements are few. Figure 12.1 shows the results of comparing the predictions of the PROFILE soil geochemistry model (Warfvinge and Sverdrup 1992) to measurements at the C2 catchment at Plynlimon, mid-Wales (Zak and

Beven 1995). PROFILE is an equilibrium geochemistry model that treats the soil as a sequence of soil horizons in series from top to bottom. It requires 35 user-defined parameter values, 26 of which are allowed to vary by soil horizon, normally resulting in the order of 100 parameter values to be specified. A Monte Carlo experiment was carried out using 10 000 randomly selected sets of parameter values, chosen from qualitatively feasible ranges for a restricted number of parameters. The other parameters were kept constant at reasonable values. Each simulation was compared to the measured values of the integrated weathering rate, and the pH and BC/Al ratio of the Bs3 horizon (which dominates the soil geochemistry at this site).

Each plot in Figure 12.1 demonstrates the combined goodness of fit to all three measures for some of the varied parameters. Each point on the plots represents one of the 10 000 Monte Carlo sets of parameter values. It is clear that for most of the parameters, there are combinations of parameter values that give better fits to the observations across the whole of the parameter range considered. The only parameter in the model that shows any strong sensitivity is the reaction coefficient for gibbsite which controls aluminium solubility in the model with a consequent strong effect on pH. This reaction is used widely in geochemical models, despite the fact that there is little or no gibbsite in temperate soils, including the soils at C2. In addition, it was found in this study that none of the combinations of parameter values could simulate the estimated weathering rate at C2

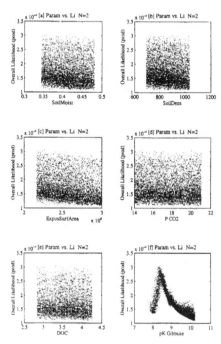

Figure 12.1   Results of fitting the PROFILE geochemical model to data from the C2 catchment in mid-Wales for six of the PROFILE model parameters. Each point represents a run of the model with randomly chosen parameter values. The higher the likelihood value for a given run, the better the fit to the observations (after Zak and Beven 1995)

adequately. The implication in this case is that the model should be rejected (but PROFILE is being used to estimate critical loads for acid deposition in many areas of Europe and North America).

EQUIFINALITY AND UNCERTAINTY IN HYDROLOGICAL MODELLING

It is suggested therefore that model equifinality may be axiomatic of environmental modelling where highly parameterised models requiring calibration are fitted to limited data that integrate the response of the system of interest over time and space. Equifinality implies that any parameter values determined by such calibration will be conditional on the other values of the model parameters such that any physical interpretation of the values must be made with care. Equifinality also implies uncertainty. Different model structures or parameter sets that are considered acceptable simulators will, in general, produce different predictions. Beven and Binley (1992) have applied a Bayesian methodology (generalised likelihood uncertainty estimation–GLUE) for estimating this predictive uncertainty based on associating a likelihood weight with each simulation. Their application uses Monte Carlo simulation of multiple randomly chosen parameter sets within a single model structure as the basis for estimating the uncertainty (see also Beven 1993; Romanowicz et al. 1994; Freer et al. 1996). Extension to multiple model structures is straightforward.

The GLUE approach rejects the idea that there may be some optimal model or parameter set. Models can only be evaluated in terms of their relative likelihood of being an acceptable simulator of the system of interest or rejected as being non-behavioural. Such a view seems to lie somewhat uneasily between several rather different philosophical viewpoints on the structure of science. Most environmental scientists will agree that model/theory confirmation is a matter of degree of empirical adequacy (van Fraasen 1980; Oreskes et al. 1994); the point here is that adequacy may be limited or conditional, requiring further tuning or modification of ancillary conditions as more or different types of data become available. This would appear at first sight to result in a purely relativist attitude to the problem of modelling complex environmental systems, in keeping with the views of Feyerabend (1975) on the development of scientific thought (see Beven 1987, for a discussion in relation to hydrology). The estimation of likelihoods is certainly consistent with a relativistic philosophical stance which does not require any necessary or strong correspondence between theory and reality.

The problem can, however, be viewed from within other traditions as a problem of model/theory falsification. It is now well recognised that falsification is fraught with difficulties, but we use it here in a weak sense in respect of the declaration of certain models as 'non-behavioural' in simulating a particular system of interest. There may indeed be many models that are behavioural or acceptable simulators, but an important part of the process of modelling is then the rejection of some of those models on the basis of existing or new evidence. This does not necessarily imply that there is any correspondence between those models retained and reality; nor that rejection of a model in one application implies that a model may not be a useful predictor elsewhere. Indeed the very fact that there may be multiple models or competing hypotheses retained does not encourage such views.

However, whilst recognising the difficulties associated with the concept of falsifiability, in this context it raises some interesting possibilities. It may be possible to design testable hypotheses and associated experiments that would allow model structures or parameter sets to be designated as non-behavioural, i.e. a certain class of models or parameter sets will be deemed falsified. The resulting studies might represent a very different approach from the experimental work associated with modelling carried out at present in which the concern tends to be with the measurement of parameters or state variables at small (but manageable) scales. Such an experimental design may not be the most cost-effective approach to refining the likelihood associated with individual models and consequently to constraining the set of behavioural models and consequent predictive uncertainty. Such an approach has much in common with the Bayesian methodology espoused by Howson and Urbach (1989). Rejection of all the models tried on the basis of some reasonable criteria will suggest a serious lack of predictive capability.

Why has this approach not already been adopted widely in hydrological modelling? One reason is that it is actually too easy to falsify the currently available models on the basis of either their assumptions or their performance relative to observations. The modelling process is then saved by the adoption of less stringent criteria of acceptability or recourse to ancillary arguments which allow that it may not be possible to predict all the observations all of the time (arguments of scale, spatial heterogeneity, lack of time variability in parameter values, uncertainty in theoretical descriptions of the processes, etc.). In this context, relativism is commonly practised – albeit using qualitative rather than quantitative measures of performance and without explicit recognition of the process. I have suggested elsewhere that the result is more akin to prophecy than to prediction (Beven 1993).

In what follows the implications of these conclusions for geomorphological studies and modelling will be considered.

EQUIFINALITY, EQUIFINALITY AND EQUIFINALITY

The use of the term 'equifinality' has had a somewhat different content in geomorphology compared with the usage above, stemming from the principles of 'general systems theory' outlined by von Bertalanffy (1951, 1962) and introduced into geomorphology by Culling (1957) and Chorley (1962) (see Haines-Young and Petch 1983). The concept in this context is used to denote the possibility of similar landforms being derived from different initial conditions in different ways by possibly different processes. Haines-Young and Petch (1983) provide a critical review of the concept, suggesting that the unthinking resort to equifinality in explanation of landforms is a failure of methodology. They suggest that if similar landforms can truly be shown to be the result of different processes then equifinality is an empty problem. They particularly object to the link with the method of multiple working hypotheses made in the work of Cooke and Reeves (1976). Cooke and Reeves interpret equifinality in the sense that it may not be possible to distinguish between several different theories for the formation of a particular landform. Haines-Young and Petch (1983, p. 466) by contrast argue that if two or more theories cannot be distinguished on the basis of the predictions that they make about landform character *then they are poor theories*. To describe those features as 'equifinal' does not detract from this situation. Use of the term merely encourages the maintence of those theories in an *ad hoc* and uncritical way. The aim of the geomorphologist should be to develop those theories so that they can be tested *against each other*. Only then, through the process of experiment and observation can the geomorphologist hope to eliminate any false conjecture.

They further suggest that the only valid use of the term is the much more restricted sense used by Culling (1957) who suggests that in open systems the operation of similar processes will, over time, tend to produce similar forms from a range of initial conditions. Culling suggests that graded streams may be considered 'equifinal'. A link with the gradualist concept of dynamic equilibrium may be discerned here, but the use of the term in this way was later criticised by Culling (1987, p. 68) himself in the light of more recent work on nonlinear dynamic systems theory and chaos. He notes:

The ubiquity of noise means that all stable systems are transient.... It is now known that transients can exhibit chaotic behaviour and that these chaotic transients may have extremely long lives ($\sim 10^6$ iterations). Chaotic transients can only compound the difficulties of recognising chaotic behaviour in the landscape. Despite all these difficulties, however, it is known that chaotic motion and strange attractors into the heartland of physical geography for turbulent flow is irregular, intermittent, self-similar and whether we like it or not ubiquitous.

Culling's (1987, p. 69) conclusion is that equifinality is a vague and transient concept that will ultimately be subsumed into the well defined apparatus of abstract dynamical systems. Geomorphological systems are nonlinear and subject to random forcings of events of different magnitudes. Similar to other nonlinear systems they should be expected to show significant sensitivity to initial conditions and random perturbations. He distinguishes between equifinality *sensu strictu*, where a perturbed system will eventually return to its original form, and weaker forms of equifinality which imply only persistence of some property, i.e. stability in some sense. He defines a number of ways in which properties may exhibit local (small perturbation only) or Lyapounov stability (return to a similar form) or, in a weaker sense, ergodic or topological persistence. The application of nonlinear dynamical theory to geomorphological systems has been further explored by Culling (1988), Malanson et al. (1992), Phillips (1993, 1994) and others.

The experience of model equifinality in hydrology suggests that there is, in fact, little incompatibility between all these views when it comes to *practical* geomorphological explanation. If there are, indeed, many models that may be compatible with the geomorphological evidence, they should include those models that exhibit equifinality in the senses outlined by Culling (1987). Haines-Young and Petch (1983) note that part of the attraction of the concept of equifinality may come from the fact that landforms present extremely difficult objects to study. As a result it may be very difficult to obtain the necessary data over sufficient periods of time and sequences of events to decide between multiple working hypotheses (or models). That does not mean that they are necessarily poor hypotheses, only that the problem is currently undecidable within the limitations of currently available models and data. If information was available to determine that the hypotheses were poor, they would normally be rejected.

If indeed, geomorphological systems are sensitive in their nonlinear dynamics to initial conditions and random forcings then it follows that much of the history of particular landforms may now be lost from view. This is not inconsistent with the fact that in many environments some effects of past geomorphological processes and climatic regimes are readily distinguished, even after long periods of time. Geomorphological systems are indeed transient, they should be expected to show the remnant results of past and present processes, but the possibility of chaotic behaviour means that the *trajectory* of their development may be undecidable on the basis of present-day evidence alone. Thus, the consequences of understanding from dynamical systems theory suggest that equifinality may not be an indication of poorly developed methodology but may be implicit in the nature of geomorphological systems.

One practical consequence of this equifinality is in the application of geomorphological models which represent a (more or less) rigorous way of formulating practical hypotheses about geomorphological systems. As in the case of hydrological models described above, geomorphological models require necessary simplifications and abstractions to be tractable and involve parameters that must be calibrated in some way. The models are nonlinear and may demonstrate chaotic behavior (Phillips 1993). Within such a model framework there may then be many combinations of initial conditions, model behaviours and parameter sets that are consistent with the limited observations available about a particular class of landform. They are then equifinal in some sense, indeed in a very similar sense to that used by Cooke and Reeves (1976) and criticised by Haines-Young and Petch (1983). This analysis would suggest that there may be very many situations in geomorphology where equifinality stems not from an inherent property of the system but from an inherent property of the process of *study* of the system.

At first sight this would appear to be a very unhealthy situation for geomorphological science, as expressed in the concerns of Haines-Young and Petch (1983). This is not necessarily the case; equifinality of hypotheses and models today, when properly recognised, can lead to the formulation of experimental and analytical methodologies that may allow rejection of some of the competing explanations in the future. One suspects, however, that there will be an irreducible set of possible explanations and that equifinality will, itself, exhibit persistence.

In summary, equifinality would appear to remain a valuable concept in geomorphological studies as a result of the inherent limitations and constraints on understanding both the genetic evolution and modelling of landforms. It expresses, in shorthand form, the impossibility of distinguishing between many possible histories from different possible initial conditions and different possible process mechanisms on the basis of the available evidence.

Qualitative reasoning to argue for one trajectory rather than another has ultimately to depend on faith. Quantitative reasoning, based on model predictions, will result in many different sets of model structures, initial and boundary conditions and parameter values that will be compatible with the available data. However, it is hoped that recognising this equifinality may lead to a more robust approach to testing the viability of different model explanations, leading to the rejection of some but, undoubtedly, to the retention of many. The class of retained models may, of course, be inherently interesting in themselves. Similarities and differences may lead to improved understanding.

The next section explores the background to model equifinality in geomorphological explanations, starting from the geomorphologist's perceptual model of the processes that are her/his concern.

EXPLANATION IN GEOMORPHOLOGY – THE PERCEPTUAL MODEL

In geomorphology, as in hydrology and all other environmental sciences, there is a difference between a scientist's perception of how the system of interest operates and what is included in the working models being used. Since the work of Popper and Bachelard, it has been recognised that both are socially conditioned; that both theory development and interpretation of experimental and other evidence are carried out within a social and historical context of interaction and competition between research groups, individual scientists, teachers and students. Geomorphology has not been subject to the detailed sociological scrutiny as some other areas of science (e.g. Knorr-Cetina 1981) but there has been a succession of reviews of the status of the subject that allow the framework for a perceptual model to be assessed, both in terms of the philosophy of the science (e.g. Haines-Young and Petch 1983; Richards 1990, 1994; Rhoads and Thorn 1993, 1994; Bassett 1994; Rhoads 1994) and the subject-matter itself (e.g. Brunsden 1985, 1990; Scheidegger 1987).

The perceptual model is not, of course, written down. It is individual to each geomorphologist depending on her/his teachers and training, his/her field experience of different environments, the literature and conference presentations s/he has been exposed to, and day-to-day discussions within a research group. Putting a perceptual model into writing will necessarily require simplification (but also perhaps useful critical review and formalism). The important thing here is that any perceptual model will recognise complexities and multiple possible explanations of landforms in a way that cannot be included in the mathematical descriptions that form the basis for any predictive capability. The perceptual model is inherently qualitative, but conditions both responses to experimental evidence and decisions about the dominant processes and representations of those processes to be included in quantitative models.

For certainly decisions must be made. Quantitative models are necessarily crude approximations of our perceptual understanding of what is important. There are many processes for which we may understand the governing principles in detail but cannot apply those principles at scales of interest because of lack of information about characteristic parameter values or boundary conditions that are only poorly known or too complex to be feasibly known. There are other processes for which we do not have an adequate description at any useful scale. These decisions are constrained by the current perceptual model and considerations of *feasibility* in terms of mathematical tractability, computing and data requirements. There is also, perhaps, a competitive edge to the process, observing and improving upon what is being done elsewhere (or at least doing something a little different).

Consider then a perceptual model for a particular area of geomorphology, the development of a hillslope/river network system. This is an area that has recently been the subject of significant (and competitive) modelling activity. A (simplified) perceptual model of the controlling processes will involve the following elements. The primary driving forces for hillslope and channel development are gravity and the hydrology, which largely controls erosion, deposition, chemical weathering and removal of material by solution. The balance of hydrological processes of surface and subsurface flows and 'losses' to evapotranspiration will affect the pattern of geomorphological development and may lead to important seasonal differences in geomorphological processes (e.g. Schumm 1956; Howard and Kerby 1983; Harvey 1994). There is a feedback between hillslope form and flow processes that will control the dynamics of surface and subsurface contributing areas for runoff and the concentration of flows and resulting shear stresses.

Over long periods of time, there may be important feedback mechanisms between vegetation cover, soil development, weathering processes, erosion and deposition with different constraints in different environments. Man can have an important impact over short periods of time. Extreme events (floods, droughts, mass movements or volcanic eruptions; e.g. Starkel 1976; Baker 1978; Newson 1980; Dunne 1991; Howarth and Ollier 1992; Nott 1992) can also have important impacts over short (and sometimes long) periods of time. We assume that uniformitarianism holds in the sense that the physical and chemical dynamics of the processes involved will not change, but the boundary conditions and values of controlling parameters may change over time. The relaxation time of the system to such short-term disturbances will control how the system is perceived as being in some 'dynamic equilibrium' and how far the magnitude–frequency of system responses can be related to the magnitude–frequency characteristics of the external forcing in terms of concepts such as the 'dominant' or 'formative' event (Wolman and Gerson 1978; Brunsden 1985, 1993; Dunne 1991). Some systems may be perceived as apparently continually in disequilibrium (e.g. Stevens et al. 1975). In some circumstances, the sequence of events may be important as well as the magnitude–frequency distribution, particularly where some threshold phenomena control the response (Anderson and Calver 1977; Beven 1981). Sediment transport depends on complex thresholds for the initiation of motion and erosion and may be transport limited or supply limited (perhaps at different times or different locations within the same system, e.g. Newson 1980; Coates and Vitek 1980; Campbell and Honsaker 1982).

Thresholds (for example, for shallow mass movements or surface erosion at a point) may evolve over time, and might also vary spatially with vegetation or soil patterns. The exceedance of thresholds may depend on spatial patterns of rainfall intensity and ante-cedent conditions that may depend on hillslope form as well as vegetation, soil and preceding weather patterns. The analysis of nonlinear dynamic systems suggests that perturbations of the system might, in some circumstances, lead to switches in the mode of behaviour and associated processes, without necessarily any relaxation back to the original system state. Geomorphological systems show some evidence of self-organisation in patterns of dendritic rill and channel networks, meandering channels, and slope–area relationships (see Hallet 1990; Rinaldo et al. 1993; Rigon et al. 1994; Rodriguez-Iturbe et al. 1994; Ijjasz-Vasquez and Bras 1995).

The operation of these processes is set within a historical context of changing external forcing associated with climate change and tectonic effects (e.g. Thornes and Brunsden 1977). Both will be expected to show irregular rates of change over time. The residual features of previous climatic regimes and geomorphological processes may still exert important controls on current landforms and processes (e.g. Fried and Smith 1992), for example in those temperate areas that were subjected to successive periods of glacial and periglacial processes in the last ice age. It is generally impossible to know the initial conditions for slope development (except for some laboratory and man-made systems). It is also impossible to know in any detail the parameters that control the process responses, particularly those of the subsurface. Biotic controls on soil permeability and soil strength through root growth and decay and the effects of soil fauna may be very difficult to assess. Soil physical characteristics can generally only be determined on a small number of samples that may be a poor representation of the soil mass as a whole. There may also be considerable heterogeneity of processes associated with the nature of the surface and its vegetation cover (e.g. Dunne et al. 1991) in ways that may be very difficult to understand (see for example Hawkins 1982; Hjelmfelt and Burwell 1984).

So much for the (simplified!) perceptual model. In summary: 'the key-words of modern geomorphology are: mobility, rhythm, flux, instability, adjustment, sensitivity, complexity and episodicity' (Brunsden 1985, p. 52). To this must now be added the possibility of chaos and strange attractors (Culling 1987, 1988; Phillips 1993, 1994). Modelling such systems is clearly very difficult. We can conclude that in addressing the modelling problem we will generally have no information about the initial conditions, little information about the changing nature of the external forcing (both climatic and tectonic) over time, poor information about the effects of man except in the very recent past, little knowledge of how the physical and biotic characteristics of the system have changed over time, and relatively poor mathematical descriptions of the processes of development at the scales of interest.

How is the modeller to proceed in the face of such uncertainty? One answer is certainly deductively; it saves having to come too close to reality and address the need for and prediction of real data.

MODELLING WITHOUT DATA – DEDUCTIVE GEOMORPHIC REASONING

Deductive reasoning has a long and prestigious history in science. It allows the consequences of a given theory and set of assumptions to be enumerated and in many cases tested. There have been a number of well-documented cases of deductive predictions in science that have later been confirmed by observation. The implication is then that the assumptions of the theory are a good approximation to reality and from a strong realist viewpoint, that the variables embodied in the theory are real variables. It is not necessary to make such claims for quantitative geomorphological theorising, which is incomplete, based on empirical expressions and recognised as approximate. The process is more normally referred to as modelling.

Quantitative geomorphological modelling has been growing rapidly in popularity as an indoor sport in recent years (e.g. Ahnert 1976, 1977, 1987; Armstrong 1976; Cordova et al. 1976; Kirkby 1985; Roth et al. 1989; Willgoose et al. 1991a, b, c, 1992; Chase 1992; Howard 1994; Moglen and Bras 1995). These models solve partial differential equations of mass conservation for water and sediment coupled to various semi-empirical erosion and transport laws, mostly of the general form

$$R = (K_1 + K_2 C)C^p D^m \sin^n \alpha - K_3) \tag{1}$$

where R is the rate of transport, D is the local value of discharge, $\sin \alpha$ is the local value of slope angle, and C is the local soil depth, K_1, K_2 and K_3 are coefficients and n, m, p are exponents which vary with the nature of the transport process (splash, viscous flow (creep), plastic flow (debris flows and slides), suspended wash transport and fluvial transport; see Ahnert 1977). For each process included in the model, the six coefficients and exponents must be specified. Even if the slope and channel process parameters are assumed to be stationary in time and space, potentially there are 30 parameters to be specified for these five processes, although many are normally set to zero. There may be additional parameters associated with weathering processes (e.g. Ahnert 1976, 1977, who uses a two-parameter formulation), while running the model requires a field of initial elevations and boundary conditions in terms of net runoff rates and a field of rates of tectonic uplift in the area of the simulation.

These 'laws' are empirical–causal idealisations which, in themselves, have little explanatory power in terms of the underlying mechanisms, and have parameters that may require calibration for particular applications. Rough ranges of these parameters are known from experimental and previous modelling experiences (e.g. Figure 12.2 from Kirkby 1990), although it has been suggested that no general agreement exists on the powers n and m (Kooi and Beaumont 1994, p. 12–207). The nonlinearity of the transport laws necessitates approximate numerical solutions, in most models using finite difference approximations on a regular square mesh, and values of $m > 1$ imply that the magnitude–frequency distribution of events may be important. All the models produce dissected landscapes that have some similarity to real landscapes, despite being gross simplifications of the perceptual model described above. In all these cases, the forcing due to external variables is continuous and steady with no allowance for extreme or catastrophic events or periods of 'relaxation' between major events. Time derivatives in the equations are usually treated explicitly and, in most cases, little study is reported of the stability constraints on the solution of these nonlinear equations.

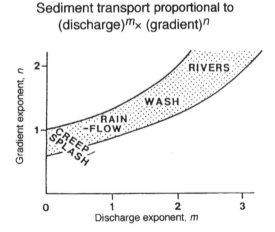

Sediment transport proportional to
(discharge)^m× (gradient)^n

Figure 12.2   Range of parameters m and n for different processors (after Kirkby 1990)

These are examples of what Morton (1993) calls 'mediating models'. They mediate between an underlying theory, which in geomorphology is developed largely in rough qualitative terms (the perceptual model), and the quantitative prediction of landscape development. They have the general characteristics revealed by Morton's analysis: they have assumptions that are false *and known to be false*: they are not, however, arbitrary but reflect physical intuition; they tend to be purpose specific with different (and possibly incompatible) sets of assumptions and auxiliary hypotheses for different purposes; they have real explanatory power but may never (nor are they expected to) develop into full theoretical structures. They also have a history, in that successful modelling techniques tend to be refined and inherited by later models (see Schrader-Frechette, 1989, for a hydrological example).

The predictions of such models are valid only within the context of the model structure itself. This will necessarily include any effects of the solution algorithms used, for example the effects of numerical dispersion within a finite difference scheme, meaning that the approximate solution may not be convergent with the original differential equations, despite the fact that the numerical solution may remain stable throughout. In addition, these are dissipative nonlinear discrete time systems; depending on the nature of the attractors of the solution, model predictions may be sensitive to initial conditions (for geomorphological examples see, for example, Willgoose et al. 1991b; Ijjasz-Vasquez et al. 1992; Howard 1994) while slightly different models applied to the same set of boundary conditions may result in significantly different predictions. Deductive inference then refers to the model; any inference about the behaviour of real systems is likely to be tenuous.

What is clear is that both landscape and models belong to a class of systems that produce dendritic structures. This arises out of the simple feedbacks between flow, erosion and sediment discharge. Various arguments have been advanced in the literature for the constraints that lead to a dendritic network for shedding water and sediment, including asymptotic efficiency arguments for the form of particular networks in particular circumstances (e.g. Woldenberg 1966; Rodriguez-Iturbe et al. 1992; Ijjasz-Vasquez et al. 1993), where the parameters of that particular system can be specified.

Consider, however, if we wish to use such models to deduce (or perhaps more correctly *abduce*) the development of particular landforms where the initial conditions, transport laws and historical boundary conditions are not well known. There are a number of problems in trying to do this. Each of the models quoted above, albeit gross simplifications of the real processes, requires the specification of a gamut of parameters. Thus, in similar fashion to the hydrological models discussed earlier, there may be many different sets of parameter values within a number of different model structures that will be equally consistent with some statistical measures of goodness of fit between modelled and real landforms. In addition, it is known that model predictions may be sensitive to their initial conditions and precise values of parameter values (see for example Moglen and Bras 1995). Thus, for any given model, there may be many different initial conditions that are equally consistent with the chosen measures of fit. In addition, the history of the boundary conditions in terms of the magnitude, frequency and sequence of events is equally unknowable; there may be many sequences that, when interacting with the parameter values and initial conditions, will be equally acceptable as simulators of today's landforms, particularly if those forms are close to an attractor in the solution space. In short, model equifinality should be an expectation in geomorphology.

The inference from this analysis is not that it is not possible to simulate today's landforms using geomorphological models. The problem is rather the contrary, it may be all too easy to produce statistically similar landforms (especially if the statistics are not too discriminating since it is very difficult to characterise precisely all the characteristics of a landform in terms of a few statistical indices or fractal dimensions which may be limited in discriminatory power). Thus there may be many model 'explanations' of the landform, equally valid given the information available but all known to be false through being based on modelling with all its limitations. Many factors have been knowingly glossed over or ignored in these first modelling studies, due to lack of knowledge and computing power.

Will further refinements of the models or data help in this respect? Again, experience from hydrological modelling suggests not a great deal. It is almost an aphorism that refinements of models within the normal development of a science tend to introduce complexity and require additional parameter values and boundary conditions. Increasing the dimensionality of the parameter space in this way leads to further identifiability problems, unless those parameter values can be estimated quite independently of the model (unlikely in geomorphology). Consideration of the nature of specific landforms requires consideration of their unique as well as generic characteristics. The unique characteristics include the heterogeneity of characteristics and parameters and previous history, all of which are essentially unknowable.

GEOMORPHOLOGICAL PREDICTABILITY – COMPARISON WITH REAL LANDFORMS

Pure deduction requires validation only for internal and numerical consistency. Models for geomorphological development, which may or may not be internally and numerically consistent, can also be compared with real landforms. Current models can produce a wide range of landforms, similar to the wide range of landforms seen in different real landscapes (e.g. Howard 1994). A variety of landform modelling studies have reported qualitative assessments of the realism of the simulated landscapes; the difficulty comes in comparing model simulations with *particular* landforms with their own unique characteristics of underlying geology and history, including the persistence of effects from past events or tectonic or climatic regimes. There have been very few studies that have attempted to do this. The examples below will illustrate some of the problems involved.

One early study was that of Ahnert (1970) who compared a simple slope profile model predicting both bedrock profile and overlying waste thickness with data on waste cover collected from three short profiles on gneiss bedrock in North Carolina. Two other, longer, profiles were eliminated from the analysis because of irregularities in slope and waste cover that are probably due to local variations of rock resistance (Ahnert 1970, p. 93). The model was purely deterministic based on a simple exponential relationship between weathering and soil depth and a transport rate that is directly proportional to slope angle. It appears to have required two parameters (a weathering rate scaling parameter and transport rate scaling parameter) and an initial slope profile. Slope development was predicted using explicit time stepping. It was found that the model could explain of the order of 86% of the field-measured waste thickness values when 'a profile similar to the field slopes was singled out for comparison'. Ahnert (1970, p. 96) concludes that 'the close agreement between the properties of the model slope and those of the field slopes indicates that the theoretical model is very probably a valid representation of conditions and processes on real slopes'.

This early study, albeit limited in scope, neatly illustrates the general problems of model validation. With the considerable benefit of a current viewpoint, it might be suggested now that, even within the framework of Ahnert's simple model, there might be a number of different representations that would produce results consistent with these field data, while a general explanatory model that would include the other slopes excluded from the validation exercise would require more complexity and parameter values and consequently data to be collected. It is worth re-emphasising that model equifinality raises problems about the physical significance of parameter values determined by calibration. Such values may only have significance within the context of the particular model structure used and will be conditional on values of other parameters. This, by extension, includes parameter values determined by calibration reported in the literature.

Since the time of Ahnert's study, the formulation of geomorphological models has indeed become generally more complex with more parameters to be specified. In addition, it is possible to compare the complete field of predicted values with the real topographic characteristics. Nobody, however, would suggest that a geomorphological model could predict a landscape in precise detail. The tendency therefore has been to compare generalised *indices* of behaviour. One example is the study of Willgoose (1994) who compares area–slope–elevation plots of both model and real landscapes (Howard 1994 shows a similar comparison). For the modelled landscapes, earlier work had shown that such plots show consistent (but different) shapes for the two cases of dynamic equilibrium (when uplift rate is equal to erosion rate) and 'declining equilibrium' (the fixed base level case when normalised hillslopes show a characteristic pattern). Effective parameter values were calibrated by fitting the field data to the characteristic relationships for each form. Some 40% of the variance of the field data was explained for the dynamic equilibrium case and 55% for the declining equilibrium. Confidence limits on the fitted parameters varied up to $\pm 10\%$ (with the base level for declining equilibrium being particulary well calibrated) while it was found that data from individual subcatchments in the field area showed slopes on the plots considerably different from the aggregated data (perhaps due to different effective base levels). Willgoose (1994, p. 158) notes that while these results show 'that the area–slope and area–slope–elevation relationships can be consistent with observed field data, this does not constitute a validation'. He suggests that a proper validation would require field data collected of total load sediment transport at a range of catchment areas at an 'undisturbed' site. Even then, assumptions of statistical homogeneity of catchment and erosion characteristics would be required. This study makes the difficulty of separating parameter calibration and model validation quite clear.

Mogelen and Bras (1995) have also tried to calibrate a model to whole landscape characteristics using an extension of the Willgoose SIBERIA model. To do so, they have assumed that the landscape is in steady state with erosion in equilibrium with uplift and that the values of m and n in equation (1) can be specified for different processes. They use the observed cumulative area distribution and slope–area curves to calibrate the parameters of the model which include a parameter that controls the heterogeneity of the resistance of the soil to erosion. Two free parameters are fitted to the cumulative area distribution for two catchment areas of different topography using nonlinear least squares. They suggest that the inclusion of heterogeneity is important in reproducing the observed cumulative area distributions. Their model uses a linear law for creep on the hillslopes; elsewhere it has been suggested that linear diffusion is inadequate to reproduce the temporal pattern of scarp degradation (Andrews and Bucknam 1987) and that a nonlinear diffusion law (with at least one extra parameter) is necessary. Although these studies are still very much in their early stages, the attraction of adding complexity and parameters to 'explain' observed landscape features is already apparent.

By concentrating on the 'equilibrium' characteristics of both modelled and field landforms the Willgoose (1994) and Moglen and Bras (1995) studies avoid the problem of persistence of features from past events and regimes of tectonic uplift or climate. In fact, Willgoose notes that his study does not address the interaction between the time of adjustment to uplift events and time between uplifts. Brunsden (1993), in a general dis-cussion of the problem of persistence, uses a framework in terms of *formative events* and *relaxation times*. Ahnert (1987) discusses the relaxation time towards dynamic equili-brium within the context of a distributed slope development model in an application to simulate the slopes of the Kall valley in the northern Eifel. The Kall valley exhibits a Tertiary denudation surface of low slope in its upper reaches, with an increasingly incised channel downstream, thought to be the result of Quaternary headward erosion with slope development affected by periglacial processes. The model used, SLOP3D, is an extension of the hillslope model cited earlier, and in this study is used to simulate the progressive development of a single slope with an initial condition taken as a current profile on the Tertiary surface. Six other profiles from further down valley were compared with the model productions assuming that the history of the valley allowed for spatial variation in the field to be replaced by temporal variation in the model. Note that this allows an additional degree of freedom in choosing which time step to compare with each field profile. In the Kall valley, profile 7 which is considerably further downstream than profile 6 is compared with a simulated profile at about half the time of that for profile 6.

The version of the SLOP3D model used appears to require nine parameters to be fitted. Of these, Ahnert (1987, p.5) notes that the four controlling suspended load wash 'keep the regolith from becoming too thick but have little effect of the shapes of the profile'. A threshold slope parameter for the occurrence of debris slides 'equals approximately the maximum angles of waste-covered slope found in the Kall valley' (Ahnert 1987, p. 5). The remaining parameters control the rate of fluvial downcutting at the base of the slope, the slow mass movement rate and the weathering rate. These are all assumed constant. Ahnert (1987) comments on the effects of climatic fluctuations that no specific mor-phological traces of the effects of fluctuations remain. 'Apparently they caused merely intensity variations during the continuing slope development but not any significant changes in the direction of that development' (Ahnert 1987, p. 6), even though Ahnert calculates that the time required for the development of the model slopes is of the order of 1 million years. The relaxation time to equilibrium for this area is estimated as of the order of 5 million years, a time scale within which both climatic and tectonic fluctuations have been significant.

Ahnert (1987, p. 6) shows that, 'after many attempts with different combinations' of parameter values, a simulation was obtained that fits the observed field profiles well. He suggests that this match is not obviously due to equifinality in the sense of similar forms arising from different process representations. Some model runs produced 'qualitatively similar forms. However, in quantitative terms all of these deviated more from the natural Kall valley slope profiles' (Ahnert 1987, p. 6). Interpreting this conclusion in terms of the equifinality concepts described above, Ahnert is clearly suggesting that the range of possible models consistent with the field data is highly constrained despite the uncertainty in the history of these slopes and the appropriate parameter values for both weathering and transport (see also Kirkby 1984). An investigation of just how constrained the feasible parameter sets are, in this and other situations, would be of great interest. Measures of model performance also require further study, but experience with hydrological models suggests that Ahnert's conclusion is optimistic. If it proves correct, however, it will be of great significance for geomorphological reasoning and prediction.

THE PROBLEM OF FUTURE HISTORIES – UNKNOWABILITY AND UNCERTAINTY

This chapter has attempted to clarify the different notions of equifinality associated with geomorphological theorising and modelling. The potential for equifinality in modelling particular landforms has been emphasised, although one of the few comparisons of modelled and field hillslopes (Ahnert 1987) suggests that the range of parameter values giving simulations consistent with field data may be highly constrained. This is likely to be optimistic, however, since equifinality should be expected as a general characteristic of the limitations of models that are false and of data that are generally inadequate for model parameter identification and in some cases unknowable. The consequences of equifinality are uncertainty in inference and prediction.

There is, however, a need and a market for geomorphological predictions in such areas as the design and near-term future development of erosion on landfill sites (e.g. Riley 1994) and making the long-term safety case for radionuclide repositories (e.g. United Kingdom Nirex 1995). The initial conditions for such predictions will be known in broad scope (either a design or actual current landform) although the heterogeneity of current slope and channel characteristics will be difficult to define precisely. Future boundary conditions, however, are clearly the stuff of speculation even in the relatively short term. Longer-term climate predictions using global climate models (GCM) cannot be considered reliable, being subject to 'flux corrections' and poorly validated at the regional scale even for mean monthly predictions. There will be even more uncertainty about the changing probabilities of extremes under changing climatic conditions which may be important in geomorphological development, particulary for threshold-controlled processes. Perhaps the best strategy towards future prediction is to consider what might happen under different possible scenarios of boundary conditions with a view to identifying those scenarios that might prove application critical.

Thus, there will be uncertainty arising from different possible models and parameter sets and uncertainty arising from different scenarios of possible boundary conditions. The possibilities are numerous and it may be difficult to assess or assign any probability of occurrence or likelihood to each possibility except in some subjective way (as in the case of the first Intergovernmental Panel on Climate Change (IPCC) report (Houghton et al. 1990) on future climate and sea level changes). This is a particular problem for application critical predictions. Decision analysis can make use of some estimate of the risk of occurrence of critical events, given the uncertainties in the prediction process, if it could be made available. It is possible to assess such risks within either Bayesian likelihood or fuzzy set frameworks but it will be clear from the discussion above that assessment of the likelihoods of both models and scenarios will be inherently subjective, even given some 'validation' of models in predicting current landforms.

There is one way in which this concept of multiple scenarios for acceptable models might be used as a proper tool for geomorphological investigations. Consider a sample set of viable models produced by Monte Carlo simulation within ranges of parameter values and conditions considered feasible in a particular situation that reproduce (to some appropriate level of similarity) the nature and historical development (as far as it is known) of real landforms. That range truly reflects the uncertain knowledge about landscape development within the limitations of the modelling process, but may also contain information about competing modes of behavior within the model structures used. If so, it suggests that a process of hypothesis testing (as discussed earlier), in which critical and perhaps novel analyses are used to eliminate certain model scenarios from the current viable set, may be a valid way of improving model structures. Limitations in knowledge, data, and lack of experimental techniques for discriminating between model scenarios, should, however, be expected to lead to a degree of irreducibility of the set of feasible models and consequently to uncertainty in inference and prediction.

This suggests two requirements for work in the future. The first is for creative experiment: collecting measurements that will allow for different hypotheses and assumptions to be tested in a way that eliminates some of the set of possible behavioral models. This is not a simple task, in that failure in a test can often be avoided by the simple addition or refinement of auxiliary assumptions (such as heterogeneity of para-meter values) that allow underlying model structures to be protected and that many of the possible measurements may not have great power in discriminating between models and parameter sets. The second is for continuing monitoring of sites so that the likelihoods associated with particular scenarios can be refined as time progresses. It probably remains an open question as to whether this strategy, as it evolves in symbiosis with model development and improvement, will increase or decrease the uncertainty in predictions of future geomorphological change.

ACKNOWLEDGEMENTS

The model simulations on which Figure 12.1 is based were made by Suan Zak with funding from the UK NERC and Department of the Environment. Field data were sup-plied by Brian Reynolds of the UK Institute of Terrestrial Ecology, Bangor; the PROFILE model by Per Warfvinge and Harald Sverdrup of the University of Lund, Sweden. The comments of the editors and referees have greatly helped improve the presentation, but if the reader still loses the thread of the argument the fault is entirely mine.

REFERENCES

Abbott, M.B., Bathurst, J.C., Cunge, J.A., O'Connell, P.E. and Rasmussen, J. 1986. An introduction to the European Hydrological System – Système Hydrologique Européen SHE. 2. Structure of a physically based distributed modelling system, *Journal of Hydrology*, **87**, 61–77.

Ahnert, F. 1970. A comparison of theoretical slope models with slopes in the field, *Zeitschrift für Geomorphologie*. Supplementband **9**, 87–101.

Ahnert, F. 1976. Brief description of a comprehensive three-dimensional process–response model of landform development, *Zeitschrift für Geomorphologie, Suppmementband* **25**, 29–49.

Ahnert, F. 1977. Some comments on the quantitative formulation of geomorphological processes in a theoretical model, *Earth Surface Processes*, **2**, 191–202.

Ahnert, F. 1987. Approaches to dynamic equilibrium in theoretical simulations of slope development, *Earth Surface Processes and Landforms*, **12**, 3–15.

Ambroise, B., Beven, K.J. and Freer, J. 1996. Towards a generalisation of the TOPMODEL concepts: topographic indices of hydrological similarity, *Water Resources Research*, in press.

Anderson, M.G. and Calver, A. 1977. On the persistance of landscape features formed by a large flood, *Transactions Institute British Geographers*, **2**, 243–254.

Andrews, D.J. and Bucknam, R.G. 1987. Fitting degradation of shoreline scarps by a nonlinear diffusion model, *Journal of Geophysical Research*, **92**, 12 857–12 867.

Armstrong, A.C. 1976. A three-dimensional simulation of slope forms, *Zeitschrift für Geomorphologie* Supplementband, **25**, 20–28.

Baker, V.R. 1978. The Spokane Flood controversy and the Martian outflow channels, *Science*, **202**, 1249–1256.

Bassett, K. 1994. Comments on Richards: the problems of real geomorphology, *Earth Surface Processes and Landforms*, **19**, 273–276.

Bates, P.D. and Anderson, M.G. 1993. A two-dimensional finite element model for river flow inundation, *Proceedings of the Royal Society of London*, Series A, **440**, 481–491.

Bathurst, J.C., Wicks, J.M. and O'Connell, P.E. 1995. The SHE/SHESED basin scale water flow and sediment transport modelling system, in *Computer Models of Watershed Hydrology*, edited by V.P. Singh, Water Resource Publication, Highlands Ranch, Colo., pp. 563–594.

Beven, K.J. 1981. The effect of ordering on the geomorphic effectiveness of hydrologic events, in *Erosion and Sediment Transport in Pacific Rim Steeplands*, IAHS Publication No. 132, pp. 510–526.

Beven, K.J. 1987. Towards a new paradigm in hydrology, in *Water for the Future: Hydrology in Perspective*, IAHS Publication No. 164, pp. 393–403.

Beven, K.J. 1989. Changing ideas in hydrology: the case of physically-based models, *Journal of Hydrology*, **1–5**, 157–172.

Beven, K.J. 1993: Prophecy, reality and uncertainty in distributed hydrological modelling, *Advances in Water Resources*, **16**, 41–51.

Beven, K.J. 1995a. Linking parameters across scales: subgrid parameterisations and scale dependent hydrological models, *Hydrological Processes*, **9**, 507–525.

Beven, K.J. 1995b. Process, heterogeneity and scale in modelling soil moisture fluxes, in *Global Environmental Change and Land Surface Processes in Hydrology: The Trials and Tribulations of Modelling and Measuring*, edited by S. Sorooshian and V.K. Gupta, Springer-Verlag, New York, in press.

Beven, K.J. and Binley, A.M. 1992. The future of distributed models: calibration and predictive uncertainty, *Hydrol. Process.*, **6**, 279–298.

Beven, K.J. and Kirkby, M.J. 1979. A physically-based variable contributing area model of basin hydrology, *Hydrological Sciences Bulletin*, **24**, 43–69.

Beven, K.J., Lamb, R., Quinn, P.F., Romanowicz, R. and Freer, J. 1995. TOPMODEL, in *Computer Models of Watershed Hydrology*, edited by V.P. Singh, Water Resource Publications, Highlands Ranch, Colo., pp. 627–668.

Blackie, J.R., and Eeles, C.W.O. 1985. Lumped catchment models, in *Hydrological Forecasting*, edited by M.G. Anderson and T.P. Burt, Wiley, Chichester, pp. 311–346.

Brunsden, D. 1985. The revolution in geomorphology: a prospect for the future, in *Geographical Futures*, edited by R. King, Geographic Association, Sheffield, pp. 30–55.

Brunsden, D. 1990. Tablets of stone, toward the ten commandments of geomorphology, *Zeitschrift für Geomorphologie*, Supplementband **79**, 1–37.

Brunsden, D. 1993. The persistence of landforms, *Zeitschrift für Geomorphologie*, Supplementband **93**, 13–28.

Campbell, I.A. and Honsaker, J.L. 1982. Variability in badlands erosion: problems of scale and threshold identification, in *Space and Time in Geomorphology*, edited by C.E. Thorn, George Allen and Unwin, London, pp. 59–80.

Chase, C.G. 1992. Fluvial landsculpting and the fractal dimension of topography, *Geomorphology*, 5, 39–58.

Chorley, R.J. 1962. *Geomorphology and General Systems Theory*, US Geological Survey Professional Paper 500-1B, Washington, DC.

Coates, D.R. and Vitek, J.D. eds 1980. *Thresholds in Geomorphology*, George Allen and Unwin, London, 498 pp.

Cooke, R.J. and Reeves, R.W. 1976. *Arroyos and Environmental Change in the American Southwest*, Oxford, 213 pp.

Cordova, J.R., Rodriguez-Iturbe, I. and Vaca, P. 1976. On the development of drainage networks, in *Recent Developments in the Explanation and Prediction of Erosion and Sediment*, edited by D.E. Walling, IASH Publication 137, pp. 239–250.

Culling, W.E.H. 1957. Multicyle streams and the equilibrium theory of grade, *Journal of Geology*, 65, 259–274.

Culling, W.E.H. 1987. Equifinality: modern approaches to dynamical systems and their potential for geographical thought, *Transactions of the Institute of British Geographers*, 12, 57–72.

Culling, W.E.H. 1988. A new view of the landscape, *Transactions of the Institute of British Geographers*, 13, 345–360.

Duan, Q., Soroshian, S. and Gupta, V.K. 1992. Effective and efficient global optimisation for conceptual rainfall–runoff models, *Water Resources Research*, 28, 1015–1031.

Duband, D., Obled, C. and Rodriguez, J.-Y. 1993. Unit hydrograph revisited: an alternate iterative approach to UH and effective precipitation estimation, *Journal of Hydrology*, 150, 115–149.

Dunne, T. 1991. Stochastic aspects of the relations between climate, hydrology and landform evolution, *Transactions of the Japanese Geomorphological Union*, 12, 1–24.

Dunne, T., Zhang, W. and Aubrey, B.F. 1991. Effects of rainfall, vegetation and microtopography on infiltration and runoff, *Water Resources Research*, 27, 2271–2286.

Feyerabend, P.K. 1975. *Against Method: Outline of an Anarchistic Theory of Knowledge*, New Left Books, London, 339 pp.

Freer, J., Beven, K.J. and Ambroise, B. 1996. Bayesian estimation of uncertainty in runoff prediction and the value of data: an application of the GLUE approach, *Water Resources Research*, in press.

Fried, A.W. and Smith, N. 1992. Timescales and the role of inheritance in long-term landscape evolution, northern New England, Australia, *Earth Surface Processes and Landforms*, 17, 375–385.

Grayson, R.B., Moore, I.D. and McMahon, T.A. 1992. Physically-based hydrologic modelling. 2. Is the concept realistic? *Water Resources Research*, 28, 2659–2666.

Haines-Young, R.H. and Petch, J.R. 1983. Multiple working hypotheses: equifinality and the study of landforms, *Transactions of the Institute of British Geographers*, 8, 458–466.

Hallet, B. 1990. Spatial self-organization in geomorphology: from periodic bedforms and patterned ground to scale-invariant topography, *Earth Science Reviews*, 29, 57–76.

Harvey, A.M. 1994. Influence of slope/stream coupling on process interactions on eroding gulley slopes: Howgill Fells, Northwest England, in *Process Models and Theoretical Geomorphology*, edited by M.J. Kirkby, Wiley, Chichester, pp. 247–270.

Hawkins, R.H. 1982. Interpretation of source-area variability in rainfall–runoff relationships, in *Rainfall–runoff relationships*, edited by V.P. Singh, Water Resource Publications, Fort Collins, Colo., pp. 303–324.

Hjemfelt, A.T. and Burwell, R.E. 1984. Spatial variability of runoff, *Journal of Irrigation and Drainage*, ASCE, 110, 46–54.

Houghton, J.T., Jenkins, G.J. and Ephraums, J.S. 1990. *Climate Change: The IPCC Scientific Assessment*, Cambridge University Press, Cambridge.

Howard, A.D. 1994. A detachment-limited model of drainage basin evolution, *Water Resources Research*, 30, 2261–2285.

Howard, A.D. and Kerby, G. 1983. Channel changes in badlands, *Geological Society of America Bulletin*, 94, 739–752.

Howarth, R.J. and Ollier, C.D. 1992. Continental rifting and drainage reversal: the Clarence River of eastern Australia, *Earth Surface Processes and Landforms*, **17**, 381–397.

Howson, C. and Urbach, P. 1989. *Scientific Reasoning: The Bayesian Approach*, Open Court Press, La Salle, Illinois.

Ijjasz-Vasquez, E.J. and Bras, R.L. 1995. Scaling regimes of local slope versus contributing area in digital elevation models, *Geomorphology*, **12**, 299–311.

Ijjasz-Vasquez, E.J., Bras, R.L. and Moglen, G.E. 1992. Sensitivity of a basin evolution model to the nature of runoff production and to initial conditions, *Water Resources Research*, **28**, 2733–2741.

Jakeman, A.J. and Hornberger, G.M. 1993. How much complexity is warranted in a rainfall–runoff model? *Water Resources Research*, **29**, 2637–2650.

Jensen, K.H. and Mantoglou, A. 1992. Application of stochastic unsaturated flow theory: numerical simulations and comparisons to field observations, *Water Resources Research*, **29**, 673–696.

Kirkby, M.J. 1975. Hydrograph modelling strategies, in *Process in Physical and Human Geography*, edited by R. Peel, M. Chisholm and P. Haggett, Heinemann, pp. 69–90.

Kirkby, M.J. 1984. Modelling cliff development in South Wales: Savigear reviewed, *Zeitschrift für Geomorphologie*, **28**, 405–426.

Kirkby, M.J. 1985. A two-dimensional simulation model for slope and stream evolution, in *Hillslope Processes*, edited by A.D. Abrahams, George Allen and Unwin, London, pp. 203–222.

Kirkby, M.J. 1990. The landscape viewed through models, *Zeitschrift für Geomorphologie*, Supplementband 79, 63–81.

Knorr-Cetina, K.D. 1981. *The Manufacture of Knowledge*, Pergamon, Oxford, 189 pp.

Konikow, L.F. and Bredehoeft, J.D. 1992. Groundwater models cannot be validated, *Adv. Water Resour.*, **15**, 75–83.

Kooi, H. and Beaumont, C. 1994. Escarpment evolution of southwestern Africa: insights from a surface processes model that combines diffusion, advection and reaction, *Journal Geophysical Research*, **99**, 12 191–12 209.

Malanson, G.P., Butler, D.R. and Georgakakos, K.P. 1992. Nonequilibrium, lags and determinstic chaos in geomorphology, *Geomorphology*, **5**, 311–322.

Moglen, G.E. and Bras, R.L. 1995. The importance of spatially heterogenous erosivity and the cumulative area distribution within a basin evolution model, *Geomorphology*, **12**, 173–185.

Morton, A. 1993. Mathematical models: questions of trustworthiness, *British Journal for the Philosophy of Science*, **44**, 659–674.

Newson, M. 1980. The geomorphological effectiveness of floods–a contribution stimulated by two recent floods in mid-Wales, *Earth Surface Processes*, **5**, 1–16.

Nott, J.F. 1992. Long term drainage evolution in the Shoalhaven catchment, southeast highlands, Australia, *Earth Surface Processes and Landforms*, **17**, 361–374.

O'Loughlin, E.M. 1981. Saturation regions in catchments and their relations to soil and topographic properties, *Journal of Hydrology*, **53**, 229–246.

Oreskes, N., Schrader-Frechette, K. and Belitz, K. 1994. Verification, validation and confirmation of numerical models in the earth sciences, *Science*, **263**, 641–644.

Phillips, J.D. 1993. Instability and chaos in hillslope evolution, *American Journal of Science*, **293**, 25–48.

Phillips, J.D. 1994. Determinstic uncertainty in landscapes, *Earth Surface Processes and Landforms*, **19**, 389–401.

Refsgaard, J.-C. and Storm, B. 1995. MIKE SHE, in *Computer Models of Watershed Hydrology*, edited by V.P. Singh, Water Resource Publications, Fort Collins, Colo., pp. 733–782.

Rhoads, B.L. 1994. On being a real geomorphologist, *Earth Surface Processes and Landforms*, **19**, 269–272.

Rhoads, B.L. and Thorn, C.E. 1993. Geomorphology as science: the role of theory, *Geomorphology*, **6**, 287–307.

Rhoads, B.L. and Thorn, C.E. 1994. Contemporary philosophical perspectives on physical geography with emphasis on geomorphology, *Geographical Review*, **84**, 90–101.

Richards, K.S. 1990. Real geomorphology, *Earth Surface Processes and Landforms*, **15**, 195–197.

Richards, K.S. 1994. Real geomorphology revisited, *Earth Surface Processes and Landforms*, **19**, 277–281.

Rigon, R., Rinaldo, A. and Rodriguez-Iturbe, I. 1994. On landscape self-organization, *Journal Geophysical Research*, **99**, 11 971–11 993.

Riley, S.J. 1994. Modelling hydrogeomorphic processes to assess the stability of rehabilitated landforms, Ranger Uranium Mine, Northern Territory, Australia–a research strategy, in *Process Models and Theoretical Geomorphology*, edited by M.J. Kirkby, Wiley, Chichester, pp. 357–388.

Rinaldo, A., Rodriguez-Iturbe, I., Rigon, R., Ijjasz-Vasquez, E. and Bras, R.L. 1993. Self-organised fractal river networks, *Physical Review Letters*, **70**, 822–826.

Rodriguez-Iturbe, I., Rinaldo, A., Rigon, R., Bras, R.L. and Ijjasz-Vasquez, E. 1992. Fractal structures as least energy patterns: the case of river networks, *Geophysical Research Letters*, **19**, 889–892.

Rodriguez-Iturbe, I., Marani, M., Rigon, R. and Rinaldo, A. 1994. Self-organised river landscapes: fractal and multifractal characteristics, *Water Resources Research*, **30**, 3531–3539.

Romanowicz, R., Beven, K.J. and Tawn, J. 1994. Evaluation of predictive uncertainty in nonlinear hydrological models using a Bayesian approach, in *Statistics for the Environment. II. Water Related Issues*, edited by V. Barnett and K.F. Turkman, Wiley, Chichester, pp. 297–317.

Roth, G., Siccardi, R. and Rosso, R. 1989. Hydrodynamic description of the erosional development of drainage patterns, *Water Resources Research*, **25**, 319–332.

Scheidegger, A.E. 1987. The fundamental principles of landscape evolution, *Catena* Supplement, **10**, 199–210.

Schrader-Frechette, K.S. 1989. Idealised laws, antirealism and applied science: a case in hydrogeology, *Synthese*, **81**, 329–352.

Schumm, S.A. 1956. Evolution of drainage systems and slopes in badlands at Perth Amboy, New Jersey, *Geological Society of America Bulletin*, **67**, 597–646.

Singh, V.P. (ed.) 1955. *Computer Models of Watershed Hydrology*, Water Resource Publications, Highlands Ranch, Colo.

Sorooshian, S. and Gupta, V.K. 1995. Model calibration, in *Computer Models of Watershed Hydrology*, edited by V.P. Singh, Water Resource Publications, Highlands Ranch, Colo., pp. 23–68.

Starkel, L. 1976. The role of extreme (catastrophic) meteorological events in contemporary evolution of slopes, in *Geomorphology and Climate*, edited by E. Derbyshire, Wiley, London, pp. 204–246.

Stephenson, G.R. and Freeze, R.A. 1974. Mathematical simulation of subsurface flow contributions to snowmelt runoff, Reynolds Creek watershed, Idaho, *Water Resources Research*, **10**, 284–298.

Stevens, M.A., Simons, D.B. and Richardson, E.V. 1975. Nonequilibrium river form, *Journal of Hydraulics Division* ASCE, **101**, 558–566.

Thornes, J.B. and Brunsden, D. 1977. *Geomorphology and Time*, Methuen, London, 208 pp.

United Kingdom Nirex Ltd 1995. *Nirex Biosphere Research: Report on Current Status in 1994*, Report No. S/95/003, Harwell.

Van Fraasen, B.C. 1980. *The Scientific Image*, Clarendon Press, Oxford, 235 pp.

Von Bertalanffy, L. 1951. An outline of general systems theory, *British Journal for the Philosophy of Science*, **1**, 134–165.

Von Bertalanffy, L. 1962. *General Systems Theory*, Brazilier, New York.

Warfvinge, P. and Sverdrup, H. 1992. Calculating critical loads of acid depositions with PROFILE–a steady-state soil chemistry model, *Water, Air and Soil Pollution*, **63**, 119–143.

Wheater, H.S., Jakeman, A.J. and Beven, K.J. 1993. Progress and directions in rainfall-runoff modelling, in Jakeman, A.J., Beck, M.B. and McAleer, M.J. (eds) *Modelling Changes in Environmental Systems*, Wiley, Chichester, pp. 101–132.

Willgoose, G.R. 1994. A statistic for testing the elevation characteristics of landscape simulation models, *Journal Geophysical Research*, **99**, 13 987–13 996.

Willgoose, G.R., Bras, R.L. and Rodriguez-Iturbe, I. 1991a. A physically-based coupled network growth and hillslope evolution model. 1. Theory, *Water Resources Research*, **27**, 1671–1684.

Willgoose, G.R., Bras, R.L. and Rodriguez-Iturbe, I. 1991b. A physically-based coupled network growth and hillslope evolution model. 2. Nondimensionalisation and applications, *Water Resource Research*, **27**, 1685–1696.

Willgoose, G.R., Bras, R.L. and Rodriguez-Iturbe, I. 1991c. Results from a new model of river basin evolution, *Earth Surface Processes and Landforms*, **16**, 237–254.

Willgoose, G.R., Bras, R.L. and Rodriguez-Iturbe, I. 1992. The relationship between catchment and hillslope properties: explanation of a catchment evolution model, *Geomorphology*, **5**, 21–37.

Woldenberg, M.J. 1966. Horton's laws justified in terms of allometric growth and steady state in open systems, *Geological Society of America Bulletin*, **77**, 431–434.

Wolman, M.G. and Gerson, R. 1978. Relative scales of time and effectiveness of climate in watershed geomorphology, *Earth Surface Processes*, **3**, 189–208.

Young, P.C. and Beven, K.J. 1994. Data-based mechanistic modelling and the rainfall-flow non-linearity, *Environmetrics*, **5**, 335–363.

Zak, S.K. and Beven, K.J. 1995. *Uncertainty in the Estimation of Critical Loads: A Practical Methodology*, CRES Technical Report TR129, Lancaster University, Lancaster, UK, 35 pp.

13 Deterministic Complexity, Explanation, and Predictability in Geomorphic Systems

Jonathan D. Phillips

Department of Geography, East Carolina University

ABSTRACT

Increasingly, geomorphic systems are viewed as nonlinear dynamical systems (NDS) and are examined using the tools and concepts of NDS theory. Many geomorphic systems have recently been shown to exhibit the more complex traits of some NDS, including deterministic chaos and self-organization. The exercise of determining that a geomorphic system has, or may have, deterministic complexity is quite troubling to many geomorphologists, usually for one or more of four general reasons:

1. Expectations of what NDS theory can or should reveal are unrealistic.
2. Many NDS concepts and analytical techniques are imported from mathematics and physics, where the simple, abstract systems bear little resemblance to complicated real-world landscapes.
3. NDS analyses are often based on mathematical models which lack rigorous field tests.
4. Merely showing that a geomorphic system exhibits determinstic complexity, while providing a plausible explanation for unexplained real-world complexity, provides no mechanistic explanations of geomorphic process or evolution.

This chapter addresses these issues. It is argued that many concepts and definitions of traditional or mainstream NDS theory are indeed inappropriate for earth sciences. There is a need for alternative concepts and definitions devised or adapted specifically for the earth sciences. This is worth doing, as there are examples of specific, testable hypotheses generated by NDS theory which suggest that the latter has some legitimate explanatory value in geomorphology. Finally, a case study of the evolution of soil landscapes is used to show that NDS theory can generate insight into mechanisms of landscape evolution. NDS

The Scientific Nature of Geomorphology: Proceedings of the 27th Binghamton Symposium in Geomorphology held 27–29 September 1996. Edited by Bruce L. Rhoads and Colin E. Thorn. © 1996 John Wiley & Sons Ltd.

theory, concepts, and methods are not a panacea for geomorphology; neither are they simply a bandwagon fad. Rather, they provide another useful tool in the geomorphologist's kit.

INTRODUCTION

There is chaos in geomorphology. Many geomorphic systems show clear evidence of chaotic dynamics and deterministic complexity. These phenomena cause many nonlinear dynamical systems to behave unpredictably (at certain scales), to exhibit extraordinary sensitivity to initial conditions, and to show complicated, pseudorandom patterns even in the absence of environmental heterogeneity and stochastic forcings. There is also chaos among geomorphologists, in the more traditional sense. Some have hailed chaos theory and other aspects of NDS theory as a revolutionary perspective. They see a potential to fundamentally change our view of earth surface processes and landscape evolution, and to provide answers to previously unanswerable questions. At the far end of the continuum, NDS theory receives derisive sneers as just another scientific fad, with little relevance to geomorphology beyond providing more toys for computer modelers. Near both ends of the continuum there is considerable hand-wringing that geomorphologists are either missing the nonlinear boat, or riding a nonlinear bandwagon.

Given the widespread impacts of NDS theory, not only in mathematics and physics, but in our neighbour discplines of climatology, meteorology, and geophysics, it was inevitable that geomorphologists would investigate chaos, fractals, and other aspects of NDS, and attemp to apply them to geomorphic problems. The result is that a number of studies have shown that geomorphic systems may exhibit chaos, complex self-organization, and other forms of deterministic complexity. Having done so, we face the question: So what? To have value beyond the pedagogic, applications of NDS theory in geomorphology must provide testable hypotheses, and/or answers to significant geomorphic questions. In this chapter I will attempt to determine the extent to which that is the case.

DETERMINISTIC CHAOS IN GEOMORPHIC SYSTEMS

Instability, Chaos, and Entropy

There is neither need nor space for a review of NDS theory as it applies to earth surface systems. Rather, I will briefly review the methodological basis of my arguments, opting in this case for self-citation rather than self-plagiarism (Phillips 1992a, 1993b, 1994, 1995a, b). Essentially, if you can describe a geomorphic system as a system of partial differential equations, as a box-and-arrow diagram, or as an interaction matrix specifying the components and whether they have positive, negative, or negligible impacts on each other, the stability of the system can be determined using the Routh–Hurwitz criteria (RHC). It does not matter if the system is nonlinear – it probably is – because the stability properties of the original nonlinear system and the interaction-matrix linearized version of it are identical. The RHC allow one to determine whether or not the system has any positive Lyapunov exponents. If it does, the system is unstable to small perturbations and potentially chaotic. An n-dimensional (where the number of dimensions equals the number of components) system has n Lyapunov exponents, which determine the rate of convergence or divergence of initially similar system states in the system phase space, and thus the sensitivity to perturbations or to variations in initial conditions.

Sensitivity to initial conditions is depicted by this standard relationship from chaos theory, where the Δ represents the difference between two system states at the start (time zero) and at some future time t:

$$\Delta_t \sim \Delta_0 e^{\lambda t}. \tag{1}$$

The separation at time t is a function of the Lyapunov exponent. The system is not, and cannot be, chaotic unless there is at least one positive Lyapunov exponent. Because an unstable system has at least one $\lambda > 0$, dynamic instability is tantamount to a chaotic system. In much of the NDS literature, a chaotic system is in fact defined as one which has a positive Lyapunov exponent.

Deterministic chaos is a property of some NDS whereby even simple deterministic systems can produce complex, pseudorandom patterns, independently of stochastic forcings or environmental heterogeneity. In chaotic systems complexity and unpredictability are inherent in system dynamics. Such systems are strongly sensitive to initial conditions, in that initially similar states diverge exponentially, on average, and become increasingly different over time. Chaotic systems are also sensitive to perturbations of all magnitudes.

The Kolmogorov (K-) entropy of a NDS measures its 'chaoticity', because K-entropy is equal to the sum of the positive Lyapunov exponents. In real landscapes, measured entropy can be due to deterministic complexity, or to 'colored noise', the combination of randomness and deterministic order. Culling (1988b) was apparently the first to suggest exploiting the relationship between K-entropy (estimated using standard statistical or information theoretic entropy measures) and chaos in geomorphic systems. In this chapter it is used to show the relationship between self-organization, instability, and chaos. Note that there are three forms of entropy referred to in geomorphology. Thermodynamic entropy is a measure of the amount of thermal energy unavailable to do work, or the disorder in a closed system. Statistical (information theoretic) entropy measures the loss of information in a transmission, or the degree of disorder in a statistical distribution. Kolmogorov (K-) entropy measures the expansion of a system's phase space (the n-dimensional space defining all possible system states or combinations of values of the n components). The mathematical equivalence of these entropy measures is evidence of their interrelatedness (Brooks and Wiley 1988). The argument here deals specifically with K-entropy. Brooks and Wiley (1988) argue from an analogy between thermodynamic and K-entropy, but Culling (1988b) shows that when the complexity of a topographic surface increases, both statistical and K-entropy also increase. Zdenovic and Scheidegger (1989) demonstrate the change in statistical entropy during landscape degradation. Ibanez et al. (1990, 1994) and Phillips (1995b) show that modes of landscape evolution whereby the complexity or diversity of the landscape increases (for example, fluvial dissection or progressive pedogenesis) result in increasing K-entropy (and vice versa). The statistical entropy of a spatial or temporal distribution produced by a nonlinear dynamical system is an estimate of its K-entropy (Culling 1988b; Kapitaniak 1988).

Self-organization is common in geomorphology (Hallet 1990), and is seen in, among other things, slope morphology, bedforms, patterned ground, beach cusps, and drainage networks. Self-organization is linked to instability and chaos, and can be described using the K-entropy. The formal mathematical arguments are spelled out elsewhere (Phillips 1995a, b), but can be summarized thus:

1. If a geomorphic system is organizing itself, initially similar forms are becoming differentiated, as for example when a planar bed develops ripples or dunes; a landscape is dissected by fluvial erosion; or weathered debris develops soil horizons.
2. This differentiation represents increasing divergence over time, on average.
3. Increasing divergence over time requires a positive Lyapunov exponent, and thus finite positive K-entropy.
4. The increasing divergence cannot continue indefinitely (cf. the finite amplitude of ripples and dunes; fluvial erosion to base level; soil profile maturity). This means that while the phase space stretches exponentially in one direction according to the largest positive Lyapunov exponent, the overall volume of the phase space must contract.
5. If phase space contraction is to occur in a nonlinear dynamical system the sum of all n Lyapunov exponents must be negative.

The positive λ reflect the K-entropy or 'chaoticity', and the rate of disorganization; the negative λ give the rate of organization. If an open, dissipative geomorphic system is to organize itself, there must be at least one positive Lyapunov exponent, but the sum of λ must be negative. The sum of the diagonal elements of the system interaction matrix are equal to the sum of real parts of the complex eigenvalues, and to the Lyapunov exponents. If all diagonal elements are $\leqslant 0$ or $\geqslant 0$ the result is clear; otherwise the relative magnitude of positive or negative terms must be known. These diagonal elements are self-effects, i.e. self-limiting or self-reinforcing feedback mechanisms for components of the geomorphic system.

There are two criteria for determining whether a geomorphic system is self-organizing, subject to two assumptions:

1. The system is a (probably nonlinear) dynamical system of the form

$$dx_i/dt = f_i(x_1, x_2, \ldots, x_n), (c_1, c_2, \ldots, c_n) \quad i = 1, 2, \ldots, n. \tag{2}$$

The x's are the n components of the geomorphic system and the c's coefficients.
2. The system can be represented by an $n \times n$ interaction matrix $\mathbf{A}$. The equations need not be fully specified, and in fact a box-and-arrow model or qualitative interaction matrix may be the working tool.

In practice the assumptions can be satisfied if the critical internal components of a geomorphic system can be identified, and if the system components, subject to their external inputs and constraints, can be described as functions of each other. The criteria are:

- The matrix $\mathbf{A}$ is unstable according to the Routh–Hurwitz criteria.
- The sum of the diagonal of $\mathbf{A}$ is negative ($\sum \mathbf{a}_{ii} < 0$).

The mechanics of this analysis are illustrated in the case study later in the chapter.

Then, subject to the assumptions, putting the arguments into reverse shows that a self-organizing open, dissipative geomorphic system must be unstable according to the RHC.

If it is unstable, it is also chaotic, as it must have at least one $\lambda > 0$. Self-organization indicates deterministic chaos.

Evidence of Chaos

Deterministic chaos is present in turbulent flows, and therefore plays a role in the mechanics of many geophysical phenomena which include turbulent flows (see Turcotte

Table 13.1 Geomorphic systems found to be potentially chaotic, or asymptotically unstable, which is tantamount to chaos (see text)

Geomorphic system or phenomenon	Method (see notes)	Reference
Infiltration–excess runoff generation	1, 2	Phillips (1992b)
Stream flow	3, 5	Jayawardena and Lai (1994)
Marsh response to sea level rise	1, 2	Phillips (1989[a], 1992a)
At-a-station hydraulic geometry	1, 2	Slingerland (1981)
		Phillips (1990, 1992b)
Downstream hydraulic geometry	1, 2, 6	Callander (1969)
		Ergenzinger (1987)
Evolution of fluvially dissected landscapes	2	Ibanez (1994)
Evolution of soil landscapes	2	Ibanez et al. (1990)
Evolution of regolith and soil thickness	6	Arlinghaus et al. (1992)
		Phillips (1993a, e)
Soil development	1, 2, 6	Phillisps (1993b, c, d)
Drainage basin evolution	6	Willgoose et al. (1991)
		Ijjasz-Vasquez et al. (1992)
Topographic evolution (relief increasing)	4	Phillips (1995a)
Semiarid soil–landform–vegetation–climate systems	1	Thornes (1988)
		Phillips (1993f)
Microtopographic roughness of glacial deposits	2	Elliott (1989)[a]
River platform change	2, 5	Hooke and Redmond (1992)
River longitudinal profiles	6	Slingerland and Snow (1988)
		Renwick (1992)
Solute (Ca) runoff	5	Kempel-Eggenberger (1993)
Coatal onlap stratigraphy	6	Gaffin and Maasch (1991)
Inititiation of channelized surface drainage	6	Smith and Bretherton (1972)
		Loewenherz (1991)
Microtopography – soil property relationships	2	Miller et al. (1994)[a]
Turbulent fluid flows	1, 2, 3, 4, 5, 6	Numerous authors; see Turcotte (1992) for an introduction
Generalized geomorphic mass flux systems	1, 6	Mayer (1992)
		Phillips (1992c)
Hydrothermal eruptions	2, 5	Nicholl et al (1994)
Earthquake activity	3	Li and Nyland (1984)
Fluvial bedforms	1	Mendoza-Cabrales (1994)

Notes: (1) Stability analysis to test for positive eigenvalue or Lyapunov exponent. (2) Test of empirical data for sensitivity to initial conditions. (3) Correlation dimension of singular spectrum analysis of time series. (4) Largest Lyapunov exponent or Lyapunov spectrum analysis of time series. (5) Phase portrait or attractor reconstruction. (6) Numerical simulation models. All methods except (6) rely (in the references cited) on field data or conceptual models derived from field observations.
[a] These authors did not explicitly address chaos or stability, but present field evidence of either increasing divergence over time, or disproportionately large landscape variations arising from very small variation in some controlling factor.

Table 13.2 Studies finding evidence of self-organization in geomorphic systems

Gemorphic system or phenomeon	Method (see notes)	Reference
Formation of periglacial patterned ground	1, 3	Hallet (1990 Werner and Hallet (1993)
Formation of nonperiglacial patterned ground	1, 3	Ahnert (1994)
Evolution of beach cusps	3	Werner and Fink (1994)
Lateritic weathering	1	Nahon (1991)
Fluvial riffle-pool sequences	1, 3	Clifford (1993)
Fluvial bedforms	1, 3	Nelson (1990) McLean (1990)
Landslides	1	Haigh (1988)
Sedimentary organization of gravel barrier beaches	1	Carter and Orford (1991)
Sandpile models	3	Bak et al (1987) Carlson et al. (1990)
Drainage network evolution	2, 3	Woldenberg (1969) Rinaldo et al. (1993) Takayasu and Inaoka (1992) Kramer and Marder (1992) Masek and Turcotte (1993) Stark (1991)
Evolution of fluvially dissected terrain	2, 3	Rigon et al. (1994) Stark (1994)
Increasing topographic relief	3	Phillips (1995b)
Scale-invariant topography	2, 3	Hallet (1990) Turcotte (1990)

Notes: (1) Field or laboratory observations. (2) Analysis of map, digital elevation, or remotely sensed data. (3) Numerical models.

1992; Newman et al. 1994). It is logical to suspect that chaotic behavior may exist in other earth surface processes, and that chaotic flows and geophysical dynamics might leave chaotic imprints on the landscape (Culling 1987, 1988a; Slingerland 1989; Malanson et al. 1990, 1992). Consequently, in recent years a number of studies have sought evidence of chaotic behavior in geomorphic systems. Table 13.1 summarizes those studies, along with work explicitly dealing with system stability, which have shown or found evidence of instability and deterministic chaos in geomorphic systems.

Chaos and instability are not, of course, found in every search. For example, Wilcox et al. (1991) found no evidence of deterministic chaos in complex snowmelt runoff time series, and Zeng and Pielke (1993) have suggested, with respect to atmospheric dynamics, that evidence of a chaotic attractor in time series can be misleading. Results are sometimes equivocal, as well: Montgomery (1993) found evidence of chaotic behavior in a model of river meander formation, but could neither support nor falsify those findings with field data. Results are also process-, site-, or situation-specific in many cases. Phillips (1992b), for instance, found that infiltration–excess runoff generation was likely to be chaotic, but not saturation–excess runoff. Other analyses make it clear that while chaos may or does occur under realistic situations in a particular geomorphic system, it does not occur at all places or at all times (Phillips 1993e).

Table 13.2 lists the studies which have found evidence of self-organization in geomorphic systems. Again, this is not meant to suggest that all phenomena in Table 13.2 are always or invariably self-organizing and chaotic. Many, in fact, have both self-organizing and nonself-organizing modes.

Emboldened by the evidence from the studies cited in Tables 13.1 and 13.2, let us declare, for the sake of argument, that some geomorphic systems are chaotic under some circumstances. So what?

PROBLEMS APPLYING NDS THEORY TO GEOMORPHOLOGY

Culling (1987, 1988a) and Malanson et al. (1990, 1992) held that chaos theory provides a powerful pedagogic framework for geomorphology, but that its utility for problem-solving is limited. This is because real landscapes nearly always exhibit some environmental heterogeneity – stochastic complexity – in addition to any deterministic, chaotic, complexity which may be present. It is extremely difficult to distinguish chaos from noise when both are present, and more difficult still to isolate the two. Further, methods for detecting and analyzing chaos in empirical data requires large data sets – 5000 observations is considered a small data set in the NDS literature, but many geomorphic data sets are far smaller. Considerable methodological progress is being made in overcoming and circumventing these problems, though they remain serious issues. But even if these methodological hurdles are cleared, there are more fundamental difficulties in applying NDS theory to geomorphology.

Abstract Concepts and Concrete Landscapes

NDS theory and methods often depend on phase portraits and trajectories, conceptually and sometimes operationally. In the n-dimensional phase space defined by the n components of the geomorphic system, the system state at any given time can be mapped into the phase space. Over time, changes in system state are represented by a sequence of points which define a trajectory. This poses few intrinsic problems for mathematical models or rapidly varying phenomena. However, in most cases landscapes change too slowly to allow a phase portrait to be constructed from empirical information–there are simply too few points to be mapped into the phase space.

The geologic record may allow the identification of previous system states, and can provide more points for the phase space. However, the record is incomplete, and (to put it mildly) can be confusing. For instance, does a paleosol represent a single set of environmental controls, or is it overprinted with evidence from several different soil-forming events or episodes – i.e. how many system states does it reflect? The record is also likely to be incomplete and biased. Coastal sedimentary sequences, for example, are likely to preserve some evidence of system states associated with sea level transgression, and little evidence of regression.

Deterministic chaos is characterized and defined on the basis of sensitivity to initial conditions. In numerical models we know and can control variations in initial states. In geomorphology initial states, even in a general sense, are often unknown. Minor variations in initial conditions are not just unknown, but also unknowable.

It is unfortunate that much of the chaos literature (including, alas, some of my own writings) has used the term sensitive *dependence* on initial conditions, rather than simply sensitivity to initial conditions. Dependence is correct insofar as mathematics goes: the most minuscule variations in starting values produce quite different, and unique, values at any given future time after the initial transients die out. However, this phrasing leads to the misguided hope that in a chaotic system the initial state can be deduced from the current state. If that were the case, what a boon it would be for paleoenvironmental reconstructions! Rather, chaos implies the opposite, from the geoscientist's perspective – nearly identical initial conditions, which differ in unmeasurable and geomorphically insignificant ways, could produce quite different results in a chaotic earth surface system. In this sense the latter exhibit sensitive *independence* of initial conditions.

Another problem arises from our field orientation. Even mathematical modellers unfailingly present their results in the form of graphic depictions of landforms. Geomorphologists invariably raise the question: Is chaos a property of geomorphic systems, or merely a property of mathematical models of geomorphic systems? The answer is that chaos is not strictly a mathematical artifact – of the 52 studies cited in Tables 13.1 and 13.2, less than a fifth rely exclusively on numerical models or mathematical arguments. Nevertheless, the question will continue to arise until we develop: (1) more and better concrete, pedagogical examples of chaos in real landscapes, and (2) *geomorphic* rather than mathematical concepts and definitions of complex nonlinear behaviors.

Interpretations and Expectations

There has been much emphasis in the NDS literature, appropriately enough, on the deterministic complexity in nonlinear systems. There are, clearly, important implications regarding predictability, important precautions regarding numerical modelling, and important opportunities to explain complex, irregular temporal and spatial patterns. Real landscapes usually contain considerable quantities of both simplicity, regularity, and order on the one hand and complexity, irregularity, and disorder on the other. Depending on one's outlook, interests, or purposes, one or the other may be emphasized. For those whose goals and inclinations favor the former, deterministic complexity can be quite troubling.

But irregularity and weirdness are only one side of the nonlinear coin. Beyond the order and determinism inherent in nonlinear systems, many unstable, chaotic, and complex self-organizing systems exhibit trends quite familiar to geomorphologists. The tendency of perturbations to persist and grow, for example, has long been well known in the development of rills, and the growth of dune blowouts and nivation hollows (for example). Instability and chaos merely (in these cases) provide frameworks for making generalizations about such phenomena and placing them in a broader context (Scheidegger 1983). The increasing divergence over time characteristic of unstable and chaotic systems need not imply incomprehensible results at broader scales, either. With respect to topographic evolution, increasing divergence in elevation simply implies an increase in relief. The result may be quite irregular, fractal topography, but such trends are certainly not unknown, rare, or particularly daunting to geomorphologists.

Because much work in chaos theory and nonlinear dynamics has dealt with physical processes such as turbulent flows, process geomorphologists in particular may have developed some unreasonable expectations that NDS theory should or could provide insight into process mechanics. This would be true only when NDS theory is applied to process-mechanical systems. The advantages of NDS approaches – like those of systems approaches in general – lie in holistic understanding. To the extent NDS theory has advantages over reductionist approaches, it lies in understanding how the components of geomorphic systems fit together, and the likely outcomes of a number of process–response relationships operating simultaneously and sequentially. An understanding of mechanics, or identification of specific process–response relationships, is more likely to be the starting point for an NDS analysis than the outcome. Sometimes NDS theory, applied to broader-scale systems, does indeed provide insight into specific geomorphic mechanisms, but it is not reasonable to expect or demand that it routinely do so, or judge its value on that basis. Likewise, one generally does not expect reductionist studies of process mechanics in aeolian or fluvial saltation to explain the evolution of ergs or drainage basins.

Deterministic Uncertainty

Chaos means that observed stochasticity and randomness may be apparent, not real, i.e. the complex patterns are completely deterministic even though they appear random. In real landscapes, however, chaos may be apparent rather than real. Recalling the relationship between entropy and chaos, the observed entropy of a spatial pattern of a geomorphic phenomenon A, which has one of $i = 1, 2, 3, \ldots, n$ discrete occurrences at each of N locations is

$$H_{\mathrm{obs}}(A) = 1/N \ln(xN! / \prod^{n} f_i), \tag{3}$$

where f_i is the number of locations where the ith outcomes of A occurs, and $1/x$ is a factor by which the number of possible arrangements of A has been reduced by some underlying control. In a truly random arrangement, any outcomes of A can occur anywhere, so $0 < x \leq 1$. If the pattern of A were completely deterministic and nonchaotic, then $H_{\mathrm{obs}}(A) = 0$. Denoting the maximum possible entropy of the pattern of A (that associated with a purely random arrangement) $H(A)$,

$$H_{\mathrm{obs}}(A) = H(A) + \ln x. \tag{4}$$

The underlying constraint x thus yields finite positive entropy, which is produced either by deterministic chaos or by colored noise.

If x is known and measurable (for example, the effects of lithologic constraints on channel networks), then $H_{\mathrm{obs}}(A)$ is colored noise. Gaining more information about x would reduce uncertainty and increase predictability. If x is known but unmeasurable (for example, effects of past bioturbations on sediment properties), x represents deterministic uncertainty, because an underlying deterministic cause increases entropy, uncertainty, and unpredictability. It is chaotic in the sense of a deterministic cause and an inability to reduce uncertainty by deterministic means. In many cases, of course, x could represent several factors, known and unknown; measurable and immeasurable.

The distinction between colored noise and chaos, then, may be a function of the extent of knowledge about constraints of A, and the technology available to measure them. Chaos, as well as randomness, can be apparent, and thus deterministic *uncertainty* is a more appropriate term for real landscapes (McBratney 1992; Phillips 1994).

Deterministic uncertainty would seem to lead us back to the traditional reductionist scientific dogma – that we can or could, ultimately explain everything if we just have more and better measurements. The distinction is that while the deterministic uncertainty concept allows that improved measurements might reduce uncertainty (consistent with the traditional, reductionist view), it also recognizes the presence of complex, nonlinear dynamics inherent to the system which can explain two phenomena which are not otherwise explained:

1. Variability which is far out of proportion with that of the environmental controls, i.e. the growth rather than the mere presence or persistence of variations in initial conditions or perturbations – for example, variations in final infiltration rates on runoff plots many orders of magnitude greater than any variability in rainfall application rates or soil hydrologic properties (Phillips 1992b).
2. Variability which increases over time with no corresponding increase in the variability of the environmental controls – for example, increasing irregularity in the configuration of estuarine shorelines over a 40-year period (Phillips 1989).

Predictability and Explanation

It can be quite useful to know that a geomorphic system is potentially chaotic. Chaos means long-term deterministic prediction is impossible, but also that short-term deterministic prediction of even the most complicated patterns is possible, and that there is some broad-scale order. Knowing a system is chaotic, we can focus efforts at those scales. In between, we know we should use stochastic methods – which work equally well whether the randomness is real or apparent – for prediction. Chaos is sometimes said to be the death knell for reductionist approaches, as it implies irreducible complexity that is immune to resolution via gathering more and more detailed data. An alternate view is that chaos allows us to focus our reductionist studies on those phenomena (or scales) where they will be most fruitful

Chaos also implies the presence of a strange attractor. This means, in principle, that even a very complicated pattern arising from a system, with many apparent degrees of freedom can be described on the basis of just a few variables. This is quite attractive, but unfortunately chaos theory provides no way to tell which of the variables or degrees of freedom are the few critical ones! But chaos also implies self-organization. Seeking and finding the manifestations of self-organization can provide considerable insight into geomorphic evolution (Hallet 1990).

The means to distinguish between chaos and stochasticity, to detect strange attractors, and to provide evidence for self-organization are quite useful in the collective toolkit of geomorphologists. These findings represent only a general step, however, toward understanding landscape evolution, and explaining landforms and surface processes, and process–response relationships.

In general, we can recognize four situations with respect to the utility of NDS theory in explanation and prediction. First, in some situations NDS approaches are useless. Generally, this applies to problems that are linear, or where systems approaches in general are inappropriate. Second, NDS theory is sometimes redundant – it tells us things we already knew or could obtain some other way. For example, many of the recent NDS-based analyses of fluvial network development (see Table 13.2) are raising the same questions and yielding the same answers as earlier analyses based on topologic randomness or least-work principles. A third situation is where NDS theory has descriptive value, in that it accurately describes the behavior of geomorphic systems, and allows an understanding or interpretation of that behavior not available otherwise. Finally, the most desirable and useful situation is one in which NDS theory has explanatory value. This is discussed in detail below.

EXPLANATORY VALUE

To be attractive as an explanation or potential source of explanations to empiricist geomorphologists, a construct must either:

- Explain observations not otherwise satisfactorily explained; or provide a plausible explanation which either fits the observed facts better than, or fits the facts equally as well as, and is simpler than, alternative explanations.
- Provide hypotheses about landscape evolution or the functioning of geomorphic systems which are testable based on field observations, and which are unlikely to have arisen otherwise.

Explanation manifests itself differently in the two dominant geomorphological paradigms. In the historical paradigm, explanations should describe the origin of geomorphic features, and postulate a plausible process or mechanism for their development. Tests are based on comparisons between field observations and the implications of the proposed explanations, and on attempts to falsify hypotheses. The latter are based on some notion of determining what should be observed in the landscape or the stratigraphic record if the NDS-generated hypothesis is false (to constitute strict falsification this should be more rigorous than the exercise of finding field evidence consistent with the hypothesis). In the process paradigm, explanations are called upon to identify processes or mechanisms responsible for particular features or phenomena, or to describe process mechanics. Tests are based on field or laboratory experiments.

The explanatory value of NDS is perhaps better suited to the historical paradigm, simply because its questions are more likely to be holistic and less likely to be reductionist than those of the process paradigm. Thus while the modern 'discovery' of chaos occurred in the field of atmospheric dynamics, its explanatory value thus far is far greater in paleoclimatology than in weather forecasting. This does not mean NDS theory has no explanatory value in process geomorphology, however, as shown below.

Explaining Variability

Complexity and irregularity in the form of extensive and complicated spatial variability over short distances and small areas are well known in geomorphology. This is often conceptualized as local-scale heterogeneity overlaid on broad-scale order and/or as the product of multiple controls over process–response relationships acting simultaneously over a range of spatial and temporal scales (cf. Burrough 1983; Chase 1992; de Boer 1992). This is certainly the case in many situations. However, there are at least two types of spatial complexity which are not satisfactorily explained by the environmental heterogeneity and multiple-scale arguments: those where the spatial variability is disproportionately high compared to the variability of controlling or influencing factors, and those where the variability increases over time without any increase in the variability of controlling factors.

For example, infiltration and percolation into homogeneous sand typically occur in the form of a spatially complex pattern of 'fingered flow'. Despite the homogeneity of the medium, and even with no measurable variation in moisture supply or application, a complex pattern of wetting front depths, soil moisture content, and moisture flux rates develops. Presumably, there are tiny, unnoticed, and unmeasurable variations within the sand, or in moisture supplies at the surface. However, the variations in moisture fluxes and storage are many orders of magnitude larger than any variability in hydrologic controls. Selker et al. (1992) were able to explain these phenomena on the basis of the growth of wetting front instabilities, using a model based on the Richards equation, and confirmed with a series of experiments. In this case NDS theory provides a reasonable explanation, within the process paradigm, where no other currently exists. Another example is Nahon's (1991) study of complex self-organization in lateritic weathering, which identifies specific geochemical processes responsible for the observed weathering features.

In the historical paradigm, there are questions where NDS-based explanations have advantages over alternatives. Saltzman and Verbitsky's (1994) phase-space model of ocean state, global ice, atmospheric CO_2, and bedrock depression beneath ice sheets is one example. Internal chaotic instabilities in the model provide a plausible explanation of Pleistocene climate change that fits the $\delta^{18}O$ record as well as competing explanations, and is conceptually simpler. Other examples involve soil spatial variability in the absence of observable variation in soil-forming factors (Phillips 1993b, c, d), and increasing irregularity of eroding marsh shorelines over time (Phillips 1989, 1992a).

Producing Testable Hypotheses

NDS theory can, and does, produce hypotheses that particular geomorphic systems do, or do not, exhibit chaos. These can then, sometimes, be tested using empirical data sets. This is fine, as far as it goes, but provides little direct insight into how landscapes work. Likewise, hypotheses that a geomorphic system will (or will not) produce complex, irregular spatial or temporal patterns may be testable, but the outcomes provide only the most general of insights. However, in some cases hypotheses of this general nature imply specific mechanisms, or suggest specific, concrete modes of landscape evolution which are, in turn, testable.

Werner and Fink (1994) developed a model of beach cusp formation based on complex, self-organizing interactions between wave processes and beach morphology. The alternative explanation of cusp formation involves edge waves. Because the competing explanations are mutually exclusive, the Werner/Fink model results in a testable hypothesis: the formation of beach cusps can be observed, along with the presence or absence of edge waves (test results available to date are inconclusive). Determining the extent to which beach cusp formation results from edge waves or wave–beach interactions will provide considerable insight into the evolution and dynamics of beach forms and processes.

Ibanez et al. (1994) applied NDS concepts to the evolution of soil landscapes and fluvially dissected terrain in Spain. This yielded the observation – which can be interpreted as a hypothesis – that as fluvial dissection proceeds, there is increasing diversity of landforms, soils, and ecosystems, thus greatly enlarging the state space of the earth surface system as a whole. This is testable, at least in the aggregate and to a first approximation, by comparing measures of landform, soil, and ecosystem diversity in basins of different stages or degree of dissection. Results of the test would be quite useful in understanding the relationship between soil landscape development and the evolution of ecological diversity.

SOIL LANDSCAPE EVOLUTION

I present here a case study of soil landscape evolution in the North Carolina coastal plain to show how NDS approaches can provide testable hypotheses relevant to both historical and process paradigms. The intimate relationships between pedological and geomorphological evolution are well described elsewhere (Birkeland 1984; Retallack 1990; McFadden and Kneupfer 1990; Gerrard 1992). The landscape–scale implications of chaotic pedogenesis and soil–landscape evolution have been discussed before (Arlinghaus et al. 1992; Culling 1988b; Ibanez et al. 1990; Phillips 1993b, c). The intent here is to show that an NDS-based anlaysis can generate testable hypotheses on specific soil-geomorphic phenomena.

Textural Differentiation Model

On the uplands of the North Carolina coastal plain (Figure 13.1), the most important factors defining the differences between soil types are the texture and thickness of A-, E-, and B-horizons. In the unconsolidated coastal plain sediments, under the prevailing humid subtropical climate, Ultisols are produced in all but the youngest sites, those subjected to dominantly regressive pedogenesis, or nearly pure sands. Textural differentiation is produced by chemical weathering and clay mineral synthesis in surficial (A- and E-horizons) and B-horizons, and by translocation via chemical dissolution–precipitation and lessivage. Differences in other soil properties, such as chemistry, structure, color, and consistency, arise as secondary impacts of textural differentiation, due to differences in drainage and degree of development, and in response to local variations in other soil-forming factors.

A conceptual model of textural differentiation can be constructed to account for differences in the texture and thickness of A- and E- versus B-horizons. Critical components are the rate of clay synthesis in the A- and E-horizons, A, internal clay synthesis in the B-horizon, B, the rate of eluvial loss from the surficial horizons, E, the rate of illuvial gain in the B-horizon, I, and the rate of moisture flux Ω. Units of clay synthesis and translocations

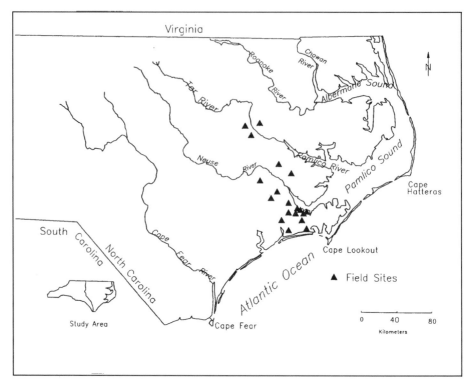

Figure 13.1 The North Carolina coastal plain. Triangles indicate sites of field soil geomorphology investigations used to develop or test the textural differential model

have dimensions of MT^{-1}. Moisture flux in this case is the transmissitivity of the soil profile, with dimensions LT^{-1}.

$$A = c_i e^{-k} 1^{A\Delta t} \tag{5}$$

$$E = \Omega c_2 A \, \Delta t \tag{6}$$

$$I = E \tag{7}$$

$$B = c_3 e^{-k} 3^{B\Delta t} \tag{8}$$

$$\Omega = c_4 - f(B \, \Delta t + I \Delta t) \tag{9}$$

The c's are constants reflecting, respectively, the maximum A- and E-horizon clay mineral synthesis rate as controlled by climate and geochemistry; the removal rate from surface horizons associated with a given moisture flux; the maximum clay mineral synthesis rate in the B-horizon; and the moisture flux rate in the unaltered soil. The k's are coefficients describing, respectively, the decline in clay synthesis as clay accumulates in the surface and B-horizons.

The equations reflect illuviation as a direct function of eluviation, clay accumulation in the surface horizons as synthesis minus eluvial loss, and accumulation in the subsurface as illuviation plus *in situ* synthesis. Eluvial loss from B-horizons occurs in the region, but is insignificant compared to the other processes. Weathering and clay synthesis rates slow as clay accumulates, due to depletion of weatherable minerals, though it might be argued for other situations that the increased water-holding capacity associated with accumulation of fines would increase weathering rates.

Decline in clay synthesis is shown as a negative exponential function of clay accumulation, in accordance with earlier models of feedback effects in weathering. The moisture flux is shown as a generalized, unspecified negative function of subsoil clay accumulation. In eastern North Carolina, hydraulic conductivities of argillic horizons are typically half to less than a tenth of those in surface horizons. The specific form of these feedback functions is not important in this context; only that they are indeed negative functions.

The relationships between the major components of the system are shown in Table 13.3. This interaction matrix gives only the positive, negative, or negligible (zero) influences of A, B, E, I, and Ω on each other, as reflected in the equations. The signs of the matrix elements are all that is necessary to determine whether there are any positive Lyapunov exponents. This method can therefore be used when exact (or generally applicable) relationships between system components cannot be specified, but the general nature of their interactions can be. Also, results are not dependent on any particular parameter values, or any particular form of the governing equations (Mendoza-Cabrales 1994; Phillips 1992a; Scheidegger 1993; Slingerland 1981). Variables a and β are not included, as they are linear combinations of other components. Links shown in the interaction matrix – a linearized version of the highly nonlinear equation system – include only the direct links – i.e. those that do not work through any other component.

The coefficients of the characteristic polynomial of the matrix can be computed based on the feedback (F), where

$$F_k = \sum (-1)^{m+1} L(m, k) \tag{10}$$

where k is the level or order and $L(m, k)$ signifies m disjunct loops of total length k. Disjunct loops are sequences of a_{ij} (matrix elements) with no i or j in common. For example, feedback at level four (the fourth coefficient in the polynomial) would include all combinations of disjunct loops whose total length (number of components) is four. The

Table 13.3 Interaction matrix showing the positive, negative, and negligible mutual influence in the soil textural differentiation model (A, E, I, B, Ω, respectively, represent rates of A- and E-horizon clay synthesis, eluviation of clay from surficial horizons, illuviation in B-horizons, B-horizon clay synthesis, and moisture flux

	A	E	I	B	Ω
A	$-a_{11}$	a_{12}	0	0	0
E	0	0	a_{23}	0	0
I	0	0	0	0	$-a_{35}$
B	0	0	0	$-a_{44}$	$-a_{45}$
Ω	0	a_{52}	0	0	0

RHC criteria are that the system is stable (i.e. all eigenvalues have negative real parts and all $\lambda < 0$ if and only if $F_k < 0$ for all k, and (for $n = 5$):

$$F_1^2 F_4 + F_1 F_5 - F_1 F_2 F_3 - F_3^2 > 0. \tag{11}$$

In the textural differentiation system all $F_k < 0$, and the second stability criterion is also likely to be met. This implies that the system is stable to small perturbations and should not behave chaotically. However, this stability is contingent upon all relationships operating as shown in Table 13.3.

If the role of clay accumulation in moisture storage to promote chemical weathering is stronger than the effects of mineral depletion, then equations (5) and (8) and the signs of a_{11} and a_{44} change, and the system becomes unstable. This could occur in early stages of weathering where the supply of weatherable minerals is not limiting, or where there is an influx of new weatherable minerals (for example, via deposition).

There is also evidence that clay accumulation in the B-horizon does not necessarily limit eluviation or moisture flux, despite the clearly lower hydraulic conductivities in argillic horizons. The latter are typically on the order of 1.5 to 15 cm h^{-1} in coastal plain Ultisols. While this can limit movement during storm runoff, it is unlikely to inhibit water movement and eluviation over longer time scales. This would alter equation (9) and interaction matrix elements a_{34} and a_{45}, and results in instability.

Implications

There are two major implications with respect to processes. The NDS analysis shows that two particular process mechanisms determine whether or not the system behaves chaotically. First, is the chemical weathering rate limited by the reaction rates or weatherable minerals? Second, do the hydraulic conductivity contrasts between the illuvial and overlying horizons limit eluviation (or the depth of illuviation)? These implications could clearly be tested in the field, and results could be linked to related hypotheses involving instability or spatial chaos–for example, where weathering is reaction-limited and depth of illuviation is not inhibited by B-horizon development, the spatial variability of soil profile morphologies and soil types in comparison to observable variation in controlling factors should be greater than where weathering is mineral-limited and eluviation–illuviation is inhibited. However, the process hypotheses do not need to be linked to NDS hypotheses to be useful, as they have intrinsic significance for landscape evolution.

There is another major implication for soil landscape evolution. Where reaction rates are limiting and/or clay accumulation does not inhibit eluviation, the textural differentiation model is unstable and chaotic. If this is the case, then the spatial variability of the soil cover, in terms of the presence, texture, and thickness of A-, E-, and B-horizons in individual pedons, should increase over time. Soil cover on older soil landscapes should be more variable than that on younger ones, even where climate, biotic effects, parent material, and topography are very similar.

This hypothesis has already been tested in an area of the lower North Carolina coastal plain where illuvial clay accumulation does not limit eluviation. The number of soil types, at the series level, was compared at two otherwise identical sites on adjacent geomorphic surfaces differing in age by about 150 000 years. The younger, late Pleistocene site had only one soil type while the older Pleistocene site had at least seven (Phillips 1993d). The hypothesis was thus not rejected.

Given the fact that exponential divergence cannot last indefinitely, and the order and self-organization which must emerge at some scale in a chaotic system, there should be some broad-scale regularities in the soil landscape, and a limit to the age of soil landscapes where increasing divergence is observed. The former is at least intuitively observed in the aggregate, coarse-scale predictability of the soil landscape model in the coastal plain. The latter can be tested if longer chronosequences can be developed.

DISCUSSION AND CONCLUSIONS

It is now clear that geomorphic systems are often complex, nonlinear, dynamical systems. That involves, in some cases, deterministic chaos and self-organization. However, applications of NDS theory to geomorphology have been hampered by three major problems:

1. Concepts and analytical techniques imported from mathematics and physics, where the simple, abstract systems bear little resemblance to real landscapes;
2. A lack of field tests;
3. Limited ability to provide explanations of geomorphic processes or evolution.

The first problem is beginning to be solved by devising and adapting terminology and methodology appropriate to the geosciences (Ibanez et al. 1994; McBratney 1992; Phillips 1994; Zeng and Pielke 1993). The second can also apparently be overcome, as NDS theory has produced some field-testable hypotheses relevant to understanding surface processes and landscape evolution. The third will be–and I argue, is being–overcome to the extent NDS theory can explain geomorphic phenomena not otherwise explained, or produce testable hypotheses unlikely to be produced otherwise. In addition to examples from the literature, a case study shows NDS theory can be used to identify critical process mechanisms in soil formation and to provide a plausible explanation for soil landscapes where broad-scale regularities are overprinted with dramatic local-scale variability disproportionate to any variability in controlling factors.

In general, while there are certainly inappropriate and unedifying applications of NDS theory in geomorphology, it has been shown in at least some cases to provide explanation, and to increase rather than decrease predictability. Whether that utility ultimately makes it just one of many useful perspectives for the geomorphologist or a perspective of preeminent importance remains to be seen. I believe that, whether or not geomorphologists explicitly embrace NDS theory and methods, the discipline is likely to continue to evolve:

- Away from efforts to identify single, inevitable ultimate steady-state equilibria, and toward the recognition of inherently unstable and non-adjusting as well as steady-state 'climax' forms (see Renwich 1992).
- Away from a linear cause-and-effect viewpoint whereby a given set of environmental controls produces a given landscape, and toward the recognition of multiple landscape responses and modes of adjustment.

- Away from perspectives which emphasize either the regularities or complexities in the landscape, and toward those which deal with both simultaneously, and which recognize order and stability as an emergent property of spatial and temporal scale.
- Away from a view of unstable and nonsteady-state forms as anomalies or deviations from the norm, and toward a view where these are seen as norms in their own right.

If these predictions are correct, then NDS theory is clearly an appropriate conceptual framework, but not necessarily the only appropriate one.

Nonlinear systems approaches have at least one thing in common with any other conceptual framework, theory, or methodology: they do not, and cannot, answer or even address every question asked by geomorphologists. Given the prominence of NDS theory in science and its growing use in geomorphology, I argue that geomorphologists should have at least a passing acquaintance with NDS concepts. But those who do not 'do' NDS are not missing the boat, any more than those who do not do cosmogenic isotope dating, geoarcheology, or microprocess mechanics (for example).

Nonlinear dynamical systems is unlikely to be the methodological or conceptual banner behind which all geomorphologists can rally, or the rubric under which all of geomorphology can be interpreted. It seems clear by now that there is probably no such banner or rubric (Rhoads and Thorn 1993). Whether NDS theory lives up to its promise depends in large measure on whether we promise, or expect, too much.

REFERENCES

Ahnert, F. 1994. Modelling the development of non-periglacial sorted nets, *Catena*, **23**, 53–63.
Arlinghaus, S.L., Nystuen, J.D. and Woldenberg, M.J. 1992. An application of graphical anlaysis to semidesert soils, *Geographical Review*, **82**, 244–252.
Bak, P., Tang, C. and Wiesenfeld, K. 1987. Self-organized criticality: an explanation of $1/f$ noise, *Physical Review Letters*, **59**, 381–384.
Birkeland, P.W. 1984. *Soils and Geomorphology*, Oxford University Press, New York, 372 pp.
Brooks, D.R. and Wiley, E.O. 1988. *Evolution as Entropy*, 2nd edn., University of Chicago Press, 415 pp.
Burrough, P.A. 1983. Multiscale sources of spatial variation in soils, *Journal of Soil Science*, **34**, 577–620.
Callander, R.A. 1969. Instability and river channels, *Journal of Fluid Mechanics*, **36**, 465–480.
Carlson, J.M., Chayes, J.T., Grannan, E.R. and Swindle, G.H. 1990. Self-organized critically and singular diffusion, *Physical Review Letters*, **65**, 2547–2550.
Carter, R.W.G. and Orford, J. 1991. The sedimentary organization and behavior of drift-aligned barriers, *Coastal Sediments '91*, American Society Civil Engineers, New York, pp. 934–948.
Chase, C.G. 1992. Fluvial landsculpting and the fractal dimension of topography, *Geomorphology*, **5**, 39–57.
Clifford, N.J. 1993. Formation of riffle-pool sequences: field evidence for an autogenic process, *Sedimentary Geology*, **85**, 39–51.
Culling, W.E.H. 1987. Equifinality: modern approaches to dynamical systems and their potential for geographical thought, *Transactions, Institute of British Geographers*, **12**, 57–72.
Culling, W.E.H. 1988a. A new view of the landscape, *Transactions, Institute of British Geographers*, **13**, 345–360.
Culling, W.E.H. 1988b. Dimension and entropy in the soil covered landscape, *Earth Surface Processes and Landforms*, **13**, 619–648.

DeBoer, D. 1992. Hierarchies and scale in process geomorphology: a review, *Geomorphology*, **4**, 303–318.

Elliott, J.K. 1989. An investigation of the change in surface roughness through time on the foreland of Austre Okstindbreen, North Norway, *Computers and Geosciences*, **15**, 209–217.

Ergenzinger, P. 1987. Chaos and order–the channel geometry of gravel bed braided rivers, *Catena*, supplement **10**, 85–98.

Gaffin, S.R. and Maasch, K.A. 1991. Anomalous cyclicity in climate and stratigraphy and modelling nonlinear oscillations, *Journal of Geophysical Research*, **96B**, 6701–6711.

Gerrard, A.J. 1992. *Soil Geomorphology*, Chapman and Hall, London, 269 pp.

Haigh, M.J. 1988. Dynamic systems approaches in landslide hazard research, *Zeitschrift für Geomorphologie*, Supplement **67**, 79–91.

Hallet, B. 1990. Spatial self-organization in geomorphology: from periodic bedforms and patterned ground to scale-invariant topography, *Earth-Science Reviews*, **29**, 57–75.

Hooke, J.M. and Redmond, C.E. 1992. Causes and nature of river planform change, in *Dynamics of Gravel-Bed Rivers*, edited by P. Billi et al., Wiley, Chichester, pp. 559–571.

Ibanez, J.J., Bellexta, R.J. and Alvarez, A.G. 1990. Soil landscapes and drainage basins in Mediterranean mountain areas, *Catena*, **17**, 573–583.

Ibáñez, J.J., Perez-Gonzalez, A., Jimenez-Ballesta, R., Saldana, A. and Gallardo-Diaz, J. 1994. Evolution of fluvial dissection landscapes in Mediterranean environments: quantitative estimates and geomorphic, pedologic, and phytocenotic repercussions, *Zeitschrift für Geomorphologie*, **38**, 105–119.

Ijjasz-Vasquez, E.J., Bras, R.L. and Moglen, G.E. 1992. Sensitivity of a basin evolution model to the nature of runoff production and to initial conditions, *Water Resources Research*, **28**, 2733–2741.

Jayawardena, A.W. and Lai, F. 1994. Analysis and prediction of chaos in rainfall and streamflow time series, *Journal of Hydrology*, **153**, 23–52.

Kapitaniak, T. 1988. *Chaos in Systems with Noise*, World Scientific, Singapore, 189 pp.

Kempel-Eggenberger, C. 1993. Risse in der geoökologischen Realität: Chaos und Ordnung in geoökologischen systemen, *Erdkunde*, **47**, 1–11.

Kramer, S. and Marder, M. 1992. Evolution of river networks, *Physical Review Letters*, **68**, 205–208.

Li, Q. and Nyland, E. 1994. Is the dynamics of the lithosphere chaotic? In *Nonlinear Dynamics and Predictability of Geophysical Phenomena*, edited by W.I. Newman et al., American Geophysical Union Geophysics Monograph, **83**, pp. 37–41.

Loewenherz, D.S. 1991. Stability and the initiation of channelized surface drainage: a reassessment of the short wavelength limit, *Journal of Geophysical Research*, **96B**, 8453–8464.

McBratney, A.B. 1992. On variation, uncertainty, and informatics in environmental soil management, *Australian Journal of Soil Research*, **30**, 913–936.

McFadden, L.D. and Knuepfer, P.L.K. (eds) 1990. *Soils and Landscape Evolution*, Elsevier, Amsterdam, 340 pp.

McLean, S.R. 1990. The stability of ripples and dunes, *Earth-Science Reviews*, **29**, 131–144.

Malanson, G.P., Butler, D.R. and Georgakakos, K.P. 1992. Nonequilibrium geomorphic processes and deterministic chaos, *Geomorphology*, **5**, 311–322.

Malanson, G.P., Butler, D.R. and Walsh, S.J. 1990. Chaos in physical geography, *Physical Geography*, **11**, 293–304.

Masek, J.G. and Turcote, D.L. 1993. A diffusion-limited aggregation model for the evolution of drainage networks, *Earth & Planetary Science Letters*, **119**, 379–386.

Mayer, L. 1992. Some comments on equilibrium concepts and geomorphic systems, *Geomorphology*, **5**, 277–295.

Mendoza-Cabrales, C. 1994. Is bedform development chaotic? *Hydraulic Engineering '94*, American Society Civil Engineers, New York, pp. 78–81.

Miller, S.T., Brinn, P.J., Fry, G.J. and Harris, D. 1994. Microtopography and agriculture in semi-arid Botswana. 1. Soil variability, *Agricultural Water Management*, **26**, 107–132.

Montgomery, K. 1993. Non-linear dynamics and river meandering, *Area*, **25**, 97–108.

Nahon, D.B. 1991. Self-organization in chemical lateritic weathering, *Geoderma*, **51**, 5–13.

Nelson, J.M. 1990. The initial instability and finite-amplitude stability of alternate bars in straight channels. *Earth-Science Reviews*, **29**, 97–115.

Newman, W.I., Gabrielsou, A. and Turcotte, D. (eds) 1994. *Nonlinear Dynamics and Predictability of Geophysical Phenomena*, American Geophysical Union Geophysical Monograph 83.

Nicoll, M.J., Wheatcroft, S.W., Tyler, S.W. and Berkowitz, B. 1994. Is Old Faithful a strange attractor? *Journal of Geophysical Research*, **99B**, 4495–4503.

Phillips, J.D. 1989. Erosion and planform irregularity of an estuarine shoreline, *Zeitschrift für Geomorphologie*, Supplement **73**, 59–71.

Phillips, J.D. 1990. The instability of hydraulic geometry, *Water Resources Research*, **26**, 739–744.

Phillips, J.D. 1992a. Qualitative chaos in geomorphic systems, with an example from wetland response to sea level rise, *Journal of Geology*, **100**, 365–374.

Phillips, J.D. 1992b. Deterministic chaos in surface runoff, in *Overland Flow: Hydraulic and Erosion Mechanics*, edited by A.J. Parson and A.D. Abrahams, University College London Press, London, pp. 177–197.

Phillips, J.D. 1992c. Nonlinear dynamical systems in geomorphology: revolution or evolution? In *Geomorphic Systems, Proceedings of the Binghamton Geomorphology Symposium*, edited by J.D. Phillips and W.H. Renwick, Elsevier, Amsterdam, pp. 219–229.

Phillips, J.D. 1993a. Spatial-domain chaos in landscapes, *Geographical Analysis*, **25**, 101–117.

Phillips, J.D. 1993b. Stability implications of the state factor model of soils as a nonlinear dynamical system, *Geoderma*, **58**, 1–15.

Phillips, J.D. 1993c. Progressive and regressive pedogenesis and complex soil evolution, *Quaternary Research*, **40**, 169–176.

Phillips, J.D. 1993d. Chaotic evolution of some coastal plain soils, *Physical Geography*, **14**, 566–580.

Phillips, J.D. 1993d. Instability and chaos in hillslope evolution, *American Journal of Science*, **283**, 25–48.

Phillips, J.D. 1993f. Biophysical feedbacks and risk of desertification, *Annals of the Association of American Geographers*, **83**, 630–640.

Phillips, J.D. 1994. Deterministic uncertainty in landscapes, *Earth Surface Processes and Landforms*, **19**, 389–401.

Phillips, J.D. 1995a. Nonlinear dynamics and the evolution of relief, *Geomorphology*, **14**, 57–64.

Phillips, J.D. 1995b. Self-organization and landscape evolution, *Progress in Physical Geography*, **19**, 309–321.

Renwick, W.H. 1992. Equilibrium, disequilibrium, and nonequilibrium landforms in the landscape, in *Geomorphic Systems, Proceedings of the Binghamton Geomorphology Symposium*, edited by J.D. Phillips and W.H. Renwick, Elsevier, Amsterdam, pp. 265–276.

Retallack, G. 1990. *Soils of the Past, An Introduction to Paleopedology*, Unwin-Hyman, London, 520 pp.

Rhoads, B.L. and Thorn, C.E. 1993. Geomorphology as science: the role of theory, *Geomorphology*, **6**, 287–307.

Rigon, R., Rinaldo, A. and Rodriguez-Iturbe, I. 1994. On landscape self-organization, *Journal of Geophysical Research*, **99B**, 11 971–11 993.

Rinaldo, A., Rodriguez-Iturbe, I., Bras, R.L. and Ijjasz-Vasquez, E.J. 1993. Self-organized fractal river networks, *Physical Review Letters*, **70**, 822–826.

Saltzman. B. and Verbitsky, M. 1994. Late Pleistocene climatic trajectory in phase space of global ice, ocean state, and CO_2: observationsand theory. *Paleoceanography*, **90**, 767–779.

Sheidegger, A.E. 1983. The instability principle in geomorphic equilibriu, *Zeitschrift für Geomorphologie*, **27**, 1–19.

Selker, J.S., Steenhuis, T.S. and Parlange, Y.-S. 1992. Wetting front instability in homogeneous sandy soils under continuous infiltration, *Soil Science Society of America Journal*, **56**, 1346–1350.

Slingerland, R. 1981. Qualitative stability analysis of geologic systems with an example from river hydraulic geometry, *Geology*, **9**, 491–493.

Slingerland, R. 1989. Predictability and chaos in quantitative dynamic stratigraphy, in *Quantitative Dynamic Stratigraphy*, edited by T.A. Cross, Prentice-Hall, Englewood Cliffs, NJ, pp. 45–53.

Slingerland, R. and Snow, R.S. 1988. Stability analysis of a rejuvenated fluvial system, *Zeitschrift für Geomorphologie*, Supplement **67**, 93–102.

Smith, T.R. and Bretherton, F.P. 1972. Stability and the conservation of mass in drainage basin evolution, *Water Resources Research*, **8**, 1506–1529.

Stark, C.P. 1991. An invasion percolation model of drainage network evolution, *Nature*, **352**, 423–425.

Stark, C.P. 1994. Cluster growth modelling of plateau erosion, *Journal of Geophysical Research*, **99B**, 13 957–13 960.

Takayasu, H. and Inaoka, H. 1992. New type of self-organized criticality in a model of erosion, *Physical Review Letters*, **68**, 966–969.

Thornes, J.B. 1988. Erosional equilibria under grazing, in *Conceptual Issues in Archeology*, edited by J.L. Bintliff et al., Edinburgh University Press, pp. 193–210.

Turcotte, D.L. 1990. Implications of chaos, scale-invariance, and fractal statistics in geology, *Global and Planetary Change*, **89**, 301–308.

Turcotte, D.L. 1992. *Fractals and Chaos in Geology and Geophysics*, Cambridge University Press, New York, 221 pp.

Werner, B.T. and Fink, T.M. 1994. Beach cusps as self-organized patterns, *Science*, **260**, 968–971.

Werner, B.T. and Hallet, B. 1993. Numerical simulation of self-organized stone stripes, *Nature*, **361**, 142–145.

Wilcox, B.P., Seyried, M.S. and Matison, T.H. 1991. Searching for chaotic dynamics of snowmelt runoff, *Water Resources Research*, **27**, 1005–1010.

Willgoose, G.R., Bras, R.L. and Rodriguez-Iturbe, I. 1991. A coupled channel network growth and hillslope evolution model, *Water Resources Research*, **27**, 1671–1696.

Woldenberg, M.J. 1969. Spatial order in fluvial systems: Horton's laws derived from mixed hexagonal hierarchies of drainage basin areas, *Geological Society of America Bulletin*, **80**, 97–112.

Zdenkovic, M.L. and Scheidegger, A.E. 1989. Entropy of landscapes, *Zeitschrift für Geomorphologie*, **33**, 361–371.

Zeng, X. and Pielke, R.A. 1993. What does a low-dimensional weather attractor mean? *Physics Letters A*, **175**, 299–304.

14 Limitations on Predictive Modeling in Geomorphology

Peter K. Haff

Department of Geology, Center for Hydrologic Science, Duke University

ABSTRACT

Sources of uncertainty or error that arise in attempting to scale up the results of laboratory-scale sediment transport studies for predictive modeling of geomorphic systems
include: (i) model imperfection, (ii) omission of important processes, (iii) lack of
knowledge of initial conditions, (iv) sensitivity to initial conditions, (v) unresolved heterogeneity, (vi) occurrence of external forcing, and (vii) inapplicability of the factor of
safety concept. Sources of uncertainty that are unimportant or that can be controlled at
small scales and over short times become important in large-scale applications and over
long time scales. Control and repeatability, hallmarks of laboratory-scale experiments, are
usually lacking at the large scales characteristic of geomorphology. Heterogeneity is an
important concomitant of size, and tends to make large systems unique. Uniqueness
implies that prediction cannot be based upon first-principles quantitative modeling alone,
but must be a function of system history as well. Periodic data collection, feedback, and
model updating are essential where site-specific prediction is required. In large geomorphic systems, the construction of successful predictive models is likely to be based
upon discovery of emergent variables and a corresponding dynamics, rather than upon
scaling up the results of well-controlled laboratory-scale studies.

INTRODUCTION

Recent efforts at simulating the evolution of large geomorphic systems such as alluvial
fans (Koltermann and Gorelick 1992), deltas (Tetzlaff and Harbaugh 1989), hillslopes
(Ahnert 1987; Kirkby 1986), fluvial drainage systems (Montgomery and Dietrich 1992;
Howard et al. 1994; Willgoose et al. 1991, 1994), and badlands (Howard 1994) rely on the

*The Scientific Nature of Geomorphology: Proceedings of the 27th Binghamton Symposium in Geomorphology held 27–29
September 1996.* Edited by Bruce L. Rhoads and Colin E. Thorn. © 1996 John Wiley & Sons Ltd.

implementation of large-scale quantitative sediment transport models. The need for environmental forecasting and for prediction of the future behavior of engineered sediment systems (e.g. Toy et al. 1987; Hanson and Kraus 1989; Riley 1994) is another potential application of quantitative sediment transport models to landscape evolution. Engineering prediction differs from geomorphic or geologic reconstruction in that risks and costs are associated with prediction, that prediction targets the unknown future while reconstruction targets the past, and that time- and space-specific prediction of future landscape configuration may be required, as opposed to generic reconstruction of landforms and landscapes. Reliable prediction would allow us to foresee how landscape features such as watercourses, soils, topography, and vegetation may evolve by their own dynamics, or in response to climate change or tectonic activity, or as the result of modern disturbances such as grazing, bulldozing, clearing of vegetation, military operations, construction of highways and utility corridors, and strip mining. Engineering prediction often involves shorter time scales than geologic reconstruction, but predictive time spans of thousands of years or more are of interest, especially where large-scale surficial disturbances occur (e.g. on military bases), or where hazardous materials are involved (as at the Ward Valley, California, nuclear waste site). Increasing use of landscape evolution codes in the geomorphic literature may presage widespread use of these methods for engineering and environmental purposes. The risk-associated context in which these methods will be used suggests that the time is appropriate to examine limitations that are likely to apply to predictive geomorphic modeling.

Several authors have recently addressed critically the nature of prediction in the geosciences and in geological engineering (Tetzlaff 1989; Oreskes et al. 1994), with special attention to fluvial geomorphology (Baker 1988, 1994a, b), coastal engineering (Smith 1994; Pilkey et al. 1994), and hydrology (Konikow 1986; Anderson and Woessner 1992; Konikow and Bredehoeft 1992; Rojstaczer 1994). Schumm (1985, 1991) has discussed distinctive features of geomorphic systems such as singularity (uniqueness) and sensitivity that make prediction of future behavior difficult in geomorphology. The present chapter focuses on the connection between quantitative constitutive models or relations and prediction of specific sediment transport behavior in large-scale (e.g. fluvial, coastal, and hillslope) geomorphic systems. Constitutive relations are defined at the smallest resolved scale of the problem and summarize our knowledge (or assumptions) about system behavior below the chosen level of resolution. One can attempt to derive constitutive rules from controlled experiments (laboratory-scale) in terms of basic physical quantities like grain size and shear stress that appear to be important determinants of transport. Alternatively, one can invoke rules that are not directly based on the underlying physics and then attempt to justify their selection by showing that certain aspects of the dynamics of geomorphic systems appear to be consistent with the choice of such 'emergent' rules. Both of these approaches are discussed below. The issue of predictability cannot, ultimately, be resolved by logic, mathematics, or computer simulation. It can only be addressed by making predictions and then comparing those predictions against the future behavior of real-world geomorphic systems. The small amount of effort given in current geomorphic research to prediction of the future is understandable in view of the long time scales and large spatial scales that are often of interest. Nonetheless, the relative absence of predictive studies, with subsequent confirmation or refutation of results, represents a significant gap in the geomorphic agenda.

Bedload transport is an example of a specific process that has been studied extensively in the laboratory and that also plays an important role in the evolution of many geomorphic systems. Over a period of many decades, small-scale experiments in flumes have been carried out and semiempirical and empirical models developed and refined on the basis of those experiments (ASCE 1975). These models reflect properties such as grain size and surface slope that are underlying determinants of the physical behavior of the system. For example, the Meyer-Peter formula can be written (Meyer-Peter and Müller 1948) (in SI units) as

$$\frac{q^{2/3}S}{d_{50}} = 17 + 0.4\frac{q_s^{2/3}}{d_{50}},$$

where q_s and q are, respectively, sediment and water discharge in kg m^{-1} s^{-1}, S is the local slope, and d_{50} is the median grain size of the bed material in meters. Other models or 'formulas' are expressed directly in terms of bed shear stress. Although empirical, these expressions are derived from experiments where the independent variables are well-controlled. The form of these equations is therefore directly connected to measured physical behavior under specific experimental conditions. Figure 14.1 shows the tightly clustered experimental results upon which the Meyer-Peter formula is based.

When the predictions of formulae due to different authors are compared, however, agreement between them is typically poor. Figure 14.2 shows predictions of bedload transport rates based upon a number of well-known transport equations. These curves were derived under diverse experimental conditions, and experimental variability may account for some of the differences in prediction shown in the figure. However, the transport equations illustrated are intended to be applicable for a range of grain sizes from medium sand to granules. It thus seems fair to make a general comparison, as shown in Figure 14.2, of the transport rates predicted by these equations. This comparison is based upon results of a comprehensive review of sedimentation sponsored by the American Society of Civil engineers (ASCE 1975). Variation over several orders of magnitude between predictions for different models suggests that application of such formulas to large-scale geomorphic systems, where local conditions are often poorly known, will result in significant uncertainty.

In fluvial problems, empirical rating curves that bypass any reference to important underlying physical variables (grain size, shear stress, etc.) have often been used in practice. River transport of suspended and contact load can be described by power-law rating curves that relate transport rate to total discharge (ASCE 1975; Richards 1982). Rating curves are also commonly used to estimate reservoir sedimentation (Singh and Durgunoglu 1992) and soil erosion (Wischmeier 1976). Such rating curves do not reflect underlying small-scale properties of sediment transport, but are keyed to measurable large-scale properties such as total discharge and average slope.

These examples reflect engineering attempts to make predictions of geomorphic processes. Such attempts may seem crude by the standards of scientific geomorphology, but geomorphic prediction is in practice nearly always based on empiricism. This is partly due to immaturity of the scientific basis of geomorphology, but it also reflects the fact that complex systems such as those characteristic of geomorphology tend to be resistant to reductionism. Because of its overwhelming success in physics, one is accustomed to

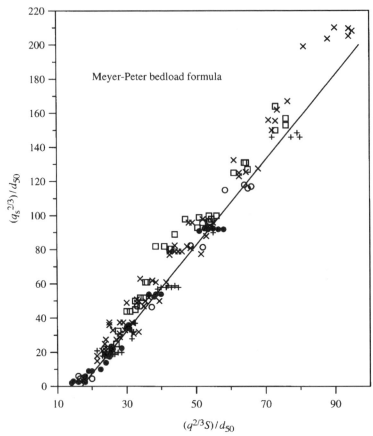

Figure 14.1 Sediment transport rate $q_s^{2/3}/d_{50}$ versus water discharge rate $q^{2/3}S/d_{50}$. Data compiled by Meyer-Peter and Müller (1948). S is water surface slope, q_s and q are, respectively, sediment and water discharge in kg m^{-1} s^{-1}. The line is the original Meyer-Peter formula

$$\frac{q^{2/3}S}{d_{50}} = 17 + 0.4\frac{q_s^{2/3}}{d_{50}} \text{ (SI units).}$$

The symbols represent experiments with particles of different mean size d_{50}: 0.0286 m (closed circles); 0.00505 m (open circles); 0.00702 m (plus signs); 0.00494 m (crosses); 0.00317 m (squares)

connecting reductionism with scientific investigation in general. In geomorphic systems, 'empirical' variables that are found to be useful for prediction may in fact be related to emergent variables of the system. In such cases, searching for emergent variables, and the constitutive rules that connect them, should be a central focus of activity of geomorphological science. This point of view has been recently articulated by Werner (1995) in terms of geomorphic attractors.

To return to the case of bedload transport, if a laboratory-tested model such as that due to Meyer-Peter (Figure 14.1) were an accurate constitutive model, it might in principle be used as the basis for scaling up to large-scale geomorphic applications. Several decades

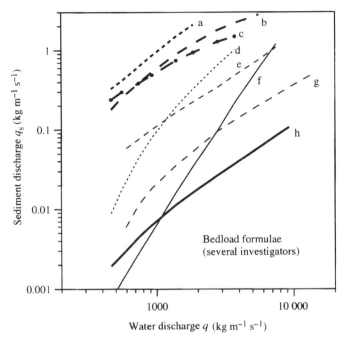

Figure 14.2 Sediment transport rate q_s versus water discharge q, for a sandy bed, based on several transport formulas: Shields (a), Einstein–Brown (b) DuBoys (c), Engelund-Hansen (d), Blench (e), Laursen (f), Schoklitsch (g), Meyer-Peter (h). Based on information compiled by the American Society of Civil Engineers (ASCE 1975)

ago, scaling up meant, at most, using local constitutive expressions of sediment discharge rates to predict total stream sediment discharge by multiplying by the stream width (e.g. Graf 1971; ASCE 1975; Yalin 1977). Today, the calculational limitation to upward-scaling is being lifted as numerical simulation approaches based on cell or grid methods (e.g. Tetzlaff and Harbaugh 1989) allow the integration of a large number of pieces of local information. The simulation studies mentioned above of fans, drainage networks, and other geomorphic features are direct consequences of this relaxation of earlier constraints on modeling. An example in engineering is the impending replacement by the US Department of Agriculture of the Universal Soil Loss Equation (Wischmeier 1976) (a rating curve approach) as the basis for estimating potential erosion on agricultural fields and rangelands. A detailed mechanistic, computer-based soil loss model, the Watershed Erosion Prediction Project (WEPP) (Lane et al. 1993) will be used instead. The use of computer-based methods in attempts to predict the future behavior of sediment transport systems can be expected to continue to increase. Potential limitations on the predictive power of such approaches, especially where long time scales are involved, is the focus of this chapter.

In moving from studies of sediment transport at small spatial scales and short time scales to large-scale geomorphic applications, several sources of uncertainty or error arise that affect accuracy of prediction. These uncertainties provide limits on how effectively one can expect to use knowledge of the basic physical processes of sediment transport to make useful large-scale predictions in geomorphology. Sources of uncertainty and error are discussed below.

MODEL IMPERFECTION

Incremental 'improvement' in sediment transport models at the laboratory scale will not necessarily add to our ability to make predictions of sediment transport at a large scale. For example, the explicit introduction of grain-size fractions as variables to be used in place of mean grain size would seem to allow in principle greater predictability of bed evolution in poorly sorted sediment. But the uniqueness of each natural sediment bed in terms of vertical and lateral variations in grading, and the difficulty of measuring such variability, make model implementation increasingly difficult as the model becomes more 'realistic'. For such reasons, quantitative geomorphic studies of large-scale evolution are not usually based on models derived directly from laboratory-scale studies of sediment transport rates, such as those shown in Figures 14.1 and 14.2. Instead, expressions for sediment flux q_s are parameterized in terms of more easily determined variables. Willgoose et al. (1991) choose $q_s \propto q^m S^n$, where q is local water discharge (averaged over a cell), S is local slope (also averaged over a cell), and m and n are parameters that in principle can vary spatially to represent heterogeneity in surface conditions.

Although in some cases large-scale power-law rules can be shown mathematically to result from suitable averaging of small-scale power-law transport formulas, such a procedure cannot be carried out in practice for each large-scale application because of lack of knowledge of the fine-scale detail necessary to implement the averaging. Moreover, in general, averaging of a small-scale power-law rule will not result in a power-law rule for averaged variables, except in special circumstances, because the transport rules are non-linear. Essentially one must search for new, higher-level rules that emerge at the large scale. In practice, the applicability of a sediment transport rule to be used at the large scale must be determined on the basis of its effectiveness in specific applications at that scale. One cannot expect that these rules are normally reducible to or explicitly derivable from laboratory-scale transport formulas. It may turn out that large-scale power-law rules, such as that given above, are suitable emergent rules for specific applications, but this fact must be established by the large-scale utility of these rules, not by appeal to the form of the small-scale transport formula.

In any case, for bedload transport, both engineering experience and critical studies show that small-scale transport formulas are unreliable, as suggested by Figure 14.2. A report of the American Society of Civil Engineers (ASCE 1975, p. 229) concludes an extended discussion of bedload transport by noting that 'sediment discharge formulas, at best, can be expected to give only estimates'. Graf (1971, p. 156) comments, following an extensive analysis of many bedload formulas, that 'an application of bedload equations to field determinations remains but an educated guess'. More recently, Gomez and Church (1989, p. 1182), on the basis of a detailed statistical analysis of 12 bedload formulas, conclude that 'on the basis of the tests performed by us ... none of the selected formula, and, we guess no formula, is capable of generally predicting bed load transport in gravel bed rivers'. This state of affairs pertains to using bedload models locally, where bed and discharge conditions ought to be best known. Use of such models as the basis for prediction in large-scale applications is even more problematic. Moreover, the models themselves represent only one of a number of sources of uncertainty in modeling sediment transport in large-scale systems, and not necessarily the most important source. Consequently, attempts to reduce model imperfection by introduction of an 'improved' small-scale sediment transport 'formula' are likely to be of only limited effectiveness in improving one's ability to predict at the large scale. Prediction and understanding at a small scale, on the other hand, may well benefit from refinement and development of small-scale models.

OMISSION OF SIGNIFICANT PROCESSES

Incremental improvement of small-scale models of erosion and deposition cannot correct the defect of omitting significant physical processes. The larger the spatial scale of the environmental system and the longer the time scale of interest, the greater is the chance that more than one important process will be present. In their monograph on large-scale simulation of clastic sediment transport, Tetzlaff and Harbaugh (1989) employ a fluvial picture of sediment erosion and deposition. They solve depth-averaged equations for water velocity over a given topographic surface, invoke a sediment transport model (similar to the Meyer-Peter and Müller bedload formula (1948) used in river sedimentation studies), pass sediment from one cell to another, and follow the evolution of topography. One possible application of such models is to alluvial fan building (Tetzlaff and Harbaugh 1989). However, alluvial fans may develop in the presence of nonfluvial transport mechanisms. In a given location, fluvial processes may dominate fan construction, but elsewhere debris flows may be important contributors to the fan building process (e.g. Whipple and Dunne 1992). Debris flows can contain up to 80–90% solids concentration

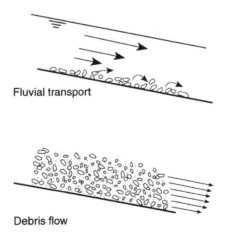

Fluvial transport

Debris flow

Figure 14.3 Bedload normally is composed of a thin carpet of sediment driven forward by the flow of the overlying water, top panel. Debris flows, whose rheology is similar to that of a Bingham plastic, are comprised of an intimate mixture of poorly sorted clastic particles and water, with water content as low as 10%, bottom panel by weight (Johnson and Rodine 1984). These flows move under rheological conditions (Figure 14.3) that are distinct (Johnson and Rodine 1984; Iverson and Denlinger 1987) from the fluvial conditions that characterize flume studies such as those from which most bedload formulas are derived. If a fluvial erosion and deposition model were applied at a location where debris flow processes are important, then no amount of attention to improvement of the fluvial model can account for the effects of a physically distinct process.

UNKNOWN INITIAL CONDITIONS

Initial conditions are statements about a system that must be made before a model can be implemented. In fluvial transport, these conditions might include the distribution of grain sizes, the cohesiveness of bank material, topographic details of the bed, distribution of channel vegetation, as well as information about stream flow characteristics. Initial conditions also include the distribution of sediment characteristics with depth below the bed surface, since erosion can expose previously buried material. These conditions are known only approximately, or, in some regions of the system, not at all. In studies of coastal sediment transport, essentially all models predicting the shoreface profile along barrier-island dominated coastlines assume that the shoreface is composed entirely of loose sand transportable in bedload or suspension. However, field studies along the North Carolina coast (e.g. Riggs et al. 1995, and references therein) show the existence of numerous premodern features of variable cohesiveness and transportability that underlie or breach the thin veneer of sand (typically < 1 m thick) on the shoreface. Such sediments range from marsh peats and tidal flat muds to indurated sandstones and gravels. The map shown in Figure 14.4, based on side-scan sonar imaging, shows the distribution of silty sand, sandy gravel, and bedrock outcrops along a section of the coast of North Carolina (Thieler et al. 1995). Predictive capabilities of sediment transport models based on the assumption of a uniform, loose, sandy shoreface are critically impacted wherever exposure of cohesive units occurs, or where significant changes in grain size occur, as in the figure. Information of this type has generally not been available for model studies of the shoreface.

Role of Data Collection

These observations suggest that as far as predictive power is concerned, local site-specific data collection can be at least as important as model choice or model refinement. Improved data collection, of course, improves knowledge of initial conditions. More important is that periodic data collection can be combined with model feedback as a practical strategy to generate a running prediction. In general, prediction of specific system behavior diverges over time from actual system behavior. Divergence occurs not only because of incompletely known initial conditions, but also because of the influence of all other sources of uncertainty. Periodic monitoring of the state of the physical system provides the information needed to correct the model (reset the initial conditions) as required. Reliable predictions of specific behavior for essentially all large-scale systems (not just sediment transport systems) require some combination of data collection and feedback. As the values of system variables begin to diverge from predicted values, data

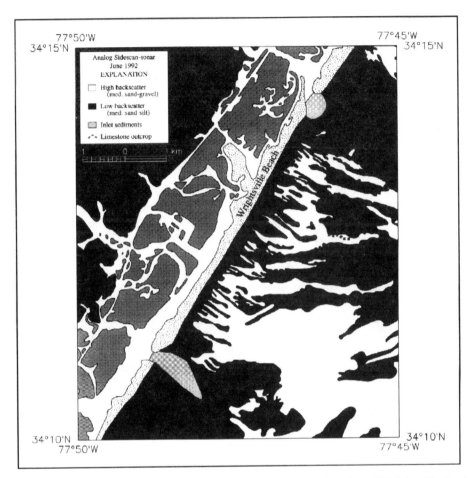

Figure 14.4 Map of sediment cover and bedrock outcrops on the shoreface of Wrightsville Beach, North Carolina. Based on sidescan sonar studies of Thieler et al. (1995). Dark color represents fine sand and silt; light color represents medium sand and gravel; rectangular symbols represent outcrops of resistant Pleistocene deposits. Adapted with permission of Elsevier Science

collection provides information that can be used to correct the model, thus producing a running prediction. Sometimes this procedure is highly quantitative. Kalman filtering techniques have been used successfully to update predictions of water levels for the purpose of optimizing power generation on the Niagara River (Crissman et al. 1993). Here ice jams in the river lead to water level fluctuations that require frequent feedback to 'correct' the model. Frequent measurements of water levels at selected observation stations are used within a hydraulic routing model to update values of discharge and water elevation at each section of the river. Often the procedure is more qualitative. Periodic inspection of dams, aircraft, and other engineered structures for cracks and corrosion or other defects are used as checks to monitor predicted performance over the lifetime of the system. Here, the 'model' is not necessarily a formal mathematical model, but the set of assumptions and relations that are used to predict system behavior. Likewise, large sediment transport systems require data collection and model updating if accurate prediction is to be continuously extended into the future. Thus, the decision to replenish a beach by pumping sand onto the shoreface is determined on the basis of periodic 'data' collection (observation of the state of the beach), not on the basis of long-term model prediction, no matter how 'realistic' or sophisticated the model might be.

SENSITIVITY TO INITIAL CONDITIONS

In nonlinear systems like those that characterize sediment transport there can exist a sensitivity to initial conditions that effectively prohibits detailed prediction of system evolution. A strong dependence on initial conditions is a highlight of chaotic behavior (Lorentz 1993). Several recent studies have focused on chaotic behavior in geomorphic systems (e.g. Slingerland 1989; Phillips 1992).

Whether a given environmental system is technically chaotic is not always easy to establish (Ruelle 1994). However, models of debris transport and deposition (Tetzlaff 1989), drainage network evolution (Howard 1994; Howard et al. 1994) and stream braiding (Murray and Paola 1994) all show that recalculation of detailed configurations of these systems after a long enough period of time is not usually possible if initial conditions are changed even slightly. Since actual initial conditions are usually known in practice only poorly, errors in prediction are always present in such systems.

These limitations can be important if the required prediction is sufficiently specific. The future pattern of occupation of channel distributaries across a river delta depends strongly on small details of present topography and channel form, and specific prediction is difficult. But, avulsion of the Mississippi River into the channel of the Atchafalaya (Richards 1982; McPhee 1990), a specific event, would have enormous consequences for the city of New Orleans and downstream industrial activity (which would no longer be on the river), and a specific predictive capability for sediment transport and erosion in that reach of the river is highly desirable. Formation of a tidal inlet in the low and narrow region north of Buxton on the North Carolina Outer Banks (Pilkey et al. 1982; Figure 14.5), would lead to isolation of the lower part of Hatteras Island and significantly impinge the economics of the region. A specific predictive capability is desired, but inlet formation by a strong overwash event may depend sensitively on island topography and shoreface bathymetry existing at the time of a given storm, and hence be difficult to predict. On the other hand, if it is the statistical occurrence of avulsion, or tidal inlet formation, or general rather than specific aspects of their dynamics, geometry, or stratigraphy that is the main interest, then sensitivity to initial conditions would not necessarily be an impediment to prediction.

Many features of geomorphic systems are relatively insensitive to initial conditions. Certain channel characteristics of rivers survive large variations in initial conditions, though precise channel patterns can be very sensitive to these conditions. Braided rivers tend to develop on steep alluvial slopes or in watercourses characterized by high sediment load. Meandering streams develop on more gentle slopes and where bank stability is provided by vegetation or cohesive sediment. If model predictions can distinguish quantitatively between conditions required for braiding or meandering, then those pre-

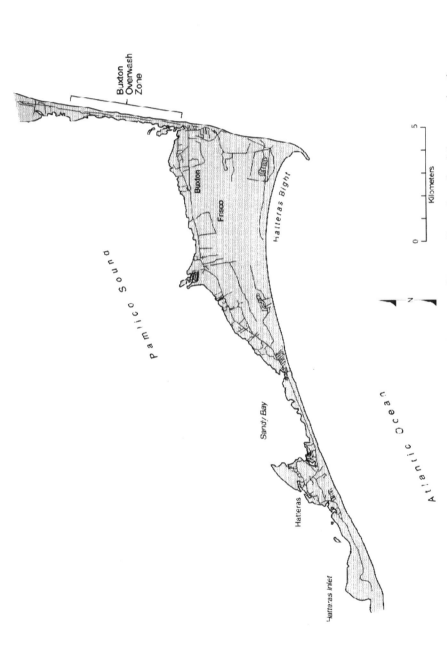

Figure 14.5 Map showing Hatteras Island, North Carolina in the vicinity of the town of Buxton. Formation of an inlet connecting the open ocean on the east to the sound on the west side of the island, in the narrow region north of Buxton (Buxton Overwash Zone), would isolate the town and the southern part of Hatteras Island. Here a site-specific prediction of sediment transport, and of inlet formation, is desirable

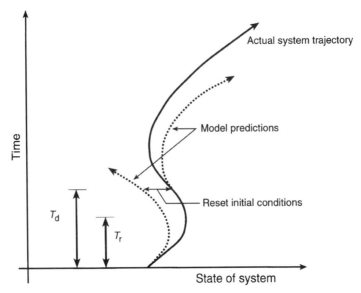

Figure 14.6 The solid curve represents evolution of system through time. Dotted lines show evolution of model prediction, which diverges unacceptably (as determined by the purpose of the prediction) from actual system behavior after a characteristic time T_d. A running prediction will be useful if the divergence time T_d, after which an update of the model is required, exceeds the time T_r needed to respond to the prediction

dictions, even though inaccurate in the sense that detailed channel forms cannot be reproduced, would clearly be useful. Thus, recent simulation studies of evolution of drainage basins and drainage networks (e.g. Howard 1994; Willgoose et al. 1991) focus appropriately on generic aspects of landscape evolution such as patterns of slope retreat and drainage basin geometry.

The previously noted strategy of data collection and feedback to periodically correct system initial conditions can be applied to help remedy sensitivity to initial conditions. An important factor in such 'prediction–correction' schemes is the relation of the time scale T_d, over which the prediction diverges unacceptably from the actual system trajectory, to the time scale T_r describing the period of time needed to respond in an appropriate way to the prediction (Figure 14.6). Here T_r might be the time required to clear a channel by dredging, or to replenish a recreational beach by pumping sand onto the shoreface; T_d is determined by experience with many cycles of prediction and model correction. If $T_r < T_d$ then the model updating scheme can be successful. If $T_r > T_d$ then the divergence time scale is too short to be useful.

UNRESOLVED HETEROGENEITY

In some physical systems, predictive capability increases with increasing size, because at larger scales one averages over many of the 'details' which make the system so heterogeneous at small scales. In fluid or solid mechanics, the unknown and unknowable detailed motion of molecular constituents can be replaced by appropriate, well-behaved averages. However, not all systems are guaranteed to have useful average properties of this kind. Within the interior of a structural beam in a building, a small averaging volume can be used to define the local state of stress in the beam. But, as the averaging volume is taken larger and larger, eventually exceeding the size of the beam, this approach breaks down. Because a building normally contains heterogeneities of all sizes up to the size of the structure itself, there is no 'typical' averaging volume of the kind one invokes in fluid or solid mechanics, small with respect to the large-scale system, but large compared with the inherent system heterogeneities (e.g. molecules). There is no small-scale physical model (constitutive law) that can be used to predict the detailed behavior of 'buildings'.

Likewise, in large-scale geomorphic systems, it may also be impossible to define a meaningful averaging volume. Heterogeneities appear in the form of variations from place to place in factors such as vegetative cover, soil type and bedrock exposure, and in the presence of discrete entities such as rills and streams. For small enough cells, the basic underlying transport model (such as one of those in Figure 14.2) may directly provide rules for changing the state of the cell (e.g. gain or loss of sediment). However, in larger cells, if the dynamics internal to a cell are sufficiently heterogeneous, then cell evolution rules are not reducible to constitutive rules derived from basic physical (laboratory-scale) principles of sediment transport.

Even small cells can be heterogeneous. Figure 14.7 shows several potential sources of heterogeneity in a dry watercourse. The field of view is about 10 m across in the middle ground. Local variations of flow depth caused by surface irregularities lead to a corresponding distribution of shear stress. Since the bedload sediment transport rate q_s is a strong function of shear stress τ (e.g. the Einstein–Brown formula (ASCE 1975) gives approximately $q_s \sim \tau^3$), the mean transport rate is not equal to the transport rate calculated on the basis of the mean depth or stress (Baird et al. 1993). Figure 14.8 shows the sediment transport rate q_s, calculated on the basis of the Einstein–Brown formula, as a function of slope for flow in a channel of given width and depth. This greatly exceeds the sediment transport rate $q_s(\bar{\tau})$ computed on the basis of the average stress $\bar{\tau}$, where in this case averaging is performed over a swath about twice as wide as the channel.

Other heterogeneities appear in Figure 14.7. Vegetation, whose occurrence is correlated with small changes in elevation, grows preferentially on low terraces lying a few decimeters above the active watercourse. Low flows are little influenced by the presence of plants. At higher flows, the presence of vegetation creates extra flow resistance, roots resist erosion, and sediment and other debris are captured behind vegetative obstructions. Moreover, the land surface above the low-flow channel has a different roughness, a larger mean clast size, and in places a soil crust that is lacking in the low-flow channel. On the left side of the figure, a low bluff several meters high deflects flow, and undergoes parallel retreat by spalling thin slabs of consolidated alluvium into the channel. Significant heterogeneity within an area less than 10 m on a side points to the challenge of eliciting simple emergent constitutive rules for cells in which several processes contribute to intracell dynamics. Bedload transport rules of the kind discussed earlier would be applicable (if at all, given Figure 14.2) on scales much smaller than that of Figure 14.7. Moreover, if intracell heterogeneities are present, the time evolution of the internal state of each cell becomes itself a nonlinear problem that needs to be solved. Internal cell evo-

Figure 14.7 Heterogeneity in surface elevation, clast size, and vegetative cover is present at several scales in this dry watercourse (Greenwater Valley, California). Field of view is about 10 m in the middle ground

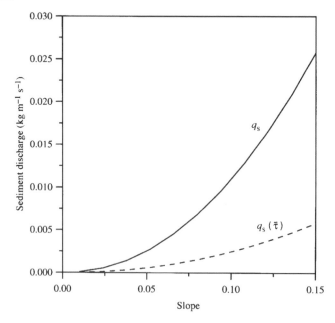

Figure 14.8 Comparison of sediment transport rate q_s in kg m^{-1} s^{-1} computed on the basis of Einstein–Brown formula for flow in a channel of width 0.475 m, depth 0.0225 m, and discharge $Q = 6.8 \times 10^{-3}$ m^2 s^{-1} with sediment transport rate $q_s(\bar{\tau})$ computed for the same flow rate but by averaging the bed shear stress over a width of 1 m

lution, however, must be captured in simple rules that can be stated a priori. The rules must be simple because no significant computational resources can be devoted to calculation of intracell dynamics. Such resources have already been used up in creating fine-scale system resolution.

This does not necessarily mean that appropriate constitutive rules cannot be found for heterogeneous cells, but the rules need to be 'discovered' at the cell level, rather than derived from fundamental principles of sediment transport. This is the origin of the power-law-type transport rules used in geological reconstructive modeling and landscape simulation, as discussed above. In such studies, cell rules are invoked that are effective in creating final-state landscapes and stratigraphy that resemble existing landscapes and stratigraphy. These rules are not derived directly by averaging over fundamental physical-process models. For example, the sediment discharge model used as a basic cell rule by Tetzlaff and Harbaugh (1989) in their simulation of large-scale fluvial erosion and deposition resembles the well-known Meyer-Peter and Müller (1948) bedload formulas, expressible as a power law of mean water discharge or excess shear stress. However, water discharge or shear stress variations within a cell, not to mention variations in surface roughness and grain size, will clearly be substantial for most applications, and 'averaging' is never carried out explicitly. The simulations of Koltermann and Gorelick (1992), who studied alluvial fan deposition over a period of 600 000 years using the Tetzlaff and Harbaugh (1989) approach, were performed with a horizontal cell resolution of 120 m. The cell-level sediment transport model used in such studies, therefore, represents essentially a new rule postulated, and hopefully confirmed, at a scale much greater than that at which any physical bedload formula has been, or could be, derived and tested. Such transport rules are not 'averaged' results of laboratory-scale bedload formulas, but should be regarded as new discovered or emergent rules, to be tested, and then used or discarded on their own merits. Thus, rules for large-scale applications should be chosen on the basis of known constraints, such as mass conservation, dependence on variables thought to be important (such as local discharge rates) and on the basis of calibration (fit) to specific field-scale studies. Such rules are not based upon fundamental transport physics, but upon observations (field collection of data) and experience with the modeling requirements of the landscapes of interest.

EXTERNAL FORCING

External forcing arises in an open system where mass, energy, and momentum can enter and be discharged through the system boundaries. Like some of the other sources of uncertainty, external forcing becomes an increasingly important factor in prediction as system size increases. While laboratory experiments are usually carried out at a scale where isolation from external events is possible, large systems are always exposed to the vagaries of nature such as storms and climate change. In fluvial sediment transport, external forcing may be due to increases of discharge resulting from storms or dam releases, to injection into the mainstream of quantities of water and sediment from side channels and slopes falling outside the model boundaries, to backwater effects due to stream impoundment or rising sea level, to tectonic uplift, and to base level lowering. The

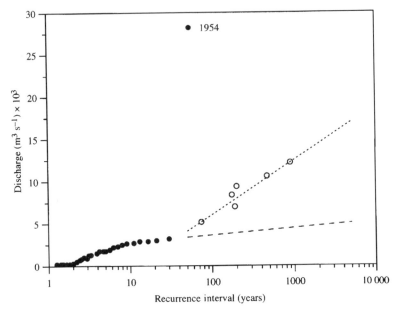

Figure 14.9   Flood frequency data for the Pecos River near Comstock, Texas, (adapted from Patton and Baker 1977) as an example of stochastic external forcing. Extrapolation (dashed line) of historical flood-frequency record (solid circles) suggests a rather different pattern of long recurrence-time events than does geological evidence based on reconstruction of paleofloods (open circles). The isolated 1954 flood data point on the upper part of the figure underscores the risk of basing statistical projections on a limited sample size

predictive capacities of any model are limited if unpredictable external forcing can occur. Sometimes, as in a dam release or impoundment, forcing may be anticipated. In other cases, as in forcing due to storms, it may be anticipated only statistically. A statistical treatment of forcing is possible only if the distribution of events is known. Patton and Baker (1977) discussed historic flood frequencies on the Pecos River in Texas (Figure 14.9). The historical flood record (solid circles) in Figure 14.9 might be extrapolated according to the dashed line to provide an assumed long-term distribution of events upon which prediction of future flood and sediment transport might be based. The data point for the massive flood of 1954, and the reconstructed flood discharges determined by paleo-hydrological studies (open circles) indicate the pitfalls of this approach. To the extent that a suitable distribution of forcing events is available (or a range of possible outcomes is specified), prediction may be possible, with the frequency of forcing, $1/T_f$, playing a role somewhat similar to the divergence frequency, $1/T_d$, described earlier. However, if forcing is due to occurrence of an event of a kind not anticipated, for example a shift in vegetative cover due to drought climate change, or change in land use, then one loses even a statistical basis for prediction.

INAPPLICABILITY OF THE FACTOR OF SAFETY CONCEPT

In engineering projects that affect property, life, or incur large costs, the risk of a serious mismatch between prediction and outcome can often be protected against by incorporation of a factor of safety in engineering design. The factor of safety helps to protect against a faulty design prediction and against larger than expected excursions of external forcing parameters. Engineered systems are generally constructed of synthesized or carefully selected materials and elements whose properties and interrelationships with each other the engineer (ideally) understands. Thus, elements of a building or bridge are selected and assembled to desired specifications, in a manner over which the engineer has control. If certain elements of the system are not well-understood, or fail to meet specifications, they are rejected and elements or materials substituted whose behavior is acceptable. In short, engineered systems are designed and assembled, and design and human assembly imply a high level of understanding and control. Consequently, model predictions can be effective for such systems, and factors of safety can be included in the design.

For geomorphic systems, however, the material properties and 'construction' of the system (an alluvial surface subject to erosion or deposition, a river channel, a tidal inlet) are not usually well-specified or well-controlled. In their natural state, these systems are not designed and assembled. Where efforts are made to engineer these systems, as in flood control projects, success is due in large measure to the addition of engineered structures (e.g. levees and dams) to the natural system – structures that incorporate a factor of safety. Thus the courses of concrete-lined rivers in the Los Angeles basin are predictable over many flood seasons, and they seldom overflow their banks. Models that are used to predict behavior of a natural system (e.g. how an alluvial surface responds to the forces of erosion following clearing of vegetation) have much less predictive capability than models of typical engineered systems, because the elements of the natural system are not chosen and constructed to predetermined specifications. Environmental systems are highly heterogeneous, material properties are often little understood, and initial conditions are usually poorly known. It is not surprising that a number of famous dam failures, exemplified by the Saint Francis dam (Wiley et al. 1928) and the Vaiont dam (Kiersch 1964) disasters, occurred primarily as the result of initial 'failure' in geologic material adjacent to or upstream of the dam structure, not because the physical (engineered) dam itself failed.

The factor of safety concept is basically a linear scaling method. Where relationships between variables are highly nonlinear, and a clear understanding of the controlling variables does not exist, as in many natural and environmental systems, the factor of safety concept is inapplicable. Lack of a factor of safety results in a reduced probability that a system will behave in a desired (or known) way in the future, and hence leads to a reduced predictive capability.

A substitute for a factor of safety in environmental prediction is the worst-case scenario (Schumm 1985). While the factor of safety depends for its effectiveness on designed and tested construction methods (making a dam thicker), the effectiveness of the worst-case scenario in geomorphic applications depends on the reliability of a model of an undesigned natural system. If the worst-case scenario represents only a modest deviation from the conditions under which the model was derived and calibrated, worst-case predictions are likely to be more useful than when the worst-case scenario represents extreme conditions, for example storm wave attack on a beach. This is where the model will be most susceptible to failure, because of a relative lack of calibration data and because of the increased likelihood of occurrence of unanticipated or neglected processes.

DISCUSSION

The extent to which small-scale sediment transport models that are defined in terms of basic physical quantities such as particle size and bed stress can be scaled up and used as a basis for specific geomorphic predictions is limited by several sources of uncertainty. These include model imperfection, omission of important process, lack of knowledge of initial conditions, sensitivity to initial conditions, unresolved heterogeneity, occurrence of external forcing, and inapplicability of the factor of safety concept. One or more of these sources of uncertainty is likely to arise in any attempt to predict large-scale geomorphic behavior on the basis of scaled-up laboratory-scale studies of sediment transport. Consequently, large-scale geomorphic prediction is difficult to implement in terms of our understanding of the physical behavior of sediment transport at the small scale.

System size seems to be the most fundamental factor that limits predictability in geomorphic modeling. The occurrence of unanticipated processes, the lack of knowledge of initial conditions, the occurrence of external forcing, and the presence of unresolved heterogeneity all become increasingly important with increases in system size. They can often be avoided in systems of small size. Correspondingly, control and repeatability, usually thought of as hallmarks of science in general, are more accurately hallmarks of small size. The fact that large geomorphic systems are often unique, and that control and repeatability are limited or absent, suggests that standard methods of analysis (such as reductionism) that often work well at laboratory scales, may be inapplicable.

Lack of information on initial conditions is a direct consequence of the large size of many geomorphic systems. This lack may be countered to some extent by increases in data collection. Data collection may occur through simple observation, or by input from deployed instrumentation. Updated or corrected predictions can then be obtained as collected data is fed back to the model. From this procedure a characteristic time scale emerges, the divergence time T_d, characterizing the typical period over which the prediction is valid without correction. Frequent data collection will maximize the time over which the prediction is actually usable. Usefulness implies that a characteristic response time T_r must be less than T_d. Data collection, and model updating, are also necessary if one wishes to maximize predictive capabilities in situations where other resources of uncertainty degrade model prediction.

A running prediction is useful if the objective is to produce short-time scale predictions of the evolution of a geomorphic system. Data collection and feedback become less effective as longer-term predictions are sought, since the utility of data collection and feedback in updating the model depends on the experience one gains in applying the model over a number of divergence times. This is how the value of the divergence time is discovered. If the system veers 'prematurely' from the predicted trajectory to an undesirable state, such as excessive erosion at a waste-burial site, an engineered response may be difficult or impossible because of the scale or cost of the problem. Then the prediction has failed its purpose, even though the model can be periodically 'corrected' and a new, more accurate prediction made.

It has been argued that models developed to predict large-scale geomorphic behavior cannot be based simply on upscaled versions of the laboratory-scale sediment transport laws. Reductionism, i.e. explaining behavior at the large scale by appeal to small-scale phenomena, works well with uniform systems, which are usually small themselves. But at large scale, emergent variables that are not (in practice) derivable from basic physics are more likely to be the useful building blocks of predictive modeling. Although not derivable from basic physics, the emergent rules must still obey fundamental physical constraints such as mass conservation. Emergent variables are often used to describe the behavior of large heterogeneous systems in areas other than geomorphology. In biology and ecology, the existence of discrete organisms immediately suggests possible emergent variables. A model of fish schooling might start with the fish itself as the fundamental entity of interest and then attempt to predict schooling patterns from rules of individual fish behavior (Niwa 1994) rather than on the basis of smaller-scale physically based rules such as those describing the internal workings of an individual fish. In geomorphology, identification of emergent variables may be more difficult. In studies of landscape evolution, average water discharge and average slope are two commonly used variables. These are implicitly used as emergent variables, since explicit derivation from fundamental transport laws is not possible, except in the unusual circumstance where local conditions are especially uniform. Put another way, it is a mistake to (necessarily) equate consistency of landforms – for example that shown by the tendency of rivers to follow the rules of hydraulic geometry – with simplicity of the underlying physics. There is great physical consistency in human anatomy and function (e.g. two arms, two eyes, upright posture), but this simplicity has nothing to do directly with the underlying physics. The consistency is instead a manifestation of emergent behavior in extraordinarily complex systems. This is the nature of nonlinear processes that lead to complexity – that simple variables arise at the large scale which are not in practice derivable from the small scale where the simple physics does operate. As far as prediction is concerned, these arguments suggest that in complex geomorphic systems an investment of research effort in improving our knowledge of emergent variables through large-scale modeling and comparison with large-scale phenomena may be more useful than a similar investment of effort in developing small-scale, physically based constitutive models.

A further consequence of size is the uniqueness of geomorphic systems. Large systems are not replicable, as laboratory and most engineered systems are, nor are their constitution, composition, and structure designed or controlled. Two landscapes may be similar in many respects, but will always differ in particulars. Particulars can often be averaged away in small uniform systems (such as molecular positions in a glass of water), but in large systems they persist as heterogeneities (resistant soil layers, patterns of vegetation) that may significantly influence the course of geomorphic evolution. As a consequence, each geomorphic system must be approached anew and analyzed for its specifics and peculiarities, as well as for its similarities to other systems. One should expect no automatic predictive formula to apply to these systems. Instead, experience with similar systems and attention to the historical (geological) record can help inform of conditions and behavior that need to be accounted for in any attempt at modeling. The presence of abandoned braided channels in regions where modern streams and rivers are meandering may indicate the presence of an environmental driving force (such as climate or vegetative cover) that must be accounted for in long-term landscape evolution. These historical factors are largely absent in laboratory-scale experiments, where physical determinants are dominant, but such factors do (or should) provide the starting point from which large-scale geomorphic modeling is initiated and the framework within which it is performed. An important and practical, if obvious, conclusion is that large-scale geomorphic modeling cannot be effectively pursued on the basis of engineering and physical knowledge alone, but must be conditioned by historical and geological understanding. This observation has important implications for the training of engineers and scientists whose future activities will involve use of large-sale geomorphic models.

Finally, many of the conclusions regarding the predictive nature of modeling in geomorphology appear applicable to modeling in the environmental sciences in general. The problems that affect prediction in geomorphology, the ubiquitous occurrence of unknown initial conditions, external forcing and unresolved heterogeneity, and the corresponding inability of reductionism to produce large-scale predictions, are generic problems of size, not of a particular system. The same general approaches that pertain to geomorphic modeling – data collection, model updating through feedback, discovery of emergent variables, avoidance of reductionism, and attention to history – may also be expected to be necessary components of large-scale environmental modeling in general.

ACKNOWLEDGEMENTS

This work was supported by the US Army Research Office, Grants No. DAAL03-92-G-0370 and DAAH04-94-G-0067. I thank J. Raghuraman for her assistance with the MS.

REFERENCES

Ahnert, F. 1987. Approaches to dynamical equilibrium in theoretical simulations of slope development, *Earth Surface Processes and Landforms*, **12**, 3–15.

Anderson, M.P. and Woessner, W.W. 1992. The role of the postaudit in model validation, *Advances in Water Resources*, **15**, 167–173.

ASCE 1975. *Sedimentation Engineering*, edited by V.A. Vanoni, American Society of Civil Engineers, Manuals and Reports on Engineering Practice, No. 54, American Society of Civil Engineers, New York, 1975.

Baird, A.J., Thornes, J.B. and Watts, G.P. 1993. Extending overland flow models to problems of slope evolution and the representation of complex slope-surface topographies, in *Overland Flow*, edited by A.J. Parsons and A.D. Abrahams, Chapman and Hall, New York, pp. 199–223.

Baker, V.R. 1988. Geological fluvial geomorphology, *Bulletin of the Geological Society of America*, **100**, 1157–1167.

Baker, V.R. 1994a. Geological understanding and the changing environment, *Transactions of the Gulf Coast Association of Geological Societies*, **44**, 13–20.

Baker, V.R. 1994b. Geomorphological understanding of floods, *Geomorphology*, **10**, 139–156.

Crissman, R.D., Chu, C-L., Yu, W., Mizumura, K. and Corbu, I. 1993. Uncertainties in flow modeling and forecasting for Niagara River, *Journal of Hydraulic Engineering*, **119**, 1231–1250.

Graf, W.H. 1971. *Hydraulics of Sediment Transport*, McGraw-Hill, New York, 513 pp.

Gomez, B. and Church, M. 1989. An assessment of bed load sediment transport formulae for gravel bed rivers, *Water Resources Research*, **25**, 1161–1186.

Hanson, H. and Kraus, N.C. 1989. *GENESIS: Generalized Model for Simulating Shoreline Change,* Tech. Report CERC-89-19, Coastal Engineering Research Center, Waterways Experiment Station, US Army Corps of Engineers, Vicksburg, Miss.

Howard, A.D. 1994. A detachment-limited model of drainage basin evolution, *Water Resources Research,* 30, 2261–2285.

Howard, A.D., Dietrich, W.E. and Seidl, M.A. 1994. Modeling fluvial erosion on regional to continental scales, *Journal of Geophysical Research,* 99, 13971–13986.

Iverson, R.M. and Denlinger, R.P. 1987. The physics of debris flow – a conceptual assessment, in *Erosion and Sedimentation in the Pacific Rim, Proceedings of the Corvallis Symposium,* edited by R.L. Beschta, International Association of Hydrological Sciences Publ. No. 165, Wallingford, Oxfordshire, UK, pp. 155–165.

Johnson, A.M. and Rodine, J.R. 1984. Debris flow, in *Slope Instability,* edited by D. Brunsden and D.B. Prior, Wiley, Chichester, pp. 257–361.

Kiersch, G.A. 1964. Vaiont reservoir disaster, *Civil Engineering,* 34, 32–39.

Kirkby, M.J. 1986. A two-dimensional simulation model for slope and stream evolution, in *Hillslope Processes,* edited by A.D. Abrahams, Allen and Unwin, London, pp. 203–222.

Koltermann, C.E. and Gorelick, S.M. 1992. Paleoclimatic signature in terrestrial flood deposits, *Science,* 256, 1775–1782.

Konikow, L.F. 1986. Predictive accuracy of a ground-water model – lessons from a postaudit, *Ground Water,* 24, 173–184.

Konikow, L.F. and Bredehoeft, J.D. 1992. Ground-water models cannot be validated, *Advances in Water Resources,* 15, 75–83.

Lane, L.J., Nearing, M.A., Laflen, J.M., Foster, G.R. and Nichols, M.H. 1993. Description of the US Department of Agriculture water erosion prediction project (WEPP) model, in *Overland Flow,* edited by A.J. Parsons and A.D. Abrahams, Chapman and Hall, New York, pp. 377–391.

Lorentz, E. 1993. *The Essence of Chaos,* University of Washington, Seattle, 227 pp.

McPhee, J. 1990. *The Control of Nature,* Noonday Press, New York, 272 pp.

Meyer-Peter, E. and Müller, R. 1948. Formulas for bed-load transport, in *Proceedings of the 2nd Meeting of the International Association for Hydraulic Structures Research,* Stockholm, pp. 39–64.

Montgomery, D.R. and Dietrich, W.E. 1992. Channel initiation and the problem of landscape scale, *Science,* 225, 826–830.

Murray, A.B. and Paola, C. 1994. A cellular model of braided rivers, *Nature,* 371, 54–56.

Niwa, H.-S. 1994. Self-organizing dynamic model of fish schooling, *Journal of Theoretical Biology,* 171, 123–136.

Oreskes, N., Shrader-Frechette, K. and Belitz, K. 1994. Verification, validation and confirmation of numerical models in the earth sciences, *Science,* 263, 641–646.

Patton, P.C. and Baker, V.R. 1977. Geomorphic response of central Texas stream channels to catastrophic rainfall and runoff, in *Geomorphology in Arid and Semi-Arid Regions,* edited by D.O. Doehring, Publications in Geomorphology, State University of New York, Binghamton, NY, pp. 189–217.

Phillips, J.D. 1992. Qualitative chaos in geomorphic systems, with an example from wetland response to sea level rise, *Journal of Geology,* 100, 365–374.

Pilkey, O.H. Jr, Neal, W.J., Pilkey, O.H. Sr and Riggs, S.R. 1982. *From Currituck to Calabash,* Duke University Press, Durham, NC, 91 pp.

Pilkey, O.H. Jr, Young, R.S., Bush, D.M. and Thieler, E.R. 1994. Predicting the behavior of beaches: alternatives to models, *Littoral 94 Proceedings,* Vol. 1, edited by S. de Carvalho and V. Gomes, European Coastal Zone Association of Science and Technology, Lisbon, pp. 53–60.

Richards, K. 1982. *Rivers: Form and Process in Alluvial Channels,* Methuen, London, pp. 90–91.

Riggs, S.R., Cleary, W.J. and Synder, S.W. 1995. Influences of inherited geologic framework on barrier shoreface morphology and dynamics, *Marine Geology,* 126, 213–234.

Riley, S.J. 1994. Modelling hydrogeomorphic processes to assess the stability of rehabilitated landforms, Ranger Uranium Mine, Northern Territory, Australia – a research strategy, in *Process Models and Theoretical Geomorphology,* edited by M.J. Kirkby, Wiley, New York, pp. 357–388.

Rojstaczer, S.A. 1994. The limitations of ground water models, *Journal of Geological Education*, **42**, 362–368.

Ruelle, D. 1994. Where can one hope to profitably apply the ideas of chaos?, *Physics Today*, **47**, 24–30.

Schummn, S.A. 1985. Explanation and extrapolation in geomorphology: seven reasons for geologic uncertainty, *Transactions of the Japanese Geomorphological Union*, **6**, 1–18.

Schumm, S.A. 1991. *To Interpret the Earth: Ten Ways to Be Wrong*, Cambridge University Press, Cambridge, England, 133 pp.

Singh, K.P. and Durgunoglu, A. 1992. Predicting sediment loads, *Civil Engineering*, **62**, 64–65.

Slingerland, R. 1989. Predictability and chaos in quantitative dynamic stratigraphy, in *Quantitative Dynamic Stratigraphy*, edited by T.A. Cross, Prentice-Hall, Englewood Cliffs, NJ, pp. 45–53.

Smith, A.W.S. 1994. The coastal engineering literature and the field engineer, *Journal of Coastal Research*, **10**, iii–viii.

Tetzlaff, D.M. 1989. Limits to the predictive stability of dynamic models that simulate clastic sedimentation, in *Quantitative Dynamic Stratigraphy*, edited by T.A. Cross, Prentice-Hall, Englewood Cliffs, NJ, pp. 55–65.

Tetzlaff, D.M. and Harbaugh, J.W. 1989. *Simulating Clastic Sedimentation*, van Nostrand Reinhold, New York, 202 pp.

Thieler, E.R., Brill, A.L., Cleary, W.J., Hobbs III, C.H. and Gammisch, R.A. 1995. Geology of the Wrightsville Beach, North Carolina shoreface: implications for the concept of shoreface profile of equilibrium, *Marine Geology*, **126**, 271–287.

Toy, T.J. and Hadley, R.F. 1987. *Geomorphology and Reclamation of Disturbed Lands*, Academic Press, London, 480 pp.

Werner, B.T. 1995. Eolian dunes: computer simulations and attractor interpretation, *Geology*, **23**, 1107–1110.

Whipple, K.X. and Dunne, T. 1992. The influence of debris-flow rheology on fan morphology, Owens Valley, California, *Bulletin of the Geological Society of America*, **104**, 887–900.

Wiley, A.J., Louderback, G.D., Ransome, F.L., Bonner, F.E., Cory, H.T. and Fowler, F.H. 1928. *Report of Commission Appointed by Gov. C.C. Young to Investigate the Causes Leading to the Failure of the St Francis Dam, California*, State Printing Office, Sacramento.

Willgoose, G., Bras, R.L. and Rodriquez-Iturbe, I. 1991. A coupled channel network growth and hillslope evolution model, 1, Theory, *Water Resources Research*, **27**, 1671–1684.

Willgoose, G., Bras, R.L. and Rodriquez-Iturbe, I. 1994. Hydrogeomorphology modelling with a physically based river basin evolution model, in *Process Models and Theoretical Geomorphology*, edited by M.J. Kirkby, Wiley, New York, pp. 271–294.

Wischmeier, W.H. 1976. Use and misuse of the universal soil loss equation, *Journal of Soil and Water Conservation*, **31**, 5–9.

Yalin, M.S. 1977. *Mechanics of Sediment Transport*, Pergamon Press, Oxford, 289 pp.

INTERDISCIPLINARY AND INTRADISCIPLINARY CONTEXTS

Geomorphology is clearly a scientific discipline embedded in the earth sciences, but it is, even more formally, an academic subdiscipline of geography and geology. Intellectually, this position provides the enrichment of diversity; academically, it often produces tension. Where diversity promotes substantively disparate views of the discipline or contrasting immediate objectives (often accompanied by sharply differing techniques) there is clearly great scope for fragmentation. Conflicting loyalties emerge as geomorphologists identify with geomorphology, geography, or geology; or increasingly with environmental science, geophysics, or engineering of one sort of another. As a practical matter of survival the practitioners of an academic (sub)discipline as small as geomorphology cannot dismiss such issues as trivial. One of the primary roles that a strong sense of scientific identity and purpose can play, after that of quality control, is disciplinary unification.

Rowland Twidale uses his long career both as an academic and as a consulting geomorphologist as a vantage point for examining the role of geomorphology in contemporary geology. Using numerous examples from Australia he demonstrates that geomorphology may be viewed as a powerful component of geology undeserving of the short shift which it frequently receives from the majority of geologists. Bernard Bauer traces the development of geomorphology in relation to geography. In doing so he champions a sophisticated perspective on the relationships among geomorphology, geography, and geology in which he demonstrates that attempts to assign modern geomorphology to a single academic discipline based on historical heritage are inappropriate.

Waite Osterkamp and Cliff Hupp cast the disciplinary net wider by comparing the development of geomorphology to that of ecology within the context of Thomas Kuhn's conception of scientific development. They highlight that even though both disciplines experienced periods when concepts derived from Darwinian evolution, equilibrium theory, and both evolution and equilibrium were important, the role that these various schools of thought played in the 'complex' disciplines of geomorphology and ecology differs fundamentally from Kuhn's notion of a paradigm or exemplar, which he developed for the 'basic' sciences (e.g. physics, chemistry). Based on this analysis they conclude that the development of geomorphology does not conform with Kuhn's model of scientific development.

The section concludes with what might be categorized as an applied view of geomorphology by William Graf. However, in this context it also serves as an exemplar of a future role for geomorphology. In examining the role that geomorphology and geomorphologists have, and may play, in management of American rivers, Graf draws attention to something that geomorphologists have generally preferred to ignore – the fundamental influence of humanity on the landforms/landscapes that are studied. There is little doubt that most geomorphologists focus their attention on understanding what they identify as a 'natural' or unsullied world. Even where this reality is understood to be a myth (almost everywhere?), the significance of human intervention tends to be downplayed. To understand the landscapes before us, as well as to manage them for society, geomorphologists must plunge quickly and vigorously into the task of pursuing a course in which human impact is overtly and systematically integrated into the fabric of the discipline.

15 Derivation and Innovation in Improper Geology, aka Geomorphology

C. R. Twidale,

Department of Geology and Geophysics, University of Adelaide

ABSTRACT

Geomorphologists have contributed to geological understanding in several ways. First, the form of the land surface provides an introduction to various geological phenomena and the first indications of subsurface structure and events, past, present and future. Second, knowledge of contemporary processes, to which geomorphologists have contributed, facilitates the interpretation of palaeosurfaces, of past deposits and of past events. Third, the land surface is part of the record of earth history, which is the core and focus of geology. Over considerable areas of each of the continents morphology provides the only clues to Phanerozoic events. Fourth, geomorphological concepts are of considerable practical or applied significance, for example in the search for minerals, in engineering geology and in planning generally.

Yet geomorphology has lost and continues to lose status in the geological profession, both academic and non-academic (governmental, industrial). Some geologists regard geomorphology as not proper or not real geology. The possible reasons for this are broached and responses discussed. The feasibility of geomorphology severing its ties with geography and geology and 'going it alone' is considered. Though attractive in principle the move is regretfully rejected as impractical. The loss of the geological link is also seen as deleterious, for geomorphology is basically chronological. Geomorphology has been particularly hard hit by recent educational and research trends, and by the shortage of job openings which utilise geomorphological education. The best hope for the future of geomorphology lies in a general return to and acceptance of curiosity-driven education, learning and research, together with success in convincing our geological colleagues of the rigour, value and relevance of our discipline.

The Scientific Nature of Geomorphology: Proceedings of the 27th Binghamton Symposium in Geomorphology held 27–29 September 1996. Edited by Bruce L. Rhoads and Colin E. Thorn. © 1996 John Wiley & Sons Ltd.

THE ESSENTIAL *MÉNAGE À TROIS*

Geomorphology is, at once, an integral part of both geology and geography. In the opinion of some, and because of its essentially chronological character, it is more closely related to and integrated with the former than the latter. Geology can be regarded as the sum of an infinite number of geographies of past ages, and geography as the geology of the present. But the present is in a sense illusory, for by the time various environmental events have been noted, recorded and analysed, they are part of the past: the recent past, but nevertheless the past. Those events will never be reproduced, for every moment, like every site and every individual, is unique. It is, however, legitimate to generalise, and to group and interpret similar events and features in terms of laws, principles or models. This is the practice in all the sciences, including the supposedly more exact sciences which, despite a determined effort to achieve precision, are also dominated by the idea of approximations.

For geographers, the physical and biological worlds form an integrated whole which forms the backdrop and basis for human activities and which in turn is directly or indirectly modified by those actions. The land surface is a visible part of the physical stage on which biological and human activities take place. As in the theatre, events on stage are the focus of attention, but are profoundly influenced by what goes on backstage, and above and below the boards, in the gridiron and the traps. The land surface is not only an important factor in the complex and spatially varied interrelations that are the essence of geographical study, but also provides evidence of events and processes in the atmosphere above and the Earth beneath. Some geographers are content to take that physico-biological world, of which the form of the land surface is a visible, measurable and, in human terms, comparatively stable foundation, as a passive factor which interacts with others only in terms of responses to natural catastrophic events, such as floods, and in so far as there are reactions to human errors and excesses (e.g. accelerated soil erosion).

Genetic explanation, understanding the land surface, has no part in such anthropocentric, empirical and static views of geography. What does it matter if a surface is Pleistocene or Permian in age? What is important is surely its morphology (relief amplitude, altitude, etc.) and weathering characteristics, and hence soil producing and agricultural potential (e.g. Thrower 1960); though how potential can be predicted without an understanding of the processes involved is not clear. What does it matter whether a volcano is associated with a convergent or a divergent plate junction or a hot spot? What is important is the character of the extruded lava and the character and frequency of eruption. The land surface can be taken as read, without reference to its evolution. Like some practitioners in other disciplines, some geographers are interested in data rather than explanations, in results rather than causes, in the quasi-static present rather than the evolutionary past. They are able, for example, to accept the intrinsic poverty of most Australian soils without linking that condition to the great age of much of the land surface and the implied long periods of leaching.

But other geographers, experienced in the field realities, driven by intellectual curiosity and aware of future possibilities, find questions of origin both interesting and essential to geographical synthesis and understanding. This attitude is of course based in academic interest, and in a conviction of the interrelations of the totality of factors at work at and near the Earth's surface. It is far removed from the cost–benefit accountability presently so much in vogue with those, call them cynics, realists or pragmatists, so many of whom are now involved in the universities, the politics of science and politics, and who, as Oscar Wilde expressed it, know the price of everything but the value of nothing.

Ironically, however, spatial distribution is an essential research tool in many of the natural sciences, but especially in multi-factor studies such as geography and ecology. Geomorphology must also be included in this category. Mapping is a common and obvious way of establishing and investigating distributional patterns. Yet this procedure makes a genetic approach mandatory. In order to produce a map of natural complexes, paradoxically, and if only from a purely pragmatic standpoint, interrelationships must be understood. Every site cannot be investigated, so that having, for instance, established bedrock–landform–soils–vegetation relationships at one site, it is necessary to extrapolate, and then check, at others. This is the basis both of single factor and complex, of detailed and regional, mapping, as exemplified by the well-known and successful broad-scale landscape and land-use mapping procedures adopted by the Commonwealth Scientific and Industrial Research Organisation (CSIRO) Land Research Division, and of detailed soils maps produced by the CSIRO Division of Soils (see e.g. Taylor and Hooper 1938; Christian 1952; Butler 1979). In order to extrapolate with any confidence, it is necessary at least to begin to understand the interrelations of various factors. The rule also applies in geology, for though 'walking the outcrop' is still desirable, time frequently does not permit this laudable practice. Photogeology is increasingly the basis of mapping, calling for genetic linking of form and foundation, of the visible surface and the hidden sub-surface.

Bauer (1995, also Chapter 16, this volume) points out that because the discipline has interfaces with most of the other natural or applied sciences, there are many logical and useful types of geomorphology. Thus, a subdiscipline of engineering geomorphology, for example, could well develop; indeed, some of the more fundamental geomorphological work – on drainage network densities – presently being carried out in Australia is due to civil engineers (see e.g. Willgoose et al. 1991). Engineering is, in the 'civil' sense, per-force concerned with the behaviour of materials, an exercise which is invaluable in the interpretation of landscape, as well as in other areas. One has only to recall the con-tributions of the likes of Bagnold, Leopold and Langbein, Terzaghi and Skempton, to appreciate the distinguished contributions in matters germane to geomorphology of col-leagues from engineering, or with engineering backgrounds. And the introduction of con-cepts and approaches from other of the natural sciences can also only be beneficial, as for instance in the study of rock glaciers (Wahrhaftig and Cox 1959; Wahrhaftig 1987).

Yet, it is not astigmatic but merely realistic, to perceive geomorphological investigations in a temporal framework, and to recognise links with the past and the constraints on our interpretations of the present imposed by the time element (Baker and Twidale 1991). Geomorphology, like geology and geography, is not only a natural science but an Earth science. We are concerned with a planet almost 5 billion years old and many of our landscapes, regional and local, mega and micro, have their origins in the distant past. Thus, though the Lochiel Landslip, located on the Bumbunga Range, about 115 km north of Adelaide, developed in gently dipping Proterozoic quartzites on 9 August 1974 (Twidale 1986), its origins can be traced back some 700–1000 million years. Its devel-opment was, in an immediate sense, due to heavy winter rains, but ultimately and critically to thin lenses of hydrophilic clays which are interbedded with the quartzite, and which provide the lubrication that allowed the quartzite beds to slide downdip over one another. Thus, its origins can be traced to the muds deposited in shallow pools formed behind the beaches of the region in later Proterozoic times. All landscapes are in some measure relic and palimpsest, and to understand the present, account must needs be taken of the geographies of past eras. Similarly, many familiar landforms and landscapes have their origins in magmatic, tectonic or thermal events, or in climatic, weathering, erosional or depositional episodes of the more or less distant past (see e.g. Twidale and Vidal Romani 1994a; Twidale 1994). Whether concerned with the significance of accelerated soil erosion, or the implications of plate motions, a chronological approach imposes an essential perspective on contemporary processes and events that is beneficial not only for professional Earth scientists but also for informed laypersons.

The multifaceted character of geomorphology implies collaboration with colleagues from other, cognate, disciplines (which links are best arrived at not by formal arrangements or regulations, but informally, through friendship and mutual respect and interests). Such cooperative efforts have the potential not only to introduce new concepts, knowledge, skills and perspectives to the study of landforms, but also to bring new concepts and perspectives to those cognate disciplines. Collaboration implies mutual benefits, symbiosis rather than parasitism.

Geomorphology has links with many cognate disciplines, but is at present philosophically and methodologically closest to geology and geography (environmental studies is taken as essentially geography with an applied bent; titles like 'Department of Geography and Environmental Studies' are surely tautological and opportunistic?). Just as where cricket is played is linked to British settlement, where geomorphology resided has varied according to tradition, tempered by personal interest. Thus in the European and British worlds academic geomorphology has been linked administratively with geography. There have been notable exceptions in men like Cotton in New Zealand, Hills in Australia and King in South Africa, Lagasquie in France, each a distinguished geomorphologist based in geology. In the United States, on the other hand, both academic and professional geologists have long been prominent in advancing the discipline, though again there are many exceptions, with geographical geomorphologists responsible for signal advances; and quantitatively the balance may well have changed, for geographical geomorphology is to the fore in the USA, especially in the west. Elsewhere the arrangement varies according to background and historical accident. In Spain, for example, individual interest seems to override administrative base, with geomorphological work flowing from both geologists and geographers; which is as it ought to be.

All three disciplines are retrospective, and in this they are perhaps more overt in their backward look than most; for all sciences are in some degree retrospective rather than predictive, in the sense that they are quite good at explaining what has happened, or what might have happened, in the recent or distant past, but falter when anticipating events. For instance, geophysicists, perhaps the most numerate (and therefore exact?) of solid Earth scientists (in contradistinction to our colleagues whose main concerns are with fluid geophysics, with the oceans and the atmosphere) confidently predicted the depth at which a borehole would intersect the Mohorovičić discontinuity beneath the Kola Peninsula, but were proved in error (Kirjuchin and Hetzer 1989; Kazansky 1992).

In theory there ought to be a constant exchange of data and ideas between geomorphologists on the one hand and geologists and geographers on the other. In practice such reciprocal exchanges do not always and everywhere take place. There are many possible reasons for this, but in part it is due to the failure of geomorphologists to capture the attention and interest of their colleagues; there are many exceptions, past and present, but many, perhaps most, of our earth science colleagues remain unaware or unconvinced of the interest and potential significance of geomorphological data and concepts. For whatever reason or reasons, many see geomorphology as a separate and distinct discipline, certainly with historical and practical links with geography and geology, but not essentially either geology or geography.

But all three earth science disciplines considered here are integrational as well as analytical. They are derivative in that they draw heavily for basic concepts, data and techniques on a wide range of cognate natural sciences; and indeed many major advances in the geosciences, as for example in physical dating, have followed from basic discoveries elsewhere. All three disciplines claim distinction for their synthetic, holistic or rounded views that add, and in some instances provide contradictory views, to the narrower reductionist interpretations of physics and chemistry, anthropology and economics; in a sense, geology, geomorphology and geography are testing grounds in the real world for the findings of the systematic disciplines. All three suffer in the eyes of our colleagues from being intrinsically generalist. In addition, as can also be said of both geography and geology, and because of their all-embracing character, geomorphology offers a construction of the mind, as well as a discipline concerned with specific phenomena; for the three Earth sciences endeavour to understand features and events together in their spatial and temporal contexts.

The three have much else in common, but, for the sake of economy, and only for this reason, and *pro tempore*, in this discussion geomorphology, and geomorphologists, are treated as separate and distinct from geology and geography, and from geologists and geographers. Geographical links are considered by Bauer (Chapter 16, this volume). Here the relations of geomorphology and geology are discussed. Obviously this is a personal statement. Essentially factual data form the basis of much of the discussion, as for example in the following section, but it is equally clear that impressions, subjective opinions and speculations loom large in the discussions of the final section. Such is the nature of this particular beast.

LINKS WITH GEOLOGY

Geomorphology is concerned with the evolution of the Earth's surface. Traditionally it has been based in field investigations, supported by various laboratory techniques the nature of which have varied through time; fundamentally, however, geomorphologists have depended on field observations and reasoning therefrom: look and think. Very different interpretations have frequently flowed from the same data, reflecting not so much flawed methodologies as the ingenuity of the human mind. Many of the most enduring of geomorphological discoveries have been based in and stimulated by observations and analyses in the field, where paradoxes and anomalies not only generate ideas and arouse the imagination, but also where those ideas can be tested against reality.

Geomorphological data and concepts have many applications in geology. Some are academic, others practical or applied, and yet others concerned with testing ideas or creating a receptive intellectual climate. The academic contributions of geomorphology to geology can be considered at three levels. First, the land surface provides the budding geologist with his or her initial impression of the Earth in all its variety – continents and oceans, mountains and plains, ice sheets and deserts, volcanic craters and limestone caves. Even to begin to understand the reasons for such variations is to broach basic questions concerning the explanation of past events in terms of present processes, the extent to which subsurface structure finds expression in morphology, and the degree to which past events are reflected in the contemporary landscape. These first impressions are important, for they have the potential to open new vistas for students; yet they are frequently ignored or mistakenly taken as read by modern specialists.

Second, even for an experienced geologist, an understanding of the land surface provides invaluable clues to subsurface structure and to past events. For instance, a knowledge of contemporary processes is essential to any understanding of ancient deposits, whether sedimentary or volcanic. Again, an awareness of forms and controversies is invaluable, so that palaeoplanation surfaces of contrasted morphologies do not imply different climatic environments, and palaeopediments do not necessarily imply aridity or semi-aridity. The study of features resulting from modern earthquakes permits better interpretation of palaeoseismic forms and stress fields. The present is the key to the past.

The greater the degree of understanding, the broader the background brought to the study of landforms, the greater is the potential for germane criticism of ideas and the development of new interpretations. As Pasteur pointed out, chance favours the prepared mind. Certainly, the best field men I have known and worked with, people like Hills and Öpik, Wopfner and Jennings, Wahrhaftig and Hutton, each had (and in one instance happily still has) a marvellous 'eye for country', reflecting an eclectic view of the Earth sciences, and a holistic view of the planet. Hills for example made signal contributions to virtually all aspects of geological science, being perhaps best known for his work in structural geology and physiography, yet achieving his Fellowship of the Royal Society of London for his work on fossil fish. And many others, like Gilbert and Gilluly, Rubey and Rodgers, have demonstrated that breadth of interest need not be incompatible with profundity of expertise. Thus, Rubey was able not only to speculate rigorously and imaginatively on the origin of the oceans but also to make critical contributions to the understanding of river activity. These men were generalists whose broad interests and intellectual capacities implied not superficiality, but an enviably coherent view of the Earth.

Third, geology is concerned to understand and interpret the history of the Earth. Historical geology or stratigraphy *is* geology and is concerned with an infinite number of palaeogeographies, superimposed one upon the other to produce the present palimpsest landscape. Yet most commonly stratigraphers restrict themselves to the sedimentary, igneous and metamorphic records and neglect evidence from the erosional side of the coin and concepts derived from such investigations. They are the poorer for such self-imposed restrictions, for to ignore or neglect the evidence manifested in the land surface is deliberately to overlook a major source of information and ideas.

Some Examples of Academic and Applied Interactions

Though geomorphologists are in some countries employed in environmental capacities, the generalist and non-particular nature of our interests and approaches reduces the opportunities for employment. Thus in many countries flood hazards and coastal protection are the provinces of the engineer rather than the fluvial or coastal geomorphologist. Geomorphologists can provide an unequalled general setting of a particular problem, but, in most instances, not specific data and solutions. Particular areas of interest such as palaeosurfaces have been explained so clearly that if and when they become of economic importance (see below) they are readily understood and become part of 'proper' geology!

Surfaces and time

The morphology of the Earth's surface reflects past events, and most of the processes responsible for denudation in the past are still active, so that an understanding of the genesis of contemporary landforms offers evidence germane to the interpretation of past events. Indeed, in some areas, as for instance much of southern Africa, the Yilgarn Craton and the Hamersley Ranges of Western Australia, and the Gawler Ranges in South Australia, the land surface provides much of the evidence of events and conditions in these regions over vast periods of geological time. Such cratonic regions have been stable and unaffected by marine transgressions over most of Phanerozoic time, so that it is the correlation of planation surfaces and associated weathering and fluvial forms with valley deposits and the sedimentary sequences of adjacent basins, plus the chronology of fluvial and aeolian forms and sequences, that allows a chronology of events to be determined (see e.g. Jutson 1914; van der Graaff et al. 1977; Twidale et al. 1985; Campbell and Twidale 1991). It is for good reason that Lester King stated that '... the great plains and pla-teaux ... record in a relatively simple manner the geomorphological history of the continents' (King 1950, p. 101).

In addition to Earth history, however, several geomorphological concepts impinge on geological interpretation. For example, the links between the erosional and depositional records are not always clear, though as Kennedy (1962) has suggested, there ought to be direct correlations between volume and type of sediments contributed to depositional basins and the character of the source areas on the one hand and the erosional style – the relationship of uplift, stream dissection and wasting – on the other.

The paradox and problems presented by contrasted rates of geomorphological activity evidenced in the landscape, pose intriguing intellectual dilemmas with possible impacts, for example in basin studies and in planning. Some areas change quickly, but others are evidently stable, or essentially so, over periods of scores, or even hundreds, of millions of years (e.g. Twidale 1976a, 1994; Twidale and Vidal Romani 1994b). How stable is stable, and how frequently, both spatially and temporally, can rapid change be anticipated, for instance in the coastal setting and in areas prone to mass movements?

Unconformities denote significant stratigraphic events, and are also of interest to geomorphologists, for they are widely re-exposed as exhumed surfaces and forms (see e.g. Falconer 1911; Ambrose 1964; Twidale 1994). Their characteristics vary according to the nature of the preserving event. Regoliths rarely survive marine transgressions, lacustrine or fluvial burial, or glaciation, but they are preserved by aeolian or volcanic deposits. Even fragile forms like barchans are preserved beneath lava flows (e.g. Almeida 1953), thus permitting the reconstruction of palaeowind directions. Many erstwhile unconformities are exposed as exhumed forms and surfaces, particularly marginal to cratons. Exhumed surfaces can be confused both conceptually and in the field with two-stage or etch forms, and it is critical to stratigraphic interpretation to distinguish between the two, for one implies a hiatus during which there was burial and re-exposure, the other the formation and stripping of a regolith. Two-stage, or etch, forms have long been recognised (e.g. Hassenfratz 1791; Falconer 1911), and they carry several additional implications. For example, they have two ages, one relating to the period of initiation, the other to exposure (e.g. Twidale 1990). Also, they are azonal, for though different processes may well be involved, they are determined by conditions in the regolith, not in great measure by atmospheric climate.

In some areas, the nature and age of surfaces have economic implications. Thus, many years ago (Jack 1931; see also Wopfner 1964) it was shown that the planation surface eroded in Proterozoic and Palaeozoic strata west of the Eromanga Basin in northern South Australia is exhumed from beneath a cover of late Jurassic and early Cretaceous strata. The former extent of that cover, i.e. of the present exhumed surface, has now become of economic significance, for opal appears to be associated with the base of the Mesozoic sequence. New opal fields, like that at Mintabie, are now being sought and developed on Precambrian and Palaeozoic terranes from which the previous Mesozoic cover has been stripped.

Weathering and the regolith

The regolith is regarded by some geologists as a nuisance concealing the solid or 'real' geology, but investigations of the weathered mantle, including duricrusts and alluvial deposits, have proved rewarding in the search for gold, diamonds and nickel (e.g. Whiting and Bowen 1976; Marshall 1988; Clarke 1994). Also, processes active in the regolith result not only in continuing landform development but also in engineering hazards. Thus, during the planning of the Alice Springs–Darwin railway link in the early 1980s numerous dolines or sinkholes developed in the laterite of the Sturt Plateau, in the northern, monsoonal, Northern Territory, were located. They are still developing and it is of critical importance to determine their origin so that a least-risk route, avoiding fracture swarms and palaeovalleys, both of which sites are conducive to the silica solution responsible for the features, can be found over the Plateau (Twidale 1987a).

Nature of landforms

The nature of a landform or surface, whether erosional or depositional, is important when considering underground water supplies and oil and gas potential. Thus, whether a piedmont fan feature is an erosional pediment underlain by bedrock, with at most a few metres of regolithic cover, or an alluvial feature underlain by a thick wedge of sediments, is of some practical significance, as well as being of academic interest. The two morphologically similar features can be distinguished using various surficial criteria, such as remote sensing imagery, drainage density, and catchment characteristics (Bourne 1996; Bourne and Twidale 1996).

Fractures

The relationship between structure and surface expression is all important in photogeology and reconnaissance exploration. Many of the relationships are obvious, others subtle. As Hills' work on lineaments demonstrated (e.g. Hills 1961), many of them are fractures and the surface expression provides important clues not only to subsurface structure but also to targeting deeply buried potential mineral deposits, as for example the Olympic Dam deposits in the arid interior of South Australia (O'Driscoll 1986; Woodall 1994). Again, though at a vastly different scale, the Norseman gold discoveries of the years immediately preceding and following the Second World War were greatly assisted by Cloos's (1931) experiments on deformation and the development of sets of shears (Campbell 1990). Other examples are detailed by Heidecker, Moore and Campbell in the *Festschrift* dedicated to Hills (Le Maitre 1989).

Fracture patterns have a profound influence on various facets of landform development at scales ranging from continental to site (see e.g. Hills 1961; Zernitz 1931) and can be related to the morphology of continents, sedimentary basins, massifs and bornhardts, as well as occurrence, shape and alignment of such features as boulders, flared slopes, rock basins and gutters. Yet, commonplace though they are, fractures in general are poorly understood and none more so than those usually referred to as offloading or pressure release joints. That this terminology stands is a measure of the confidence geologists have in the implied explanation. And it is seemingly logical and persuasive (Gilbert 1904). Yet there is more than one logic, and various geologists and geomorphologists (e.g. Merrill 1897; Twidale 1964) have over the years drawn attention to anomalies between the alleged expansive and tensional environment implied in Gilbert's hypothesis and the common, indeed characteristic, occurrence of sheet fractures in bornhardts.

Many of the latter are of the same lithology as that in which the adjacent plains are developed. The residuals are most likely preserved because the rock compartments on which they are developed, are in compression, causing fractures to be suppressed and thus scarce or absent; in contrast to the high fracture densities of adjacent compartments which are thus vulnerable to groundwater penetration, to weathering and to erosion (Twidale 1982a, b). Holzhausen (1989) has shown experimentally that stress trajectories in a compressed partly confined block describe a convex upwards pattern. Various other lines of morphological and structural evidence, as well as general argument, can be brought to bear on the question at the very least calling into question the universal validity of the offloading hypothesis (e.g. Vidal Romani et al. 1995; Twidale et al. 1996), and highlighting possible stress conditions that are of practical importance to engineers and planners (e.g. Twidale and Sved 1978; Wallach et al. 1993). Thus studies of palaeoseismicity, fracture patterns and landforms, and their implications for plate tectonics during both the present migrations and former cycles (e.g. Nance et al. 1988), have considerable practical applications. Also, the tendency of fractures to adjust to landsurfaces, the plane of least principal stress, means that the geometry of palaeosurfaces is of practical interest to engineers engaged in deep excavations.

Again, investigations of fracture patterns and densities at the surface and at depth in relation to the storage of liquid nuclear waste (Blès 1986) have revealed that surface patterns provide a reliable indication of pattern at depth. Incidentally and inadvertently they also strengthened the interpretation of inselbergs based in variations in fracture density (e.g. Mennell 1904; Linton 1955) by allowing reliable extrapolation of surface patterns to higher compartments that have been eliminated by erosion (Twidale 1987b).

Drainage patterns

Fractures also have a marked influence on drainage patterns, which thus provide invaluable clues to rock type and structure (e.g. Zernitz 1931). The patterns developed by river systems are, however, frequently misunderstood, even by astute field geologists, as for example the significance of entrenched meanders (e.g. Campana 1958). Structurally anomalous drainage patterns are likewise commonly misconstrued (e.g. Madigan 1931). Some transverse drainage can be referred to catastrophic events such as diversion by faulting or folding, volcanism, ice sheets or glaciers. Some, and in particular antecedence, inheritance and superimposition, carry considerable implications for stratigraphic history (see e.g. Marr 1906; Harris 1939; Cotton 1948, p. 56; Lees 1955; Bowler and Harford 1966). But others reflect the basic locational stability of rivers. Rivers are prime examples not only of the impacts of unequal activity (Crickmay 1932, 1976) but also of positive feedback or reinforcement mechanisms; for, once established, their growth implies the gathering of more and more water, both surface and subsurface, and the increasing and enhanced dominance of the river as a master element in the regional pattern. Thus, many drainage anomalies reflect the deep erosion by master streams of fold structures, the geometry of which changes with depth: they can be attributed to stream persistence and valley impression (Oberlander 1965; Twidale 1966, 1972).

Drainage patterns and their evolution are significant in the search for minerals such as diamonds, uranium and gold, and the investigation of palaeochannels can be similarly rewarding (e.g. van der Graaff et al. 1977; Clarke 1994). Thus the Yeelarrie uranium discovery of the late 1960s and early 1970s was based on the realisation that the ore was concentrated in a buried but extensive palaeodrainage channel (Cameron 1991), and a similar genesis is advocated for the uranium deposits of the Narlaby Channel, on northern Eyre Peninsula (Bourne et al. 1974; Binks and Hooper 1984).

Verification

Geomorphological field investigations have brought to light problems of interest to other scientists and have also devised means of testing theories and problems posed by others. For example, during routine morphological mapping of the Beda Valley, southern Arcoona Plateau, South Australia, silcrete was located in two settings, valley floor and scarp foot. The former occurred in sheets, the latter as skins rich in titanium oxide (anatase), and the chemistry and mineralogy of the two differed. This led not only to the development of various new ideas on the origin of silcretes, but also raised the question of the solubility of titanium in natural conditions (Hutton et al. 1972, 1978; Milnes and Hutton 1974; Hutton 1977). The conventional chemical wisdom is that crystalline silica is of relatively low solubility in the ambient temperatures and chemical environments found at and near the Earth's surface (e.g. Krauskopf 1956), but its occurrence in river waters (e.g. Davis 1964), the widespread development of siliceous speleothems (e.g. Vidal Romani and Vilaplana 1984), and the obvious dissolution of silcrete and siliceous skins demonstrate that in some natural conditions silica is soluble. The solution of opaline forms of the mineral resulting from hydrolysis of quartz is especially important and, as in many other weathering processes, the presence of bacteria may be critical.

In a series of papers Hunter and Rubin (e.g. Rubin and Hunter 1985) have highlighted an apparent anomaly between past and present deserts. In the stratigraphic record there are thick sequences of sands evidently deposited in desert dunes of the longitudinal (linear, seif) type, which also dominate modern dunefields. At present, in some deserts at any rate, the dunes are separate entities resting on genetically unrelated substrates (see e.g. Wopfner and Twidale 1967, 1988; Mabbutt and Sullivan 1968). In order to produce sequences like those preserved in the stratigraphic record it is necessary to postulate that linear dunes migrate laterally, but in central Australia the field evidence, as well as a consideration of reinforcement principles, suggests that they do not (e.g. Nanson et al. 1992). Other explanations must be sought for the origin of the ancient dune deposits. For example, did the thick dune sequences form in actively and rapidly subsiding basins?

Geomorphological concepts are also useful in either testing or applying concepts devised by scientists in other areas. For instance, it has been suggested that the rate of cosmogenic nuclide accumulation in surface exposures can provide absolute dates for those surfaces (Lal 1991). Some early results are at odds with either stratigraphy (e.g. Phillips et al. 1990) or dates produced by other physical methods (e.g. Wells et al. 1990, 1992). The reasons may lie in the retention rates of the nuclides, which may reflect the weathering and permeability of the host rock, or the date of exposure of the bedrock surface, i.e. whether the surface is of etch origin or whether it has had a soil cover or covers since its essential initiation. For these reasons, the stepped inselbergs of north-western Eyre Peninsula and other regions (Twidale and Bourne 1975; Twidale 1982a, c) are particularly suitable for testing the cosmogenic nuclide method, for the steps separating the treads are flared slopes which are sufficiently steep to have been devoid of a regolithic cover since their exposure. The treads on the other hand, though altitudinally and temporally distinct and now mostly devoid of soil, may have carried a cover in the past and dates from them could well be anomalous or misleading. In addition, rates of erosion can be misleading, for there is no indication of any variation through time. Also, geomorphological theory suggests that most surfaces are diachronic or palimpsest, so that any dates obtained for a particular section would most likely be misleading. On the other hand the method is suitable for dating palaeoseismic and other catastrophic events.

Intellectual climate

Scientific advance comes through the conception and development of outrageous ideas, and it is essential that, just as a free society, as defined by Adlai Stevenson, is one in which it is safe to be unpopular, so unconventional explanations must not be ridiculed and automatically ruled out of court. The value of a principle is the number of things it explains, and many useful concepts have begun life in controversy and extreme scepticism, if not disrepute. If the data and argument following from an idea seem valid, and the cause involved is worth while, tenacity is justified: but tenacity and enthusiasm, not dogmatism. There is surely call for an intellectual climate which allows for rational consideration of concepts alien to the conventional wisdom, for reasoned debate rather than entrenched confrontation.

Thus, the suggestion that some landscape elements are of great antiquity conditions the mind at least to contemplate the possibility of a great age for seemingly vulnerable, and hence, by implication, youthful, deposits and surfaces. To take a simple example: on emerging from the Mt Lofty Ranges, the River Torrens near Adelaide has deposited a train of boulders, cobbles and gravels. These deposits remain essentially unlithified. Moreover, during the Miocene the immediately adjacent area to the south was a marine embayment. The location and character of these sediments suggest a youthful, later Cainozoic, and possibly Pleistocene age. The coarse, fluviatile deposits can, on another interpretation, however, be traced laterally into fossiliferous Eocene paludal or lacustrine deposits, in which case the sediments and the ancient Torrens to which they are genetically related are also Eocene. This is consistent with a widely, though not unanimously, held view of the regional geology and geomorphology, involving planation and weathering of the upland followed by block faulting beginning in the Eocene (for review and earlier references see Twidale 1976b).

Once possible antiquity is allowed, evidence can be taken at face value rather than circumvented or rejected out of hand. For instance, it may seem impossible for coastal dunes, even dunes protected by calcrete, to survive at least 20 million years, yet this is the conclusion dictated by the field evidence for the Ooldea and Barton ranges, in the arid interior of South Australia (Benbow 1990).

While not neglecting detail, it is also important to retain some perspective concerning global scale features and problems. Megageomorphology, which implies a global view of landscape patterns and evolution, will in due course become recognised as an essential adjunct to the study of many aspects of geology. I have in mind the behaviour of planation surfaces and their implications for plate tectonics, for isostatic principles, and for models of landscape evolution additional to those already suggested (e.g. Davis 1899; King 1942, 1953; Hack 1960; Crickmay 1974, 1976; Twidale 1991). Consideration of the nature and limitations of river erosion is also germane to models of landscape evolution (e.g. Crickmay 1976; Twidale 1991) as well as for the duration and ultimate significance of accelerated soil erosion. The significance of two-stage development remains under-estimated. The relationships between weathering and erosion, and in particular the erosion of regoliths, with its implications for the character of basin deposits (e.g. Nahon and Trompette 1982), have still fully to be appreciated. The solar system cannot be ignored in the quest for understanding the Earth, for it provides indispensable pictures of the planet as it was billions of years ago, as well as providing examples of landscapes in extreme aridity and processes operating at scales rarely, if ever, attained on this planet (see e.g. Baker et al. 1983; Wilhelm 1987).

Thus, geomorphology benefits enormously from geological mapping and analysis, and from the application of geological concepts. But geomorphology has in turn contributed much to geology through its role in mapping, in its analyses of the erosional as well as depositional aspects of earth history, of the reconstruction of subsurface conditions from surface morphology, of the meaning of planation surfaces, and by process studies in terms of the present providing the key to the past. Many of its contributions are even of applied significance; they are 'relevant'! Like the biological, medical and engineering sciences, all of them concerned with various aspects of nature in all its complexities, geomorphology is in considerable measure derivative. Geomorphologists have, like others, taken ideas and tested, developed and enhanced them in the field, in their real-world settings. They have been able to place them in temporal context, for instance to analyse the significance of long-period cataclysmic events as compared to the gradual effects of various processes (e.g. Wolman and Miller 1960; Baker 1973).

THE PROBLEM: THE ORPHAN ANNIE SYNDROME

Geomorphologists have offered and delivered data and concepts of obvious academic and practical value to geologists. Yet, despite this, there are problems. Geomorphology can at once be construed as the study from which both geology and geography emerged and also as derived from both disciplines. Following this second analogy, both putative parents frequently disclaim their offspring with the result that in places, and from time to time, geomorphology is orphaned. And despite Oliver Twist being regarded by some as being fortunate to live in such an exciting orphanage, the geomorphologist's lot, like the policeman's, is not an 'appy one.

The problem can be examined in two contexts. First, geomorphology is regarded by some academic colleagues as lacking rigour. It is, like geology and geography, disparagingly regarded as a 'soft' or derivative science in contradistinction to such 'hard' or basic disciplines as chemistry and physics. Geomorphologists are perceived by their peers in the systematic natural sciences as derivative generalists, concerned with the collection and coordination of data derived from the systematic sciences and in a series of spatial and temporal contexts. These critics are either unaware of, or totally misunderstand, the intellectual effort represented by, say, a geomorphological map (and producing a legend!). They regard fieldwork as an excuse for a holiday. They have no comprehension either of the difficulties and problems that arise in correlating raw data in space and in the immensity of time, or of the implications for the systematic sciences of anomalies that come to light in the complex real world.

Second, though geomorphology does not lack definition, identity, or interest and support, both in the academic world and generally, its usefulness is perceived, and in the main with good reason, to be academic and general. In hard economic times, like those many countries and universities are now experiencing (though governmental priorities also come into play), geomorphology is seen by some, in terms of cost–benefits, i.e. numbers and employment opportunities, as academically and economically non-viable, and as warranting only low priority. Geomorphology is no longer a formal or required component of several distinguished geology schools. In the United States the rejection has been most pronounced in the Ivy League institutions, but several other eminent geology schools do not now offer formal courses in geomorphology.

Geographers, too, can find geomorphology superfluous to their needs and interests. This has happened in some departments in the United States, and in Australia the distinguished biogeographical and geomorphological group at the Australian National University, once graced by such as Donald Walker, Joe Jennings and Jim Bowler, has now been subsumed in the Department of Archaeology, with the rider that any geomorphological research shall be related to matters human and thus be germane to the overall archaeological functions of the institution. In my own university, geomorphology was initially, and for historical reasons, based in geography. Then, in the 1970s, the geographers decided that geomorphology, and other physical disciplines, were not part of mainstream geography, and geomorphology was transferred to and delivered out of geology (Bowie 1982). In the early 1990s, geography found a renewed interest in things physical, including geomorphology, while geology, in financially pressing times, developed higher priorities. On such whimsies is the fate of geomorphology – and geomorphologists – determined.

That poor performance is not the reason for geomorphology being regarded as dispensable is demonstrated, beyond any possible doubt, by the distinction of many of our American colleagues who, once retired or otherwise out of sight and mind, have not been replaced; giving an innovative twist to the meaning of 'irreplaceable' (R.W. Young, personal communication, August 1995)! Some geologists still perceive geomorphology as not quite scientific; as more akin to natural history, a term used pejoratively of what are seen to be amateurish pursuits which are unsuitable and inappropriate to a rigorous intellectual discipline (shades of Newton, White, von Humboldt and Darwin!).

At present, the acceptance of geomorphology as a discipline varies directly with the economic well-being of academia and with the desirability and perceived viability of cultural–educational pursuits. Given ready funding, geomorphologists will be tolerated in the geological household, but it takes a geologist of eclectic views to welcome the discipline in hard times. Though, like George Eliot, I personally do not desire a future that will break ties with the past, and think it impractical and imprudent to do so, it might be asked why geomorphology does not simply ignore and abandon its past, and, to resume an earlier metaphor, set up house on its own, instead of cohabiting, frequently in some discomfort, with either geology or geography? Could geomorphology stand independently as a bridge, as distinct from a link, between not only geology and geography but also between several other disciplines, a concept voiced by Dusty Ritter (personal communication, November 1995)? This is an attractive idea, one that could readily be sustained in academic terms, and one I had entertained before Dusty revived my interest in its possibilities. But I fear I must still, reluctantly, rule it out of court as desirable but impractical.

A comparison may be drawn with another branch of Earth science that is of mixed parentage, has from time to time found itself understandably confused as to its place and purpose, and has occasionally asserted its independence and gone its own way, namely palaeontology or paleobiology. Because of its commercial connections, palaeontology has a stronger funding base than does geomorphology. On the other hand, separation from other geological sciences robs palaeontology of its essential stratigraphic background, as well as depriving geology *sensu lato* of essential inputs to and critical aspects of Earth history.

Many of the same arguments apply to geomorphology, though they are perhaps less obvious and less pressing. Quite apart from the problem of numbers, nowadays so crucial in university thinking and funding, geomorphology shares interests with many other disciplines in the natural and social sciences; a large 'gene pool' of ideas and concepts is advantageous, if not essential. Of these cognate disciplines, geology and geography are historically and intellectually the closest. Geomorphology has obvious and genuine links to environmental studies, but the latter is basically applied and in such a context, and in the worst-case scenario, basic science tends to be neglected and even to be subordinated to political and economic requirements. Thus, though in theory an independent geomorphology has much in its favour, it is a Utopian dream, for the realities of academic life argue against such an arrangement.

Geomorphologists are concerned to explain the present landscape. Yet, taking a coldly rational viewpoint, and thus ignoring the factors of tradition and personal relationships (either of which can have an overriding local influence), some favour geology as the chosen partner, first because of the chronological imperative, but also, and from a purely practical point of view, because many necessary or desirable facilities are common to geology and geomorphology. Though it does not sound either spectacular or obviously useful, geomorphology's main claim to a secure home in geology or geography, or both, rests on its being concerned with an integral part of Earth history, and with providing perspectives on time, space and process. These are vital, yet tenuous and in many instances intangible, links; but, then, many of our intellectual, cultural, historical and ethical underpinnings and beliefs are similarly tenuous. They are no less real for being intangible, though they are vulnerable to being shed, at least temporarily, and if convenient or advantageous to do so. A chronological and spatial perspective is to geologists as air is to humans; essential, critical, but taken for granted, until it either runs short or is threateningly polluted. This is not to suggest that geomorphology has a monopoly on understanding Earth history – far from it – but geomorphologists have made significant contributions. They have, perhaps, been too successful in explaining their thinking in so far as complex ideas have been presented in language understandable by all, in contrast to the jargon-laden obscurities of others. If to be incomprehensible is to be intellectually respectable, most geomorphologists have failed.

The past decades have been dominated by specialists, many trained in physics or chemistry and applying their knowledge and expertise to geological and geomorphological problems. The various methods of physical dating of rocks come readily to mind. Their contributions have resulted in enormous advances in knowledge of the Earth. But the need for a multidisciplinary attack is demonstrated by the team approach to research that is nowadays so commonplace. Even with such arrangements, however, there is still a need for a broader perspective, not only because of the multiplicity and variety of causative factors, but also arising from the need to see those complexities and factors in the context of an entity. Broader and more generalist integrative perspectives are or ought to be an integral part of scientific inquiry. It is to be hoped (and I am optimistic – super-optimistic according to Dusty! – that common sense will eventually prevail) that current intellectual and academic climates will change. My optimism is based in the likelihood that sooner or later it will be accepted, even by politicians, that long-term investment in research and education is essential to any developed and civilised (they are not necessarily the same thing) nation's well-being.

Many of the problems afflicting geomorphology, and other basically academic, curiosity-driven disciplines like classical studies and philosophy, stem from the present intellectual, or rather anti-intellectual, climate. In particular the attitudes born and nurtured in political correctness (which is so much in vogue but which has perverted language and logic, thought and justice, and which is anti-intellectual and anti-common sense: see e.g. Bloom 1987; Howard 1994; Hughes 1994), with the emphasis on immediate relevance and cost–benefits, must surely soon be seen for what they are, and abandoned. Political correctness is at once responsible for the diminution and humiliation of such prestigious research institutions as the United States Geological Survey (USGS) and Australia's CSIRO, and for the utter debasement of the university ethos. The politically correct find it convenient not to appreciate either the need for long-term investment in science and education, or the need for an educated, as opposed to a trained, electorate; or is it, to paraphrase Voltaire, that once the electorate becomes capable of critical analysis and thought, all is lost? Of the earth sciences, geomorphology has suffered as much as any from this philistine attitude. Appreciation of the wide-ranging perspective, a return to scholarship, learning and education, and the acceptance of curiosity as a valid reason for inquiry, can only benefit geomorphology. Eventually, necessity will compel a reversion to sound educational practices and to a recognition of integrity as an integral requirement of academic life.

The structural foundation of landscape studies is so all-pervasive and obvious that it has been taken for granted and neglected. Similarly, the role of geomorphology in the totality of geology was so obvious to, and accepted by, the great figures of the past that there was never any need to justify it – until the present. And it is, despite the growing appreciation of the role of geomorphological investigations in mineral exploration, a point of view difficult to justify to an accountant. We must persevere with a holistic and non-reductionist view of the evolution of the Earth's surface. The wide-ranging general outlook is our strength. As Eiseley (1961, p. 91), put it, for James Hutton: 'A landscape is not a given thing, shaped once and forgotten, but rather a page from a continuing biography of the planet.' Whether notionally geologists, geographers or geomorphologists, and presuming that we seek to understand, we surely ought to read *all* the pages, not just that which appeals to our sectional and vested interests.

ACKNOWLEDGEMENTS

I wish to thank Liz Campbell and Jennie Bourne for useful comments on an early draft of this chapter. As is apparent, Dusty Ritter, in his role as reviewer, contributed much to the end product by judicious questioning and by making me reconsider some views I was hesitant to embrace. This is, however, a personal statement, and the views expressed are my responsibility.

REFERENCES

Almeida, F.F.M. 1953. Botucatú, a Triassic desert of South America, in *Proceedings, XIX International Geological Congress, Algiers, 1952*, **VII**, 9–24.
Ambrose, J.W. 1964. Exhumed palaeoplains of the Precambrian Shield of North America, *American Journal of Science*, **262**, 817–857.
Baker, V. R. 1973. Palaeohydrology and sedimentology of Lake Missoula flooding in eastern Washington, *Geological Society of America Special Paper*, **144**, 79 pp.
Baker, V. R. (On behalf of Mars Channel Working Group) 1983. Channels and valleys on Mars, *Geological Society of America Bulletin* **94**, 1035–1054.
Baker, V. R. and Twidale, C.R. 1991. The reenchantment of geomorphology, *Geomorphology*, **4**, 73–100.
Bauer, B.O. 1995. Geography's contribution to geomorphology: a sense of breadth and awareness, *Preliminary Abstracts for the 1996 Binghamton Symposium in Geomorphology*.
Benbow, M.C. 1990. Tertiary coastal dunes of the Eucla Basin, Australia, *Geomorphology*, **3**, 9–29.
Binks, P.J. and Hooper, G.J. 1984. Uranium in Tertiary palaeochannels 'West Coast' area, South Australia, *Proceedings of the Australasian Institute of Mining and Metallurgy*, **289**, 271–275.

Blès, J.L. 1986. Fracturation profonde des massifs rocheuses granitiques, *Documents du Bureau de Recherches Géologiques et Minières*, **120**. 316 pp.

Bloom, A. 1987. *The Closing of the American Mind*, Penguin, London, 392 pp.

Bourne, J.A. 1996. Landform development and stream behaviour in the piedmont zone of the Flinders Ranges of South Australia, unpublished Ph.D. thesis, University of Adelaide, Adelaide.

Bourne, J.A. and Twidale, C.R. 1996. Recognition and relationship of alluvial fans and pediments in the western piedmont of the Flinders Ranges, South Australia, in preparation.

Bourne, J.A., Twidale, C.R. and Smith, D.M. 1974. The Corrobinnie Depression, Eyre Peninsula, South Australia, *Transactions of the Royal Society of South Australia*, **98**, 139–152.

Bowie, J.H. (chairman) 1982. *Review of the Department of Geology and Mineralogy and of the Department of Economic Geology*, University of Adelaide, Adelaide, 54 pp.

Bowler, J.M. and Harford, L.B. 1966. Quaternary tectonics and the evolution of the riverine plain near Echuca, Victoria, *Journal of the Geological Society of Australia*, **13**, 339–354.

Butler, B.E. 1979. A soil survey of the horticultural soils in the Murrumbidgee Irrigation Areas, New South Wales (a revised edition of Bulletin No. 118 (1938) by J.K. Taylor and P.D. Hooper), *CSIRO Bulletin*, **289**, 80 pp.

Cameron, E. 1991. The Yeelirrie uranium deposit, Western Australia, *Case Histories of Mineral Deposits*, **3**, 223–231.

Campana, B. 1958. The Flinders Ranges, in *Geology of South Australia*, edited by M.F. Glaessner and L.W. Parkin, Melbourne University Press/Geological Society of Australia, Melbourne, pp. 28–45.

Campbell, E.M. and Twidale, C.R. 1991. The evolution of bornhardts in silicic volcanic rocks, Gawler Ranges, South Australia, *Australian Journal of Earth Sciences*, **38**, 79–93.

Campbell, J.D. 1990. Hidden gold: the Central Norseman story. An account of structural geological studies and ore search at Norseman, *Australian Institute of Mining and Metallurgy Monograph*, **6**, 150 pp.

Christian, C.S. 1952. Regional land surveys, *Journal of the Australian Institute of Agricultural Science*, **18**, 140–146.

Clarke, J.D.A. 1994. Evolution of the Lefroy and Cowan palaeondrainage channels, Western Australia, *Australian Journal of Earth Sciences*, **41**, 55–68.

Cloos, H. 1931. Zur experimentellen Tektonik. Brüche und Faltung, *Naturwissenschaften*, **19**, 242–247.

Cotton, C.A. 1948. *Landscape*, Whitcombe and Tombs, Wellington, 509 pp.

Crickmay, C.H. 1932. The significance of the physiography of the Cypress Hills, *Canadian Field Naturalist*, **46**, 185–186.

Crickmay, C.H. 1974. *The Work of the River*, Macmillan, London, 271 pp.

Crickmay, C.H. 1976. The hypothesis of unequal activity, in *Theories of Landform Development*, edited by W.M. Melhorn and R.C. Flemal, SUNY, Binghamton, pp. 103–109.

Davis, S.N. 1964. Silica in streams and groundwater, *American Journal of Science*, **262**, 870–891.

Davis, W.M. 1899. The geographical cycle, *Geographical Journal*, **14**, 481–504.

Eiseley, L. 1961. *Darwin's Century*, Anchor Books, Garden City, New York, 378 pp.

Falconer, J.D. 1911. *The Geology and Geography of Northern Nigeria*, Macmillan, London, 295 pp.

Gilbert, G.K. 1904. Domes and dome structures of the High Sierra, *Geological Society of America Bulletin*, **15**, 29–36.

Hack, J.T. 1960. Interpretation of erosional topography in humid temperate regions, *American Journal of Science*, **238A**, 80–97.

Harris, W.J. 1939. Physiography of the Echuca district, *Proceedings of the Royal Society of Victoria*, **51**, 45–60.

Hassenfratz, J.-H. 1791. Sur l'arrangement de plusieurs gros blocs de différentes pierres que l'on observe dans les montagnes, *Annales de Chimie*, **11**, 95–107.

Hills, E.S. 1961. Morphotectonics and the geomorphological sciences with special reference to Australia (15th William Smith Lecture), *Quarterly Journal of the Geological Society of London*, **117**, 77–89.

Holzhausen, G.R. 1989. Origin of sheet structure. Morphology and boundary conditions, *Engineering Geology*, **27**, 225–278.

Howard, P.K. 1994. *The Death of Common Sense*, Random House, New York, 202 pp.

Hughes, R. 1994. *Culture of Complaint*, Harvill, London, 177 pp.

Hutton, J.T. 1977. Titanium and zirconium minerals, in *Minerals in Soil Environments*, edited by J.B. Dixon and S.B. Weed, Soil Science Society of America, Madison, Wisconsin, pp. 673–688.

Hutton, J.T., Twidale, C.R. and Milnes, A.R. 1978. Characteristics and origin of some Australian silcretes, in *Silcrete in Australia*, edited by T. Langford-Smith, University of New England Press, Armidale, pp. 19–30.

Hutton, J.T., Twidale, C.R, Milnes, A.R. and Rosser, H. 1972. Composition and genesis of silcretes and silcrete skins from the Beda Valley, southern Arcoona Plateau, South Australia, *Journal of the Geological Society of Australia*, **19**, 31–39.

Jack, R.L. 1931. Report on the geology of the region north and northwest of Tarcoola, *Geological Survey of South Australia Bulletin*, **5**, 31 pp.

Jutson, J.T. 1914. An outline of the physiographical geology (physiography) of Western Australia *Geological Survey of Western Australia Bulletin*, **61**, 229 pp.

Kazansky, V.I. 1992. Deep structure and metallogeny of Early Proterozoic mobile belts in light of superdeep drilling in Russia, *Precambrian Research*, **58**, 289–303.

Kennedy, W.Q. 1962. Some theoretical factors in geomorphological analysis, *Geological Magazine*, **99**, 305–312.

King, L.C. 1942. *South African Scenery*, Oliver and Boyd, Edinburgh, 308 pp.

King, L.C. 1950. A study of the world's plainlands, *Quarterly Journal of the Geological Society of London*, **106**, 101–131.

King, L.C. 1953. Canons of landscape evolution, *Geological Society of America Bulletin*, **64**, 721–752.

Kirjuchin, L.G. and Hetzer, H. 1989. Nationale Programme zum Niederbringen übertiefer Bohrungen und ihre ersten Ergebnisse, *Zeitschrift für Angewandte Geologie*, **35**, 325–332.

Krauskopf, K.B. 1956. Dissolution and precipitation of silica at low temperatures, *Geochimica et Cosmochimica Acta*, **10**, 1–26.

Lal, D. 1991. Cosmic ray labeling of erosion surfaces: in situ nuclide production rates and erosion models, *Earth and Planetary Science Letters*, **104**, 424–439.

Lees, G.M. 1955. Recent earth movements in the Middle East, *Geologische Rundschau*, **42**, 221–226.

Le Maitre, R.W. (ed.) 1989. *Pathways in Geology. Essays in Honour of Edwin Sherbon Hills*, Hills Memorial Volume Committee/Blackwell, Melbourne, 463 pp.

Linton, D.L. 1955. The problem of tors, *Geographical Journal*, **121**, 470–487.

Mabbutt, J.A. and Sullivan, M.E. 1968. The formation of longitudinal sand dunes: evidence from the Simpson Desert, *Australian Geographer*, **10**, 483–487.

Madigan, C.T. 1931. The physiography of the western MacDonnell Ranges, central Australia, *Geographical Journal*, **78**, 417–433.

Marr, J.E. 1906. The influence of the geological structure of English Lakeland upon its present features – a study in physiography, *Quarterly Journal of the Geological Society of London*, **58**, 207–222.

Marshall, T. 1988. The diamondiferous gravel deposits of the Bamboesspruit, southwestern Transvaal, South Africa, in *Geomorphological Studies in southern Africa*, edited by G.F. Dardis and B.P. Moon, Balkema, Rotterdam, pp. 495–505.

Mennell, F.P. 1904. Some aspects of the Matopos. 1. Geological and physical features, *Proceedings of the Rhodesian Science Association*, **4**, 72–76.

Merrill, G.P. 1897. *A Treatise on Rocks, Weathering and Soils*, Macmillan, New York, 411 pp.

Milnes, A.R. and Hutton, J.T. 1974. The nature of micro-cryptocrystalline titania in 'silcrete' skins from the Beda Hill area of South Australia, *Search*, **5**, 153–154.

Nahon, D. and Trompette, R. 1982. Origin of siltstones: glacial grinding versus weathering, *Sedimentology*, **29**, 25–35.

Nance, R.D., Worsley, T.R. and Moody, J.B. 1988. The supercontinent cycle, *Scientific American*, **July**, 44–51.

Nanson, G.C., Chen, X.Y. and Price, D.M. 1992. Lateral migration, thermoluminscence chronology and colour variation of longitudinal dunes near Birdsville, Simpson Desert, central Australia, *Earth Surface Processes and Landforms*, **17**, 807–819.

Oberlander, T. 1965. The Zagros streams, *Syracuse Geographical Series*, **1**, 168 pp.

O'Driscoll, E.S.T. 1986. Observations of the lineament–ore relation, *Philosophical Transactions of Royal Society of London*, **317A**, 195–218.

Phillips, F.M., Zreda, M.G., Smith, S.S., Elmore, D., Kubik, P.W. and Sharma, P. 1990. Cosmogenic chlorine-36 chronology for glacial deposits at Bloody Canyon, eastern Sierra Nevada, *Science*, **248**, 1529–1532.

Rubin, D.M. and Hunter, R.E. 1985. Why deposits of longitudinal dunes are rarely recognised in the geologic record, *Sedimentology*, **32**, 147–157.

Taylor, J.K. and Hooper, P.D. 1938. Soil survey of the horticultural soils in the Murrumbidgee irrigation areas, NSW, *Council for Industrial Research Australia Bulletin*, **118**, 108 pp.

Thrower, N.J.W. 1960. Cyprus – a landform study, *Annals of the Association of American Geographers*, **50**, 84.

Twidale, C.R. 1964. Contribution to the general theory of domed inselbergs. Conclusions derived from observations in South Australia, *Transactions and Papers of the Institute of British Geographers*, **34**, 91–113.

Twidale, C.R. 1966. Chronology of denudation in the southern Flinders Ranges, South Australia, *Transactions of the Royal Society of South Australia*, **90**, 3–28.

Twidale, C.R. 1972. The neglected third dimension, *Zeitschrift für Geomorphologie*, **16**, 283–300.

Twidale, C.R. 1976a. On the survival of palaeoforms, *American Journal of Science*, **276**, 1138–1176.

Twidale, C.R. 1976b. Geomorphological evolution, in *Natural History of the Adelaide Region*, edited by C.R. Twidale, M.J. Tyler and B.P. Webb, Royal Society of South Australia, Adelaide, pp. 43–59.

Twidale, C.R. 1982a. *Granite Landforms*, Elsevier, Amsterdam, 372 pp.

Twidale, C.R. 1982b. The evolution of bornhardts, *American Scientist*, **70**, 268–276.

Twidale, C.R. 1982c. Les inselbergs à gradins et leur signification: l'exemple de l'Australie, *Annales de Géographie*, **91**, 657–678.

Twidale, C.R. 1986. The Lochiel Landslip, South Australia, *Australian Geographer*, **17**, 35–39.

Twidale, C.R. 1987a. Sinkholes (dolines) in lateritised sediments, western Sturt Plateau, Northern Territory, Australia, *Geomorphology*, **1**, 33–52.

Twidale, C.R. 1987b. Review of J.L. Blès' 'Fracturation profonde ... ' (1986), *Progress in Physical Geography*, **11**, 464.

Twidale, C.R. 1990. The origin and implications of some erosional landforms, *Journal of Geology*, **98**, 343–364.

Twidale, C.R. 1991. A model of landscape evolution involving increased and increasing relief amplitude, *Zeitschrift für Geomorphologie*, **35**, 85–109.

Twidale, C.R. 1994. Gondwanan (Late Jurassic and Cretaceous) palaeosurfaces of the Australian Craton, *Palaeogeography, Palaeoclimatology, Palaeoecology*, **112**, 157–186.

Twidale, C.R. and Bourne, J.A. 1975. Episodic exposure of inselbergs, *Geological Society of America Bulletin*, **86**, 1473–1481.

Twidale, C.R., Horwitz, R.C. and Campbell, E.M. 1985. Hamersley landscapes of Western Australia, *Revue de Géographie Physique et Géologie Dynamique*, **26**, 173–186.

Twidale, C.R. and Sved, G. 1978. Minor granite landforms associated with the release of compressive stress, *Australian Geographical Studies*, **16**, 161–174.

Twidale, C.R. and Vidal Romani, J.R. 1994a. On the multistage development of etch forms, *Geomorphology*, **11**, 107–124.

Twidale, C.R. and Vidal Romani, J.R. 1994b. The Pangaean inheritance, *Cuadernos do Laboratorio Xeolóxico de Laxe*, **19**, 7–36.

Twidale, C.R., Vidal Romani, J.R., Campbell, E.M. and Centeno, J.D. 1996. Sheet fractures: response to erosional offloading or tectonic stress? *Zeitschrift für Geomorphologie*, in press.

Van der Graaff, W.J.E., Crowe, R.W.A., Bunting, J.A. and Jackson, M.J. 1977. Relict Early Cainozoic drainages in arid Western Australia, *Zeitschrift für Geomorphologie*, **21**, 379–400.

Vidal Romani, J.R., Twidale, C.R., Campbell, E.M. and Centeno, J.D. 1995. Pruebas morfologicas y estructurales sobre el origen de las fracturas de descamacion, *Cuadernos do Laboratorio Xeolóxico de Laxe*, **20**, in press.

Vidal Romani, J.R. and Vilaplana, J.M. 1984. Datos preliminares para le estudio espeleotemas en cavidas graniticos, *Cuadernos Laboratorio Xeolóxico de Laxe*, **7**, 305–324.

Wahrhaftig, C. 1987. Foreword, in *Rock Glaciers*, edited by J.R. Giardino, J.F. Schroder and J.D. Vitek, Allen and Unwin, Boston, pp. vii–xii.

Wahrhaftig, C. and Cox, A. 1959. Rock glaciers in the Alaska Range, *Geological Society of America Bulletin*, **70**, 383–436.

Wallach, J.L., Arsalan, A.H., McFall, G.H., Bowlby, J.R., Pearce, M. and McKay, D.A. 1993. Pop-ups as geological indicators of earthquake-prone areas in intraplate eastern North America, in *Neotectonics: Recent Advances*, edited by L.A. Owen, I. Stewart, and C. Vita Finza *Quaternary Proceedings*, **3**, 67–83.

Wells, S.G., Crowe, B.M. and McFadden, L.D. 1992. Measuring the age of the Lathrop Wells volcanic center at Yucca Mountain, *Science*, **257**, 555–558.

Wells, S.G., McFadden, L.D., Renault, C.E. and Crowe, B.M. 1990. Geomorphic assessment of late Quaternary volcanism in the Yucca Mountain area, southern Nevada: implications for the proposed high-level radioactive waste repository, *Geology*, **18**, 549–553.

Whiting, R.G. and Bowen, K.G. 1976. Gold, in *Geology of Victoria*, edited by J.G. Douglas and J.A. Ferguson, *Geological Society of Australia Special Publication*, **5**, 434–451.

Wilhelm, D. 1987. The geologic history of the Moon, *United States Geological Survey Professional Paper*, **1348**, 302 pp.

Willgoose, G., Bras, R.L. and Rodriguez-Iturbe, I. 1991. Results from a new model of river basin evolution, *Earth Surface Processes and Landforms*, **16**, 237–254.

Wolman, M.G. and Miller, J.P. 1960. Magnitude and frequency of forces in geomorphic processes, *Journal of Geology*, **68**, 54–72.

Woodall, R. 1994. Empiricism and concept in successful mineral exploration, *Australian Journal of Earth Sciences*, **41**, 1–10.

Wopfner, H. 1964. Permian–Jurassic history of the western Great Artesian basin, *Transactions of the Royal Society of South Australia*, **88**, 117–128.

Wopfner, H. and Twidale, C.R. 1967. Geomorphological history of the Lake Eyre Basin, in *Landform Studies from Australia and New Guinea*, edited by J.N. Jennings and J.A. Mabbutt, Australian National University Press, Canberra, pp. 118–143.

Wopfner, H. and Twidale, C.R. 1988. Formation and age of desert dunes in the Lake Eyre depocentres in central Australia, *Geologische Rundschau*, **77**, 815–834.

Zernitz, E.R. 1931. Drainage patterns and their significance, *Journal of Geology*, **40**, 498–521.

16 Geomorphology, Geography, and Science

Bernard O. Bauer

Department of Geography, University of Southern California

ABSTRACT

The centrality and importance of geography to the disciplinary development of geomorphology has been undervalued historically because: (1) geography's intellectual core is not easily identified nor circumscribed; (2) geography's establishment as an academically distinct discipline in North America did not occur until the late nineteenth century; and (3) geography's scientific foundations are not widely appreciated. The relationship between geology and geomorphology, in contrast, is often portrayed as being more substantive and intimate. Nevertheless, the intellectual roots of geography, geology, and geomorphology are closely intertwined and traceable to common Greek origins. Only the institutional infrastructures that have evolved to support these contemporary academic disciplines are distinct and separate.

Geomorphology's contemporary academic status as a subdiscipline of both geography and geology has often been viewed as detrimental, and several geographers and geologists over the last 75 years have advocated academic realignment that favors one or the other. This has resulted in needless caricaturing that serves an injustice upon geographical and geological practitioners, past and present. The situation is especially unfortunate and potentially damaging when the host disciplines are portrayed as having only singular, intrinsic methodologies, and by logical necessity, unique and unifying philosophies. In the case of geography, an interdisciplinary discipline that borrows from several affiliated physical, life, and social sciences (and occasionally the arts and humanities), myriad methodologies and philosophies are employed, encouraged, and critically challenged in attempts to provide integrating and synthesizing perspectives on human–environment interrelations. Geography's most important contribution to geomorphology may well be the breadth and diversity it brings to geomorphological thinking. Such breadth is manifest

The Scientific Nature of Geomorphology: Proceedings of the 27th Binghamton Symposium in Geomorphology held 27–29 September 1996. Edited by Bruce L. Rhoads and Colin E. Thorn. © 1996 John Wiley & Sons Ltd.

in: (1) an increasing number of subspecializations in geomorphology beyond the traditional cores; (2) an increasing number of methodological and philosophical perspectives
being brought to bear on geomorphic problems; (3) an increasing concern with the
integrity of geomorphological claims to knowledge, especially those that assume scientific
postures; and (4) an increasing appreciation for the necessity of utalitarian research,
especially in the face of inexorable alteration of earth's surface by the profound activity of
humans, and for the express purpose of ensuring disciplinary survival by demonstrating
contemporary relevance. In these contexts, geography serves to inform and heighten
geomorphology's awareness of the physical, intellectual, and social pulses of the world
around us.

OBJECTIVES AND CAVEATS

The original objective of this chapter was to provide an evaluation and elucidation of 'the
ways in which geographic theory and methods have influenced or are currently influencing the development of geomorphology as a science' (Rhoads and Thorn, personal
communication). Such an undertaking turns out to be unrealistic for several reasons. First,
it accepts, a priori, the existence of theories and methods that are distinctly geographical,
their readily identifiable character, their acceptance and use by the geographic community,
and their transplantation into the geomorphological corpus. Many have taken exception to
such assertions, and Schaefer (1953, p. 227), for one, contends that the '... existence of a
field ... needs no "methodological" justification'. Yatsu (1992, p. 92) concurs and suggests that 'for the development of scientific knowledge, researchers must use any method
available'. Although certain disciplines might easily be characterized on the basis of
distinctive theories and methods (e.g. mathematics or engineering), this is neither a
necessary nor sufficient condition. Geography, in particular, encompasses a broad spectrum of theories and methods, many of which have evolved in association with
developments in other disciplines, and thus, the donor–recipient relationships are not
evident.

Second, it implicitly assumes that there is widespread agreement as to the meaning and
implications of 'science', that geomorphology is considered to be a science by the broader
community of academics, and that we, as geomorphologists, find it desirable for geomorphology to be(come) scientific. These issues are at the very core of geomorphology,
and one need only scan the recent geomorphological and geographical literature to get a
sense of the prevailing confusion and ambiguity surrounding them (e.g. Richards 1990,
1994; Baker and Twidale 1991; Yatsu 1992; Rhoads and Thorn 1993, 1994; Bassett 1994;
Rhoads 1994). Ontological and epistemological concerns are central to these debates but
are ordinarily the domain of the philosopher or historian of science. Are geomorphologists
prepared to engage these debates or will they reach, once again, for the soil auger (cf.
Chorley 1978)?

Third, it is inherently confrontational because it is tantamount to geographical representation in the 'Championship of the Disciplines' (i.e. disciplines most influential to the
development of geomorphology). In such endeavors it is often convenient and effective to
place in opposition the merits of one discipline against those of another. Supporting
arguments and expositions are often based on extremist, opinionated, historically super-

ficial, and logically nondefensible assertions strengthened by hegemonic posturing. They generally fail to appreciate that disciplines are human institutions that have evolved as convenience structures, and that their intellectual, methodological, theoretical, and practical character and concerns are often inextricably intertwined. Furthermore, viewpoints are often divergent, egos are large, consensus is unlikely, and there is no absolute truth or authority to which one can turn for arbitration or resolution. A 'Championship of the Disciplines' may be good sport, but its deeper purpose is far from clear.

In view of these difficulties, this chapter will not follow a conventional tack that updates the comprehensive works of others who have assessed the important contributions of geographers to geomorphology through inventories of publications, citations, or society memberships (e.g. Graf et al. 1980; Costa and Graf 1984; Graf 1984; Marston 1989). Neither will it assert that a particular geomorphic principle, law, theory, or method has a distinct and uniquely geographical origin – this might be the case, but the supporting arguments would be hard to make. The reader is referred to any of several comprehensive histories of geography and geomorphology to search out such truths (e.g. Hartshorne 1939; Chorley et al. 1964, 1973; Dury 1983; Tinkler 1985; Beckinsale and Chorley 1991; Unwin 1992; James and Martin 1978; Walker and Grabau 1993). Instead, the chapter will argue that:

1. The intellectual and academic roots of geomorphology, geography, and geology are inextricably interwoven, and that it is therefore inappropriate and misleading to characterize key historical figures and events as exclusively 'geographic' or 'geologic' or to suggest that they were seminal in the evolution of geomorphology;
2. Geomorphology has evolved into an academic subdiscipline, despite its long-standing intellectual tradition, and has become practically dependent on its host disciplines (primarily geography and geology) for academic survival;
3. Geomorphology has benefited and will continue to benefit intellectually from the breadth and diversity that geography embraces and fosters;
4. Geomorphology's 'scientific' stance may be difficult to substantiate, and such posturing may not readily admit alternative perspectives of practical and intellectual utility to the discipline.

These conditions have profound implications for the future evolution of geomorphology since they lead to the conclusion that geomorphology stands to benefit from a strategy that advocates integration rather than separation. This is true with respect to both its interdisciplinary associations and its fundamental concerns with scale. The bulk of the chapter is devoted to providing the necessary background leading to this assessment.

A CARICATURE OF GEOGRAPHY

Throughout the history of geography, its practitioners have been variously perceived as very scientific, pseudoscientific, or antiscientific. The range of divergent viewpoints about geography's nature and utility are epitomized by the following quotes:

> Geography is queen of the sciences, parent to chemistry, geology, physics, and biology, parent also to history and economics. Without a clear grounding in the known characteristics of the

earth, the physical sciences are mere game-playing, the social sciences mere ideology (*The Times*, 7 June 1990, p. 13 quoted in Unwin 1992, p. 1).

During my recent stay in northwestern Europe I could not escape the conclusion that the position of geographers generally is not one of high esteem. I found the field criticized sharply on all sides. Most of the criticism related to a tendency for geographers to attempt research in fields they had insufficient background to enter. One critic flatly denied that geography is a field of knowledge at all, for the reason that it offers nothing unique which may be regarded as its own peculiar technique or method. He denied an appeal to cartographic expression as stoutly as he denied the proposition that all things printed in words belong to the field of literature. He claimed that our techniques are really those of the mathematician, historian, economist, demographer, geologist, engineer, or other specialist, according to the demands of the problem under consideration. He denied flatly that geographers have powers of synthesis that differ from those employed in other disciplines, or special license to stray into the domain of others. My abilities in debate were taxed severely at times (Russell 1949, p. 10–11).

The mystery (or ignorance) surrounding geography stems largely from within the discipline – geography has always had difficulty identifying its central concerns and its boundaries. Several authors have made the claim that 'the core of geography is the set of assumptions, concepts, models, and theories that geographers bring to their research and teaching' (Abler et al. 1992, p. 5), yet rarely do these authors provide an explicit listing of these methods and theories. In contrast, Tuan (1991, p. 101) suggests that the central theme of geography is the 'earth as home of Man', and that geographers are unified in their perspective on reality 'which is not so much a conscious program as a temperament or natural disposition' (Tuan 1991, p. 106). As realistic as these assessments may be, they are hardly tangible. As a result, geography has been identified more with its varying contemporary interests and practice than with any enduring, cohesive, and well-delineated subject matter. Whereas the geography of the postwar decades was imagined to involve exploration of exotic lands and cultures, regional syntheses, cartographic expression, and *National Geographic*, the discipline today has been described as multidisciplinary, interdisciplinary, integrative, and even schizophrenic. The recent inclusion of geography as a core subject in the *Goals 2000: Educate America Act* (Public Law 103- 227) coupled with the development of national geography standards (National Geographic Research and Exploration 1994) as a framework within which to achieve these educational goals may provide the nexus for alleviating much of the mystery and ignobility attached to geography.

GEOMORPHOLOGY'S ROOTS: GEOGRAPHY OR GEOLOGY?

Biased Histories of the Early Years (pre-1850)

Several geographers and geologists have asserted that geomorphology falls within the domain of geology (e.g. Sauer 1924, p. 22; Johnson 1929, p. 211; Russell 1949, p. 4; Bryan 1950, p. 198; Dury 1983, p. 92; Tinkler 1985, p. 3). If held in earnest, a myopic perspective such as this is apt to lend more significance to the role of the marble than the role of the sculptor in the production of the Venus de Milo. More often than not, such assertions have been made in pragmatic attempts to seek expedient solutions to inter-disciplinary tensions. Douglas W. Johnson, for example, suggested that,

> geomorphology itself has suffered, and will continue to suffer, from attempts to include it in the geographic realm. In the history of its development, in its methods, and in its affiliations it is a part of geology (Johnson 1929, p. 211).

However, this statement was made in the context of Johnson's 1928 presidential address to the Association of American Geographers through which he was engaging, by necessity, in a disciplinary-wide debate about the circumscription and future of an academically distinct geography. More recently, Worsley (1979) argued for the pragmatic separation of geomorphology from geography and suggested that its placement into any of several geosciences would enhance its utility to society largely because there would be access to better facilities, equipment, well-trained students, and geoscientific respectability. On intellectual grounds, such assertions are difficult to defend, and bold statements about geomorphology's detachment from geography are often followed by convoluted and hidden references to the contrary. Dury (1983, p. 92), for example, suggests that 'if geomorphology should have been located within a single discipline, that discipline should have been geology'. But, he also notes that

> the question of where geomorphology belongs ... is badly structured. Formally, it belongs where practitioners are attached for payroll purposes. Functionally, it belongs on the surface/subsurface interface (Dury 1972, p. 201).

Tinkler (1985) states quite emphatically that '... geomorphology is indisputably a part of geology ...' (p. xii), but then goes on to: (1) acknowledge the existence of institutional affiliations and academic connections between geomorphology and geography (p. xiv) and other cognate disciplines (p. 5); and (2) admit that 'processes of the atmosphere acting on the earth's surface over both the short term and the long term provide the essential catalysts that mediate the geomorphic system' (p. 5), and that an 'intimate relation exists between geomorphology and vegetation, with soil and climate as important mediating agents' (p. 5). Even William Morris Davis, while speaking before the Geological Society of America, professed that 'all geography belongs under geology, since geography is neither more nor less than the geology of today' (Davis 1912, p. 121). Davis's statement seems paradoxical unless one appreciates that: (1) Davis's intellectual allegiances to geography and geology were united; (2) geography had not yet established itself as a full-fledged academic discipline in North America; and (3) Davis's views about the nature of geographical inquiry were largely restricted to physiography but were rapidly evolving with the discipline itself (see Johnson 1929, p. 209, footnote 12). It is to these generally interrelated conditions of geography, geology, and geomorphology that I now turn my attention.

The relationship between geology and geomorphology is admittedly very intimate. During the eighteenth and nineteenth centuries, the dominant concern of geology, both as an intellectual pursuit and an academic enterprise, was understanding the character and evolution of earth's surface (Chorley et al. 1964; Davies 1969; Tinkler 1985). It is note-worthy that the term 'geomorphology' is traceable only as far back as the mid-1800s (Tinker 1985, p. 4), suggesting that there was no practical need to separate the essence of geology from that which was geomorphology – they were one and the same. Geography, on the other hand, had not yet attained prominence as an academic discipline. Consequently, the importance and intimacy of the geography–geomorphology relationship have suffered from a historical transparency. Accounts of geography's development have tended to consider geography's geomorphological concerns, quite erroneously, as short-lived or marginal. This perspective concentrates more on geography's short academic lifespan than its long-standing intellectual traditions. The situation is further complicated by geography's wide-ranging concerns that extend beyond the earth's surface *per se* into the realm of human nature and behavior. The contemporary situation is rather different. Geomorphology has become a central specialization in academic geography, whereas marginalization of geomorphology and geomorphologists on the part of some mainstream geologists is not uncommon. Even as early as 1958, the retiring President of the Geological Society of America observed that 'Quaternary studies [geomorphology] gradually lost an aura of respectability which is attached to "hard-rock" geology' (Russell 1958, p. 1).

Geography recognizes its beginnings in the writings and speculations of the ancient Greeks such as Homer, Thales, Anaximander, Hecataeus, Herodotus, Plato, Aristotle, Alexander the Great, Pytheas, Eratosthenes, and Ptolemy (James and Martin 1981). Early geographical ideas were largely physical-geographic, if not geomorphologic, and they survive because Strabo's writings on geography were found intact. Strabo described the role of the geographer as explaining 'our inhabitated world – its size, shape, and character, and its relations to the earth as a whole' (quoted in James and Martin 1981, p. 36 from a translation by Jones 1917, pp. 429–431). This perspective of earth as consisting of two domains – the habitable and the uninhabitable – can be traced to Strabo's predecessors, Aristotle and Eratosthenes. Aristotle had been concerned with the *ekumene* or inhabited part of the earth (James and Martin 1981, p. 28) which he associated with the temperate zones in the Mediterranean region. Habitability, he suggested, decreased with latitudinal distance toward the equator and toward the poles. Such speculation accorded well with observation, and the remaining task was to explore the reasons for these associations. This laid the foundation for Eratosthenes to coin the term 'geography' and establish its *raison d'être* as the study of earth as the home of man (James and Martin 1981, p. 31). This theme has been retained in varying form and degree through to the present, and Aristotle's concept of the *ekumene* eerily foreshadowed the paradigms of geographical influence and environmental determinism in geography and other related sciences at the turn of the twentieth century. The focus on earth's surface as the object of study by the Greeks is at the same time geographical and geological, and there is little evidence to suggest that these early philosophers contemplated humans as anything more than passive elements on the landscape. Humans as agents of environmental change is a theme not espoused until much later, first by Georges Louis Leclerc, Comte de Buffon in the late 1700s, and then by many geographers in the late 1800s. Geographers ultimately championed this paradigm in the 1950s (e.g. Thomas 1956), and it remains a central concern in most earth sciences and social sciences.

Up until the mid-1800s, most scholars had interdisciplinary backgrounds and concerns – they were naturalists, scientists, and philosophers. Several individuals during this classical period could claim mastery of the sum of accumulated scientific and philosophical knowledge, and it would be inappropriate in most cases to attach a single contemporary disciplinary label to them. The storehouse of knowledge had been growing exponentially, however, and disciplinary specialization became inevitable. It is at this phase of transition from a generalized to a specialized academy (roughly from the late 1700s to the mid-1800s) that most modern accounts of the foundational bases of geomorphology, geography, and geology search for their ancestry. Geographers identify individuals such as Immanuel Kant (1724–1804), Alexander von Humboldt (1769–1859), Carl Ritter (1779–1859), Arnold Guyot (1807–1884), George Perkins Marsh (1801–1882), Daniel Coit Gilman (1831–1908), Ferdinand von Richthofen (1833–1905), and Friedrich Ratzel (1844–1904), among many others, as academic forefathers. Geologists, on the other hand, are more likely to point to James Hutton (1726–1797), John Playfair (1747–1819), Charles Lyell (1797–1875), James Dwight Dana (1813–1895), John Wesley Powell (1834–1902), Clarence E. Dutton (1841–1912), and George M. Wheeler (1842–1905) as foundational figures. Are these reasonable and illuminating choices, and why is there little overlap between the lists?

Not surprisingly, such foreshortened retrospective searches for disciplinary roots end up pointing to only a select few that differ depending on disciplinary orthodoxies. This phenomenon has been recognized elsewhere and has been called the Whig interpretation of history (e.g. Livingstone 1984, p. 271), a process by which disciplinary historians judge the merits of affiliation with certain widely recognized scholars according to the contributions they are believed to have made toward establishing modern theories or paradigms (e.g. Brush 1974, p. 1169). Such retrospective reconstructions of academic lineages 'extend from personal viewpoint and experience to selection of material, the time dependence of ideas, the scientific context at a particular time, and the temptation to suggest that consensus of opinion exists where this may not indeed have been the case' (Gregory 1985, p. 2). An inherent danger to Whig historiography is that it 'looks at the past in terms of present ideas and values, rather than trying to understand the complete context of problems and preconceptions with which the earlier scientist himself had to work' (Brush 1974, p. 1169). It searches for a seed when no such beginning may realistically exist. Most 'founding fathers' are usually little more than symbolic figureheads because, in most cases, they founded neither the contemporary academic structures (e.g. departments, institutions, societies) nor the intellectual heritage of the discipline. Disciplinary histories and heroes so created can be very influential because the superficial logic is easy to grasp and because they are often the perceived essence and legitimation of a discipline or a disciplinary paradigm (see Tinkler 1985, pp. 229–230 or Sack 1992, pp. 258–259 for views about the shift from a 'Davisian' to a process-oriented school of thought; or Herries Davies 1989, pp. 7–10 for a broader geomorphological perspective). These perceived essences, rightly or wrongly, appear to have direct bearing on the future evolution of a discipline (cf. Sherman, Chapter 4 this volume).

An equally viable, if not more realistic, interpretation of disciplinary histories would recognize a common intellectual heritage, much like branches sprouting from the trunk of a tree that is supported by a diffuse root system representing the distant, less visible past. In this way, geographers, geologists, and geomorphologists alike ought to be able to trace their academic branches to Greek roots through a common trunk that spans the post-Renaissance era. However, contemporary academic ties to these distant figures are rather loose, and the notion of a common academic heritage does not sit comfortably with some disciplinary stalwarts. Thus, it is difficult for geographers to look past von Humboldt, or for geologists to see beyond Hutton, into the seventeenth century to discover a common ancestry in scholars such as Nathaneal Carpenter, Bernhard Varenius, Thomas Burnet, John Woodward, Abraham Gottlob Werner, and Horace de Saussure. These individuals, apparently, do not conform neatly with contemporary images of disciplinary practitioners and paradigms – many of their ideas and beliefs seem simplistic if not foolish by today's standards, and history seems to judge them with prejudice (see discussion by Tinkler 1985, pp. 9–12). Physical and human geographers in particular would have difficulty reaching consensus about which of these seventeenth-century scholars was the most geographic. This is, of course, a direct outcome of our Whig interpretation of history (a forward-looking perspective that recognizes that geography and geology did not exist as formal academic disciplines prior to the 1700s avoids this difficulty), but it has obvious implications for the perceived transparency of the historical link between geography and geomorphology.

Turn-of-the-century Developments

The modern period, beginning in the mid-1800s, was an era of increasing academic specialization and professionalization. Geology departments were already widely established by this time – John Woodward (1665– 1728) had endowed and named a chair of geology at Cambridge over one century before (Tinkler 1985, p. 38). But full-fledged departments of geography did not come into existence until the 1870s in Europe and until the turn of the century in North America (James and Martin 1978, p. 3). The relative timing was important to North American geomorphology for three reasons. First, established geographical scholars specializing in geomorphological subject matter had little choice but to affiliate themselves professionally with geology departments. Second, students of geography seeking advanced degrees in North America had similarly few options, and were restricted to geology departments for formal education. In both cases, physical geographers were received openly, collegially, and as peers by their geological colleagues. Third, the academic discipline of geology, which had predominant concerns for the character and evolution of the earth's surface until the 1850s, was fragmenting into specialty areas, such as mining, structural mapping, mineralogy, petrology, paleontology, seismology, and geophysics. These areas of study were gaining increasing prominence in the late 1800s and were beginning to dominate as subdisciplines. As Tinkler (1985, p. 4) describes it,

> ... geology as a subject exploded in much the same way as, for example, biochemistry has exploded in this century. The explosion left geomorphology as a small part of a vast subject and with its emphasis on, or towards, the present it tended to lose touch with a parent subject so committed to exploring the past and unveiling the origin of the earth.

These three conditions were particularly important to geomorphology because they forced an integration of geological and geographical thought. Into the academic world of geology, stepped several scholars (including Guyot, Agassiz, Gilman, and Davis) who had received formal training in European schools where the intellectual development of geography was considerably advanced over its North American counterpart. In this way, German (as well as French and British) ideas about a 'new geography' were introduced into geomorphological thinking in geology. This 'new geography' favored inductive methods based on empirical observations over theoretical deduction, sought interpretations and explanations rather than mere descriptions, and was intensely interested in the interaction of humans with their natural environments. For a brief time (loosely, 1880 to 1910) the geomorphological branches of geography and geology grew together – that is, the substance, method, and theory of geography, as they pertain to geomorphology, were those of geology as well.

Harvard was the first North American university to offer specialized training in physical geography (physiography) within a geology department. The key geographer–geologists at Harvard were Nathaniel Southgate Shaler, who was a student of Louis Agassiz (Agassiz was educated in Switzerland, taught at Harvard as a zoologist, and is noted by geomorphologists for his glacial theories), and William Morris Davis (an 'understudy' of Shaler and frequent visitor to Europe). This is not to say that North American geography relied solely on geology for intellectual stimulation. Indeed, it retained specializations and developed subdisciplinary interests outside of the geomorphological realm. At the Wharton School of Finance and Commerce (University of Pennsylvania), Emory R. Johnson, J. Paul Goode, and J. Russell Smith offered advanced education in economic and transportation geography at the turn of the century. Yale University also had a geographic tradition that extended back to 1786 with the appointment of Jedidiah Morse. Geographic instruction at Yale during the latter part of the nineteenth century focused more on ontographical subject matter, and courses were offered by Daniel Coit Gilman, William H. Brewer, Francis A. Walker, and Herbert E. Gregory. Nevertheless, there was relatively little communication between these three groups, and as far as geomorphology (physiography) was concerned, the geological–geographical union at Harvard and elsewhere was natural and unquestioned.

The activities and contributions of the key geomorphological figures at the turn of this century are good indicators of how intimate the linkage between academic geography and academic geology was in the realm of geomorphological subject matter. It is widely acknowledged that John Wesley Powell (1834–1902), Grove Karl Gilbert (1843–1918), and William Morris Davis (1850–1934) played pivotal roles in the development of North American geomorphology. Powell and Gilbert, in particular, are often pointed to as epitomizations of the geological practitioner of geomorphology (e.g. Baker and Pyne 1978) whereas geographers, in retort, claim Davis as their geomorphological champion:

> For American geological geomorphologists, the most important scientific trinity was not structure, process, and stage, emerging from the heuristic synthesis of a Harvard scholar. Rather the critical trinity was Gilbert, Powell, and Dutton ... (Baker 1988, p. 1157).

Such antithetical caricatures are, of course, a variation of the 'Championship of the Disciplines' and it is important to recognize they are not easily substantiated, nor even historically accurate (cf. Sack 1991, 1992).

William Morris Davis was indeed very much a geographer. He was the primary influence behind organization of the Association of the American Geographers (AAG), he presided over the first meeting of the AAG in Philadelphia in December of 1904, was twice the elected President of the AAG (1905, 1909), he authored *The Geographical Cycle* (Davis 1899) and several other essays on geographical research and teaching, and he was a professor of physical geography who was devoted to geographical perspectives in his teachings and his research. But, he also had a classic scientific and engineering training, worked briefly for a coal mining company, was employed as a summer field geologist by the United States Geological Survey, taught in a geology department, and served as President of the Geological Society of America (GSA). Gilbert, in contrast, was employed as a geologist throughout his professional life, wrote several geological monographs whilst employed by the United States Geological Survey, is revered for his conception and application of the scientific method, and was founding member and twice GSA President. Nevertheless, he was also a founding member of the National Geographic Society, acting President of the National Geographic Society (1904), a founding member and President of the AAG (1908), and considered himself to be a geographer as well as a geologist (Sack 1992, p. 252). Further, he coauthored physical geography textbooks with Albert Perry Brigham, one of Davis's earliest and most influential graduate students. Similarly, John Wesley Powell is best known for his explorations of the canyons of the Green and Colorado rivers under the auspices of the early Geographical, Topographical, and Geological Surveys, and he served as Director of the United States Geological Survey. Yet, he too was a founding member of the National Geographic Society and published extensively on geographical topics including several reports on Indian cultures, changing settlement patterns in the arid lands, and physiographic provinces of the United States. Powell even asserted that 'Sound geological research is based on geography. Without a good topographic map geology cannot even be thoroughly studied ...' (Powell 1885, quoted in James and Martin 1978, p. 5). James and Martin (1981, p. 160) contend that Powell might have made considerably more contributions to geographical scholarship had he not encountered official resistance to his work from people in positions of political and financial power intent on selling and developing land 'sight unseen' in the arid west.

    One could play the game of ascribing relative merit to the many accomplishments of Davis, Gilbert, and Powell, and then tallying the scorecard to see which side of the geography–geology line they fall. This serves little purpose and belittles the profound contributions that these individuals have made to both disciplines. Indeed, many of their secondary and tertiary contributions have had far greater impact on these disciplines than most researchers' primary contributions. After all, we do not refer to Leonardo da Vinci as just a painter.

    The intertwined nature of the disciplinary roots of geomorphology in geography and geology are most apparent in the early histories of the professional societies formed at the beginning of the twentieth century. The birth of the AAG in 1904 is especially revealing. Davis was eager to form an organization such as the AAG because he recognized that American geography could not come of age until a professional society existed in which true geographical scholarship was the main criterion for membership. Although other organizations such as the American Geographical Society and the National Geographic Society were already well established, they catered mostly to philanthropists, explorers, and geographically inclined aristocrats. There was as yet no forum through which academic geographers could speak to each other and to geographically informed audiences about scholarship at the forefront of the discipline. The AAG was to serve this purpose. If one looks at the characteristics of the original 48 members, however, one finds that 19 held positions as geologists, and 15 of these were Davis's past students. In fact, the membership criteria were so heavily skewed that admission in the first year was denied to

J. Russell Smith (a student of Ratzel at Munich, assistant to Emory R. Johnson at the Wharton School, and an economic-transportation geographer by contemporary standards) on the grounds that Smith had not been adequately trained in physical geography (James and Martin 1978, p. 36)!

Despite Davis's best intentions to provide opportunities for scholars with varied academic training to participate and interact within the structure of the AAG, the initial years maintained a strong physical-geographic, if not geologic, presence. Of the 22 papers presented at the first annual meeting in 1904, 13 were on topics that could be considered to be physical geography and an additional 4 on biogeography. In 1908, H.E. Gregory, Chair of the Geology Department at Yale, expressed his concern about the large proportion of geologists in the fledgling society:

> ... I am becoming exercised over the fact that each year the official staff of the Association consists chiefly of men who, to my mind, are geographers only by a stretch of that term .... If the organization is large, I see no reason why geologists with geographical leanings would not be enrolled as members; but I think that only rarely should they occupy positions as officers ...
> ... And would it not be wise for me and Fenneman and certain others who are pretty clearly geologists to resign from this organization, so as to make the cleavage between geology and geography even more distinct? (Gregory letter to Brigham 1912, quoted in James and Martin 1978, p. 47, cit. 7).

It is not unreasonable to conclude, then, that from a historical perspective, geomorphology's roots were, at the same time, geographical and geological. Distinguishing the contributions of geography from those of geology is an artificial and meaningless exercise because:

1. Contemporary disciplinary definitions and demarcations do not apply to past eras;
2. Geography did not exist as a separate academic discipline until the late nineteenth century;
3. Geology was the only academic discipline that had geomorphological concerns and it was dominated by them; and yet,
4. A strong intellectual geographic tradition is recognizable and traceable through academic geology from about the turn of the twentieth century back to the early Greeks.

GEOMORPHOLOGICAL GROWING PAINS

Academic Geography Comes of Age

Geography's intellectual and practical contributions to geomorphology during the twentieth century are commonly perceived as weak. In part, this condition can be ascribed to geology holding a privileged academic position over geography among the scientific disciplines. As noted in the previous section, geology evolved at a time when knowledge was expanding and disciplinary boundaries arose naturally – these boundaries were flexible, translucent, and ill defined. Geography, on the other hand, evolved during a period of academic specialization and interdisciplinary competition in which disciplinary boundaries were contemplated consciously and defended vigorously. Debates about the nature and purpose of new disciplines relative to their established peers observed an unspoken formalism, not unlike the unquestioned authority of parental roles within a familial structure. The roles of rival siblings however, require definition and redefinition as the members age and the family evolves. In this way, the status and subject matter of geology have not received scrutiny to the same degree as those of geography, neither from within nor without. Johnson, in the early part of this century, noted that

> ... geology, and certain other sciences took their rise in a day when knowledge was more limited, methods were more crude, and standards were lower. Their youthful errors were less harshly judged by less competent critics than exist today. Geography suffers the penalty of late development in the midst of sciences already advanced to maturity, and in the presence of experienced judges admirably documented in a variety of related fields. Assuredly the test is a severe one, and we must expect for a time to suffer in comparison with our elder associates. But we need not be unduly anxious in respect to this particular difficulty. Youth is a disease which cures itself (Johnson 1929, p. 205).

In the first decades of the twentieth century, geography began to assert itself as a true academic discipline with established university departments and several geographical societies. North American departments of geography were graduating substantial numbers of Ph.D.s, and this provided a new membership pool for the AAG. The physiographic tradition was beginning to be usurped by ontographic concerns, and geologists and physiographers became the minority by the 1920s. This was, at once, an exhilarating and frustrating stage in the evolution of geography – the future direction of the discipline was not at all clear, but the possibilities were manifold. The age of specialization had allowed other disciplines, especially those in the physical and biological sciences, to flourish through a research strategy that purposefully isolated processes under study from the complicated interactions inherent to natural systems. These complexities, however, are the essence of geography. It is not surprising, then, that there was great confusion and uncertainty within the discipline. The search for the answer to 'What is geography?' was leading geographers down many different paths. Nevertheless, they had one common bond in their widespread rejection of things geological. It was recognized by all that academic separation from geology was necessary so that the fledgling discipline could assert itself on its own terms. Even as late as 1939, Hartshorne (1939, p. 29) in his clarion call for a 'regional geography', felt it necessary to reassert that 'Geography is not an infant subject, born out of the womb of American geology a few decades ago, which each new generation of American students may change around at will.'

The need to demarcate geography from geology required a new definition for the discipline, one that established its core and its bounds, and incorporated humans as something more than a casual afterthought. Debate focused not so much on whether humans were to be central to the newly evolving and vulnerable discipline, but in what way humans were to form the nexus for study relative to their environmental platform (James and Martin 1978, p. 51). Some scholars even argued that to make the separation of geography from geology complete, it was necessary to reject intellectually the physical grounding of geography on earth's surface and not to 'cling to the peripheral specialisms to which reference has been made – to physiography, climatology, plant ecology, and animal ecology – but ... relinquish them gladly to geology, meteorology, botany, and zoology, or to careers as independent sciences' (Barrows 1923, p. 4). Such debates about the true (and desired) nature and substance of geography were not always articulate nor assertive (consider, for example, the ineffectiveness of the arguments put forth to avert the elimination of geography at Harvard as described by Smith 1987), but they obviously were necessary. There was some reluctance on the part of the majority of academic geographers to disown geomorphology completely because earth's physical surface was recognized as an essential component of geographical processes. Nevertheless, in the wake of widespread rejection of 'environmental determinism' and 'geographical influences' (e.g. Brigham 1903; Semple 1911), there was negative reaction to the historically privileged position of geomorphology and there was reluctance to admit it to positions of academic and intellectual power within geography. Further, the expansionist era associated with the American frontier and European colonization was coming to a close, and the new explorations were into social rather than physical spaces (Smith 1987, p. 168). Geomorphology was ignored, much like a child left to self-amusement in the midst of parental disputes about family finances. This was not unfamiliar territory for geomorphology – several decades earlier, geologists had 'left [geomorphology] behind, like a hapless rural milkmaid at the pit head, as the miner climbed below' (Tinkler 1985, p. 80). To compound matters, there was a distinct absence of charismatic leaders with novel approaches to the subject, or at least, a reluctance to embrace such figures and their ideas professionally. As a consequence, North American geomorphology faded into the academic background of both geography and geology (Dury 1983). Most geomorphologists retained a Davisian outlook and were silently searching for evolutionary order in the ever-increasing stock of concordant surfaces, denudation chronologies, climatic anomalies, and variants thereof. Exceptions, of course, are many and have been remarked upon extensively (e.g. Tinkler 1985; Chorley et al. 1973; Beckinsale and Chorley 1991; Yatsu 1992).

Human geographers, on the other hand, were engaged in heated and protracted debates about how to interface more closely with the social sciences and the humanities and about what constitutes appropriate geographical subject matter, methods, and theories. Geographers participated in (and defended fervently) three types of activities through the war decades:

1. They returned to their traditional roles involving the sterile, but careful and elaborate, collection, classification, and cartographic representation of worldly data especially pertaining to those places not yet explored.
2. They embarked on holistic studies of particular places as unique and interesting entities.
3. They engaged in generalizing and theorizing about earth-surface processes with the goal of formulating widely applicable laws to recurring events.

Ensconced in these debates were hidden tensions regarding spatial versus temporal studies, idiographic versus nomothetic objectives, scientific versus humanistic methods, and the role of humans as passive versus active agents of change. At various times, then, geographers adopted concerns for human ecology (e.g. Barrows 1923), chorology (e.g. Sauer 1924, 1925), regionalism (e.g. Hartshorne 1939; Finch 1939), historical geography (e.g. Brown 1948; Sauer 1941), geomorphography (Kesseli 1946), antiexceptionalism (e.g. Schaefer 1953), and applied geography (Ackerman 1945).

Ultimately, geography followed the trend of most other earth and social sciences during the 1960s and 1970s and entered into an era dominated by quantification. An unflattering assessment of this trend suggests that it was fueled by a desire to 'look more scientific'. Geographers quickly discovered, however, that a 'scientific geography' was not what everyone thought the discipline should be, and it has since entered into an age of diversity that advocates exploring alternative, if not unusual or radical, approaches to understanding the world (cf. Dear 1988). Some have caricatured it as the age of 'isms' and examples include positivism, humanism, realism, possibilism, feminism, existentialism, scientism, relativism, idealism, materialism, structuralism, and of course, postmodernism.

Finding an Academic Home

What happened to geomorphology in the meantime? The hiatus in geomorphological activity during the war decades facilitated a certain intellectual separation from the past, especially from the geomorphology of Davis (Tinkler 1985, p. 198). New geographical domains were being explored and this led to advances in coastal, karst, aeolian, tropical, periglacial, glacial, island, tectonic, climatic, and soils geomorphology. Intellectually, geomorphology was about to benefit immensely from stimuli derived from innovations external to both geology and geography (e.g. Dury 1983). In addition, geomorphology, which had been neglected, if not orphaned, by geography during the war years, was about to be readopted by geology. Unfortunately, this readoption was more in spirit than in devotion, and it is ironic that it arose as a by-product of geography's continuing debates about degree of attachment to earth's physical landscapes. The strengthening of the academic linkages between geomorphology and geology in North America was therefore not a renaissance in the sense that practicing geologists were reinvigorated with geomorphological spirit. Rather, it was the outcome of the public posturing of two widely respected earth scientists: Richard J. Russell and Kirk Bryan.

Russell and Bryan served as presidents of both the AAG and GSA, positions of considerable authority and influence. In their presidential addresses to the AAG in 1948 and 1949, respectively, both men remarked on the growing interest in geomorphology that was taking place outside of geography (Russell 1949; Bryan 1950). Bryan (1950, p. 197) suggested that the 'revival would warm the heart of Davis and would also yield him many misgivings ... [because] ... there is among the geographers much indifference and even a modicum of hostility'. More importantly, both Russell and Bryan made emphatic claims about geomorphology having its historical roots and intellectual/academic home in geology – the bulk of their addresses were concerned with how geomorphology might provide better service to the geographer. In Russell's subsequent presidential address to the Geological Society of America (Russell 1958), he even went so far as to imply that physiographers following Davis's tradition (meaning 'physical geographers') hampered the development of geomorphology because they lacked the requisite training in geophysics and geology (Russell 1958, p. 2). He then offered a prescription for how geomorphological– geologists might improve on this lamentable condition. Russell and Bryan admitted that there was (or could be) a close relationship between geography and geomorphology, but their vision clearly had geography on the receiving end of geomorphological inquiry – the former had little to contribute to the latter.

In retrospect, it seems that these two presidential addresses, although altruistic in intent, did much violence to the image of the geographer–geomorphologist. The perceived weakness of the historical linkages between geography and geomorphology became officially entrenched in the literature through their musings. Fortunately, some of the more significant advances in twentieth century geomorphological thought were coming not from retrospective introspection about the prescribed subject matter of geography or geology, but from sources that were completely unaware of or impartial to such inter- and intra-disciplinary reconditioning. Strahler (1950, 1952) is to be acknowledged for warning the geomorphological community that much of the relevant and substantive research on erosional and dynamical systems was being conducted by engineers, and that '... few geologists seem aware of this progress and there has been little evidence of geomorphologists adapting the information and methods to landform research' (Strahler 1950, p. 211). Useful innovations were coming not only from engineering, but also from hydraulics, biological systems, hydrology, thermodynamics, and statistical mathematics. The course of developments since the 1950s should be familiar to most geomorphologists, and it includes phrases and concepts such as tectonic and isostatic uplift, timebound and timeless models, reductionism, morphometric analysis, mechanics and dynamics, hydraulic geometry, magnitude and frequency, systems theory, allometry, equifinality, entropy, indeterminacy, equilibrium and thresholds, characteristic forms, process–response suites, and numerical modeling. It is also appropriate to acknowledge the advanced and insightful works of researchers such as O'Brien, Hjulstrom, Rubey, Shields, Bagnold, Leighly, Sundborg, and Yatsu who were somewhat peripheral to the mainstream of geomorphology of their time. Bagnold, for example, is now widely acknowledged for his exploration of the North African deserts and for his many contributions to our understanding of the nature of aeolian dune systems and the mechanics of sediment transport. His academic contributions span a period of almost six decades, beginning with the publication of many seminal works in the 1930s and 1940s, after having retired from a distinguished army career! The Royal Geographical Society of London awarded him the Gold Medal in 1934, the US Academy of Sciences awarded him the G. K. Warren Prize in 1969, whereas the Geological Society of America waited until 1970 to honor him with its Penrose Medal. These awards were followed by the Wollaston Medal from the Geological Society of London in 1971 and the Sorby Medal from the International Association of Sedimentologists in 1978. The geomorphological community, on the other hand, failed to recognize Bagnold's achievements officially until 1981, at which time he was awarded the David Linton Award from the British Geomorphological Research Group.

The evolution of geomorphology during this period was affected not only by intra-disciplinary debates within geography and geology, but also by social, political, and economic climates – that is, the 'internal' and 'external' histories (e.g. Livingstone 1984, p. 271; Yatsu 1992, p. 94). The latter tend to facilitate or hamper developments in certain disciplines both directly and indirectly. The opportunities afforded geology by the boom in petroleum and mineral exploration in the 1960s and 1970s, for example, eventually steered geological interests away from geomorphological subject matter. In a 1971 survey of geomorphological offerings in North American geology departments, White and Malcolm (1972, p. 146) warned that

... the number of departments having dropped or still having not offered geomorphology is surprising. This eventually may prove to be ill-timed, not only for geology as a science but also for man on this planet.

Geography, by contrast, was at this time concerned with its scientific image and became infatuated with quantitative and statistical enumeration. It availed itself easily of geomorphological research, especially the quantitative aspects of hydraulic geometry, morphometry, hypsometry, network analysis, and general systems. A minority of geographers even went so far as to suggest that the description and classification of landscape attributes in precise and nongenetic ways ought to be the primary activity of geomorphologists in geography departments, thereby arguing for a distinctive and mutually exclusive division of geomorphic labor between geography and geology (e.g. Zakrzewska 1967, 1971). Most physical geographers, fortunately, were dissatisfied with such sterile descriptive treatments of the earth's surface and were becoming more concerned with Man–land [sic] interactions and with the internal adjustments and dynamics of physical systems (e.g. Chorley 1962). In this way, the transition to a process paradigm in geomorphology seems to have been facilitated, if not encouraged, by a close association with geography (cf. Robinson 1963; Dury 1983).

It is interesting to note that claims to specific geomorphological heritages within geography or geology have invariably led to contradictory or paradoxical associations in the historical evolution of geomorphic thought. Geomorphological research by geographers is often characterized as being quantitative and process-oriented (e.g. Baker 1988; Baker and Twidale 1991). Yet, the roots of the process paradigm are claimed to be geological and are usually traced along a branch that comprises the activities of various United States Geological Survey employees such as Wolman, Langbein, and Leopold, and ultimately Gilbert using a 50-year graft. In contrast, the strong claim that many Quaternary geologists often make to a historical–genetic perspective or method in their geomorphological research (e.g. Baker and Twidale 1991) 'found its most eloquent expression in the writings of a geographer, William Morris Davis' (Baker 1988, p. 1158). These are somewhat inaccurate and inconsistent portrayals of contemporary thought in these disciplines.

CONTEMPORARY GEOMORPHOLOGY: A TAIL ON THE GEOGRAPHICAL DOG?

Although the intellectual and academic roots of geomorphology are difficult to disentangle, it must be acknowledged that the academic disciplines of geography and geology, by centering themselves over specific intellectual terrain and by setting bounds on their domains, have had a pronounced influence on what geomorphology has become. What it has become is a subdiscipline. James and Martin (1978, p. 7) suggest that four conditions need to be created before a field of learning (i.e. an intellectual discipline) is able to assert itself as a learned profession (i.e. an academic discipline):

1. Interaction of a significant number of scholars with an accepted body of concepts, images, and rules of professional behavior;

2. Establishment of university departments offering advanced (graduate) training in concepts and methods;

3. Creation of opportunities for qualified scholars to find paid employment in their areas of expertise;

4. Establishment of an organization or professional society to serve the interests of the profession and to provide focus for professional activities.

The academic discipline, therefore, encompasses both the subject matter that constitutes the basis for disciplinary inquiry and the formal and informal structures that facilitate and perpetuate that inquiry in academe.

As we have seen, scholars throughout history have speculated about and discussed earth's dynamic surface – the intellectual discipline boasts a long and noble tradition. Employment opportunities for geomorphologists seem to be adequate, if not plentiful. Geomorphologically oriented venues for scholarly exchange and professional interaction have arisen recently, and they include the British Geomorphological Research Group, Guelph Biannual Symposia, Binghamton Annual Symposia, Quaternary Geology and Geomorphology Division of the GSA, Geomorphology Specialty Group of the AAG, American Geomorphological Field Group, Friends of the Pleistocene, International Conference on Geomorphology Series, Catalina Island Workshop Series, and the Canadian Geomorphology Research Group. Nevertheless, geomorphology has not yet demonstrated the centrality, utility, popularity, or novelty that might warrant the widespread establishment of independent academic departments focusing exclusively on geomorphological subject matter, notable exceptions such as the Department of Biogeography and Geomorphology at the Australian National University aside. Geomorphology is not presently a viable academic discipline, and it remains dependent on host disciplines for academic survival.

In this context, terms such as 'geographical geomorphology' and 'geological geomorphology' seem astigmatic and inappropriate because they have the 'tail wagging the dog'. They invert the primacy of the host disciplines' role relative to that of the geomorphological subdiscipline. Further, they ignore the claims that other host disciplines, such as geophysics, oceanography, or engineering, might advance for a specialization in geomorphology – there are potentially many dogs, all with geomorphological tails. And most importantly, terms such as these are inaccurate and potentially damaging because of their implicit association of method or theory with the host discipline. 'Geographical geomorphology' and 'geological geomorphology' imply that it is accurate, indeed reasonable, to portray geography and geology as having only singular, intrinsic methodologies, and by logical necessity, unique and unifying philosophies. As we have seen, these are grossly oversimplified caricatures of geographic and geologic inquiry that serve an injustice on their practitioners, past and present. Terms such as 'geomorphological geography', 'geomorphological geology' (the latter term is credited to Powell by McGee; see Tinkler 1985, p. 4), or 'geographer–geomorphologist' (following the syntax of Robinson 1963, p. 16; Baker 1988, p. 1158; Butzer 1989, p. 48; Johnson 1929, p. 209) are more appropriate, although they seem awkward and perhaps unnecessary in the contemporary academic environment (see Campbell 1928, for views on the importance of logical terminology).

If we acknowledge geomorphology's position as an academic subdiscipline, we must also realize that neither the geomorphological tail nor the disciplinary dogs are particularly large relative to other tails and dogs. Surveys of the status of geomorphology within geography, conducted in the 1980s (Graf et al. 1980; Graf 1984; Costa and Graf 1984; Marston 1989), showed that geomorphologists accounted for only 4.3% (283 members in 1983) of the total AAG membership and 9% (1140 members in 1981) of the total GSA membership, with many of the latter considering themselves to be Quaternary geologists rather than geomorphologists. Recent figures (Graf, personal communication) show that the total number of North American geomorphologists has increased overall, with the relative percentage in the AAG increasing to 6.7% (473 members in 1994), while the relative percentage in the GSA has decreased to 7.5% (1255 members in 1994). These changes are consistent with trends identified in the early 1980s which suggest that geomorphology is increasing in relative importance in geography, but decreasing in relative importance in geology (e.g. White 1982; Costa and Graf 1984). A survey of 17 journals likely to publish articles on geomorphological subject matter (covering the period 1976–86 and listed in the *Science Citation Index* and *Social Science Index*) showed that slightly more than 50% of the contributions were from geographers even though geomorphological-geographers only account for about 22% of the geomorphological community (Marston, 1989). More importantly, only 20% of the articles published in these journals were on geomorphological subject matter, suggesting that our geomorphological 'bark' is rather muted.

Concerns about the status and importance of one's subdiscipline among recognized academic disciplines are, of course, not unique to geomorphology. Klemes (1986, p. 177S) laments the

> unsatisfactory state of hydrology [which] is, in the final analysis, the result of the dichotomy between the theoretical recognition of hydrology as a science in its own right and the practical impossibility of studying it as a primary discipline but only as an appendage of hydraulic engineering, geography, geology, etc. As a consequence, the perspectives of hydrologists tend to be heavily biased in the direction of their nonhydrologic primary disciplines and their hydrologic backgrounds have wide gaps which breed a large variety of misconceptions ... with consequent dangers both to scientific development of hydrology and to its practical utility.

Klemes goes on to present a convincing case for the establishment of separate university departments or units of hydrology. Eagleson (1991) similarly argues for recognition of the hydrological sciences as a distinct geoscience. Echoes of such sentiments are often heard within the geomorphological community. Perhaps the recent discussions about the creation of an American geomorphological society or group (much akin to the British and Canadian examples) to which geographers and geologists alike could pledge allegiance are a first step toward such a separate academic identity for geomorphology. Nevertheless, there may be lessons to be heeded from the hydrology example. Little progress has been achieved in establishing a distinct academic identity partly because of widespread dissension among the ranks of hydrologists who prefer affiliation with their host disciplines (Klemes 1986, p. 177S).

WHERE'S THE SCIENCE?

Geomorphology's status as a subdiscipline forces it to carry the same public personae as its hosts. That geology is a science is not ordinarily open to debate. That geography is a science seems less certain, especially in the contemporary academic climate. In the eyes of the public, geology is a science and geography is about maps, capital cities, and longest rivers. In the eyes of the physicist or chemist, geology is a 'soft' science and geography is about maps, capital cities, and longest rivers. In the eyes of the economist or sociologist, geology is a 'hard' science and geography is sometimes useful. In the eyes of the artist, geology and geography are fascinating. Most geologists, in contrast, consider themselves to be fundamentally scientific. Most geographers also consider themselves to be scientific, although some choose to differentiate between natural and social science, whereas others reject science and all that it stands for. Are all these views accurate, and what are the implications for geomorphology?

Most geomorphologists believe that geomorphology is a science. The title of this symposium presumes this very stance, and Rhoads and Thorn (1993) offer an informative framework within which to examine its rationality. Nevertheless, the difficult work of arguing geomorphology's status as a scientific discipline remains to be done. I doubt whether many geomorphologists have actually contemplated their research within the framework of a scientific method or even interrogated their implicit claims to being scientific. Fortunately, just 'appearing scientific' can be advantageous because the existing intellectual and social structure of academe rewards this position with respectability. Adopting a scientific stance is a matter of convention rather than conscious choice in most disciplines, and it may not reflect actual practice.

Prescriptions for scientific practice have been examined in detail by many philosophers including Kuhn, Popper, Feyerabend, Bhaskar, Lakatos, and Chalmers, and I do not wish to engage these complicated debates further. The reader is referred to Haines-Young and Petch (1986) and Rhoads and Thorn (1994) for cogent summaries of philosophical debates as they relate to geomorphology and physical geography. It is fair to say, however, that out of these debates comes one conclusion: it is extraordinarily difficult to come up with a definition of science that is, at the same time, mutually agreeable to all scientists, exclusionary of nonscientific endeavors, prescriptive of the practice of science, *and* a realistic portrayal of the history of scientific investigation in all disciplines. An example will serve the point well. Several years ago, while arguing the case for inclusion of an introductory physical geography course within the curriculum of the 'Natural World' subcategory of the General Education program, I found myself defending against repeated attacks against a few of the laboratory exercises on the grounds that they were not sufficiently scientific. I questioned the criteria on which these judgments were being made, but none of the panelists were able to provide an informative response. Their weak criticism was that these geography exercises had no 'hands-on' component that required students to be in a laboratory setting using scientific instruments. Despite the absence of articulated criteria defining a laboratory exercise, they held firm to their belief that it was generally understood what constitutes a scientific laboratory exercise and that they could easily pass judgment. Among the examples they offered was the 'classic' biology experiment of frog dissection. This intrigued me, and I asked them how it was that cutting open a frog and looking inside its belly constituted a scientific experiment as opposed to simple observation. Where was the theory? What was the hypothesis? What was being tested? How were the results evaluated critically? And if such dissection constituted science, how was it different from a geographer or geologist examining a topographic map or stereoscopic air-photo pair and taking measurements of terrain attributes? They finally conceded, somewhat reluctantly, that the frog dissection experiment required elaboration and that map analysis might be interpreted as scientific in the same vein as measuring the size, mass, and relative positioning of the internal organs of the frog. What struck me the most about this exchange, however, was that these truly well-respected, 'hard-core' scientists had never contemplated what the scientific enterprise was about. They were basing their judgments of curricular materials on preconceived notions and grade-school definitions of science rather than a full appreciation for the underlying principles guiding their day-to-day practice. They were hiding beneath the cloak of scientific hegemony and had no awareness of it.

After these meetings, it occurred to me that my training and employment as a geographer had served me well. Contemporary philosophical debates in geography, mostly by human geographers, obligate the physical geographer to defend the legitimacy of scientific claims to knowledge (e.g. Eliot Hurst 1985; Dear 1988; Marcus et al. 1992). These debates force an awareness of a broad range of philosophies including those pertaining to 'hard' sciences (e.g. positivism, rationalism, realism) and those that distance themselves from it or would undermine it (e.g. existentialism, idealism, relativism, postmodernism). The challenge to science by the 'onslaught of the isms' is predicated on the position that science is hard to define and difficult to differentiate from other knowledge-seeking enterprises, that it is fallible and subjective, that its logic has been refuted for decades, that its methodology fails miserably in the understanding of societally relevant issues, and therefore it should hold no preferred status within the division of academic labor nor in the eyes of society. The challenges against science are challenges against geomorphology if one accepts that geomorphology is a science. Why are these challenges important? In short, they encourage us to understand something of ourselves and our practice. They force us to distinguish good science from bad science. They guard against academic complacency, and demand us to question whether geomorphological research is indeed scientific, to justify these claims, and to interrogate the activities of the discipline. Geography helps in this effort because it is not completely and comfortably nestled within the sciences. It straddles both the natural and social sciences, and even borrows and develops an appreciation for artistic views of the world. While human geographers tend to be highly critical of physical geographers and vice versa, geographers in general are receptive to alternative viewpoints and perspectives.

Several recent papers by geographers have initiated an examination of the scientific bases of geomorphology, especially the role of theory (Chorley 1978; Rhoads and Thorn 1993) and the philosophical perspectives that might be most appropriate to contemporary geomorphic practice (Richards 1990, 1994; Yatsu 1992; Bassett 1994; Rhoads 1994; Rhoads and Thorn 1994). Most academics have rejected a science predicated on the extreme edicts of Baconian-style logical positivism, and many also remain unconvinced by Popperian prescriptions of critical rationalism (Haines-Young and Petch 1986), although these philosophies come closest to a popular image of scientific investigation. Some philosophers of science, such as Chalmers (1990), have attempted to define a middle ground by reconciling the orthodoxy of classical science with anarchical critiques of science (e.g. Feyerabend 1975). Chalmers (1990) suggests that the aim of science should be the 'establishment of generalizations governing the behavior of the world' (p. 29) with 'some means of substantiating those generalizations' (p. 38) and an 'emphasis on the growth and improvement of knowledge' (p. 37). As regards method and standard, Chalmers (1990, p. 39) simply requires that

> candidates for scientific laws and theories should be vindicated by pitching them against the world in a demanding way in an attempt to establish their superiority over rival claims ... [and in the physical sciences, this] will usually involve artificial experimentation and that the successful prediction of novel phenomena will be especially significant.

Unfortunately, Chalmers' middle ground cannot be used to evaluate whether disciplinary practice within geomorphology is scientific (1990, p. 116):

> While the aim of science can be *distinguished* from other aims and epistemological appraisals distinguished from other appraisals, the scientific practice involved in the pursuit of that aim cannot be *separated* from other practices serving other aims.

Richards (1990, 1994), Yatsu (1992), Bassett (1994), and Rhoads (1994) call for the adoption of a realist perspective by geomorphologists to guide their practice. Richards (1990, p. 195) even contends that 'Many geomorphologists will perhaps be unsurprised to discover that their geomorphology is essentially realist.' In part, this may be due to the broad range of colorations within the realist spectrum. One common theme among these brands of realism is that scientific observation is theory-laden and that theories tend to become more truth-like as the scientific enterprise continues. That is, former knowledge may be modified and improved, but it is rarely totally usurped as in the Kuhnian model of scientific revolutions. Little effort in geomorphology has been devoted to determining the extent to which geomorphic practice conforms to any philosophy of science, and Rhoads and Thorn (1994, p. 99) suggest that 'The challenge for realists is to show how many theoretical constructs embodied in the Davisian view of geomorphology, including references to unobservables, have been preserved in contemporary geomorphic theories.' Bishop (1980, p. 310) attempted to analyze the scientific quality of Davisian theory and concluded that:

> ... in Popperian terms, the cycle is not a scientific theory on at least two grounds. Firstly, the theory is irrefutable in a central, essential concept, that of stage; secondly, the theory has been modified in an ad hoc manner to account for those objections that could be brought against it ... Davis's 'outrageous' hypothesis might have been of more value had it been expressed in such a way as to permit its testing by falsification.

Perhaps geomorphology will turn out to be less scientific than most geomorphologists would like (cf. Sherman, Chapter 4 this volume).

Recently, we have been informed that postmodernism has 'hit geography like a tidal wave' and that it 'has flourished, because it constitutes the most profound challenge to three hundred years of post-Enlightenment thinking' (Dear and Wassmansdorf 1993, p. 321). Proponents of postmodernism claim that rationalism is not a viable philosophy, that theoretical argument and self-evident truths are invalid, and that the search for universal metanarratives is hopeless (Dear and Wassmansdorf 1993). Although the rhetoric is most impressive, postmodern arguments are hardly unique – they have been voiced by many others under less formal philosophical labels. What is disturbingly common to post-modern discourse, however, is the consistent and relentless attack on science. The postmodern left is particularly blatant in its attempts at self-legitimation through decon-struction of the scientific enterprise, and these seem ill-directed and driven by passion rather than reason. Ironically, the broader message of postmodernism – that competing claims to privilege or authority are not capable of being reconciled or resolved and therefore should be avoided and renounced (Dear and Wassmansdorf 1993) – seems to have been conveniently overlooked by these extremists. Although I am unable to imagine a useful and practical 'postmodern geomorphology', I am sympathetic to the postmodern appreciation for alternative perspectives. This position is, in fact, perfectly consistent with a 'scientific geomorphology', as pointed out by several geomorphological-geographers, and it merely requires that geomorphologists resist the temptation to don the cloak of scientific hegemony. Yatsu (1992, p. 115), for example, calls for the 'elimination of authoritarianism and the like' and suggests that, 'totally free discussions, and thus lib-ertarianism are the first step toward the advancement of the science [of geomorphology]'. Similarly, Rhoads and Thorn (1994, p. 99) state that, '... adopting a realist perspective does not necessitate that a truly scientific approach implies that all geomorphological problems must be described in the language of physics. Contemporary scientific realism explicitly acknowledges that no scientific discipline has privileged status with regard to the truth.' A postmodern view simply extends the realm to all disciplines, whether sci-entific or not.

Several earth scientists had expressed postmodern sentiments even before postmodern movements were constituted. Leighly (1955, p. 318), for example, was concerned with the academic and intellectual constraints that Davisian dictates had imposed on geomor-phological thinking for over 50 years, and while arguing for a more process-oriented physical geography, he suggested that:

> It would be good if we could again approach the earth with unhampered curiosity and attempt to satisfy that curiosity by whatever means the problems we encounter suggest. In particular, we should discard a restriction that has long been laid upon us: the prohibition of concern with processes. Let processes be restored to the central position they deserve ... The land, the sky, and the water confront us with questions whenever we look at them with open eyes. These questions, and the privilege of sharing in the quest of answers to them, are a part of our birthright.

At least one recent paper arguing for the reenchantment of geomorphology (Baker and Twidale 1991) is also surprisingly postmodern in its outlook when it speaks of the 'awe and wonder' of landscape appreciation (p. 89), when it encourages geomorphologists to become 'mavericks' (p. 90) and to 'hypothesize outrageously' (p. 96), and when it argues that an '... overemphasis on methodology, either for theoretical abstraction or "objective" measurement has made geomorphological study increasingly remote from the realities of the Earth's surface which constitutes its *raison d'être*' (Baker and Twidale 1991, p. 74). This is at once a plea for a more personal 'oneness' with the land and a rejection of scientific conservatism within geomorphology. Unfortunately, the enlightenment advo-cated by Baker and Twidale (1991) is dimmed considerably by the authoritarian and often contradictory statements that imbue the text. Readers are encouraged to think open-mindedly about alternative geomorphological perspectives (outrageous hypotheses) as long as they appreciate that 'The survival of old forms and surfaces serves as a timely reminder that Geomorphology is historical in nature and that it is an integral part of Geology' (Baker and Twidale 1991, p. 92). With regard to the value of geographical perspectives and research, Baker and Twidale (1991) opine that 'It had been demonstrated that academic life could continue and indeed flourish without Geography' (p. 76), and that geography's *raison d'être* should be 'to form a bridge between the natural and social sciences, and as an attempt to see the world in the round' (p. 76). Apparently, Baker and Twidale (1991) would have geographers act as lowly carpenters in the service of the sciences! In this, they have missed a crucial point. The bridge does not need to be built – it already exists. The challenge is to travel that bridge in order to appreciate the view from the other side, which may be considerably different. To sit on one side and ignore the other, is to deny knowledge. Even worse, to deconstruct the bridge and deepen the chasm, is to deny the existence of other worlds. This is the postmodern critique of disciplinary hegemony, and it is unfortunate that some geomorphologists remain blissfully uninformed of it.

Geomorphology, it could be argued, has been inconsistent in its treatment of the scientific enterprise. It has accepted the benefits of appearing and 'acting' scientific while failing to uphold the obligations and standards of rigorous scientific practice. Baker and Twidale (1991), for example, never question whether geomorphology is a science, yet they deplore the insidiousness of 'the substitution of elegantly structured methodology and theory for spontaneity, serendipity, and common sense' (p. 73) in geomorphological research. In so doing, they call for a rejection of methodology, theoretical abstraction, and objective measurement and interpret them as symptoms of intellectual sterility and indifference to nature when, in fact, these are at the very heart of the scientific enterprise (cf. Rhoads and Thorn 1994). Baker and Twidale (1991) uphold the primacy of a *geologically* rooted historical/genetic approach to geomorphological investigation, of which Davis's *geographical* works are the epitomization, and they do so in the face of Strahler's (1950) and Bishop's (1980) challenges to its scientific quality while at the same time reverencing the 'process-oriented' concepts of G. K. Gilbert. Are these deliberate obfuscations meant as postmodern prods to our conceptual, methodological, and theoretical complacency and sedentariness, or are the inconsistencies and contradictions truly revealing of the conceptual state of geomorphology? In either case, it may be beneficial to question whether geomorphology is best served by adopting an academic posture situated squarely and exclusively within the sciences. Much of what geomorphologists undertake in their research amounts to little more than specialized observation and speculation, as difficult as this may be. Experimentation and hypothesis testing are all but impossible, except perhaps, under reductionist strategies such as those in flumes, wind tunnels, or small experimental plots. This recognition, however, should not to be misconstrued as license to practice poor science, but rather, as an appreciation for the limitations of the scientific enterprise. We should never compromise on the quality of our science, but we should also not be reluctant to admit that there are alternative routes to knowledge. The field of medicine is a good example of a discipline that has its foundation firmly rooted in a scientific platform even though its practitioners readily admit to a dependency on skillful artistry in medical practice.

FUTURE PROSPECTS: INTEGRATION OR ELIMINATION?

In the 1950s, Bryan (1950, p. 198) remarked that, 'Natural as it may be for exponents of a subject to claim independence and to extol the unique virtue of their favorite field of effort, geomorphology can hardly claim the pre-eminence of separateness'. Some 20 years later Dury (1972, p. 200) advised that, 'This would appear to be no time to carve up the geomorphic frontier.' The twenty-first century is quickly approaching and it is worth taking stock to see if much has changed over the last 50 years. Is geomorphology's academic status sufficiently improved to entertain a complete academic separation from geography and geology, or would it be in our best interests to retain our subdisciplinary status and exploit this position?

Geomorphology may be defined as the area of study that leads to an understanding of and appreciation for landforms and landscapes including their geometry, structure (internal and external), coexistence with other forms (biotic and otherwise), and dynamics (mode of evolution and processes integral to their existence and evolution). A separate academic discipline that focuses on mere description and classification of landforms and landscapes has already proven to be too inert to be palatable. A discipline concerned with landform changes alone would find many of its practitioners idle waiting for change to occur, or, even worse, making inferences based on secondary data sources. A discipline that examines processes in the absence of the landforms *per se* loses touch with the essence of geomorphology. Conversely, a separate discipline that provides comprehensive coverage of all the many facets embodied in the definition would likely be untenable because the current demarcation of academic disciplines is too entrenched and because there are too many disciplinary claims on geomorphological spheres of knowledge. The example of the hydrological sciences is revealing in this context. Thus, the challenge for the future of geomorphology, at least in the short term, may be one of integration and synthesis.

Interdisciplinary Academic Associations

Dury (1983, p. 90) remarked that:

> discussions of disciplinary location have been largely by-passed by the general quantitative revolution of the natural sciences (and of part of the arts), by the growth of interdisciplinary research, and by the knowledge explosion, the result of which is that traditional departments are now too small for what they must do.

Many of society's contemporary concerns are focused on environmental problems such as those involving pollution, resource utilization, urbanization, spread of disease, sustainable agriculture, species-habitat preservation, and alteration of regional or global conditions. These are broad areas of research in which geomorphologists could and should play a role. Although the need for a more societally relevant geomorphology has been recognized for a long time (e.g. the 1970 Binghamton Symposium was on environmental geomorphology) and is being met by a select few (e.g. Goudie 1990; Graf 1994, and Chapter 18 this volume), there has been reluctance on the part of the broader geomorphological community to engage in this type of research. Perhaps this is because these fields of interest are nontraditional and have an applied flavor or because they require us to interact with researchers from other disciplines. Disciplinary traditions are subject to change, however, and there may be distinct advantages to working in multidisciplinary and collaborative groups during periods of economic hardship (e.g.

Church et al. 1985; Smith 1993). The shift to such a flexible academic 'culture' seems to be taking place in other disciplines, and one need only consider the number and popularity of environmental studies/sciences programs, schools, and institutes that have been established recently in universities and colleges worldwide to get a sense of future trends. Funding agencies such as the National Science Foundation have also expressed increasing interest in long-term, multi-institutional projects that are interdisciplinary and concerned with issues of environmental and social relevance and significance (e.g. Human Dimensions of Global Change program).

Several years ago, Coates (1971, p. 6) noted that:

> the science of geology has become increasingly compartmentalized at a time when many of the intriguing problems have become interdisciplinary. The geomorphologist, as the surviving generalist in earth science, can play a vital role in bridging this communication and intellectual gap.

In this regard, geomorphology's association with geography would seem to be advantageous, but only in so far as geographers are able to deliver on their claims of being synthesizing, integrative, and interdisciplinary. This may not be easy given the factionalism that is as pervasive in geography as in other disciplines – the AAG currently recognizes almost 50 specialty groups within its structure. It has been suggested that human geographers have generally failed to recognize the importance of the physical environment in their research:

> How long our colleagues in human geography will be able to maintain their distaste for the proposition that people can be strongly (and usually adversely) affected by the behaviour of the physical environment I am not prepared to guess; but until they give at least some ground, I should expect to find the separation of geomorphology from geography, in the intellectual if not in the locational sense, to widen (Dury 1983, p. 97).

It is also true that geomorphologists are generally reluctant to bestow the same privilege to humans as agents of change. Geomorphological research is still largely concerned with 'natural' systems despite a general awareness that pristine environments are the exception and that humans will continue to contribute to, if not dominate, environmental change (e.g. Goudie 1990; Nordstrom 1992). It has been estimated, for example, that humans now move more sediment on an annual basis than any other geomorphic agent (see Monastersky 1994). The notion of distinctive 'urban' climatologies or hydrologies has been recognized for decades, and it is thought that the hydrologic balance of the North American continent may have been altered perceptibly by increased surface-water storage in reservoirs created by dam closures. The interaction of individuals or societies with their geomorphological environments has found its most profound expression in natural-hazards research which will require geomorphologists to interface more directly with social scientists in the future (e.g. Gares et al. 1994). The potential contributions of other disciplinary practitioners to these efforts cannot be denied, and interactions with them should be encouraged. We need to be proactive about such interaction, and it is revealing to note that, of an estimated 1430 geomorphologists or Quaternary specialists in North America in 1983 (Costa and Graf 1984), only 48 belonged to both the GSA and the AAG (13 geologists belonged to the AAG, whereas 35 geographers belonged to the GSA).

Scale Integration

The challenge with regard to scale will be to transcend artificially constructed boundaries delineating spatial and temporal domains that are a logical necessity of our attempts to understand a complex world in a simplified way. Invariably these compartmentalizations reflect disciplinary strictures rather than natural divisions because the spatial and temporal scales of a problem are dictated by the methods used to attack them. This is a fundamental issue that may have to be addressed at the philosophical level before any methodological or practical advances can be made:

> Contention in contemporary geomorphology centers on differences in regulative principles, types of scientific arguments, and characteristics of theory employed by scientists, all of whom consider themselves geomorphologists (Rhoads and Thorn 1993). Although the point of contention is scientific methodology, the contention itself is clearly philosophical in nature. In other words, it is not possible to resolve conflicts between competing methodologies within science itself; instead, resolution of these differences must occur within philosophy (Montgomery 1991) (Rhoads and Thorn 1994, p. 99).

Contemporary research in geomorphology has been characterized as either 'historical' or 'process-oriented' (e.g. Baker and Twidale 1992; Kennedy 1992). Associated with these terms are tacit spatial and temporal scales. 'Historical' geomorphological research tends to be concerned with understanding landscape evolution as a sequence of events, and therefore it requires explicit identification of distinct time (space) elements. 'Process-oriented' geomorphological research, in its purest form, has no well-defined spatial or temporal reference points – that is, the processes can take place anywhere at any time. Nevertheless, generalizations about processes are often derived from case studies, and there will always be some idiographic component to geomorphological studies. Unfortunately, this latter reality is not easily admitted by the use of terms such as 'timelessness' and 'equilibrium', and this has caused considerable confusion and has led to needless posturing within the discipline (cf. Schumm and Lichty 1965; Kennedy 1992).

Practically, it should be clear that every process, by definition, must play itself out across space and through time. Process-oriented research tends to be reductionist because small spatial and temporal scales allow the researcher to reduce the degrees of freedom in any particular problem through control strategies. However, reductionism does not constitute diametric opposition to historicism. Kennedy (1992) offers an especially penetrating and insightful analysis of this fallacy. She argues that it is impossible to divorce the historical from the process approaches because to do so

> reduces the historical event to a boundary or initial condition ... But if one does so often enough, there is the risk of forgetting that boundary conditions must inevitably change, as part of the expectable, but inherently unpredictable sequence of endogenetic and exogenetic events ... the search for 'dynamic equilibrium' advocated by Strahler and by Chorley thus has the paradoxical effect of turning perfectly straightforward high-magnitude/low-frequency events into catastrophes (Kennedy 1992, p. 248).

In this context, our uncritical application of general systems theory has failed us miserably. The 'systems manual' has inspired us to a reductionist disassembling of Hutton's Earth Machine – to identify and classify components; to observe and describe their workings; to further dismantle these components into individual parts; to examine and describe those parts and their workings, and so on. Unfortunately, it is silent on methods of reassembly, and the essential chapter revealing to us what we truly seek – the internal workings of the Earth Machine – remains missing.

Other disciplines, such as chemistry, astronomy, and physics, have begun to take steps toward reconstructive efforts. Schweber (1993, p. 34), for example, points out that 'the reductionist approach that has been the hallmark of theoretical physics in the 20th century is being superseded by the investigation of emergent phenomena, the study of properties of complexes whose "elementary" constituents and their interactions are known'. The term 'emergent properties' refers, not to the theoretical or mathematical constructs that represent the fundamental laws which most reductionists seek, but rather, to the characteristic behavior of solutions to the mathematical expressions that describe natural phenomena. Recent research on nonlinear dynamical systems in geomorphology (e.g. Phillips 1992 and references therein) seems particularly germane in this respect. The identification of emergent properties is not a straightforward task because it involves more than simple reconstruction (i.e. putting the deconstructed pieces together). This was recognized by Anderson (1972, p. 393) who suggested that:

> the more the elementary-particle physicists tell us about the nature of the fundamental laws, the less relevance they seem to have to the very real problems of the rest of science, much less to those of society ... The constructionist hypothesis breaks down when confronted with the twin difficulties of scale and complexity. The behavior of large and complex aggregates of elementary particles, it turns out, is not to be understood in terms of a simple extrapolation of the properties of a few particles.

Anderson (1972) also argued that each scale level should have its own fundamental laws (i.e. its own ontology) and therefore, its own complexities and emergent properties.

These ideas about distinctly different scale levels have found partial support in contemporary geomorphological practice (e.g. de Boer 1992; Sherman and Bauer 1993; Phillips 1995). At the smallest scales, it is not uncommon for scientists to favor concisely expressed, deterministic relations that invoke force balances or conservation principles (of mass, momentum, vorticity, entropy, or energy). At intermediate scales, much of the mathematical formality is retained, but some of the deterministic physics or chemistry are replaced by parameterizations that invoke phenomenological or constitutive coefficients such as conductivity, diffusivity, viscosity, elasticity, erodibility, porosity, and permeability, or the various coefficients of drag, friction, cohesion, and strength. These coefficients are used to relate the principal variables in a given process–response interaction when accounting for individual particles and their behavior is no longer possible. Ensemble averaging over domains of interest (representative elemental volumes) becomes necessary. At the largest scales, descriptions of system behavior usually assume probabilistic (indeterministic) properties or an idiographic and historical character. One of the challenges for geomorphologists, then, is to identify the existence and domain of these scale levels – their bounds, fundamental laws, and emergent properties.

It is worth remembering, however, that some 'laws' may transcend scale in the sense that they will be characteristic of several contiguous levels. The law of gravity, for example, applies to most scales of geomorphological interest, as might the principles of least action, sufficient reason, and uniformity of nature (Yatsu 1992, p. 88). Geomorphologists should continue the search for general or 'universal' laws or principles, but with the proviso that they be scientific rather than methaphysical. The formal analysis of Church and Mark (1980) into the character of proportional relations in geomorphology and their interpretation as allometric or self-similar 'growth laws' is exemplary in this regard. Research addressing both emergent and universal properties is necessary if geomorphology is to evolve toward a societally relevant discipline with nomothetic explanations and predictive powers.

SUMMARY REMARKS

Traditional accounts of the development of geomorphology have suffered from Whig historiography which adopts a backward-looking perspective that interprets the relevance and importance of past figures and events in the context of contemporary values and paradigms. Because contemporary geomorphology considers itself to be scientific, it often traces its roots selectively to those figures and events that lend credence to this scientific image. Such posturing is usually based less on the realities of current practice or reasoned direction than on misrepresented images and value-laden emphases. Although there are tangible benefits to such posturing, especially since scientific activities are generally held in high esteem by society and within academe, there are accompanying costs. It is paramount that we identify explicitly these costs and benefits because there are implications for the future of the discipline (cf. Raguraman 1994).

Popular histories of North American geomorphology are biased in that they fail to recognize the relative importance of geography to geomorphological disciplinary development. In part, this may be ascribed to geography's relatively recent arrival as an academic discipline in North America, despite its long-standing intellectual traditions that extend to the ancient Greeks. The important contributions of geology and geologists, although central, are unduly stressed, and such preferential representation may be misleading in several ways. First, it is not evident that the intellectual discipline of geomorphology, with its focus on earth-surface phenomena, can or should be traced to any single geological or geographical seed or root, whether a person, institution, event, or activity. Intellectual interest in the world around us is a human trait that has persisted throughout history, even before there were disciplines. Disciplinary demarcations are socially constructed institutions that have become necessary only recently because individuals can no longer 'know it all'. Imposing contemporary disciplinary structures on academic thought during historical periods is therefore inaccurate, inappropriate, and unwarranted. Further, it ignores the malleable and evolving nature of disciplines – their cores, their boundaries, and the knowledge contained therein are subject to change through time.

Second, the academic roots of geomorphology are intertwined with those of many disciplines, not just geology or geography, and it is not self-evident that one discipline can be asserted to have had a more profound role than another – the 'Championship of the Disciplines' is unwinnable. Moreover, the distinctions made between geology and geography, and between geologists and physical geographers, are often matters of trivial caricaturing. North American departments of geology during the nineteenth century, for example, retained the services of geographers, biologists, and anthropologists, and only by one very pragmatic definition would they be labeled as 'geologists'. Geomorphology has been (and continues to be) practiced in geology departments by geographically oriented researchers, and vice versa. To label key geomorphological figures such as Gilbert, Davis, and Powell as either 'geological' or 'geographical' is a gross misrepresentation of their character and an unfortunate oversimplification of their research interests and contributions. Similarly, commonplace assertions that the process-oriented and historical approaches to geomorphological research are characteristically 'geographical or geological', 'scientific or nonscientific', and substantively antithetical are not substantiable. Such caricaturing damages the integrity of geomorphology and polarizes its practitioners during a period when academe is facing some of the most profound challenges to its intellectual and theoretical foundation, disciplinary and administrative structure, social relevance and utilitarian worth, and long-term financial survival.

Finally, whether geomorphology views itself as being dominantly geological or geographical in character has implications for how it is perceived by other academicians and how its practitioners conduct their research, and indeed, what problems are deemed to be of importance. The geological persona of geomorphology seems to be more 'scientific' and also more academically privileged. The geographical persona is more diverse philosophically and methodologically, and potentially more interdisciplinary and integrative. Geomorphology and society would best be served by encouraging future development of both. I end this chapter by reiterating the sentiments of R.J. Russell in his presidential address to the AAG almost 50 years ago:

> Whether geomorphology belongs to geography or geology seems to be a question unworthy of the debate it has occasioned ... I feel somewhat happy when I notice attempts of either geographers or geologists to claim the subject, and somewhat dejected when I see either trying to pass it over to the other (Russell 1949, p. 11).

ACKNOWLEDGEMENTS

I gratefully acknowledge all those authors, cited in the bibliography, who have compiled extensive and comprehensive histories of the geographic and geologic disciplines. The fruits of their labor were plentiful, yet sufficiently ambiguous to allow for the interpretive, stylized historiography you have just read. I take all responsibility for historical untruths since I was negligent in consulting original articles prior to 1900 and therefore failed to develop a full appreciation of the nuances inherent to these early writings. I extend thanks to Curt Roseman, Dick Marston, Jeff Lee, Doug Sherman, and Melissa Gilbert for providing helpful comments, and to the National Science Foundation (SBR-9158230) for providing continuing research support.

REFERENCES

Abler, R.F., Marcus, M.G. and Olson, J.M. (eds) 1992. *Geography's Inner Worlds: Pervasive Themes in Contemporary American Geography*, Rutgers University Press, New Brunswick, NJ, 412 pp.

Ackerman, E.A. 1945. Geographic training, wartime research, and immediate professional objectives, *Annals of the Association of American Geographers*, **35**, 121–143.

Anderson, P.W. 1972. More is different, *Science*, **177**, 393–396.

Baker, V.R. 1988. Geological fluvial geomorphology, *Geological Society of America Bulletin*, **100**, 1157–1167.

Baker, V.R. and Pyne, S. 1978. G.K. Gilbert and modern geomorphology, *American Journal of Science*, **278**, 97–123.

Baker, V.R. and Twidale, C.R. 1991. The reenchantment of geomorphology, *Geomorphology*, **4**, 73–100.

Barrows, H.H. 1923. Geography as human ecology, *Annals of the Association of American Geographers*, **13**, 1–14.

Bassett, K. 1994. Comments on Richards: the problems of 'real' geomorphology, *Earth Surface Processes and Landforms*, **19**, 273–276.

Beckinsale, R.P. and Chorley, R.J. 1991. *The History of the Study of Landforms or the Development of Geomorphology*, Vol. 3; *Historical and Regional Geomorphology, 1890–1950*, Routledge, London.

Bishop, P. 1980. Popper's principle of falsifiability and the irrefutability of the Davisian cycle, *Professional Geographer*, **32**, 310–315.

Brigham, A.P. 1903. *Geographic Influences in American History*, Ginn, Boston, Mass., 366 pp.

Brown, R.H. 1948. *Historical Geography of the United States*, Harcourt, Brace, New York, NY, 596 pp.

Brush, S.G. 1974. Should the history of science be rated x?, *Science*, **183**, 1164–1172.

Bryan, K. 1950. The place of geomorphology in the geographic sciences, *Annals of the Association of American Geographers*, **40**, 196–208.

Butzer, K.W. 1989. Hartshorne, Hettner, and 'The Nature of Geography', in *Reflections on Richard Hartshorne's 'The Nature of Geography'*, edited by J.N. Entrikin and S.D. Brunn, Occasional Publication, Association of American Geographers, Washington, DC, pp. 35–52.

Campbell, M.R. 1928. Geographic terminology, *Annals of the Association of American Geographers*, **18**, 25–40.

Chalmers, A. 1990. *Science and its Fabrication*, University of Minnesota Press, Minneapolis, 142 pp.

Chorley, R.J. 1962. Geomorphology and general systems theory, *United States Geological Survey Professional Paper*, *500-B*.

Chorley, R.J. 1978. Bases for theory in geomorphology, in *Geomorphology: Present Problems and Future Prospects*, edited by C. Embleton, D. Brunsden and D.K.C. Jones, Oxford University Press, Oxford, pp. 1–13.

Chorley, R.J., Beckinsale, R.P. and Dunn, A.J. 1973. *The History of the Study of Landforms*, Vol. 2: *The Life and Work of W.M. Davis*, Methuen, London, 874 pp.

Chorley, R.J., Dunn, A.J. and Beckinsale, R.P. 1964. *The History of the Study of Landforms*, Vol. 1: *Geomorphology before Davis*, Methuen, London, 678 pp.

Church, M.A., Gomez, B., Hicken, E.A. and Slaymaker, O. 1985. Geomorphological sociology, *Earth Surface Processes and Landforms*, **10**, 539–540.

Church, M.A. and Mark, D.A. 1980. On size and scale in geomorphology, *Progress in Physical Geography*, **4**, 342–390.

Coates, D.R. 1971. Introduction to environmental geomorphology, in *Environmental Geomorphology*, edited by D.R. Coates, State University of New York, Binghamton, NY, pp. 5–6.

Costa, J.E. and Graf, W.L. 1984. The geography of geomorphologists in the United States, *Professional Geographer*, **36**, 82–92.

Davies, G.L. 1989. *The Earth in Decay: A History of British Geomorphology, 1578–1878*, Oldbourne, Macdonald Technical and Scientific, London, 326 pp.

Davis, W.M. 1899. The geographical cycle, *Geographical Journal*, **14**, 481–504.

Davis, W.M. 1912. Relation of geography to geology, *Bulletin of the Geological Society of America*, **23**, 93–124.

Dear, M. 1988. The postmodern challenge: reconstructing human geography, *Transactions, Institute of British Geographers*, **13**, 262–274.

Dear, M. and Wassmansdorf, G. 1993. Postmodern consequences, *Geographical Review*, **83**, 321–325.

De Boer, D.H. 1992. Hierarchies and spatial scale in process geomorphology: a review, *Geomorphology*, **4**, 303–318.

Dury, G.H. 1972. Some recent views on the nature, location, needs, and potential of geomorphology, *Professional Geographer*, **24**, 199–202.

Dury, G.H. 1983. Geography and geomorphology: the last fifty years, *Transactions, Institute of British Geographers*, NS, **8**, 90–99.

Eagleson, P.S. 1991. Hydrologic science: a distinct geoscience. *Reviews of Geophysics*, **29**, 237–248.

Eliot Hurst, M.E. 1985. Geography has neither existence nor future, in *The Future of Geography*, edited by R.J. Johnston, Methuen, New York, pp. 59–91.

Feyerabend, P.K. 1975. *Against Method*, New Left Books, London.

Finch, V.C. 1939. Geographical science and social philosophy, *Annals of the Association of American Geographers*, **29**, 1–28.

Gares, P.A., Sherman, D.J. and Nordstrom, K.F. 1994. Geomorphology and natural hazards, in *Geomorphology and Natural Hazards*, edited by M. Morisawa, *Geomorphology*, **10**, 1–18.

Goudie, A. 1990. *The Human Impact on the Natural Environment* (3rd edn), The MIT Press, Cambridge, Mass., 388 pp.

Graf, W.L. 1984. The geography of American field geomorphology, *Professional Geographer*, **36**, 78–82.

Graf, W.L. 1994. *Plutonium and the Rio Grande*, Oxford University Press, New York, 329 pp.

Graf, W.L., Trimble, S.W., Toy, T.J. and Costa, J.E. 1980. Geographic geomorphology in the eighties, *Professional Geographer*, **32**, 279–284.

Gregory, K.J. 1985. *The Nature of Physical Geography*, Edward Arnold, London, 262 pp.

Haines-Young, R.H. and Petch, J.R. 1986. *Physical Geography: Its Nature and Methods*, Harper and Row, London, 230 pp.

Hartshorne, R. 1939. *The nature of geography*, The Association of American Geographers, Lancaster, PA.

Herries Davies, G.L. 1989. On the nature of geo-history, with reflections on the historiography of geomorphology, in *History of Geomorphology*, edited by K.J. Tinkler, Unwin Hyman, Boston, Mass., pp. 1–10.

James, P.E. and Martin, G.J. 1978. *The Association of American Geographers – The First Seventy-five Years 1904–1979*, Association of American Geographers, 279 pp.

James, P.E. and Martin, G.J. 1981. *All Possible Worlds – A History of Geographical Ideas* (2nd edn), Wiley, New York, 508 pp.

Johnson, D. 1929. The geographic prospect, *Annals of the Association of American Geographers*, **14**, 168–231.

Jones, H.L. 1917. Translation of Strabo, *The Geography of Strabo*, G.P. Putnam & Sons, New York.

Kennedy, B.A. 1992. Hutton and Horton: views of sequences, progressions and equilibrium in geomorphology, in *Geomorphic Systems*, edited by J.D. Phillips and W.H. Renwick, *Geomorphology*, **5**, 231–250.

Kesseli, J.E. 1946. A neglected field: geomorphography (published abstract), *Annals of the Association of American Geographers*, **36**, 93.

Klemes, V. 1986. Dilettantism in hydrology: transition or destiny? *Water Resources Research*, **22**, 177s–188s.

Leighly, J. 1955. What has happened to physical geography? *Annals of the Association of American Geographers*, **45**, 309–318.

Livingstone, D.N. 1984. The history of science and the history of geography: interactions and implications, *History of Science*, **22**, 271–302.

Marcus, M.G., Olson, J.M. and Abler, R.F. 1992. Humanism and science in geography, in *Geography's Inner Worlds: Pervasive Themes in Contemporary American Geography*, edited by R.F. Abler, M.G. Marcus and J.M. Olson, Rutgers University Press, New Brunswick, NJ, pp. 327–341.

Marston, R.A. 1989. Geomorphology, in *Geography in America*, edited by G.L. Gaile and C.J. Willmott, Merrill Publishing Company, Columbus, Ohio, pp. 70–90.

Monastersky, R. 1994. Earthmovers – humans take their place alongside wind, water, and ice, *Science News*, **146**, 432–433.

National Geographic Research and Exploration 1994. *Geography for Life: National Geography Standards – 1994*, National Geographic Society, Washington, D.C., 272 pp.

Nordstrom, K.F. 1992. *Estuarine Beaches: An Introduction to the Physical and Human Factors Affecting Use and Management of Beaches in Estuaries, Lagoons, Bays and Fjords*, Elsevier Applied Science, New York, 225 pp.

Phillips, J.D. 1992. Nonlinear dynamical systems in geomorphology: revolution or evolution?, in *Geomorphic Systems*, edited by J.D. Phillips and W.H. Renwick, *Geomorphology*, **5**, 219–229.

Phillips, J.D. 1995. Biogeomorphology and landscape evolutions: the problem of scale, *Geomorphology*, **13**, 337–347.

Powell, J.W. 1885. The organization and plan of the United States Geological Survey, *American Journal of Science*, **29**, 93–102.

Raguraman, K. 1994. Philosophical debates in human geography and their impact on graduate students, *Professional Geographer*, **46**, 242–249.

Rhoads, B.L. 1994. On being a 'real' geomorphologist, *Earth Surface Processes and Landforms*, **19**, 269–272.

Rhoads, B.L. and Thorn, C.E. 1993. Geomorphology as science: the role of theory, *Geomorphology*, **6**, 287–307.

Rhoads, B.L. and Thorn, C.E. 1994. Contemporary philosophical perspectives on physical geography with emphasis on geomorphology, *Geographical Review*, **84**, 91–101.

Richards, K. 1990. 'Real' geomorphology, *Earth Surface Processes and Landforms*, **15**, 195–197.

Richards, K. 1994. 'Real' geomorphology revisited, *Earth Surface Processes and Landforms*, **19**, 277–281.

Robinson, G. 1963. A consideration of the relations of geomorphology and geography, *Professional Geographer*, **15**, 13–17.

Russell, R.J. 1949. Geographical geomorphology, *Annals of the Association of American Geographers*, **39**, 1–11.

Russell, R.J. 1958. Geological geomorphology, *Bulletin of the Geological Society of America*, **69**, 1–22.

Sack, D. 1991. The trouble with antitheses: the case of G.K. Gilbert, geographer and educator, *Professional Geographer*, **43**, 28–37.

Sack, D. 1992. New wine in old bottles: the historiography of a paradigm change, in *Geomorphic Systems*, edited by J.D. Phillips and W.H. Renwick, *Geomorphology*, **5**, 251–263.

Sauer, C.O. 1924. The survey method in geography and its objectives, *Annals of the Association of American Geographers*, **14**, 17–34.

Sauer, C.O. 1925. *The Morphology of Landscape*, University of California Publications in Geography, **2**, 19–53.

Sauer, C.O. 1941. Foreword to historical geography, *Annals of the Association of American Geographers*, **31**, 1–24.

Schaefer, F.K. 1953. Exceptionalism in geography: a methodological examination, *Annals of the Association of American Geographers*, **43**, 226–249.

Schumm, S.A. and Lichty, R.W. 1965. Time, space, and causality in geomorphology, *American Journal of Science*, **263**, 110–119.

Schweber, S.S. 1993. Physics, community, and the crisis in physical theory, *Physics Today*, **46**, 34–39.

Semple, E.C. 1911. *Influences of Geographic Environment on the Basis of Ratzel's System of Anthropo-geography*, Henry Holt, New York, 683 pp.

Sherman, D.J. and Bauer, B.O. 1993. Dynamics of beach-dune systems, *Progress in Physical Geography*, **17**, 413–447.

Smith, D.G. 1993. Fluvial geomorphology: where do we go from here?, in *Geomorphology: The Research Frontier and Beyond*, edited by J.D. Vitek and J.R. Giardino, *Geomorphology*, **7**, 251–262.

Smith, N. 1987. 'Academic war over the field of geography': the elimination of geography at Harvard, 1947–1951, *Annals of the Association of American Geographers*, **77**, 155–172.

Strahler, A.N. 1950. Davis' concepts of slope development viewed in the light of recent quantitative investigations, *Annals of the Association of American Geographers*, **40**, 209–213.

Strahler, A.N. 1952. Dynamic basis for geomorphology, *Bulletin of the Geological Society of America*, **63**, 923–938.

Thomas, W.L. (ed.) 1956. *Man's Role in Changing the Face of the Earth*, University of Chicago Press, Chicago, Ill., 1193 pp.

Tinkler, K.J. 1985. *A Short History of Geomorphology*, Barnes & Noble Books, Totowa, 317 pp.

Tuan, Y.F. 1991. A view of geography, *Geographical Review*, **81**, 99–107.

Unwin, T. 1992. *The Place of Geography*, Longman Scientific & Technical, New York, 273 pp.

Walker, H.J. and Grabau, W. (eds) 1993. *The Evolution of Geomorphology: A Nation-by-nation Summary*, Wiley, New York.

White, S.E. 1982. Geomorphology linking time and space, *Geotimes*, **27**, 18.

White, S.E. and Malcolm, M.D. 1972. Geomorphology in North American geology departments, 1971, *Journal of Geological Education*, May, 143–147.

Worsley, P. 1979. Whither geomorphology?, *Area*, **11**, 97–101.

Yatsu, E. 1992. To make geomorphology more scientific, *Transactions, Japanese Geomorphological Union*, **13**, 87–124.

Zakrzewska, B. 1967. Trends and methods in land form geography, *Annals of the Association of American Geographers*, **57**, 128–165.

Zakrzewska, B. 1971. Nature of land form geography, *Professional Geographer*, **23**, 351–354.

17 The Evolution of Geomorphology, Ecology, and Other Composite Sciences

W.R. Osterkamp
US Geological Survey, Tucson, Arizona

C.R. Hupp
US Geological Survey, Reston, Virginia

ABSTRACT

Geomorphology, ecology, and related disciplines are complex composites of the basic sciences – physics, chemistry, and biology. The basic sciences have developed through paradigm (exemplar) definition and replacement, but the composite sciences, too complex to generate applicable exemplars, developed from principles borrowed from basic science. Succeeding periods of speculation and observation, composite sciences adopted exemplars of evolution from biology and equilibrium from chemistry. Effort continues to unite these conflicting approaches.

   Work of geomorphologists, ecologists, and other composite scientists suggests that periods of Darwinian evolution, equilibrium, and integration occurred in common sequence but with different timing in the composite sciences. Fundamental differences between the basic and composite sciences – simplicity versus complexity, suitability versus inappropriateness to direction by exemplars, and a generally theoretical versus applied quality – provide explanations for different patterns of development. Based on common characteristics and methods of investigation, future trends in composite science are anticipated.

The Scientific Nature of Geomorphology: Proceedings of the 27th Binghamton Symposium in Geomorphology held 27–29 September 1996. Edited by Bruce L. Rhoads and Colin E. Thorn. © 1996 John Wiley & Sons Ltd.

INTRODUCTION

Events defining a history of science and the branching and growth of disciplines within science are well-established, but the causes of the events and branching continue to inspire debate. This chapter, based on a thesis of Thomas Kuhn (1970), suggests how several branches of modern science have developed. Specifically, this overview considers geomorphology, including physical geography, and ecology within all science-related study. It is proposed that predecessors to contemporary geomorphologists and ecologists, largely observers through the mid-seventeenth century and beyond, were profoundly influenced by Darwin's *Origin of Species* in 1859. Furthermore, a predictable counterreaction to an evolutionary orientation followed. Finally, inevitable amalgamation of the poles is being adopted by recent natural scientists. The leaders of these tenures of thought are termed observationalists, Darwinists, equilibrists, and integrationists; a fifth group conceivably could be unifiers. If North American scientists are unduly emphasized, the bias is unintentional.

Numerous discussions treat the emergence of science from Greek philosophy, which typically viewed nature as an organism. The separation accelerated in the second millennium with the growth of Persian mathematics and the scientific technique. Roger Bacon, an early observationalist, was among the first, about AD 1720, to insist on observation, objectivity, and repeatable experimentation. As the speculations of Herodotus, Aristotle, Seneca, and others following them helped usher in the Renaissance, science increasingly became a separate component of philosophy (Bowler 1992).

With the Renaissance, a view of nature based on observation became predominant and rifting among philosophy, religion, and technology deepened. With turmoil of the sixteenth-century Reformation, emphasis was placed on quantitative techniques and experimentation of the scientific method. The advances did not depart from observation, but represented subtle change in technique. Hence, Renaissance and post-Renaissance observationalists were also epistomologists, bridging philosophy and science by attempting to understand the limits, validity, and methods to develop knowledge (Bowler 1992). Observationalists continue a presence, but few are engaged in quantitative experimentation without embracing an overriding doctrine.

Modern science, for this discussion, began with the separation of physics from the cosmology of Copernicus and Kepler in the early seventeenth century, and continued with Robert Boyle's studies on gases, a transition from alchemy to chemistry. Development of biology, the other basic science, was closely tied to late eighteenth- and early nineteenth-century work of William Smith, James Hutton, Rodney Murchison, and others in paleontology and stratigraphy. Systematic observations of blood circulation in animals were made by William Harvey (1628), but the timing of the first significant applications of experimental method to biology is unclear.

The Structure of Modern Science

Thomas S. Kuhn, a historian, suggested in *The Structure of Scientific Revolutions* (1970) that modern science, studies based on the experimental approach, has a history of data-gathering punctuated by shorter periods of 'paradigm' upheavals that force reevaluation and redirection. The history of modern science largely addresses events defining progress in the basic sciences, and Kuhn (1970) mostly discusses physics, chemistry, and biology. This chapter extends the paradigm concept to geomorphology, ecology, and other disciplines that combine elements of the basic sciences and technology.

A *paradigm* (Kuhn 1970) is loosely synonymous with archetype, an all-inclusive model. Dominance of one style, however, gives way to another. With acceptance of a paradigm, *normal science* holds, a 'continuation of a particular research tradition' (Kuhn 1970, p. 11). Normal science extends '. . . the knowledge of those facts that the paradigm displays as particularly revealing, by increasing the extent of the match between those facts and the paradigm's prediction, and by further articulation of the paradigm itself' (Kuhn 1970, p. 24).

Normal science supports the prevalent paradigm, but with time, some data appear anomalous. When too many conflicting data reduce paradigm utility, the science becomes unstable and subject to discovery – the revolution of paradigm replacement. Acceptance of a paradigm implies acceptance of the rules and standards that define the system; thus, consensus is established for the conduct of normal science, including the goals of continuing research. Most importantly, adoption of a paradigm characterizes the science until replacement recurs: '. . . to desert the paradigm is to cease practicing the science it defines' (Kuhn 1970, p. 34). Because a new paradigm upsets custom, older scientists are threatened and tend to reject the model. Generally, full recognition of a paradigm requires one or two decades and results in a terraced advancement of the discipline – treads of normal science interrupted by paradigms.

Following criticism of his use of paradigm, Kuhn substituted *exemplar(s)*, which 'are concrete problem solutions, accepted by the group as, in a quite usual sense, paradigmatic' (Kuhn 1977, p. 297). Through this chapter, therefore, exemplar replaces paradigm. The breadth of application that a concept classed as paradigmatic has remains in question (e.g. Haines-Young and Petch 1986); it is inferred here that Kuhn (1970, 1977) intended an exemplar to be broadly applicable to a science. Another difficulty that resulted in scant criticism was Kuhn's focus on the 'basic' sciences: physics, chemistry, and biology. Excepting geology, Kuhn gave no significant recognition to other disciplines – the 'composite' sciences. For responses to criticisms, see 'Postscript – 1969' (Kuhn 1970, p. 174–210).

Diversity in Science

The term *composite science* refers to complex disciplines such as geomorphology and ecology, acknowledging that they are composed of distinct parts of other types of study. Thus, we define composite science as a discipline with specific and generally agreed-upon goals requiring various scientific and technological approaches of investigation to meet those objectives. A goal of geomorphology, for example, is a genetic interpretation of landforms, and techniques of physics, chemistry, biology, and engineering are employed to develop interpretations. Similar statements seem fitting for other disciplines regarded here as composite, or compound, sciences.

Because the composite sciences have diverse inputs, they typically are more applied than the basic sciences, making them less dominated by established order. Physics is cleanly defined as the study of the material universe, but ecology is concerned with interrelations of organisms and their environments and must account for variables including climate, soil physics and chemistry, and plant physiology of competing species. The basic sciences, products of early observationalists and gaining identity in the seventeenth century through exemplar sequencing, provided a basis for most composite science two to three centuries later but could not provide similar traits of exemplar structure. An exception composite science is geology, a principal source from which geomorphology arose. Geology established a modern identity late in the eighteenth century, is similar to basic science in some respects, and directly benefited from exemplars of basic science, especially evolution and advances in chemistry. The genetic complexity of a composite science, including geology, largely precludes the rule of an encompassing exemplar that guides the research of its 'normal science'. In much the same manner as their parents, however, the composite sciences have exhibited a progression or evolution of development, two centuries later, but mostly without benefit of exemplar direction. The result has been the theft or appropriation of exemplars proposed for a basic science.

This chaper suggests a context for understanding the development and operation of the composite sciences. Numerous papers, including Kuhn's (1970, 1977), explore the basic sciences as a set; others treat a specific composite science (e.g. Chorley et al. 1973; Kitts 1977; McIntosh 1985; Sack 1992; Frodeman 1995). Few, however, consider the development of composite sciences as a group, which may be necessary to understand how any one member has matured. We propose that the composite sciences, lacking exemplar heredity of the basic sciences, exploited, with little modification, Darwinian evolution for use within each discipline. Charles Darwin (Figure 17.1), in *Origin of Species*, wrote two chapters on geology, largely paleontology, and the effect on geology and other disciplines was profound, self-evident, and has been discussed exhaustively. A thesis here is that a more subtle effect of evolution, largely ignored but persisting to the present, has been its dominant influence, both positive and negative, on other composite sciences.

EXEMPLARS AND THE PARENT SCIENCES

Geomorphology and ecology grew from parents of basic science and geology and share histories entwined with them. Thus, exemplars controlling physics, chemistry, and biology provided form to composite science as well. Noteworthy examples of exemplars that have had but indirect effect on composite sciences include the laws of motion by Isaac Newton in the 1680s, discovery of oxygen by Joseph Priestley about 1770, development of atomic theory by John Dalton about 1805, and observations of Gregor Mendel that genes obey probabilistic laws (1865). Of greater pertinence, however, was publication of *The Origin of Species by Means of Natural Selection; or, the Preservation of Favored Races in the Struggle for Life*, by Charles Darwin (1859), and development of equilibrium theory, common to all of science but best expressed for chemical reactions by van't Hoff (1884):

$$A + B \leftrightarrow C + D \tag{1}$$

Equation (1) quantifies Le Chatelier's principle, that a system at equilibrium adjusts to a stress (i.e. change in temperature, pressure, or concentration of matter) so as to reestablish equilibrium. Although the van't Hoff equation was new to physical science, equilibrium had long been observed in engineering and had been recognized by Latin speculators as signified by *vix medicatrix naturae* (loosely meaning the balance and effort of natural

Figure 17.1 Portrait of Charles Darwin

healing). One of many applications of the van't Hoff equation to the composite sciences is the partition coefficient, K_d (Olsen et al. 1982), which is a mass ratio of a contaminant, C_s, sorbed to soil particles, to the equilibrium concentration of the contaminant, C_c, in water:

$$K_d = C_s/C_e \qquad (2)$$

Much of twentieth-century chemistry, the study of changes of matter, has been governed by equilibrium theory of equation (1). Physical chemistry, the interface between physics and chemistry, was developed by Jacobus van't Hoff (Figure 17.2). Similarly, Darwinian evolution and its long time scales for over a century have been tenets of biology, simply defined as the study of life. Components of biology treating shorter temporal scales than evolution, biochemistry, biophysics, and ecology, among others, typically embrace equilibrium theory.

An exception to the generalization that the composite sciences evolved from and later than the basic sciences is geology. Dependent on history, tailored after physics (Kitts

Figure 17.2 Portrait of Jacobus van't Hoff (courtesy of the Nobel Foundation)

1977; Frodeman 1995), but lacking susceptibility to experimentation, geology has been termed a *derivative* science (Schumm 1991; Frodeman 1995) and *protoscience* (Kitts 1977) that is parent to other composite sciences and is itself a composite. Geology, the study of the Earth, includes subdisciplines of petrology (treating the occurrence of rocks), stratigraphy (the description of divisions of rocks and their historical significance), sedimentology (the study of sedimentary rocks), tectonics (the study of the architecture of the Earth), historical geology (the study of temporal change on earth), geophysics (the study of the Earth, Moon, and other planets), and geochemistry (the study of the distributions of elements). An extreme example, the breadth of geology epitomizes the complexities of the composite sciences and demonstrates that an exemplar relevant to geochemists, for example, may have limited application to other segments of 'the group' (Kuhn 1977) of geologists.

As did physics and chemistry, geology separated from philosophy in the eighteenth century, largely due to naturalists such as Abraham Werner, James Hutton, Rodney Murchison, Alexander von Humboldt, and Charles Lyell (Bowler 1984; Tinkler 1989). Their studies included broadly scoped observations of mineralogy, rock types, geomagnetism, climatology, and the fossil record, which in 1798 led to recognition of evolution by William Smith's *Law of Faunal Succession*, about 60 years before Darwin's *Origin of Species*. Ironically, therefore, the exemplar of evolution initially may have been constructed for a composite science too complex to use it effectively, and perhaps was incorporated into biology only after it had matured sufficiently to accommodate the immensity of the concept.

An alternative view might suggest that geology has been guided by the Neptunist, Plutonist, and catastrophist schools of landscape formation, by uniformitarianism (Hutton 1795), which was an element of Plutonism based largely on faunal succession and later embraced by Darwin, or more recently by geochemical and convection models, including plate tectonics. We suggest, however, that the former were broad speculations to explain observed rocks and landforms, and the latter, however important they may be in explaining the occurrence of continents, are applicable only to part of geology. This viewpoint agrees with Giere's (1988), that generalizations applied to all science are too broad to be useful, but differs with his preference for models (as opposed to exemplars) of narrow scope and applicable perhaps only to a portion of a science. Regardless of perspective, no present exemplar seems sufficiently broad to serve geology fully, but geologic studies, being strongly tied to time, continue to embrace doctrines of faunal succession, uniformitarianism, and evolution.

THE COMPOSITE SCIENCES

A composite science is an area of study with well-defined objectives and scope, but which requires data from and overlapping with two or more of the basic sciences. With geology, geomorphology, ecology, soil science, and hydrology are examples discussed here. By this definition, psychology, economics, and other social sciences are either questionable or excluded, but developmental interpretations proposed for geomorphology and ecology may apply to those disciplines as well.

Although exemplars in geomorphology have been suggested (e.g. Ritter 1988; Sack 1992; Rhoads and Thorn 1994), heterogeneity may preclude the emergence of exemplars from within composite science. Thus, Kuhn (1970) scarcely mentions composite sciences and does not refer to aggregated lines of scientific study. Instead, the Kuhn model is defined uniquely for the basic sciences and cannot be extrapolated easily to composites. Both in theory and practice, an inability to generate an applicable exemplar and to provide regulation of a science with it results in a vacuum. The void forces the science to improvise, to borrow, to plagiarize exemplars and accompanying techniques to establish its identity as a science, and these borrowed exemplars inevitably conflict with parts of the science.

The following examples show similar progressions that started with exemplar- deficient observation. In each case, possibly excepting hydrology, this stage was succeeded by a period dominated by Darwinian evolution – the initial borrowed exemplar. In each case but at different times, evolution was either partly or mostly replaced with a time-independent exemplar of equilibrium, which in turn was moderated by integrationists attempting to merge the extremes. Although this progression partially mirrors change in the basic sciences, it has greater resolution because the basic sciences evolved through exemplar replacement consistent with the science, whereas imposition of exemplars on the composite sciences eventually resulted in incompatibilities.

Geomorphology

The establishment of geomorphology as a discipline distinct from geology, geography, or parts of engineering was relatively recent. A graph of geomorphology documents its emergence in the late nineteenth century (Vitek and Ritter 1989), and diagrams the labyrinth of topics comprising this complex science. Owing to this recency, domination by observationalists was short but observation has extended into succeeding periods. Most pre-Darwinian examples of geomorphic observationalists are best distinguished as geologists, geographers, or naturalists. Most notable, perhaps, was Alexander von Humboldt, a Prussian explorer of broad interests in geology, mineralogy, geophysics, climatology, and botany, who helped found physical geography and was a major stimulus for explorations of the American West (Pyne 1980). Among contemporaries of Darwin or those prominent shortly after was John Wesley Powell, who recognized structural control on stream courses, stating that folded structures tend to divert water around them. Powell, second director of the US Geological Survey (USGS), classified landforms and differentiated between valleys that trend perpendicular to the strike and those that trend parallel to rock layers. Based on Colorado River expeditions, Powell (1875) defined consequent, antecedent, and superimposed channels. Clarence Dutton, a companion of Powell, extended his ideas, such as the deduction that the leveling of the landscape is a product of river-bottom corrasion and slope weathering.

Another observationalist of the Darwinian period was Louis Agassiz, a dedicated follower of Humboldt (Pyne 1980). Agassiz was a Swiss zoologist and paleontologist – and later was an antievolutionist (McIntosh 1985), apparently threatened by the concept – who, in 1837, proposed that 'a great ice period' had occurred prior to uplift of the Alps (Agassiz 1840). Agassiz came to the United States in 1846, was first in a series of renowned geomorphologists at Harvard University, gained international acceptance for glacial landscape development (Flint 1971), and (despite previous problems with chronologies) largely started in North America the subdisciplines of glacial and Quaternary geology and glacial geomorphology.

Kirk Bryan was among the last of the acclaimed geomorphic observationalists. Bryan replaced William Morris Davis at Harvard in 1926 and was mentor there to J.T. Hack. A field-oriented generalist who contributed important papers on channel changes, soil phenomena, erosion and sedimentation, alluvial chronology in the southwest United States, terraces, slope retreat, and gully gravure, Bryan never employed a specific doctrine or exemplar. He was supportive of an evolutionary approach to geomorphology, especially that of Walther Penck, but scorned emerging quantitative techniques (Higgins 1975).

Identity for geomorphology occurred with William Morris Davis. A disciple of Darwin, Davis (Figure 17.3) wrote papers, best expressed in 'The geographical cycle' (Davis 1899), that treated landforms as evolutionary, time-dependent landscape features. Essentials of Davis's geographical cycle of erosion are well-known, but are summarized as (1) initial uplift of an area or rock mass, (2) progressive wearing down of the rock mass by weathering and erosion through unequal stages of landscape youth, maturity, and old age,

Figure 17.3 Portrait of William Morris Davis (from Photo Collection, US Geological Survey Library, Lakewood, Colorado)

and (3) an ultimate condition of landform reduction by erosion, yielding a peneplain of very gentle slope.

Davis patterned his model after Darwin's concept of evolution, and therefore subscribed to uniformitarianism. Writing of antecedent valleys (as previously defined by Powell) in the Appalachian Mountains, Davis (1883, p. 357) asserted that he 'tests the past by the present'. Time was mentioned repeatedly as the principal variable of landform genesis, and referrals to the 'origin of land-forms' and 'origin of cross valleys' intentionally followed Darwin's *Origins of Species*. With direct reference to organic evolution, Davis (1899, p. 485) wrote:

The larva, the pupa, and the imago of an insect; or the acorn, the full-grown oak, and the fallen old trunk, are no more naturally associated as representing the different phases in the life-history of a single organic species, than are the young mountain block, the maturely carved mountain-peaks and valleys, and the old mountain peneplain, as representing the different stages in the life-history of a single geographic group. Like land-forms, the agencies that work upon them change their behaviour and their appearance with the passage of time.

Shortly following, Davis (1899, p. 485) wrote that the 'sequence in the developmental changes of land-forms is, in its own way, as systematic as the sequence of changes found in the more evident development of organic forms'. Earlier, Davis (1883, p. 325) had suggested that the 'many pre-existent streams in each (Appalachian) river-basin concentrated their water in a single channel of overflow, and that this one channel survives – a fine example of natural selection'.

Davis, an eloquent lecturer and writer, dominated geomorphology for a half century using Darwin's exemplar. He was responsible, for example, for establishment of the Association of American Geographers in 1904 (Chorley et al. 1973, p. 417). Through the mid-twentieth century, most geomorphologists practiced 'normal science' of Davis's cycle of erosion, but alternative viewpoints in a context of time-dependency were expressed by Walther Penck early in the century, and by Lester King in the 1950s and 1960s. Penck described *knickpunkte* and *piedmont treppen* within a system of noncyclic slope retreat and crustal movement. King modified Davis's concept of peneplain formation by suggesting that landscapes form through 'integration of pediments that are enlarged by headward recession of scarps' (Higgins 1975, p. 9).

Commenting on the appeal that Davis's system sustained during several generations of geomorphologists, Higgins (1975, pp. 12–14) listed features including simplicity, seeming applicability to prediction and interpretation, presentation, and rationality. Davis's application of organic evolution to the physical world was timely and enticing, and an evolutionary basis for geomorphic thought 'filled a void'. That void was the lack of a doctrine, an exemplar, explaining landform development conformably with uniformitarianism. Hence, the exemplar of organic evolution, following temporary rejection by some of the prior generation (e.g. Tarr 1898; Smith 1899), was readily applied to landscapes.

The enthusiastic adoption and prolonged popularity of the erosion cycle were gradually eroded by doubts of Davis's assumptions regarding structure, process, and especially time (Hack (1960, p. 87) sardonically added senility as a final stage of Davis's cycle). Analogously to basic science, data and observations accumulated placing doubt on the Davisian system, and those data necessitated an exemplar replacement. The shift occurred because few if any peneplains could be identified, slope variations could not be explained, and variable uplift and erosion rates were difficult to address. These disparities caused a generally perceived need for change, but the shift did not and probably could not occur by the introduction of a revolutionary concept as drives exemplar replacement in the basic sciences. Thus, the discipline collectively sought an alternative to the Davisian system, and the alternative exemplar, from chemistry via soil science and hydrology, was equilibrium theory.

Dynamic equilibrium was first proposed for landforms by G.K. Gilbert (1877) in 'Geology of the Henry Mountains (Utah)'. Gilbert, who was influenced strongly by Powell, presented his ideas of geomorphic equilibrium prior to publication of Davis's system and van't Hoff's concept of chemical equilibria, but the popularity of evolution relegated this part of his classic paper to relative obscurity. Many have assumed that Davis's forcefulness explains the dominance of his ideas over Gilbert's. It seems likely, however, that the mood of the emerging discipline was insufficiently sophisticated to consider equilibrium concepts (e.g. Chorley et al. 1973, pp. 196–197; Higgins 1975, pp. 12–14; Ritter 1978, pp. 4–5; Pyne 1980, pp. 254–261).

Harbingers of exemplar shift were papers by Horton (1945), describing morphometric approaches to drainage basins, and Strahler (1950, 1952, 1954, 1957), who anticipated the application of equilibrium to landscapes through quantitative techniques. Robert Horton was, like Roger Bacon, insistent on the application of quantified data and mathematical techniques to investigate process; although a meticulous engineer steeped in equilibrium techniques, he was influenced by the popularity of the Davisian system. Arther Strahler stressed process and incorporation of mechanics, fluid dynamics, and quantitative techniques into geomorphic studies, although he too was inclined to blend the erosion cycle into an equilibrium format (e.g. Strahler 1954, p. 353). It was not until J.T. Hack (1960) revived Gilbert's concepts of geomorphic equilibria (himself initially unaware of this proposal nearly a century earlier) that doubts of the Davisian system presented a suitable climate for exemplar replacement. Although Hack's paper was the catalyst, change occured more through a consensus of uneasiness than by a startling new idea. An imposed exemplar replaced an imposed exemplar, a result being, in the 1960s and 1970s, attention to equilibrium-related topics such as allometry, topology, and a variety of statistical techniques.

John Hack completed his doctorate at Harvard under Kirk Bryan in 1940. After two years of teaching, he joined the USGS and served with its Military Geology Unit. Following World War II, Hack began pursuing reseach interests along the Maryland coastal plain and in the Appalachian Ridge and Valley Province where Davis had developed many of his ideas. Hack joined a group of scientists working in the Shenandoah Valley; others were C.C. Nikiforoff, C.B. Hunt, and Harvard graduate-student friends M.G. Wolman, C.S. Denny, J.C. Goodlett (Figure 17.4), and L.B. Leopold. Hunt, Wolman, Denny, and Leopold were geomorphologists/hydrologists, whereas Nikiforoff was a Russian refugee soil scientist (Figure 17.5) and Goodlett a plant ecologist. Members of this unique band of equilibrists interacted, reinforcing equilibrium concepts expressed in numerous papers on geomorphology, ecology, pedology, and hydrology (Osterkamp 1989).

Among Hack's products were reports on longitudinal stream profiles (Hack 1957), entrenched meanders (Hack and Young 1959), and the geomorphology and plant ecology of an Appalachian watershed (Hack and Goodlett 1960). These studies revealed conflicts with erosion-cycle concepts and led to 'Interpretation of erosional topography in humid temperate regions' (Hack 1960), which explicitly offered time-independent equilibrium as an alternative to the Davisian system. Although detractors such as J. Hoover Mackin were antagonistic, acceptance occurred rapidly, and it became fashionable to be critical of Davis.

Dominance of equilibrium was short because many geomorphologists realized the futility of discarding time. Thus, integrationists – those trying to reconcile systems of evolution and equilibrium – soon offered explanations of compatibility, that a system is applicable depending on scales of space and time. Richard Chorley (1962), although preferring open-system dynamic equilibrium, concisely analyzed differences between the systems and benefits of each. Schumm and Lichty (1965) presented objective-dependent guidelines for applying the conceptual models to landscapes. Efforts to reconcile the polar

Figure 17.4 Photograph of mid-twentieth-century equilibrists J.T. Hack (geomorphology) on the left and J.C. Goodlett (plant ecology), in the Little River Basin, Virginia, 1955 (courtesy of Clare Hack)

extremes of geomorphic systems have yielded recently to models of integration including geomorphic thresholds (e.g. Schumm 1973, 1979), complex response (Schumm 1973), nonlinear dynamics (Middleton 1990), and renewed recognition of a systems approach (Ritter 1978).

Ecology

The term *ecology*, and its goal of relating organisms and environment, generally are attributed to German zoologist Ernst Haeckel (1866), who later tied the concept firmly to evolution by stating that 'ecology is the study of all complex interrelations referred to by Darwin as the conditions of the struggle for existence' (Allee et al. 1949). The first significant practice of ecology, however, may be attributable to Humboldt's pre-Darwinist studies in South America (Gendron 1961).

Created as normal science to validate the exemplar of evolution, ecology experienced a short period of unbiased yet poorly directed observation. Plant communities were assumed static, and vegetation of large areas was described simply by compiling species lists (Joyce 1993) or by assuming simple relations between vegetation and climate (Merriam 1894). In a discussion of the origins of ecology, McIntosh (1985) suggests *polymorphic* to describe ecology, and notes that prior to about 1910, ecologists were criticized for being too

Figure 17.5   Photograph of mid-twentieth-century equilibrists C.C. Nikiforoff (soil science), on the left and J.T. Hack, Christmas 1954 (courtesy of Clare Hack). Figures 17.4 and 17.5 signify the interdisciplinary interactions that have contributed to the imposition of exemplars by the composite sciences

concerned with observation and description. One culprit was V.M. Spalding (1903, p. 207), who maintained that ecologists should '... ascertain and record fully, definitely, perfectly and for all time facts'. Process or interpretation were not objectives.

The temporal aspect of evolution was introduced into ecology by Darwinists Henry Cowles, Frederick Clements, and Victor Shelford. They rejected the compilation of species lists but instead emphasized species change through time (Allen and Hoekstra 1992). Cowles studied geology and botany at the University of Chicago, where he took his doctorate in 1898 and taught. Earlier Cowles had been a student of Davis at Harvard (H.M. Raup, oral presentation, Rutgers University, 1972). In 1895 Cowles worked with the USGS and may have interacted with Davis, who conducted USGS field studies then in the New England area. As student and professor in Chicago, Cowles, who was influenced strongly by Davis's teaching and writing, worked with distributions of xeric vegetation on the sand-dune and beach deposits bordering the south shore of Lake Michigan (Cowles 1899), and applied the erosion model to vegetation patterns of the sandy shorelines. As a direct impetus 'from the contemporary studies of the cycle of erosion by the physiographer, William Morris Davis' (Raup 1952, p. 306), Cowles identified *successions* of plants relative to distance from the lake edge and the time required, following *disturbance*, for shoreline retreat to have progressed that distance. The oldest, most distant stand was dominated by beech and maple trees, termed the climax community (Colinvaux 1973). The ecology of Cowles, therefore, was fully parallel to the cycle-of-erosion model and direct analogies are evident between initial uplift and disturbance, stage of erosion and succession (the assumed sequence by which plant communities change in an orderly progression through time), and peneplain and climax community. As did the Darwinian and Davisian systems, succession relied on deduction and minimized process (Mayr 1982). Presumably, 'succession', as used in ecology, was derived directly from William Smith's 'faunal succession', as applied to the fossil record over 100 years earlier.

Frederick Clements (Figure 17.6) grew up on the prairie and attended the University of Nebraska. Clements's observations of prairie plants, and changes he saw in their distributions following disturbance by frontier wagon traffic and land development, made him responsive to Cowles's shoreline observations. Clements (1928, p. 3) viewed plants as members of a highly organized community, or *complex organism* that 'arises, grows, matures, and dies ...'; environment was given little attention. With the persuasiveness of Davis, Clements promoted Cowles's concepts of succession, and thereby imposed the exemplar of evolution on plant ecology. Clements (1916) expanded the concept of the climax community, applying it globally to *formations*, plant communities controlled by climate but exhibiting a range of seral stages of various primary successions, all subject to eventual areal climax or formation. In this manner, Clements developed classes of global formations. Whether 'formation', which was applied to plant communities of Midwestern landscapes in the 1890s (Allen and Hoekstra 1992), was extracted from similar usage in earth science is unclear.

Victor Shelford was an animal ecologist and a student/colleague of Cowles who also worked on the Lake Michigan sand dunes; predictability, and conforming to a normal-science effort to ratify an ecological use of evolution, he claimed animal succession in parallel to Cowles's observations for plants (Colinvaux 1973). Shelford worked closely with Clements, adopted his system, including convergence toward a regional climax, and collaborated on 'bio-ecology' (Clements and Shelford 1939), a combining of animal and plant ecology (McIntosh 1985).

Succession remains dominant in plant ecology owing to obvious changes in species composition that occur after disturbance and because of the same simplicity and applicability (H.M. Raup, oral presentation, Rutgers University, 1972) noted by Higgins (1975) for the Davisian system. The lack of recognized process, however, soon exposed the Clementsian system to criticism and the potential for an exemplar shift. Partial replacement during the 1920s was effected by L.G. Ramensky, in Russia, and by H.A. Gleason in Illinois. As Strahler had favored quantitative approaches to geomorphology, Ramensky and Gleason advocated techniques, largely environment-dependent rather than time-dependent, that required detailed measurements. Instead of using Clements's community concept, Ramensky and Gleason emphasized survival of individual plants, maintaining

Figure 17.6   Portrait of Frederick Clements (from Desert Laboratory, Tucson, Arizona)

that the collection of individuals in an area is a consequence of similar conditions of dispersal and environmental requirements. Moreover, Gleason stressed vegetative change as a function of environmental change along a continuous gradient (McIntosh 1975, 1983; Allen and Hoekstra 1992). Ramensky (1924), quoted by McIntosh (1983, p. 8), anticipated conversion by ecologists toward a structure founded on process and equilibrium when he wrote that ecology's future lies 'in deeper analysis of relations, acting factors and equilibrium mechanisms'.

As did Ramensky and Gleason, equilibrists William Cooper, Hugh Raup, and Robert Sigafoos recognized that climax forests are theoretical and cannot be documented. They dismissed succession, maintaining that plant associations depend strictly on migration and environmental selection. Although a student and advocate of Cowles, Cooper envisioned a mosaic of vegetation patches (McIntosh 1985), in which a 'forest as a whole remains the same, the changes in various parts balancing each other' (Cooper 1913, p. 43). Cooper too used landform as a parallel to ecologic change by cogently comparing long-term vegetation development to stream braiding (Allen and Hoekstra 1992). Raup, professor and director of the Harvard Forest, 1946 to 1967, and Sigafoos, USGS, were advisors to Hack and Goodlett, and Sigafoos and Goodlett were students of Raup and Bryan at Harvard. During a career in which he strongly questioned the steady-state ideal of succession to a climax forest, Raup (1941) preferred vegetative adjustment toward dynamic equilibrium.

Reflecting Raup's disdain of Clementsian ecology and his attention to process, Robert Sigafoos was first to detail interactions among vegetative development, flood damage, and floodplain dynamics. Sigafoos (1964) demonstrated widespread disturbance to bottomland vegetation by floods and thereby emphasized dynamic equilibrium and change on temporal scales too short to result in climax. The British animal ecologist, Charles Elton, also disregarded Darwinian concepts of adaptation and evolution, preferring process, group dynamics, and equilibria of the food chain and the food cycle – 'the sociology and economics of animals' (Elton 1927, p. vii).

Equilibrist J.C. Goodlett (Figure 17.4) studied relations of vegetation to landforms and the processes that develop landforms. Goodlett (1969, p. 35) stated that 'plant cover is a part of the open system that constitutes the landscape, and the vegetation is in a state of continuous adjustment with its environment'. Commenting on landform change and its effect on vegetation, Goodlett (1969, p. 38) extended concepts of Gleason by suggesting that 'the plant cover must adjust to these modifications, or pass on. Geomorphic processes, that act to mould the landscape, take place on, in, or through the plant cover. The plants adjust to the environmental variations produced by the geomorphic processes, and in turn they affect the processes and their products.' Thus, Goodlett (1969) subscribed strongly to equilibrium, stressing that geomorphology and biology cannot be separated, and that the individualistic concept of plant ecology is similar to dynamic equilibrium of geomorphology. The Ramensky/Gleason/Goodlett model was a process-oriented, time-independent, open-system approach based on interplay between vegetation and environmental factors responsible for its composition. The equilibrium, or individualistic, concept of ecology addresses present processes to explain features, without need for final, ideal condition (Hupp 1984).

Integrationists in ecology overlapped with equilibrists. For this discussion, integrationists are those who proposed complete, interacting systems in biology, and those who provided guidance by identifying trends within ecology. Representatives of the former group are Forrest Shreve and E.P. Odum. The latter group includes Ramon Margalef, who published *Perspectives in Ecological Theory* (1968), and Robert McIntosh, whose historical works on ecology culminated with *The Background of Ecology* (1985). For rangeland ecology, E.J. Dyksterhuis (1949) modified Clementsian succession by noting that grazing influences rangeland succession in a predictable and quantitative manner.

Forrest Shreve, at the Carnegie Institution's Desert Laboratory on Tumamoc Hill, Tucson, from 1907 into 1940 (Bowers 1988), was a colleague of Clements. Clements spent winters of 1917 through 1924 at the laboratory, when incompatibility may have led to Shreve's rejection of portions of Clementsian doctrine, especially as applied to deserts (Bowers 1988). Shreve (1936, p. 213) wanted 'to weave together the separate threads of knowledge about the plants and their natural setting into a close fabric of understanding in which it will be possible to see the whole pattern and design of desert life'. Reflecting Ramnesky and Elton, Shreve (1936, p. 213) remarked that the 'distribution of a plant species reflects its tolerance for a range of environmental conditions, thus few if any species have identical distributions; trends in establishment and mortality in plant populations tell much about the conditions necessary for growth; changes in climate and vegetation along an environmental gradient reveal the conditions that limit a plant's distribution'. In earlier work, Shreve (1919) suggested dispersal processes to account for species populations in isolated mountains, a paper that inspired integrationist models of nonequilibrium insular biogeography (e.g. Brown 1978).

Like Shreve, E.P. Odum had a holistic perspective conflicting with the more publicized ideas of Clements. Ironically, *holism* was coined by philosopher/statesman, J.C. Smuts, who credited Clements for the idea (Colinvaux 1973). Odum (1953, 1969) developed the concept of *ecosystem*, an open-system ecological approach (similar to Hack's for landscapes) for fluxes of matter and energy (McIntosh 1985). As Chorley (1962) later did for geomorphology, Bertalanffy (1951) united thought in biology and ecology by applying systems theory.

Soil Science

The history of soil studies has differed from that of geomorphology owing to agricultural and economic considerations. The study of a natural resource, the science of soil processes treats the formation, properties, classification, and mapping of soils. An observationalist period that began by the late nineteenth century was largely one of noting characteristics to classify and map soils. The first widely recognized attempts, by V.V. Dokuchaev (1879) and other Russian soil scientists, were based on climate and vegetation (Nikiforoff 1949), criteria still used worldwide in soil classification. Refinements were added by K.D. Glinka in Russia, Emil Ramann in Germany, and C.F. Marbut, M. Baldwin, and James Thorp and G.D. Smith, United States. Marbut's (1928) scheme was nearly restricted to mature soil and did not consider process. Baldwin et al. (1938) added detail and nomenclature to the Marbut system, but made little attempt to account for soil differences. Thorp and Smith (1949) added complexity to prior classifications by subdividing into Great Soil Groups. These classifications were largely based on genetic factors of climate and vegetation; not until 1960 was a nongenetic system devised in the United States, by the Soil Conservation Service, founded on quantitative physical and chemical criteria (Ritter 1978).

Soil studies, applied and stressing resource management, belatedly evolved to soil science or pedology. Although weathering and soil chemistry were investigated in Europe in the 1930s, imposition of an exemplar occurred later than for other composite sciences, and when it did, evolution and equilibrium appeared nearly simultaneously. Hans Jenny (Figure 17.7), the first noteworthy Darwinian of pedology, applied evolution to soil development in a seminal treatise on soil formation. With his 'fundamental equation of soil-forming factors', Jenny (1941, p. 16) proposed that soils (S) and soil properties (s) are results of climate (cl), biota (o), topography (r), parent material (p), and time (t):

$$S, \ s = f(cl, o, r, p, t, \ldots) \tag{3}$$

For constant climate, organisms, parent material, and topography, Jenny (1941, pp. 31, 49) asserted that the soil profile is 'solely a function of time':

$$S = f(\text{time})_{cl,o,r,p,\ldots} \tag{4}$$

Figure 17.7   Photograph of Hans Jenny (courtesy of P.W. Birkeland)

Equation (4) implies orderly soil changes through time, dependent on other independent variables (Jenny 1941, 1980); hence, the system is directly analogous to Davis's erosion cycle, Clementsian succession, and Darwinian evolution (Figure 17.8). Jenny's work led to concepts of chronosequences and chronofunctions of soils (Birkeland 1990), and referred specifically to soil evolution due to influxes of heat, rainfall, and light. End members of chronosequences suggested by Jenny were soils of iron and aluminium oxides and hydroxides, equivalent to a peneplain of Davis or climax forest of Clements.

Pedologist C.C. Nikiforoff (Figure 17.5) emigrated to the United States from Russia following World War I and joined the Soil Survey. From ideas developed earlier, and recognizing but disagreeing with Jenny's (1941) equation, Nikiforoff (1942) published 'Fundamental formula of soil formation', which, analogous to earlier work of Odum, viewed soil processes as fluxes of matter and energy. As Gilbert's equilibrium was overlooked for the Davisian system, Nikiforoff's (1942) paper attracted scant attention owing to the immediate popularity of Jenny's Darwinian model. Nikiforoff's work may never have gained recognition had not John Hack used the approach to develop his own

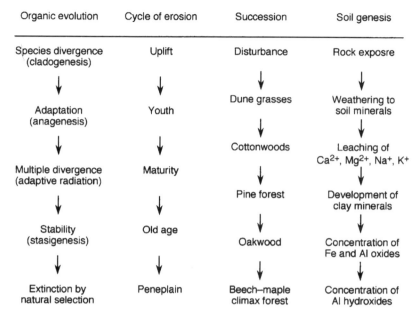

Organic evolution	Cycle of erosion	Succession	Soil genesis
Species divergence (cladogenesis)	Uplift	Disturbance	Rock exposre
↓	↓	↓	↓
Adaptation (anagenesis)	Youth	Dune grasses	Weathering to soil minerals
↓	↓	↓	↓
Multiple divergence (adaptive radiation)	Maturity	Cottonwoods	Leaching of Ca^{2+}, Mg^{2+}, Na^+, K^+
↓	↓	↓	↓
Stability (stasigenesis)	Old age	Pine forest	Development of clay minerals
↓	↓	↓	↓
Extinction by natural selection	Peneplain	Oakwood	Concentration of Fe and Al oxides
		↓	↓
		Beech–maple climax forest	Concentration of Al hydroxides

Figure 17.8 Diagram comparing generalized organic evolution with evolutionary systems imposed on geomorphology (cycle of erosion), plant ecology (succession), and pedology (soil genesis)

model of landscape dynamics. Because equilibrium, applied to pedology and other composite sciences, considers fluxes, progressive change is not emphasized and a flow diagram analogous to that for evolution (Figure 17.8) is impracticable.

Nikiforoff (1942, p. 847) maintained that if 'nothing is synthesized in the soil which does not decompose and nothing decomposes which is not synthesized', neither accumulation nor depletion of soil components can continue 'without coming to a certain equilibrium between the losses and gains'. A limiting expression for balance was derived as

$$S_n = A/r \tag{5}$$

in which S_n is the amount of soil substance after n years, when equilibrium has been attained, A is the mass, assumed constant, that is synthesized each year, and r is the annual rate of mass lost by decomposition. Thus, dynamic equilibrium is achieved after n years, when soil processes may still be intense but do not alter features of the profile.

Nikiforoff's thesis, tentatively accepted, exposed an elemental difficulty in applying evolution to Earth science. That difficulty was the assumption that a soil or landscape can be stable long enough to yield an excessively thick soil of oxides and hydroxides underlying a peneplain and climax forest. If stability could persist, the earth '... would be wrapped in a *lifeless mantle thoroughly deprived of the unstable minerals, composed entirely of those most resistant to any further changes, and, hence, essentially static*' (Nikiforoff 1949, p. 222). Recent pedologists have acknowledged this difficulty, but a consensus continues that even where geology and climate are constant, the effects of denudation necessitate consideration of time (e.g. Birkeland 1990). This view is especially strongly held by those desiring to date surfaces or to deduce paleoenvironments from soil characteristics.

Integrationists have focused on strengths of each system. Birkeland (1984, 1990) recognized that formation factors of mature soils have been variable – soils are poly-genetic. He therefore advocated that attention be placed on process and factors controlling soil development. Combining soil science and geomorphology, R.V. Ruhe emphasized climatic variables as determinants of soil structure and composition. Introductory comments by Ruhe (1975, p. 3) in his text on surficial geology stress both 'the nature and evolution of the landscapes and materials of the earth ...'. A classic paper integrating soil evolution and process in a semiarid environment resulted from years of field observations by Gile et al. (1981).

Hydrology

Development of hydrology as a composite science during the last two centuries was largely by engineers (Biswas 1970). Thus, use of an equilibrium exemplar, common to engineering, was substantial. Persuasive examples that post-1859 hydrology was con-trolled by evolution, however, are more difficult to cite. Additionally, because hydrology concerns water fluxes and therefore is reliant on engineering principles, the science is only weakly susceptible to a controlling creed. Trends have been apparent in the development of hydrology, however, especially because of interactions and overlap with geomorphol-ogy, ecology, pedology, and basic science.

Observationalism in hydrology began with the start of modern science in the seven-teenth century; an example was French lawyer Pierre Perrault who, about 1670, suggested that rainfall in the upper Seine River basin could account for runoff in the river (Biswas 1970). French physicist Edmé Mariotté confirmed Perrault's observation by measuring flows in the Seine River, and accordingly was a founder of hydrology (Todd 1967). O.E. Meinzer (1923) was late among observationalists in hydrology, recognizing geology as an important control of groundwater recharge and discharge: Meinzer was responsible for defining groundwater study as a major component of the USGS program.

As previously noted, John Wesley Powell initially was a geomorphic observationalist. He clearly showed evolutionary thinking in surface-water hydrology and erosion, however, when writing of his explorations of the Colorado River (1875) and the Unita Mountains (1876). His concept of denudation to base level anticipated the Davisian system, and his recognition that all mountains are reduced by streamflow became fundamental to the cycle of erosion (Chorley et al. 1973). Robert Horton, the scientist and engineer who insisted on quantification, published 'Erosional development of streams and their drainage basins; hydrophysical approach to quantitative morphology' (1945), a benchmark Davisian treatment of drainage evolution using quantitative approaches.

Equilibrium approaches are basic to hydrology, especially water-balance studies. A particular advance in hydrologic equilibrium, however, was 'The hydraulic geometry of stream channels and some physiographic implications', by L.B. Leopold and Thomas Maddock, Jr (1953) who extended hydraulic–geometry relations from engineering regime theory (Kennedy 1895; Lacey 1930) and the graded-stream concept (Mackin 1948). Discussing hydraulic adjustments, Leopold and Maddock (1953, p. 51) stated that a '... particular rate of increase of both velocity and depth downstream is necessary for maintenance of approximate equilibrium in a channel ...'. Within groundwater hydrology, C.V. Theis, from the University of Cincinnati, was recruited to the USGS by Meinzer. Theis (1935) proposed the 'nonequilibrium' equation, a watershed advance in ground-water theory based on equilibrium and assumptions of infinite areal extent and homogeneity of aquifers.

Hydrologic integrationists are regarded here as those combining attention on processes with rates at which processes change owing to external inputs of climate, biotic alteration, or land disturbance. Among recent integrationists, Walter Langbein and M.G. Wolman recognized effects that short-term climate change and land-use change have on stream-flow, channel morphology, and sediment discharge.

Social Science

Social sciences, including economics, are not composites as defined. Their histories, however, also show forcing of concepts, exemplars of 'cultural equilibrium', from basic science (Schumpeter 1934, p. xi). For example, response to evolution, following spec-ulations of Adam Smith, David Ricardo, and John Stuart Mill, was shown within economics by Karl Marx. If not an evolutionist, Marx reflected late-1800s Darwinism by writing of temporal factors in profits, social impoverishment, and business-cycle trends. Herbert Spencer, a British philosopher and biologist who was repeatedly cited by Clements and was the probable source of his 'complex organism' (Worster 1977), espoused 'social Darwinism', the suggestion that elite classes of wealth and power possess biological superiority and thus attain superior socioeconomic position. Combining evo-lution with Greek tradition regarding nature as an organism, French geologist/cleric Pierre Teilhard de Chardin (1959) compared the ontogeny of human society to species whose evolutionary histories necessarily direct their destinies.

Anticipating economic 'marginalists', Alfred Marshall applied 'partial equilibrium analysis' to the balance between supply and demand. F.H. Knight, an unabashed equi-librist, was among those proposing economic analogs to mass, inertia, momentum, force, and space, thereby applying Newtonian laws of motion to economic dynamics; he asserted that the 'root idea in economic statics is clearly the notion of *equilibrium*, and hence of *forces* in equilibrium' (Knight 1935, p. 162). P.A. Samuelson noted that economists, Marx and Marshall as examples, have applied biological and physical concepts to economic study '... in which evolution and organic growth is used as the antithesis to statical equilibrium analysis' (Samuelson 1961, pp. 311–312). Although Samuelson failed to appreciate the extent to which evolution and equilibrium affected the social sciences, he seemed to recognize that exemplars of another discipline could not be imposed suc-cessfully when he mused that results had been hazy and disappointing.

Early this century, unilinear sequences of social stage – evolutionary sociology – yielded to universal patterns of equilibrium and associated periods of disequilibrium. Integrationist John Maynard Keynes analyzed determinants of effective demand and levels of national income and employment rather than corporate equilibrium or allocation of resources. Thus, Keynes de-emphasized business equilibria and time. Although the enjoining of evolution, equilibrium, and integration on the social sciences cannot be documented as forcefully as for composite science, writings of leaders from Marx through the present clearly show similar and vigorous impacts by borrowed and probably inappropriate exemplars.

CONCLUSIONS

In this discussion, extending ideas of Thomas Kuhn, we propose that composite sciences have matured much differently than basic sciences. A principal reason for differences in development is that basic sciences are defined by exemplars (Kuhn 1970), whereas composite sciences are too complex to be defined by an exemplar and, lacking exemplar superstructures, normal science is undirected. The necessary and partly beneficial result is the imposition of exemplars that may rule effectively but are ill-suited and hence fail to define. Because imposed exemplars cannot guide the conduct of composite science appropriately, their acceptance may lead to doctrine if not dogma.

This thesis, with supporting examples, are interpretations (for a similar premise, see Stoddart 1986, Chapters 8 and 11), but regardless of how the composite sciences have grown, their periods of growth may be better identified by style or fad than as functions of scientific discovery. Styles of normal composite science proposed here are Darwinian evolution followed by equilibrium theory. The former was preceded by observation and the latter was followed by attempts to combine the extremes. This pattern of exemplar imposition on the composite sciences seems less similar to the use of definitive exemplars by the basic sciences than it does to styles in the arts (i.e. the baroque, classical, romantic, and impressionist periods). As the romantic period differed in time and duration for each art form, the period and duration of dominance by evolution differed within each composite science. Furthermore, the time scales of models depicted in Figure 17.8 differ, ranging from centuries for succession to millions of years for peneplanation.

If borrowed exemplars guided composite science, a secondary factor was personal interactions. As Mozart was influenced by Haydn, and Monet by contemporaries of French impressionism, style lineage has been apparent in composite science. Examples are close relationships that Nikiforoff, Hack, Leopold, and others enjoyed in the Shenandoah Valley, and contacts among Cowles, Clements, Shelford, Davis, and, in an opposite manner, Shreve. Linear influences in the USGS seem apparent for Humboldt, Powell, Dutton, Gilbert, Meinzer, Bryan, Theis, Hack, Sigafoos, and possibly even Davis and Cowles, and similarly with students and faculty at Harvard as represented by Agassiz, Davis, Cowles, Bryan, Hack, Leopold, Raup, Sigafoos, and Goodlett. During formative years of the natural sciences in the British Isles, Darwin was influenced by Hutton, who enriched ideas of Smith. Although effects of personal interactions within basic science are also obvious, they may be less pronounced when exemplar replacement helps direct the science instead of the science embracing an attractive exemplar.

The choice of evolution as an exemplar by composite scientists may have been hasty. Figure 17.8 represents models as parallel flow diagrams, but actual parallels between evolution and its use in geomorphology, ecology, and soil science are questionable. Of the models diagrammed, evolution starts with variability, but variability develops through time in the others. Extinction is not the culmination of evolution as is peneplanation or climax communities for erosion and succession. Extinction occurs at any stage of evolution, and some species persist indefinitely. Whereas evolution is linear, erosion, succession, and soil genesis are cyclic. As these discrepancies between evolution and its use in the composite sciences became apparent, the need for exemplar replacement became imperative. More as integrationist than equilibrist, these thoughts were expressed succinctly but differently by Raup (1964, p. 26):

> The geologists had their peneplain; the ecologists visualized a self- perpetuating climax; the soil scientists proposed a thoroughly mature soil profile, which eventually would lose all trace of its geological origin and become a sort of balanced organism in itself. It seems to me that social Darwinism, and the entirely competitive models that were constructed for society by the economists of the nineteenth century, were all based upon a slow development towards some kind of social equilibrium. I believe there is evidence in all of these fields that the systems are open, not closed, and that probably there is no consistent trend towards balance. Rather, in the present state of our knowledge and ability to rationalize, we should think in terms of massive uncertainty, flexibility, and adjustment.

The composite sciences continue to integrate disparate points of view. Recently the maturing process has narrowed with normal-science studies of limited scope and scale, examples being the use of fluvial dynamics to analyze river-channel islands, use of cosmogenic radioisotopes to estimate denudation rates, or documentation of numerically small occurrences of adult plants suggestive of specific ecological conditions. Conversely, an ultimate goal of any science is a sweeping exemplar, or *symbolic generalization* (Kuhn 1977, p. 297), of unified theory. If the cosmology of Copernicus and Kepler or relativity of Einstein in particle (nonquantum) physics represent unifying theory, another cycle of exemplar replacement may not be feasible, and normal-science decline may be assured. Essential problems of planetary motion have long been solved, and already many look back at the *Golden Age of Physics*.

Owing to complexity and inability of composite sciences to be defined by exemplars, unifying principles may not be possible. Holism as an ideal provides a goal, but does not present explanation. The Gaia hypothesis (Lovelock 1965) is a holistic suggestion that the dynamics of physical and biological processes on Earth are integrated and function as one evolving system (Margulis 1993). A full-circle journey from early Greek speculation, holism, as represented by the Gaia hypothesis, does seem to have application to composite science. A unifying theory may be unattainable, but future goals of composite science, being broadly scoped, may emphasize complete integration of physical and biotic processes at various scales of time and space.

Attention to integrated process studies at decadal and watershed scales increases with environmental concern. New exemplars, original or imposed, seem unlikely to direct further conduct of the composite sciences. It seems, however, that small-scale studies of integrated process, driven by borrowed exemplars of limited application, must yield to global perspectives. A unified theory may be an unrealistic objective, but a unified perspective seems desirable.

REFERENCES

Agassiz, L. 1840. *Études sur les glaciers*: Neuchâtel, privately published, 346 pp.
Allee, W.C., Emerson, A.E., Park, O., Park, T. and Schmidt, K.P. 1949. *Principles of Animal Ecology*, Saunders, Philadelphia, 837 pp.

Allen, T.F.H. and Hoekstra, T.W. 1992. *Toward a Unified Ecology*, Columbia University Press, New York, 384 pp.

Baldwin, M., Kellogg, C.E. and Thorp, J. 1938. Soil classification, in *Soils and Men*, US Department of Agriculture Yearbook, pp. 979–1001.

Betalanffy, L. von 1951. General systems theory: a new approach to unity of science, in *Problems of General Systems Theory*, Symposium of the American Philosophical Society, Toronto, 1950, pp. 302–311.

Birkeland, P.W. 1984. *Soils and Geomorphology*, Oxford University Press, New York, 372 pp.

Birkeland, P.W. 1990. Soil-geomorphic research – a selective overview, *Geomorphology*, **3**, 207–224.

Biswas, A.K. 1970. *History of Hydrology*, North-Holland, Amsterdam, 336 pp.

Bowers, J.E. 1988. *A Sense of Place – the Life and Work of Forrest Shreve*, The University of Arizona Press, Tucson, 195 pp.

Bowler, P.J. 1984. *Evolution – the History of an Idea*, University of California Press, Berkeley, 412 pp.

Bowler, P.J. 1992. *The Fontana History of the Environmental Sciences*, Fontana Press, London, 634 pp.

Brown, J.H. 1978. The theory of island biogeography and the distribution of boreal birds and mammals, *Great Basin Naturalist Memoirs*, **2**, 209–227.

Chorley, R.J. 1962. Geomorphology and general systems theory, *US Geological Survey Professional Paper 500-B*, 10 pp.

Chorley, R.J., Beckinsale, R.P. and Dunn, A.J. 1973. *The History of the Study of Landforms or the Development of Geomorphology* – Vol. 2: *The Life and Work of William Morris Davis*, Methuen, London, 874 pp.

Clements, F.E. 1916. *Plant Succession – an Analysis of the Development of Vegetation*, Carnegie Institution of Washington Publication No. 242, Washington, DC.

Clements, F.E. 1928. *Plant Succession and Indicators*, H.W. Wilson, Washington, D.C.

Clements, F.E. and Shelford, V.E. 1939. *Bio-ecology*, Wiley, New York, 425 pp.

Colinvaux, P.A. 1973. *Introduction to Ecology*, Wiley, New York, 621 pp.

Cooper, W.S. 1913. The climax forest of Isle Royale, Lake Superior, and its development, *Botanical Gazette*, **55**, 1–44, 115–140, 189–235.

Cowles, H.C. 1899. The ecological relations of the vegetation of the sand dunes of Lake Michigan, *Botanical Gazette*, **27**, 95–117, 167–202, 281–308, 361–391.

Darwin, C.R. 1859. *The Origin of Species by Means of Natural Selection; or, the Preservation of Favored Races in the Struggle for Life*, John Murray, London, 502 pp.

Davis, W.M. 1883. Origin of cross valleys, *Science*, **1**, 325–327, 356–357.

Davis, W.M. 1899. The geographical cycle, *Geographical Journal*, **14**, 481–504.

Dokuchaev, V.V. 1879. Abridged historical account and critical examination of the principal soil classifications existing, *Transactions, Saint Peterburg Society of Nature*, **10**, 64–67 (in Russian).

Dyksterhuis, E.J. 1949. Conditions and management of range land based on quantitative ecology, *Journal of Range Management*, **2**, 104–115.

Elton, Charles 1927. *Animal Ecology*, Macmillan, New York, 207 pp.

Flint, R.F. 1971. *Glacial and Quaternary Geology*, Wiley, New York, 892 pp.

Frodeman, R. 1995. Geological reasoning: geology as an interpretive and historical science, *Geological Society of America Bulletin*, **107**, 960–968.

Gendron, V. 1961. *The Dragon Tree – a Life of Alexander, Baron von Humboldt*, Longmans, Green, New York, 214 pp.

Giere, R.N. 1988. *Explaining Science, a Cognitive Approach*, University of Chicago Press, Chicago, 321 pp.

Gilbert, G.K. 1877. Geology of the Henry Mountains (Utah), *US Geographical and Geological Survey of the Rocky Mountains Region*, US Government Printing Office, Washington, DC, 160 pp.

Gile, L.H., Hawley, J.W. and Grossman, R.B. 1981. *Soils and Geomorphology in the Basin and Range Area of Southern New Mexico – Guidebook to the Desert Project*, New Mexico Bureau of Mines and Mineral Resources, Memoir 39, 222 pp.

Goodlett, J.C. 1969. Vegetation and the equilibrium concept of the landscape, in *Essays in Plant Geography and Ecology*, edited by K.N. Greenidge, Nova Scotia Museum, Halifax, pp. 33–44.

Hack, J.T. 1957. Studies of longitudinal stream profiles in Virginia and Maryland, *US Geological Survey Professional Paper 294-B*, pp. 45–97.

Hack, J.T. 1960. Interpretation of erosional topography in humid temperature regions, *American Journal of Science, Bradley Volume*, **258-A**, 80–97.

Hack, J.T. and Goodlett, J.C. 1960. Geomorphology and forest ecology of a mountain region in the central Appalachians, *US Geological Survey Professional Paper 347*, 66 pp.

Hack, J.T. and Young, R.S. 1959. Intrenched meanders of the North Fork of the Shenandoah River, Virginia, *US Geological Survey Professional Paper 354-A*, 10 pp.

Harvey, W. 1628. *De motu cordis et sanguinis in animalibus*, R. Willis (translator), 1848, London.

Haeckel, E. 1866. *Generelle Morphologie der Organismen: Allgemeine Grundzüge der organischen Formen-wissenschaft, mechanisch begründet durch die von Charles Darwin reformirte Descendenz-Theorie*, 2 vols, Reimer, Berlin.

Haines-Young, R.H. and Petch, J.R. 1986. *Physical Geography: Its Nature and Methods*, Harper & Row, London, 230 pp.

Higgins, C.G. 1975. Theories of landscape development – a perspective, in *Theories of Landscape Development*, edited by W.N. Melhorn and R.C. Flemal, Publications in Geomorphology, State University of New York, Binghamton, NY, pp. 1–28.

Horton, R.E. 1945. Erosional development of streams and their drainage basins; hydrophysical approach to quantitative morphology, *Geological Society of America Bulletin*, **56**, 275–370.

Hupp, C.R. 1984. Forest ecology and fluvial geomorphic relations in the vicinity of the Strasburg Quadrangle, Virginia, Ph.D. dissertation, George Washington University, Washington, DC, 168 pp.

Hutton, J. 1795. *Theory of the Earth, with Proofs and Illustrations*, Vols 1 and 2, Edinburgh; Vol. 3, London, 1899.

Jenny, H. 1941. *Factors of Soil Formation*, McGraw-Hill, New York, 281 pp.

Jenny, H. 1980. *The Soil Resource – Origin and Behavior*, Springer-Verlag, New York, 377 pp.

Joyce, L.A. 1993. The life cycle of the range condition concept, *Journal of Range Management*, **46**, 132–138.

Kennedy, R.C. 1895. Prevention of silting in irrigation canals, *Institute of Civil Engineers Proceedings*, **119**, 281–290.

Kitts, D.B. 1977. *The Structure of Geology*, Southern Methodist University Press, Dallas, 180 pp.

Knight, F.H. 1935. *The Ethics of Competition*, Harper and Brothers, New York, 363 pp.

Kuhn, T.S. 1970. *The Structure of Scientific Revolutions*, 2nd edn, The University of Chicago Press, Chicago, 210 pp.

Kuhn, T.S. 1977. *The Essential Tension*, The University of Chicago Press, Chicago, 366 pp.

Lacy, G. 1930. Stable channels in alluvium, *Institute of Civil Engineers Proceedings*, **229**, 259–384.

Leopold, L.B. and Maddock, T. Jr. 1953. The hydraulic geometry of stream channels and some physiographic implications, *US Geological Survey Professional Paper 252*, 57 pp.

Lovelock, J.E. 1965. A physical basis for life detection experiments, *Nature*, **207**, 568–569.

McIntosh, R.P. 1975. H.A. Gleason, 'individualistic ecologist,' 1882–1975: his contributions to ecological theory, *Bulletin of the Torrey Botanical Club*, **102**, 253–273.

McIntosh, R.P. 1983. Excerpts from the work of L.G. Ramensky, *Bulletin of the Ecological Society of America*, **64**, 7–12.

McIntosh, R.P. 1985. *The Background of Ecology – Concept and Theory*, Cambridge University Press, Cambridge, 383 pp.

Mackin, J.H. 1948. Concept of the graded river, *Geological Society of America Bulletin*, **59**, 463–512.

Marbut, C.F. 1928. A scheme for soil classification, *Proceedings and Papers of the First International Congress of Soil Science*, Vol. 4, pp. 1–31.

Margalef, R. 1968. *Perspectives in Ecological Theory*, University of Chicago Press, Chicago, 111 pp.

Margulis, L. 1993. Gaia and the colonization of Mars, *GSA Today*, **3**, 277–280, 291.

Mayr, E. 1982. *The Growth of Biological Thought*, The Belknap Press of Harvard University Press, Cambridge, Mass., 974 pp.

Meinzer, O.E. 1923. Outline of ground-water hydrology with definitions, *US Geological Survey Water-Supply Paper 494*, 71 pp.

Merriam, C.H. 1894. Laws of temperature control of the geographic distributions of terrestrial animals and plants, *National Geographic Magazine*, **6**, 229–238.

Middleton, G.V. 1990. Non-linear dynamics and chaos: potential applications in the earth sciences, *Geoscience Canada*, **17**, 3–11.

Nikiforoff, C.C. 1942. Fundamental formula of soil formation, *American Journal of Science*, **240**, 847–866.

Nikiforoff, C.C. 1949. Weathering and soil evolution, *Soil Science*, **67**, 219–230.

Odum, E.P. 1953. *Fundamentals of Ecology*, Saunders, Philadelphia, 546 pp.

Odum, E.P. 1969. The strategy of ecosystem development, *Science*, **164**, 262–270.

Olsen, C.R., Cutshall, N.H. and Larsen, I.L. 1982. Pollutant-particle associations and dynamics in coastal marine environments: a review, *Marine Chemistry*, **11**, 501–535.

Osterkamp, W.R. (compiler) 1989. A tribute to John T. Hack by his friends and colleagues, in *History of Geomorphology from James Hutton to John Hack*, edited by K. Tinkler, Allen & Unwin, Boston, pp. 283–291.

Powell, J.W. 1875. *Exploration of the Colorado River of the West (1869–72)*, US Government Printing Office, Washington, DC, 400 pp.

Powell, J.W. 1876. *Report on the Ecology of the Eastern Portion of the Unita Mountains*, US Government Printing Office, Washington, DC, 218 pp.

Pyne, S.J. 1980. *Grove Karl Gilbert, a Great Engine of Research*, University of Texas Press, Austin, Texas, 306 pp.

Ramensky, L.G. 1924. Basic regularities of vegetation covers and their study, *Vēstnik Opytnogo dêla Stredne-Chernoz. Ob.*, Voronezh, 37–73 (in Russian).

Raup, H.M. 1941. Botanical problems in boreal America, *Botanical Review*, **7**, 147–248.

Raup, H.M. 1952. Review of Braun, E.L., Deciduous forests of eastern North America, *Ecology*, **33**, 304–307.

Raup, H.M. 1964. Some problems in ecological theory and their relation to conservation, *Journal of Ecology*, **52**, supplement, 19–28.

Rhoads, B.L. and Thorn, C.E. 1994. Contemporary philosophical perspectives on physical geography with emphasis on geomorphology, *Geographical Review*, **84**, 90–101.

Ritter, D.F. 1978. *Process Geomorphology*, William C. Brown, Dubuque, Iowa, 603 pp.

Ritter, D.F. 1988. Landscape analysis and the search for geomorphic unity, *Geological Society of America Bulletin*, **100**, 160–171.

Ruhe, R.V. 1975. *Geomorphology – Geomorphic Processes and Surficial Processes*, Houghton Mifflin, Boston, 246 pp.

Sack, D. 1992. New wine in old bottles: the historiography of a paradigm change, *Geomorphology*, **5**, 251–263.

Samuelson, P.A. 1961. *Foundations of Economic Analysis* (sixth printing), Harvard University Press, Cambridge, Mass., 447 pp.

Schumm, S.A. 1973. Geomorphic thresholds and complex response of drainage systems, in *Fluvial Geomorphology*, edited by M.E. Morisawa, Allen & Unwin, London, pp. 299–310.

Schumm, S.A. 1979. Geomorphic thresholds: the concept and its applications, *Transactions, Institute of British Geographers*, **4**, 485–515.

Schumm, S.A. 1991. *To Interpret the Earth: Ten Ways to be Wrong*, Cambridge University Press, Cambridge, 133 pp.

Schumm, S.A. and Lichty, R.W. 1965. Time, space, and causality in geomorphology, *American Journal of Science*, **263**, 110–119.

Schumpeter, J.A. 1934. *The Theory of Economic Development*, Harvard University Press, Cambridge, Mass., 255 pp.

Shreve, F.S. 1919. A comparison of the vegetational features of two desert mountain ranges, *Plant World*, **22**, 291–307.

Shreve, F.S. 1936. Plant life of the Sonoran Desert, *Scientific Monthly*, **42**, 213.

Sigafoos, R.S. 1964. Botanical evidence of floods and flood-plain deposition, *US Geological Survey Professional Paper 485-A*, 35 pp.

Smith, W.S.T. 1899. Some aspects of erosion in relation to the theory of peneplains, *University of California Publications, Bulletin of the Department of Geology*, **2**, 155–178.

Spalding, V.M. 1903. The rise and progress of ecology, *Science*, **17**, 201–210.

Stoddart, D.R. 1986. *On Geography and its History*, Basil Blackwell, Oxford, 335 pp.

Strahler, A.N. 1950. Equilibrium theory of erosional slopes approached by frequency distribution analysis, *American Journal of Science*, **248**, 673–696.

Strahler, A.N. 1952. Dynamic basis of geomorphology, *Geological Society of America Bulletin*, **63**, 923–938.

Strahler, A.N. 1957. Quantitative geomorphology of erosional landscapes, *Proceedings of the 19th International Geological Congress*, Algiers, 1952, section 13, part 3, pp. 341–354.

Tarr, R.S. 1898. The peneplain, *American Geologist*, **21**, 351–370.

Teilhard de Chardin, P. 1959. *The Phenomenon of Man*, Harper & Brothers, New York, 318 pp.

Theis, C.V. 1935. The relation between the lowering of the piezometric surface and the rate and duration of discharge of a well using ground-water storage, *Transactions, American Geophysical Union*, **16**, 519–524.

Thorp, J. and Smith, G.D. 1949. Higher categories of soil classifications – order, suborder, and great soil groups, *Soil Science*, **67**, 117–126.

Tinkler, K.J. 1989. Worlds apart: eighteenth century writings on rivers, lakes, and the terraqueous globe, in *History of Geomorphology*, edited by K. Tinkler, Unwin Hyman, London, pp. 37–71.

Todd, D.K. 1967. *Ground Water Hydrology*, Wiley, New York, 336 pp.

Van't Hoff, J.H. 1884. *Études de dynamique chimique*, Frederik Muller, Amsterdam, 296 pp (in German).

Vitek, J.D. and Ritter, D.F. 1989. Geomorphology in the USA: an historical perspective, *Transactions, Japanese Geomorphological Union*, **10-B**, 225–234.

Worster, D. 1977. *Nature's Economy – a History of Ecological Ideas*, Cambridge University Press, Cambridge, 404 pp.

18 Geomorphology and Policy for Restoration of Impounded American Rivers: What is 'Natural?'

William L. Graf

Department of Geography, Arizona State University

ABSTRACT

Geomorphology as a natural science is returning to its roots of a close association with environmental resource management and public policy. The science, after an insular period emphasizing mostly basic research questions, has graduated to a more mature phase wherein there is a new emphasis on application of established theory to address issues of social concern. In this new phase, geomorphology must become more interactive with other environmental sciences, and must establish socially relevant paradigms for research and teaching. An example of geomorphology's new directions is the research activity in fluvial geomorphology directed to the restoration of rivers downstream from American dams. The era for construction of large dams is over in the United States, leaving the nation with a fragmented river system and new social values that emphasize restoration through the Federal Power Act, Endangered Species Act, and Safety of Dams Act. The nation's rivers are now divided into segments, with some parts dominated by the direct effects of dams, some by indirect effects, and some in a preserved, unaffected state. To enhance river restoration, policymakers are focusing on dams, modifying them, changing their operating rules (on the Colorado, Trinity, Gunnison, and Kissimmee rivers), and in some cases removing them (on the Elwha, Kennebec, Milwaukee, and Gila rivers). The objective of these measures is to return the downstream landscape to its original, natural condition, yet geomorphic and ecologic changes related to the dams are not completely reversible. The issue of what is natural, and how closely restored systems can approximate natural conditions downstream from dams are challenging policy and scientific questions for fluvial geomorphology. Reaches downstream from the dams have experienced departures from natural conditions by changes in their hydrology (water yield,

The Scientific Nature of Geomorphology: Proceedings of the 27th Binghamton Symposium in Geomorphology held 27–29 September 1996. Edited by Bruce L. Rhoads and Colin E. Thorn. © 1996 John Wiley & Sons Ltd.

peak flows, low flows, timing of discharge events), sedimentology (sediment discharge, size distribution of sediment), geomorphology (channel and floodplain forms and processes), and biotic systems (riparian vegetation, fish, and wildlife). The underlying geomorphology of the systems helps determine what is natural and how closely restoration can approximate original conditions. Policy analysis by geomorphologists provides predictions of outcomes from a variety of options, whereas basic science seeks general explanation. Geomorph-ological science can contribute directly to policy analysis by modification of its traditional objectives: (1) defining what is natural by Quaternary studies and historical geomorphology; (2) explaining the operations of the present system by empirical process studies; and (3) analyzing policy options by use of predictive models. To be useful to policymakers, the reductionist, analytic investigations of fluvial geomorphology must evolve into an integrative, synthesizing science.

INTRODUCTION

During its early years as a definable science, geomorphology had a close association with resource assessment and public policy for management of the environment. As economic development connected to the Industrial Revolution stimulated the investigation of the resource potential of landscapes around the globe in the nineteenth century, geomorphology simultaneously emerged as a science to explain surface processes (Tinkler 1985). By the mid-twentieth century, this connection between the practical and the theoretical had broken down, however, and emphasis in geomorphological research had shifted to highly focused theoretical work that sought to exclude the human variable from its analysis of environmental systems. While this work resulted in sophisticated theory for geomorphic systems, the implementation of those theories in problem solving was a low priority, and the connections of geomorphology with other sciences as well as with decision-makers waned. As the functional relationships among various natural systems have become increasingly apparent in the last two decades, geomorphology as a science has partly returned to its historical roots by becoming less insular, by addressing the problems of applying theoretical constructs to problems of interest to decision-makers, and connecting to scientists in other disciplines such as zoology, botany, hydrology, and environmental design.

It would be a daunting task to survey the connections between geomorphology and policy for all the varied branches of the science. Instead, it is the purpose of this chapter to explore how the science is addressing the connection in fluvial geomorphology for the problems associated with the downstream impacts of dams. These problems are instructive because they are typical of the multifaceted aspect of many environmental issues, and they illustrate how geomorphological research can inform decision-makers using empirical evidence supported by theoretical interpretations. Fluvial geomorphology and the downstream impacts of dams is also a suitable venue for exploring science and policy because the issue is a global one of major importance. There are 2357 dams with large reservoirs (those containing more than 10^8 m^3 or 10^5 acre-ft), with their waters inundating a surface area the size of California (L'vovich and White 1990). Of the total discharge of the 139 largest river systems of the northern hemisphere, 77% is directly affected by dams (Dynesius and Nilsson 1994).

Downstream geomorphological impacts of dams and associated channelization works are large scale, and include changes in channel sizes, patterns, sinuosity, and hydraulic properties as a result of changes in flow regimes and sediment loads (Brookes 1988). Changes by dams to river systems in Europe, for example, are a legacy of human-induced channel changes that began in Roman times (van Urk and Smit 1989), and that have become progressively more prominent since the Middle Ages (Decamps et al. 1994). Changes in rivers of Spain, France, Italy, and Greece brought about by humans are on a scale of Quaternary-long changes resulting from hydroclimatologic influences (Lewin et al. 1995). Large rivers in Asia, especially China, have an equally long history of the impact of dams with major changes in geomorphology as the result. In the United Kingdom the rivers are generally smaller than those in China, but they experience similar substantial adjustments (Carling 1987; Zhou and Pan, 1994). In Australia, rivers have biological, hydrological, and geomorphological configurations that owe much to the histories of their dams (Warner 1988; Benn and Erskine 1994). Tropical rivers, with their high volume throughputs of water also evidence wide-ranging impacts of dams and their operations (Pickup 1980; Pickup and Warner 1984).

The fluvial system of the United States is not a natural river network. Dams have an impact on flow in every major stream in the conterminous 48 states, and the nation has 87 dams that impound reservoirs of 1.2×10^9 m^3 (1×10^6 acre-ft) or more (Graf 1993, pp. 36–38). More than 50 000 dams of all sizes have a storage capacity that is more than three times greater than the mean annual runoff of the country's surface (Table 18.1). Channelization, bank stabilization, and artificial works line thousands of kilometres of what were once natural channels (Lagasse 1994; Simons and Simons 1994). By certain definitions, the nation has about 5.1×10^6 km (3.2×10^6 miles) of stream channels (Echeverria et al. 1989), but 17% is under reservoir waters, and of the remainder, human activities have altered all but about 2%. The Wild and Scenic Rivers System permanently protects from development only about 0.3%. The period of large dam construction is now over, but the period of river presentation is just beginning (Figure 18.1). The result of dam construction and piecemeal preservation efforts is a fragmented river system, where segments dominated by engineering works are interspersed with segments that approach natural conditions. Administration as well as physical processes in these systems are

Table 18.1 Reservoirs and dams in the United States

Reservoir capacity (acre-ft)	Number	Total capacity (acre-ft)
> 10 000 000	5	121 670 100
1 000 000–10 000 000	82	186 480 100
100 000–1 000 000	482	136 371 900
50 000–100 000	295	20 557 000
25 000–50 000	374	13 092 000
5000–25 000	1411	15 632 000
50–5000[a]	50 000[b]	5 000 000
< 50[c]	2 000 000	10 000 000
Total		508 803 100

[a] Mean reservoir size estimated to be 100 acre- ft.
[b] US Army Corps of Engineers estimates.
[c] Mean reservoir size estimated to be 5 acre-ft.
Source: Data from US Army Corps of Engineers, compilation from Graf (1993, p. 17).

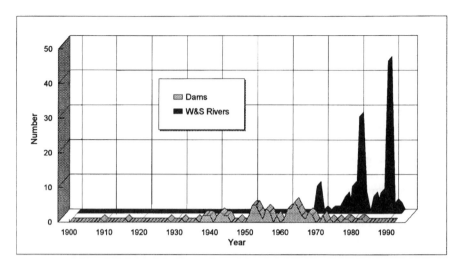

Figure 18.1 The temporal characteristics of policy changes for United States rivers as reflected by the timing of completion of large dams (those with reservoir capacity greater than 1 000 000 acre-ft) and designation of preserved rivers under the Wild and Scenic Rivers Act. Data from Graf 1993, pp. 36–42

poorly integrated, and demand new scientific theory for their understanding, and new policy perspectives for their management.

This artificial system conflicts with current American social values as expressed in laws and regulations to preserve and restore at least some fluvial environments for recreation, aesthetic purposes, and especially wildlife habitat. At the national level, the Wilderness Act (1964), Wild and Scenic Rivers Act (1968), Endangered Species Act (1973), Clean Water Act (1977), and numerous state river preservation systems are expressions of a national ethic emphasizing the desire for natural conditions associated with rivers. While preservation is a matter of legal protection of the resource as it exists, landscape or ecosystem restoration is an active intervention requiring considerable knowledge and expertise. The fundamental question in river resource management in the twenty-first century is likely to be all-encompassing: how can the nation foster economic development while at the same time preserving and enhancing environmental quality? The issue is not limited to water resources, but rather includes all physical, chemical, and biological aspects of the fluvial system (Gregory et al. 1991). The geomorphic and hydrologic components form the foundation of this complex environmental system, and any successful policy designed to manage the system must account for their behavior.

The purpose of this chapter is to explore the opportunities for interaction between science and policy in retoring the nation's rivers to more stable, more natural conditions. The chapter seeks to identify commonalities in research from several rivers, and to bring unity to a group of efforts that often have operated in isolation from each other. The following pages explore three aspects of river restoration related to dams. First, how do science and policy interact with regard to river management to restore river systems? Second, the goal of restoration is a natural system, but what do we mean by 'natural' in this application? Finally, what can we learn from specific applications of restoration science and policy?

INTERACTION OF POLICY AND SCIENCE

Policymakers are legislators, elected executive officials, and agency personnel who have been entrusted by the public to provide overall administrative leadership. They are also administrators of private business concerns, as well as participants in nongovernmental organizations and groups who seek to influence the course of events (National Research Council 1995). The researchers who provide scientific knowledge for policy include private consultants, university faculty, and government employees. The formal contacts between researchers and policymakers are usually in the form of contracts addressing particular applied problems using a body of more general basic knowledge, so that the distinction between basic and applied science is often indistinct. Formal contacts also occur in the form of expert testimony before courts, administrative law judges, or advisory panels.

In the United States, the most important federal agencies dealing with river restoration include the US Army Corps of Engineers, Bureau of Reclamation, Fish and Wildlife Service, Environmental Protection Agency, and the Department of Energy. The Corps of Engineers undertakes projects designed for flood control in a partnership arrangement with other sponsors who may be federal, state, or local agencies (Black 1987, pp. 98–116). The Corps also administers Section 404 of the Clean Water Act, which governs the issuance of permits for altering the channels of the 'waters of the United States', a legal designation that includes most major rivers in the nation. The Bureau of Reclamation is a water resource development agency charged with supplying water to agricultural and urban users (Holmes 1972; Office of the Federal Register 1983). The Fish and Wildlife Service administers the Endangered Species Act, and so is concerned with the protection and enhancement of habitats, particularly wetlands, that support wildlife. The Environmental Protection Agency is responsible for administering the Clean Water Act, a policy that focuses mostly on chemical quality of water, including sediment as a pollutant. The Department of Energy, through its administration of the national laboratories, is concerned with water and sediment pollution associated with the nation's nuclear weapons program.

The geomorphologists who may influence decisions in these agencies fall into one of three categories (Kodras and Jones 1990): researchers outside the state, researchers who are outside consultants to the state, and those directly employed by the state. Those researchers outside government, mostly university-based researchers, derive funding from a variety of sources, but maintain a critical independence from the governmental agency, because their employment does not directly depend on the government. Their agendas may be driven by the availability of funding in various topical areas, but because they have a range of choices for funding sources they adopt research subjects based partly on personal preferences. Their work may subsequently support particular positions adopted by decision-makers. The outside agents enjoy considerable independence, but at the price of relative isolation from the policymakers, and their input is usually indirect. Those researchers who act as consultants provide input for particular cases defined a priori by the funding agency. They have direct access to the decision-makers, but they rarely set their own agendas. Finally, researchers employed by government agencies are agents of the state, and though they lack the independence of other investigators, they are in positions to directly influence decision-makers.

In the past two decades, policymakers and scientists involved in river issues have undergone significant changes in emphasis that are mutually supportive. Policymakers dealing with rivers have reflected the increasing interest among the American public in environmental quality by emphasizing habitat preservation and restoration. The objective of these efforts has been to reduce the impacts of human activities, particularly on wildlife, by maintaining and enhancing entire riparian and fluvial ecosystems, through integrated basins management (Gore and Shields 1995). At the same time, fluvial geomorphology has become increasingly systems-oriented, with more emphasis on the integrative aspects of the science (Hugget 1985). Geomorphological research is now often associated with ecosystem- or watershed-scale investigations rather than isolating a single or a few key components of the river. This trend toward synthesis leads to broader conclusions and results more likely to be of direct use to decision-makers than the reductionist, analytic approaches dominant in the 1960s and 1970s.

PRESENT STATE OF GEOMORPHOLOGY AND POLICY

The rapid increase in demand for geomorphic knowledge with policy implications is especially noticeable in river restoration efforts. Projects presently under consideration for restoration efforts include the removal of the Columbia Falls Dam on Pleasant River, Maine, and modification of operational rules for Sheppard Dam on the Brazos River, Texas; Kingsley Dam on the Platte River, Nebraska; Hawk's Nest Dam on the New River, West Virginia; the San Luis Reservoir system in southern California; and the operation of several dams on the Columbia River in the Pacific Northwest (E. Hunt 1988). Within the next few years, this list may grow with several anticipated additions (Table 18.2).

Heretofore, fluvial restoration has been largely the purview of zoologists, botanists, and engineers, but geomorphology is an important component of any effort to successfully restore rivers to more natural arrangements. Basic geomorphic research, theory building, and application are beneficial in answering three questions facing all river restoration work in reaches affected by dams. *First, what is natural?* Quaternary studies and historical geomorphologic investigations can establish the characteristics of the channel and near-channel processes and landforms before the installation of dams. *Second, how does the system work with dams in place?* Empirical investigations that focus on system operation, integrating the artificially controlled hydrology, landform, sediment, and biotic components can make contributions to understanding the dynamics of the present situation. *Third, what is likely to happen if there are structural or operational changes to enhance restoration?* Predictive studies based on spatially variable, iterative models rooted in basic geomorphic and physical principles can provide improved predictions for decision-makers. Many geomorphologists are already involved in work exemplifying these questions including individuals or groups in the US Geological Survey and at several universities, including Johns Hopkins, Colorado State, University of Wyoming, Utah State University, Arizona State University, and the University of California, Berkeley.

Table 18.2 Dams proposed for removal and requiring FERC relicensing

Dam	River	State	Comments
Gillespie	Gila	Arizona	Breached by 1993 flood
Newport No. 11	Clyde	Vermont	Washed out by 1994 flood
Pine River	Pine	Minnesota	Targeted for removal in FERC
Stronich	Manistee	Michigan	relicensing process by
Several structures	Manomenee	Michigan–Wisconsin	American Rivers, Inc.
No. 160	Genessee	New York	
Condit	White Salmon	Washington	
Savage Rapids	Rogue	Oregon	Removal planned by owners
Elwha	Elwha	Washington	
Glines Canyon	"	"	
Edwards	Kennebec	Maine	Removal likely
Ice Harbor	Snake	Washington	Recommended for removal
Lower Monumental	"	"	by the Oregon Resources
Little Goose	"	"	Council
Lower Granite	"	"	
Hells Canyon	"	Oregon– Idaho	
Brownlee	"	"	
Peton	Deschutes	Oregon	
Gold Ray	Rouge	"	
Windester	North Umpqua	"	
Three Mile Falls	Umatilla	"	
Rodman	Ochlawha	Florida	Under debate in state legislature

Note: List is incomplete, with frequent changes in planning and policy.

Fluvial restoration has become an important component of public policy for rivers. The National Academy of Sciences recently established a philosophical basis for this work (National Research Council 1992). Restoration is often included in legislative directives related to specific streams such as the Trinity and Elwha rivers. In the last Congress (103rd Congress, 2nd Session), representatives introduced four major bills aimed at river restoration, and though they did not pass, they are likely to continue to be considered because they emphasize the federal, state, local, and tribal cooperative approaches presently popular in Congress. Restoration will be a component of policy regarding dams for the foreseeable future, and it represents an important demand for geomorphic knowledge.

WHAT IS NATURAL?

If fluvial restoration has as its goal the re-creation of a predisturbance, natural condition, how does one define that natural condition? More importantly, how does one measure naturalness in the most common cases, those that are partly natural and partly artificial, with a mixture of direct and indirect human influences? The construction of a continuum of natural-to-artificial systems begins with the definition of the end points, followed by the identification of intermediate states. Specification of completely artificial rivers is obvious because they are comprised of engineered or completely disrupted systems. The River

Walk along the San Antonio River in downtown San Antonio, Texas, is a famous example of such a system. With its water flows controlled by gates, its channel defined by cement walls, its vegetation consisting of imported plants, and its built landscape, the River Walk is completely unlike the ecosystem it replaced.

Artificial river reaches are those experiencing the effects of human activities, either directly or indirectly. Built environments in and near rivers obviously impact fluvial processes locally, but other impacts may propagate themselves to distant locations downstream. Modifications of flow characteristics by a dam, for example, affect the geomorphology of the river for a considerable distance, in some larger rivers for more than 100 km (Williams and Wolman 1984). Enhanced or depleted sediment supply has a compound effect downstream by changing the particle sizes and geographic distributions of deposits. The deposits, in turn, exert partial control over biotic distributions and influence the fate of contaminants in the system.

The natural end of the spectrum for ecosystems in general (not limited to rivers) is also relatively easy to define, with some attempts now established in law. Section 2 of the 1964 Wilderness Act defines a natural system (including rivers) in the formal sense of wilderness: 'A wilderness, in contrast with those areas where man and his own works dominate the landscape, is hereby recognized as an area where the earth and its community of life are untrammeled by man, where man himself is a visitor who does not remain' (Public Law 88-577; Hendee et al. 1978, p. 68).

The 1968 Wild and Scenic Rivers Act (Public Law 90-5423) codified definitions of naturalness specifically for rivers. Section 16(a) of the Act defines a river as 'a flowing body of water or estuary or a section, portion, or tributary thereof, including rivers, streams, creeks, runs, kills, rills, or small lakes'. The law contains a scale based mostly on accessibility, degree of disruption, and the presence of control structures (Coyle 1988, pp. 14–16). Section 2(b) of the 1968 Act specifies that a wild river is free of impoundments, is generally inaccessible by trail, and is essentially primitive and unpolluted (Palmer 1993). The completely natural river channel in the classification used in this chapter is analogous to the 'wild' river as defined by the Act, but it is also free from upstream dams that might indirectly alter its hydrology and geomorphology.

By beginning with these extreme conditions of completely artificial and completely natural channels, and continuing with admittedly arbitrary gradations between them, it is possible to develop a formal classification for geomorphic naturalness for rivers (Table 18.3). This classification pertains only to the geomorphology of the rivers, but similar classifications are possible to define departures from natural conditions for any aspect of the river ecosystem, such as hydrology, riparian botanical or zoological communities, or chemical characteristics. Such classifications, used in tandem with the one given here, could address the statistical characteristics of the flow regime, water temperature and pH conditions, ratios of native to introduced fish species, dominance of exotic as opposed to native vegetation, and the concentrations of indicator chemical compounds or elements in water and sediment.

A recent application of the geomorphic naturalness scale in Table 18.3 to central Arizona rivers showed that the majority of the channel reaches were in the range of partly modified to mostly modified, with essentially artificial channels common in urban areas (Graf et al. 1994, a, b). Even in the rivers subjected to extensive engineering, however, there were many segments of channel that were essentially natural. From a naturalness

Table 18.3 Geomorphic 'naturalness' classification for river channels

Channel type	Completely natural	Essentially natural	Partly modified	Substantially modified	Mostly modified	Essentially artificial	Completely artificial
Pattern, X-section shape, materials	No obvious evidence of human activities – same forms and processes as existed prior to human occupation	No obvious evidence of human activities – same forms as prior to human occupation	Altered channel patterns, x-sectional shapes, or sediment characteristics as a result of human activities	Altered channel patterns, x-sectional shapes, or sediment result of human activities	Altered channel patterns, x-sectional shapes, or sediment characteristics as a result of human activities	Altered channel patterns, x-sectional shapes, or sediment characteristics as a result of human activities	Completely engineered and/or built channel with altered processes and sediment
Minor landforms	Same forms and processes as those found prior to human occupation	Altered by human activities or changes in sediment supply	Altered by human activities or changes in sediment supply	Altered by human activities or changes in sediment supply	Altered by human activities or changes in sediment supply	Altered by human activities or changes in sediment supply	Altered by human activities or changes in sediment supply
% Channel area engineered or disturbed	0	<10%	<10%	>10% <50%	>50% <90%	>90% <100%	100%
Descriptive notes	Completely undisturbed channel, could be a 'wild river' in the Wild and Scenic River System	Minor modifications by human action, sometimes through flow regulation; in other cases, scattered structures on an otherwise undisturbed channel	Obvious modifications by flow regulation of altered sediment supply resulting in channel disturbed by mining, metamorphosis, scattered structures	Major modifications to channel forms and processes, with up to half the channel area disturbed by mining, development, or structures	Major modifications to channel forms and processes, with most of the channel area including dredging; a development, or structures	Largely artificial channel due to engineered bed and/or banks; in some cases including dredging few natural forms or processes remain	Channel completely determined by design and manipulation with no natural forms
Example	Middle Salmon River, Idaho	Colorado River in Grand Canyon, Arizona	Platte River in western Nebraska	Potomac River near Georgetown, Maryland	Santa Cruz River near Santa Cruz, California	Illinois River in central Illinois	Los Angeles River in Los Angeles, California

perspective, the river is segmented into reaches of varying characteristics, with each segment evidencing distinct physical processes and requiring specific policy considerations. Decision-makers (the Corps of Engineers and its partner agencies at the national, state, and local level) used the classification as a basis for selecting channel segments for emphasis in restoration efforts. Generally, they will invest resources in the segments that are partly modified in an attempt to change them to essentially natural. Modifications (direct or indirect) to other channel segments are so substantial that returning them to the more natural end of the scale would be too costly. Similar applications of the naturalness scale are possible in a variety of environments, though it may be much more difficult in forested areas (Gregory and Davis 1992).

The objective of fluvial restoration, the conversion of affected channels to ones that are more natural in morphology and function, is one of the primary purposes of sound environmental management for rivers. Although planners and engineers have a long history of interest in channel design, it has only been in recent years that geomorphologists have begun to apply their scientific understanding to design problems. In an early Binghamton Geomorphology Symposium, for example, Keller (1975) reviewed the significant differences between natural and designed channels, pointing out that design might easily incorporate more natural configurations. Geomorphologists have developed the theme of designing rivers to contain natural roughness elements in arrangements that mimic undisturbed natural streams (e.g. Gregory et al. 1994). Richards et al. (1987), using examples from Scotland, Saudi Arabia, and Honduras, showed how geomorphological perspectives can improve engineering and management solutions in a wide variety of conditions. Newbury (1995) pointed out that restoration efforts are more likely to be successful if traditional engineering tools such as the Chezy equation receive new interpretations to produce more natural and less artificial designs.

SPECIFIC APPLICATIONS OF RESTORATION POLICY AND SCIENCE

Since about 1980, two avenues for change have emerged for dam and river restoration. Each involves substantial policy changes, each produces changes in the fluvial landscape, and each requires new geomorphological research. First, the management of large public dams has begun to include changes in operating rules that enhance downstream restoration. Second, owners of smaller private dams are redesigning their structures or removing them for environmental enhancement in a federal relicensing procedure. Geomorphic research using basic theoretical concepts is a common approach for policy formulation to restore fluvial environments. The following pages provide examples at a variety of scales to illustrate the relationship between geomorphological research and restoration policy. The examples are in two broad categories: operating rules for large public dams, and the removal of smaller private dams.

Operating Rules for Large Public Dams

It has only been in recent decades that the impact of large dams on downstream ecosystems has become apparent. Upstream inundation and loss of riparian habitat is obvious, but the alteration of downstream geomorphic conditions is substantial and far-reaching.

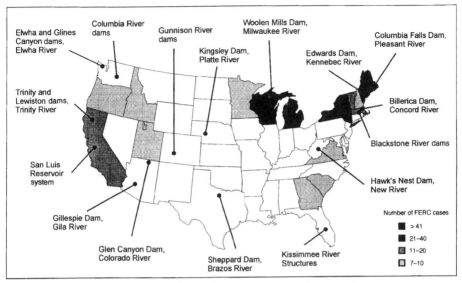

Figure 18.2   Locations of structures discussed in this chapter

Depending on the purpose of the dam and its operating rules, a variety of hydrologic changes occur in downstream flow. Most often, dam operations reduce annual flood peaks, while increasing seasonal low flows in some cases, or eliminating low flows in others. Dams operated for hydroelectric power production introduce short-term flow fluctuations that have no natural counterpart. Some of the most obvious geomorphic changes are channel scour and armoring for a limited distance below the structures, accumulation of tributary sediments in the main channel and valley, conversion of channel patterns from braided to meandering, channel shrinkage, beach and bar expansion, and floodplain expansion (Williams and Wolman 1984; Petts 1979, 1980). These geomorphic changes, in association with the changes in flow regime, result in a new substrate and hydrologic environment for riparian vegetation, with concomitant changes in wildlife populations. Fluvial restoration in these instances depends on changing the operating rules of the dams so that they more closely mimic natural flow regimes. Recent examples of this strategy include operations at Glen Canyon Dam on the Colorado River, dams on the Gunnison River, Trinity Dam on the Trinity River, and structures on the Kissimmee River (see Figure 18.2 for locations).

Glen Canyon Dam

The 1962 closure of Glen Canyon Dam on the Colorado River brought about remarkable changes in the flow of the river downstream through Grand Canyon. The dam impounds the second largest reservoir in the country, Lake Powell, with a maximum storage capacity of 33.7×10^9 m^3 (27.3×10^6 acre-ft), about two years' mean water yield of the river. The dam eliminated periodic major floods, some as great as 8500 m^3 s^{-1} (300 000 cfs), sustained low flows throughout the year at levels significantly greater than under natural conditions, and terminated the upstream contributions of sediment (Figure 18.3).

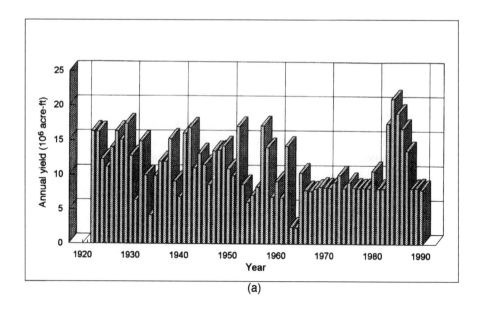

(a)

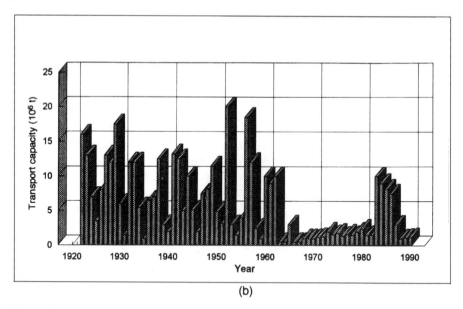

(b)

Figure 18.3   Annual water yield and sand transport in the Grand Canyon of the Colorado River below Glen Canyon Dam, Arizona, showing two impacts of Glen Canyon Dam: (a) a regularized annual water yield and (b) a reduced transport capacity as a result of reduced flood flows. Data from US Bureau of Reclamation (1994, pp. 73 and 85)

Operation of the dam for hydroelectric power production introduced the most radical changes in the flow regime. In response to daily variations in demand for electricity in the regional power grid, the dam's penstocks and turbines released as much as 850 m^3 s^{-1} (30 000 cfs) during evening hours, and as little as 28.3 m^3 s^{-1} (1000 cfs) during the morning hours. The result was a daily fluctuation in river stage in the Grand Canyon of up to 4 m (13 ft).

The new regime caused widespread geomorphic and ecologic adjustments in the Grand Canyon, which is part of a national park. The change from warm, silty waters to cold, clear waters released from the dam altered the fish population by creating conditions unfavorable for endangered native fishes such as the humpback chub and razorback sucker (Minckley 1991). Meanwhile, artificially introduced trout flourished. Before the installation of the dam, large-caliber sediment brought down to the main channel by debris flows and floods from tributaries came to rest in the main channel forming rapids, but periodic major floods moved some of these materials into downstream pools (Webb et al. 1987). After the closure of the dam, the highest flows were absent, and debris accumulated in increasingly large rapids. Additionally, the geomorphology of the channel margins changed in response to the lack of large floods. Sand deposits stranded high above the new water levels eroded and desiccated, while eroding beaches and bars near the flow margin lost material to channel pools without the replenishment that would have occurred in natural floods (Schmidt and Graf 1990). Riparian vegetation which depended upon these substrates also changed, and new plant assemblages became common. The invasion of tamarisk, an exotic shrub and tree, had begun before the closure of the dam and accelerated afterwards (Carothers and Brown 1991, pp. 111–128).

These downstream hydrologic, geomorphic, and ecologic impacts were not of concern when the dam was authorized and built. Congress authorized the dam for water storage, sediment control, and power generation, with only minor consideration given to recreation and wildlife (mostly related to the upstream reservoir). By the 1980s, however, the public perceived the downstream changes as a major problem, and the environmental quality ethic had given the geomorphology and ecology of the Grand Canyon new value (National Research Council 1987, pp. 15–20). The beaches were critical to the maintenance of the canyon's $20 million per year whitewater rafting industry, native fishes protected under the Endangered Species Act were declining, and the riparian vegetation supported a variety of desirable wildlife.

Spurred by a lawsuit by river users, the Bureau of Reclamation established the Glen Canyon Environmental Studies (GCES) in 1982. With a budget that has grown to several million dollars per year by 1995, the GCES has investigated the hydrologic, geomorphic, and ecologic processes in the river environment downstream from Glen Canyon Dam. In what was probably the largest-scale geomorphic experiment ever conducted, the operators controlled the dam discharges for a year, testing the hydrologic and geomorphic impacts of various discharge magnitudes, durations, and ramping rates. They have produced hundreds of reports that provide the most extensive knowledge base available for a canyon river. The primary geomorphologic/hydrologic conclusions are that (1) the reduced transport capacity of the river is roughly adequate to carry the available sediment inputs from tributaries below the dam, (2) under normal present conditions, beaches will slowly erode, losing their sediment to nearby pools in the channel, (3) waters below the dam are too cold for a vigorous native fishery.

Improved, scientifically based understanding of the canyon river processes has resulted in remarkable policy adjustments for operating the dam in a manner that facilitates the restoration of more natural conditions downstream. The Bureau partially simulates seasonal variation in flows, trying to strike a balance between truly natural discharges and legal requirements for water storage and delivery as outlined in a set of court decisions, treaties, and laws collectively known as the 'Law of the River'. Dam operators may allow an occasional (about once every five years) 'flood' of about 1275 m^3 s^{-1} (45 000 cfs) to scour sediment from pools and deposit it on rebuilt beaches and bars. The agency is also exploring the possibility of a multiple outlet withdrawal structure designed to take water from the reservoir at various depths, a method to increase the temperature of tail waters downstream from the dam (US Bureau of Reclamation 1994). Adaptive management allows operators to change strategy in response to changing conditions downstream as indicated by long-term monitoring.

Glen Canyon Dam is the largest structure to have modified operating rules in favor of extensive restoration, but the modifications are a result of 13 years of intensive geomorphic and other types of research with the expenditure of as much as $100 million. The outcomes of management decisions have a certain degree of predictability in the Grand Canyon. It will never be possible to restore the canyon to the natural conditions that prevailed there before the introduction of exotic species and the installation of the dam. It is possible, however, to move the canyon environment across the naturalness scale to a position less dominated by artificial hydrographs and landforms.

Gunnison River Dams

Dam operations on the Gunnison River in southwestern Colorado are also concerned with the restoration of more natural geomorphic conditions in the river channel of a protected area. Experience there shows the importance of the magnitude and timing of dam releases. The Aspinall Unit of the Colorado River Storage Project includes three dams on the Gunnison – Blue Mesa, Morrow Point, and Crystal – with a combined storage capacity of 1.3 × 10^9 m^3 (1.08 × 10^6 acre-ft), about half the mean annual water yield of the river. Immediately downstream from Crystal Dam is the Black Canyon of the Gunnison, part of a national monument that is under consideration for national park, national conservation area, and wild and scenic river status (US House of Representatives 1993). In 1991, American Rivers, Inc., a private advocacy group, designated the river as the most threatened stream in the nation because of reduced discharges. The operations of the dams have altered the hydrology of the Gunnison in Black Canyon in a manner similar to that described for Glen Canyon Dam. For the Gunnison, the primary consideration is the large flood discharges which in pre-dam conditions (prior to 1966) moved most of the debris in rapids created in the canyon by tributary discharges and debris flows (Figure 18.4). Vegetation evidence suggests that floods large enough to mobilize the rapids (310 m^3 s^{-1} or 10 950 cfs) had a pre-dam recurrence interval of 3.2 years, but a post-dam recurrence interval of more than 20 years (Auble et al. 1991; Elliot and Parker 1992). As a result, considerable buildup of material is occurring in rapids, impeding whitewater recreation and altering fish habitat.

In the early 1990s, the administrators of Black Canyon, the National Park Service, sponsored investigations of the dynamics of the rapids and large-caliber sediment in the

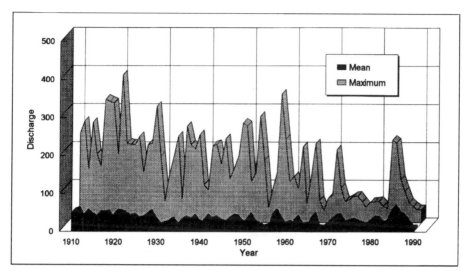

Figure 18.4 Annual flood series and the monthly mean discharges of the Gunnison River, at East Portal, Colorado, below the Gunnison River dams, showing two impacts of dam operations after 1966: more consistent low flows, and generally reduced maximum flows. US Geological Survey data, also given by Chase (1992, p. 75)

canyon with respect to flood discharges. In support of the research, the Bureau of Reclamation manipulated the discharges from Crystal Dam to create artificial 'floods' of 22 m^3 s^{-1} (775 cfs) in September 1990 and 45 m^3 s^{-1} (1500 cfs) in November 1990. Tracking of tagged boulders in rapids showed that at these discharges some debris on the downstream side of the bars was mobilized. Extension of the empirical observations using HEC-2 (a Corps of Engineers computer program) calculations for water surface profiles suggested that at the pre-dam common maximum flows of 310 m^3 s^{-1} all but the largest particles would be mobilized (Chase 1992).

In response to these findings, the National Park Service, Bureau of Reclamation, US Fish and Wildlife Service, and the state of Colorado have agreed to new operating rules for the dams of the Aspinall Unit on a four-year experimental basis (US Bureau of Reclamation 1991–93). During one year, the dams will release a spring 'flood' of 57–142 m^3 s^{-1} (2000–5000 cfs), in one year of 142–283 m^3 s^{-1} (5000–10 000 cfs), in one year of greater than 340 m^3 s^{-1} (12 000 cfs), and in one year greater than 424 m^3 s^{-1} (15 000cfs). The 'floods', released during the naturally occurring peak flow period of May 15 to June 15, will provide a restoration of more natural hydrologic and geomorphic conditions, and return the rapids to a more dynamic state.

Trinity and Lewiston Dams

Experimental discharges and changes in operation rules also play a role in fluvial restoration mandated by Congressional legislation on the Trinity River in northern California, where experience shows the irreversibility of some dam impacts. Two dams are at issue: the Trinity Dam for water storage, and Lewiston Dam, a reregulation structure (a relatively small dam downstream from a larger one designed to modify short-term releases from the large structure and to generate hydroelectric power). The Trinity Dam impounds a reservoir with a capacity of 3.02×10^9 m^3 (2.45×10^6 acre-ft) as part of the Bureau of Reclamation's Central Valley Project, while Lewiston Dam has a reservoir of minimal capacity (US Bureau of Reclamation 1981). When Congress authorized the Central Valley Project in 1955 for water resource development, it stipulated that impacts on other resources would be minimized. After the completion of Trinity Dam in 1962 and Lewiston Dam in 1963, however, the project diverted 80% of the Trinity River's discharge and eliminated the normal peak flows of the river in the spring of each year (US Senate 1984). As a result, the salmonid fishery, once the most important fishery in California for chinook and steelhead salmon, was virtually eliminated. Before the installation of the dams, the rivers had gravel beds with particular grain-size distributions which were the spawning grounds for the fish (Figure 18.5). Although fine materials occurred in the river, the spring flows flushed them downstream and maintained the gravel beds (Kondolf and Wolman 1993). After closure of the dams, the general flows declined and the lack of peak spring flows allowed the buildup of fine materials on the bed, blanketing the gravel and eliminating the spawning areas (Majors 1989). Minimum flows were 4.25 m^3 s^{-1} (150 cfs) until 1981, when they were increased to 8.5 m^3 s^{-1} (300 cfs) (Cassady et al. 1994). The

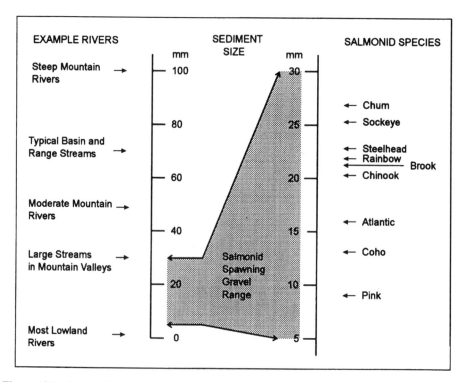

Figure 18.5 Salmonid spawning gravel sizes for various species compared to the natural sediment in groups of representative rivers showing the narrow range of useful sediment sizes. Data for spawning from Kondolf and Wolman (1993), general river data from US Geological Survey information

influx of fine material was further aggravated by logging on steep slopes in the watershed which accelerated slope erosion and increased sediment loading to the streams.

By the early 1980s, an intergovernmental task force had brought about some adjustments in an attempt to restore the fishery. A sand dredging system removed some fine sediment from one of the major sources of fine sediment, the Grass Valley Watershed, and the Bureau of Reclamation restored some streamflows from Lewiston Dam. These efforts did not restore the anadromous fish habitat to an acceptable degree, and recreational, scenic, and wildlife resources were still in a substantially modified condition. In 1984, Congress authorized $33 million to manage and restore the fishery, with another $2.4 million for monitoring (Public Law 98-541, the Trinity River Basin Restoration Act).

From the geomorphic perspective, the Trinity River presented a problem in sediment transport and channel response to changing flow patterns. As the channel became smaller and dominated by fine materials after dam closure, vegetation invaded the channel and near-channel environment. In 1992, an experimental flow of $170 \text{ m}^3 \text{ s}^{-1}$ (6000 cfs) was released from Lewiston Dam in an attempt to flush fine materials from the system and expose the underlying gravel, but the vegetation and altered channel configurations prevented a return to the original, natural conditions. Sediment mobility was indeed increased, but much of the sediment was derived from bank erosion rather than the channel floor. Conversely, many areas were surprisingly resistant to erosion because of their vegetation cover (Pitlick 1992). It thus appears that restoration of the Trinity River to anything approaching its pre-dam natural state is unlikely.

Kissimmee River Structures

Not all adjustments in operations for restoration involve large dams. In the Everglades of south Florida (Figure 18.6), a region of almost imperceptible topographic relief, the restoration of the regional southward flow of water depends on adjustments in operations of low gates and pumps along the Kissimmee River and in the Everglades area south of Lake Okeechobee (Cushman 1994). In the first major civil works project on an American river to be reversed, the Corps will also restore the straightened course of the Kissimmee to its natural meandering configuration, and return water-level fluctuations in lakes and the integrated stream network to more natural timing and ranges. The Corps' operational changes will increase the annual period of inundation for the 41 700 ha (108 000 acres) Shark River Slough addition to Everglades National Park (Figure 18.7), returning the hydrologic portion of the ecosystem to more natural arrangements (US Army Corps of Engineers 1993). The total cost of the restoration project, now under way, will be more than $500 million (US Office of Assistant Secretary of the Army 1992).

Relicensing and Removal of Private Dams

Although the federal government built and manages the largest dams in the United States, private interests have constructed thousands of smaller ones. Most of the private structures completed during the period 1930–60 were primarily for hydroelectric generation. Because the structures use the 'waters of the United States' (rivers administered by the federal government), the Federal Energy Regulatory Commission (FERC, originally the Federal Power Commission) licenses their operation. The Federal Water Power Act of

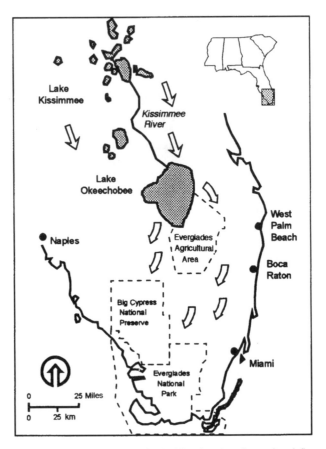

Figure 18.6 The Everglades region of south Florida, showing the regional flow of surface water with respect to Everglades National Park and the critical position of the Kissimmee River in the system. Redrawn and modified from maps by the Audubon Society and the US Army Corps of Engineers

1920 (16 USC 791a *et seq.*) gave FERC the authority to license the private dams for 50-year periods at a time when the dams appeared to be clean, environmentally sound energy sources. By the 1970s it became obvious many of the dams had deleterious effects on fluvial environments, especially on fish habitats and fish migration patterns. By posing obstacles to migrating fish that had become more important as stocks dwindled, the dams took on a central role in efforts to restore the fisheries and the habitats that support them.

During the 1980s, the original licenses of many private dams expired, and their owners began applying for relicensing (Figure 18.2 shows the distribution of FERC relicensing activity). Between 1987 and 2000, some 300 dams worth approximately $10 billion (17% of all hydroelectric projects under FERC authority) require relicensing (R.T. Hunt 1988; Boyd et al. 1990). Initially, FERC renewed licenses with hydropower as the dominant consideration, but changing public ethics placed new social values on natural rivers and the species that used them. The 1986 Electric Consumers Protection Act (Public Law 99-

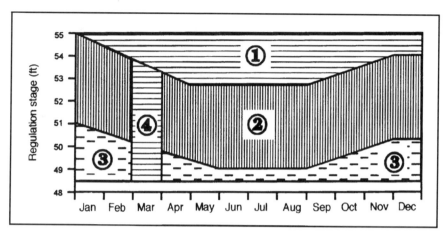

Figure 18.7 Operating rules for control structures related to Kissimmee Lake, Florida: (1) Full discharge, with releases from the lake as rapidly as possible, always preferred if possible; (2) historic stage– discharge relation, with releases from the lake made according to historical averages; (3) minimum discharge into the river maintained from the lake at 250 cfs; (4) March rule, whereby changes in lake level are limited to 0.1 ft per week and discharges are made accordingly. If lake levels fall below 48.5 ft, no discharges from the lake are permitted. Data and figure design from US Army Corps of Engineers (1993)

495) provided that the relicensing process would give equal consideration to power and nonpower uses for each project.

As a result of these policy changes, proponents of restoration successfully mounted serious challenges to relicensing, arguing that the primacy of hydropower in the decision process was not part of FERC's legal authority, and that endangered species, recreation, and aesthetics should also play a part in deciding the future of the dams. Proponents of restoration argued for changes in operating rules for some structures, but they also argued that in some cases the dams should be removed, restoring the natural hydrology and geomorphology of the rivers in question (Echeverria et al. 1989). Although this seemed to be a radical idea in the late twentieth century, it was hardly new. In the early nineteenth century, Henry David Thoreau offered the then outrageous suggestion of using a crowbar to destroy the mill dam at Billerica, Massachusetts, to restore the upstream migration of shad on the Concord River (Thoreau 1849, pp. 40–41). By the 1990s, however, removal of dams for environmental restoration of rivers had become a realistic policy. The cases of four structures illustrate the policy and hydrologic/geomorphic implications of dam removal, and the problems entangled with efforts to restore natural conditions: the anticipated removal of structures on the Elwha River, Washington; the dismantled Woolen Mills Dam on the Milwaukee River, Wisconsin; the debate about removing Edwards Dam on the Kennebec River, Maine; and the breaching of Gillespie Dam on the Gila River, Arizona.

Elwha and Glines Canyon Dams

The Elwha River drains a 818 km^2 (316 mile^2) watershed on the north slope of the Olympic Mountains of Washington (Figures 18.8 and 18.9). The lower 32 km (20 miles)

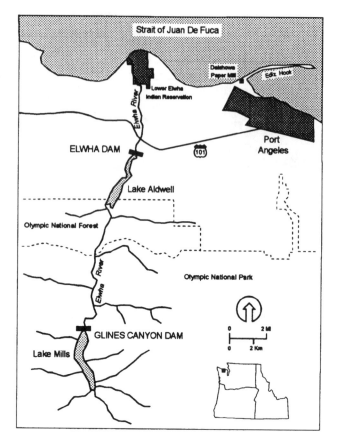

Figure 18.8 Map of the lower Elwha River, Washington, showing the locations of the two large dams with respect to Indian lands, Park Service and Forest Service lands, and the Daishowa paper mill which uses electricity produced by the dams. Based on a map by Federal Energy Regulatory Commission (1991)

of the stream above its mouth at the Strait of Juan De Fuca was once one of the state's most productive fisheries, accommodating six species of salmon as well as a variety of trout. The fish, some weighing as much as 45 kg (100 lb), provided year-round sustenance for the members of the Lower Elwha Klallam Tribe who lived along the lower reaches of the stream (US House of Representatives 1992). In 1911 the closure of Elwha Dam restricted the fishery to the lower 7.9 km (4.9 miles) of the river. While the Elwha Dam and an associated structure, the Glines Canyon Dam, provided hydroelectricity for industrial use, they also decimated the fishery, and Lake Mills, the reservoir behind Glines Canyon Dam, radically altered the river in Olympic National Park (US Senate 1992).

The relicensing application of the private corporation owning the dams triggered wide-ranging consideration of alternatives to the prevailing emphasis on hydroelectric power production. The owners of the structures proposed to partially restore natural hydrologic and biotic conditions by installing a fish ladder, protective screens, and spillway modifications on Elwha Dam, building a trap-and-haul facility at Glines Canyon Dam, and

Figure 18.9 Photograph of Elwha Dam and its reservoir, Lake Aldwell, Washington, W.L. Graf photograph 124-1, 2 August 1994

initiating annual spills of 2.8 m^3 s^{-1} (100 cfs) from Glines Canyon during the fish out-migration. The total cost of these modifications would be $14.7 million (US General Accounting Office 1991), but the restoration of the native fishery would be limited. Alternatively, after a seven-year analysis of environmental, power, and economic consideration, the two primary federal agencies advising FERC on the relicensing process, the National Park Service and the US Fish and Wildlife Service, recommended that the dams be removed (National Research Council 1992, pp. 219–220). Because FERC is legally bound to follow advice of other federal agencies, it appears that (barring fundamental changes in law) the dams will be removed within the next few years with the objective of restoring natural conditions to the entire length of the river.

FERC (1991, p. xxxiii) has recognized that the principal issue in the Elwha River is reducing uncertainties about the behavior of the large quantities of sediment stored in the two project reservoirs. The draining of the reservoirs and removal of the dams will require five years, with the likely period of adjustment for sediment transport through the system extending another two to five years. FERC estimated that most of the sediment presently stored in the reservoirs would be likely to remain in place, but fine materials would flood downstream areas. In reaches below the dams, presently armored reaches dominated by cobbles would be inundated by silt and sand, creating enlarged floodplains, meandering channels, and more wooded riparian habitats. Park Service and Fish and Wildlife Service personnel hope that within 10–20 years the 'original habitat conditions' would reappear on the lower river and that within the Lake Mills reservoir basin, 'near natural conditions' would be restored in about the same time period. In addition to eliminating about $16.5 million in electricity production per year used by the nearby Daishowa paper mill, the estimated cost of removal of the two structures is $64.3 million (US General Accounting Office 1991).

These prognostications are problematic. There are no formal empirical studies that document and quantify fluvial responses to the removal of large structures like the Elwha River dams, though US Geological Survey researchers are investigating the case. The excavation of previously stored sediment from the reservoirs after dam removal is likely to be a complex, multistage process that has not been widely documented or successfully modeled and predicted. The downstream fate of these sediments in the Elwha system is also more of a guess than a science, because although fine sediments moved through the system before the closure of the dams, they did so on a continuous basis in relatively low annual amounts. The disposition of huge masses of fine materials released into the river during a brief period is largely unknown, and immediate floodplain construction downstream is without historical precedent. Although presently available theory and modeling technology allow some reasonable estimates, it appears that the best experience is that from unintended dam breaches and natural dam failures (Jarrett and Costa 1986). These limited efforts aside, the issue of the geomorphic impact of dam removal remains a largely unstudied aspect of fluvial processes, and increases the uncertainty in managing the Elwha situation.

In addition to the scientific question, the policy considerations in the Elwha case are also at the edge of established practice. The contention that reservoir areas and downstream segments will be restored to natural conditions within two decades is unlikely if strict definitions of 'completely natural' are applied. The large amounts of sediment remaining within the Lake Mills and Lake Aldwell areas will produce channel and bank conditions never before seen in the area, and broad downstream floodplains are probably not natural. The objectives of management on the river are to return natural hydrologic conditions to benefit the fishery, but the fluvial geomorphology will probably be substantially modified (as defined in Table 18.3). Whether or not this partial restoration is acceptable from an ecosystem and landscape management standpoint is an unanswered question.

Edwards Dam

Privately owned low dams in the northeastern United States are also under consideration for removal. The best example is probably Edwards Dam, a run-of-the-river structure on the Kennebec River near Augusta, Maine (American Rivers, Inc. 1994, p. 4; Williams 1993). Constructed in 1837 for sawmill and canal operations, and later modified for hydroelectric production, the dam prevents the upstream migration of Atlantic salmon, and restricts the habitat of shad, smelt, and sturgeon (Egan 1990). Despite the modifications to the dam to improve fish passage, state officials and nongovernmental organizations concluded the changes were inadequate, and the state legislature passed a resolution in 1990 calling for the removal of the structure. In 1994, FERC began investigating of the feasibility and impact of removing the dam, and the issue is not yet settled.

From a scientific perspective, removal of Edwards Dam and other similar structures on New England rivers poses different questions than removal of dams on the Pacific coast.

Because the eastern structures are only a few metres high and are often run-of-the-river, they do not significantly alter the hydrology of the streams, and prediction of the hydrologic effects of their removal is fairly straightforward. The stored sediment and altered channel gradients pose the most important problems. The removal of the structures inevitably produces short-term downstream impacts through remobilized sediment. The masses of sediments themselves may be of only minor concern, but the chemical quality of the materials is a potential pollution problem. New England and the eastern United States was the hearth of American industrialization, and throughout the nineteenth century heavy manufacturing without environmental quality controls dominated the waterways. These industries released to the streams huge (but unmeasured) amounts of heavy metals which became attached to the sediments. Some of these contaminated sediments moved to the long-term sinks of the ocean or Great Lakes, where they now occur as easily defined, but isolated, components of the bottom sediments (Thomas 1972). Large amounts of contaminated sediment also came to rest behind the dams where they have remained interred. Removal of the structures raises the possibility of their remobilization and deposition on floodplains downstream.

An area where this issue is of enough concern to prevent dam removal is the Blackstone River system in central Massachusetts. Low dams have converted the river into a chain of lakes near the old manufacturing town of Worcester. Through the late 1800s, Worcester was a major metal machining center, and at one time was the greatest producer of metal wire in the world. Lead, zinc, copper, cadmium, and a variety of other contaminants from manufacturing are now trapped in sediments behind the dams, which also act as local sinks for runoff from abandoned factory areas. To remove the structures would be to risk hazardous remobilization of the contaminants which are more easily managed in place. In these cases, environmental restoration may be less desirable than containment of contaminated sediments.

Unlike their western counterparts, the streams of the central Atlantic and New England states offer numerous examples of dam removal that might be analyzed for the long-term impact of such efforts. On almost every stream with perennial flow, the eighteenth and early nineteenth century saw the construction of mills and associated dams for the grinding of grain and powering machinery. Between about 1870 and 1930, with the advent of a centralized industrial economy, many of these mills ceased production, and owners removed the dams. Such sites now provide analogs for analysis and prediction of the effects of anticipated additional dam removals.

Woolen Mills Dam

Some empirical evidence regarding removal is available from the case of at least one structure in the Midwest. The Woolen Mills Dam, constructed in 1919, produced a head of 4.2 m (14 ft) and a small reservoir affecting 2.4 km (1.5 miles) of the Milwaukee River in West Bend, Wisconsin (Nelson and Pajak 1990). By the late 1980s, however, the state Department of Natural Resources determined that the dam was unsafe and developed a 10-year plan for its removal. Using suitability models for establishing habitat for small-mouth bass, northern pike, and common carp, administrators restored the channel to emphasize sport fishing in an urban park setting (National Research Council 1992, p.

220). The resulting river is not even essentially natural according the scale in Table 18.2, but it is now closer to the natural side of the scale than it was previously.

Gillespie Dam

Gillespie Dam on the Gila River of central Arizona provides an unintentional example of the impacts of dam removal. The dam was built in 1921 as a 7.6 m (25 ft) high irrigation diversion structure at a constriction in the Gila River Valley west of Phoenix, Arizona (Lecce 1988). Within two years of its closure, sediment had accumulated to the 485 m (1600 ft) wide crest of the dam, and by the 1980s the wedge of stored sediment extended 11.3 km (7 miles) upstream. During flood events, water passed over the dam, but during most of year, the dam diverted all of the flow into canals. Despite surviving several floods of over 2830 m^3 s^{-1} (100 000 cfs), on 9 and 10 January 1993, the dam breached during a flood of up to 4245 m^3 s^{-1} (150 000 cfs), developing a gap 54.5 m (180 ft) wide and extending from the crest to the base of the structure (Figure 18.10). Within two weeks, channel erosion removed more than half the sediment stored behind the dam, evacuating it through the breach and distributing it along 24 km (15 miles) of the compound channel (a well-defined low-flow invert within a broader, braided high-flow channel) downstream. The deposition of fine sands elevated the bed as much as 2 m (6 ft) in some locations.

The breach caused a virtually instantaneous drop in base level of over 6.5 m (20 ft), but upstream migration of the resulting knickpoint did not produce a confined trench through the accumulated sediments. Lateral migration of the channel within the reservoir area excavated wide swaths of material and left crenelated margins in the remaining sediments. If the breach is not repaired, and restoration of the stream is possible, the long-term stability of the remaining sediments is an unresolved problem (Figure 18.11).

Public policy regarding Gillespie Dam is unsettled. Management options include repairing the dam, removing it completely, or leaving it in its present breached condition (Flood Control District of Maricopa County, Internal Project Review, 17 November 1994). The owners of the dam, a Swiss corporation with investment interests in the 10 100 ha (25 000 acres) irrigated by the water diversions from the structure, are seeking the most cost-effective solution to the problem. Because the owners installed a new pumped withdrawal system to suppy irrigation water, they are not likely to repair the dam unless they are forced to do so by regulators.

From the public perspective, there are five major issues, illustrating the administrative complexity associated with fluvial restoration connected with dams: environmental quality, dam safety, flood control, fluvial restoration, and liability. First, any alterations or work in the channel will require a permit from the Corps of Engineers under Section 404 of the Clean Water Act. The Corps is actively pursuing restoration of some segments of the river, and is likely to evaluate any option with that goal in mind. Second, the state Department of Water Resources is responsible for the safety of the structure, and engineering evaluations are required to assess the soundness of the remaining structure, especially its survivability during future floods. Third, the county flood control district is responsible for flood protection downstream from the site, as well as administering the National Floodplain Insurance Program of the Federal Emergency Management Agency. The flood control district is now employing geomorphic studies to assess the effect of channel changes downstream from the dam on flood potential. Fourth, the state

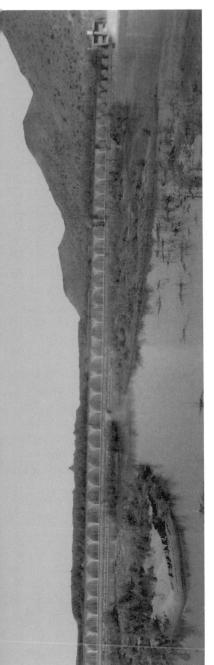

Figure 18.10 Photographs of Gillespie Dam, Gila River, Arizona, before and after its breach in a 1993 flood event. Above, showing the dam intact and operating as a run-of-the-river structure for diversion of irrigation waters, W.L. Graf photograph 51-26, 20 February 1984. Below, showing the breach in the structure, W.L. Graf photograph 110-17, 8 December 1994

Figure 18.11   Photograph of sediment above and below Gillespie Dam, Gila River, Arizona, with human figures on the dam crest for scale. The sediments behind (to the right) the dam had accumulated to a level slightly above the crest before the breach. W.L. Graf photograph 127-22, 8 December 1994

Department of Game and Fish administers an important wildlife area upstream from the dam. Restoration of the original natural river in the reach once part of the reservoir area would augment this wildlife area, but with a mixed blessing. In its natural condition, the river channel was locationally unstable, making its management for any purpose difficult. If restoration is possible, it is likely that a compromise will emerge: a less natural but more stable channel will be the ultimate objective. Finally, landowners downstream from the dam have initiated legal action, charging that mismanagement of the dam and channel led to the breach and has threatened their property with an increased flood hazard.

CONCLUSIONS

A review of science and public policy for restoration of dammed rivers along with several case examples shows that there are several consistent themes that connect science and policy:

1. As industrial societies move into postindustrial stages, social values for rivers are changing from a perspective whereby they are viewed simply as water resources alone to a perspective that is more complex. In addition to serving as water resources, rivers are coming to be viewed as multipurpose ecosystems or landscapes that serve many

objectives. Some of these objectives are compatible with each other, but it is not possible to maximize all the objectives.

2. Social values place increasing emphasis on a particular objective, the preservation or restoration of 'natural' conditions that are conductive to achieving goals related to (in order of established public preference) fish and wildlife management, recreation, and aesthetics.

3. Throughout much of the world, rivers are fragmented or segmented, divided into completely or partially controlled reaches by dams, with each reach exhibiting a geomorphological behavior that is markedly different from other nearby reaches. Theoretical models that view the river system as highly integrated must be modified to account for this segmentation.

4. It is rarely possible to restore truly natural conditions to regulated rivers, and even if dams are removed, the probability of reestablishing original pre-dam conditions is unknown, but it is likely to be variable from one case to another.

5. Dams impose hydrologic and sedimentologic changes on downstream reaches. Dams usually have one or more of the following hydrologic impacts: reduced annual water yield, reduced flood peaks, altered low flows (eliminated, or in other cases, sustained at higher than natural levels), altered seasonality of flow variation. Sedimentologic impacts include reductions in downstream sediment discharges, changes in bed material size, and concomitant upstream storage which is remobilized if the dam is removed or breached, raising problems of sedimentation and contaminant mobility.

6. Because of these hydrologic and sedimentologic changes, geomorphic conditions in channels downstream from the structures evolve into distinctly unnatural configurations. These configurations depend not only on the presence of the structures (for which we have some theory), but also on operating rules for the dams (for which we have little theory).

7. Riparian vegetation and zoological communities are intimately related to the hydrologic, sedimentologic, and geomorphic conditions, but our knowledge about these connections is only now evolving into formal understanding.

The primary consideration in establishing policy for river restoration is to address the problem of 'what is natural?' It is unlikely that long reaches of American rivers will ever return to their original, truly natural states, and they are likely to remain in their fragmented condition (Graf 1993; Dynesius and Nilsson 1994), with preserved or restored reaches interspersed with developed and dammed reaches. The rivers are therefore likely to operate as segmented systems, and though effects in one reach are likely to be transmitted downstream, different segments perform different geomorphologic functions and require different policy strategies. But it is possible to make them more natural than they are at present by selective removal of dams and alteration of operating rules for the remaining structures. Policy-making for rivers is often a perceptual issue (Gregory and Davis 1993), but the decision about how far to go in making rivers more natural is partly political (what does society want?), partly scientific (what is possible?), and partly economic (what can society afford?). Based on previously established basic understanding of the systems (e.g. Petts 1984; Williams and Wolman 1984), geomorphologists can ask meaningful research questions with understandable, convincing answers. Management can then proceed confidently on the basis of established and accepted research, with fluvial changes anticipated and taken into account. The science of geomorphology can thus help resolve the dominant philosophical conflict in management of the nation's rivers in the twenty-first century: how to maintain viable economic development of the resource for the present while simultaneously preserving a quality environment for the future.

By addressing questions of interest to policymakers, the public, and other sciences, geomorphology returns to its intellectual roots, but to do so, its practitioners must focus their efforts on subjects that others truly care about rather than on topics only of interest to a limited number of geomorphologists. Geomorphologists must more often participate in the evolving complex interactions among Earth and life scientists to address issues of ecosystem quality and restoration (Naiman et al. 1995). The recent experiences of geomorphologists as outlined in this chapter show that such interactions can be fruitful on an administrative as well as scientific basis. Whether they like it or not, the scientist and policymaker are destined to be partners.

ACKNOWLEDGEMENTS

Ginger Hinchman greatly assisted the early construction of this chapter by obtaining information and documents related to the case examples. I greatly appreciate the efforts of Ken Gregory and Doug Sherman who materially improved the chapter with their thoughtful review suggestions.

REFERENCES

American Rivers, Inc. 1994. Hydropower reform, *American Rivers*, **22**(3), 5.

Auble, G.T., Friedman, J. and Scott, M.L. 1991. *Riparian Vegetation of the Black Canyon of the Gunnison River, Colorado: Composition and Response to Selected Hydrologic Regimes Based on a Direct Gradient Assessment Model*, report by the US Fish and Wildlife Service for the National Park Service, Water Resources Division, National Park Service, Ft Collins, Colo.

Benn, P.C. and Erskine, W.D. 1994. Complex channel response to flow regulation: Cudgegong River below Windamere Dam, Australia, *Applied Geography*, **14**, 153–168.

Black, P.E. 1987. *Conservation of Water and Related Land Resources* (2nd edn), Roman & Littlefield, Totowa, NJ, 336 pp.

Boyd, D.W., Nease, R.F., Rice, J.S. and Taleb-Ibrihim, M. 1990. *Evaluating Hydro Relicensing Alternatives: Impacts on Power and Nonpower Values of Water Resources*, Contract Report GS-6922, Project 2694-7, Electric Power Research Institute, Palo Alto, Calif.

Brookes, A. 1988. *Channelized Rivers: Perspectives for Environmental Management*, Wiley, Chichester, 326 pp.

Carling, P.A. 1987. Bed stability in gravel streams, with reference to stream regulation and ecology, in *River Channels: Environment and Process*, edited by K.S. Richards, Blackwell, London, pp. 321–347.

Carothers, S.W. and Brown, B.T. 1991. *The Colorado River through the Grand Canyon: Natural History and Human Change*, University of Arizona Press, Tucson, 235 pp.

Cassady, J.C., Cross, J.B. and Calhoun, F. 1994. *Western Whitewater from the Rockies to the Pacific*, North Fork Press, Berkeley, Calif, 314 pp.

Chase, K.J. 1992. Gunnison River thresholds for gravel and cobble motion, Black Canyon of the Gunnison National Monument, MS thesis, Colorado State University, Ft Collins, Colo., 192 pp.

Coyle, K.J. 1988. *The American Rivers Guide to Wild and Scenic River Designation*, American Rivers, Inc., Washington, DC, 59 pp. Plus 10 appendices.

Cushman, J.H., Jr. 1994. Federal agencies have plans to restore Florida Everglades to a more natural state, *New York Times*, 28 September, p. A10.

Decamps, H., Fortune, M. and Gazelle, F. 1994. Historical changes of the Garonne River, Southern France, in *Historical Change of Large Alluvial Rivers: Western Europe*, edited by G.E. Petts, H. Moller and A.L. Roux, Wiley, Chichester, pp. 249–267.

Dynesius, M. and Nilsson, C. 1994. Fragmentation and flow regulation of river systems in the northern third of the world, *Science*, **266**, 753–762.

Echeverria, J.D., Barrow, P. and Ross-Collins, R. 1989. *Rivers at Risk: The Concerned Citizen's Guide to Hydropower*, Island Press, Washington, DC, 216 pp.

Egan, T. 1990. Dams may be razed so the salmon can pass, *New York Times*, 15 July, pp. 1, 14.

Elliott, J.G. and Parker, R. 1992. Potential climate-change effects on bed-material entrainment, Gunnison Gorge, Colorado, in *Proceedings, American Water Resources Association, 28th Annual Conference and Symposia*, Reno, Nevada, November, 1992, pp. 751–759.

FERC (Federal Energy Regulatory Commission) 1991. Draft environmental impact statement for Elwha and Glines Canyon Dams, Federal Energy Regulatory Commission, Washington, DC.

Gore, J.A. and Shields, F.D. Jr. 1995. Can large rivers be restored? *Bioscience*, **45**, 142–152.

Graf, W.L. 1993. Landscapes, commodities, and ecosystems: the relationship between policy and science for American Rivers, in *Sustaining Our Water Resources*, 10th Anniversary Symposium, Water Science and Technology Board, National Research Council, National Academy Press, Washington, DC, pp. 11–42.

Graf, W.L., Beyer, P.J., Rice, J. and Wasklewicz, T.A. 1994a. *Geomorphic Assessment of the Lower Gila River, West Central Arizona*, Contract Report DACW09-94-M-0494, Part 1, US Army Corps of Engineers, Arizona Area Office, Phoenix.

Graf, W.L., Beyer, P.J. and Wasklewicz, T.A. 1994b. *Geomorphic Assessment of the Lower Salt River, Central Arizona*, Contract Report DACW09-94-M-0494, Part 2, US Army Corps of Engineers, Arizona Area Office, Phoenix.

Gregory, K.J. and Davis, R.J. 1992. Coarse woody debris in stream channels in relation to river channel management in woodland areas, *Regulated Rivers*, **7**, 117–137.

Gregory, K.J. and Davis, R.J. 1993. The perception of riverscape aesthetics: an example from two Hampshire rivers, *Journal of Environmental Management*, **39**, 171–182.

Gregory, K.J., Gurnell, A.M., Hill, C.T. and Tooth, S. 1994. Stability of the pool-riffle sequence in changing river channels, *Regulated Rivers*, **9**, 35–43.

Gregory, S.U., Swanson, F.J., McKee, W.A. and Cummings, K.W. 1991. An ecosystem perspective of riparian zones, *Bioscience*, **41**, 540–551.

Hendee, J.C., Stankey, G.H. and Lucas, R.C. 1978. *Wilderness Management*, Miscellaneous Publication 1365, US Department of Agriculture, Washington, DC.

Holmes, B.H. 1972. *A History of Federal Water Resources Programs*, Miscellaneous Publication 1233, US Department of Agriculture, Washington, DC.

Hugget, R. 1985. *Earth Surface Systems*, Springer-Verlag, Berlin, 270 pp.

Hunt, E. 1988. *Down by the River: The Impact of Federal Water Projects and Policies on Biological Diversity*, Island Press, Washington, DC.

Hunt, R.T. 1988. *Guide for Developing a Hydro Plant Relicensing Strategy*, Vol. 1, *Major Issues and R&D Needs*, Research Report AP- 6038, Project 2694-5, Electric Power Research Institute, Palo Alto, Calif.

Jarrett, R.D. and Costa, J.E. 1986. Hydrology, geomorphology, and dam- break modeling of the July 15, 1982 Lawn Lake Dam and Cascade Lake Dam failures, Larimer County, Colorado. *US Geological Survey Professional Paper 1369*.

Keller, E.A. 1975. Channelization: environmental, geomorphic, and engineering aspects, in *Geomorphology and Engineering*, edited by D.E. Coates, Dowden, Hutchinson, and Ross, Stroudsburg, Penn., pp. 115–140.

Kodras, J.E. and Jones, J.P. 1990. Academic research and social policy, in *Geographical Dimensions of US Social Policy*, edited by J.E. Kodras and J.P. Jones, Edward Arnold, London, pp. 237–248.

Kondolf, G.G. and Wolman, M.G. 1993. The sizes of salmonid spawning gravels, *Water Resources Research*, **9**, 2275–2285.

Lagasse, P.F. 1994. Variable response of the Rio Grande to dam construction, in *The Variability of Large Alluvial Rivers*, edited by S.A. Schumm and B.R. Winkley, American Society of Civil Engineers, New York, pp. 395–422.

Lecce, S.A. 1988. Gillespie Dam, in *The Salt and Gila Rivers in Central Arizona: A Geographic Field Trip Guide*, edited by W.L. Graf, Department of Geography Publication 3, Arizona State University, Department of Geography, Tempe, pp. 169–180.

Lewin, J., Macklin, M.G. and Woodward, J.C. (eds) 1995. *Mediterranean Quaternary River Environments*, A.A. Balkema, Rotterdam, 292 pp.

L'vovich, M.I. and White, G.F. 1990. Use and transformation of terrestrial water systems, in *The Earth as Transformed by Human Action: Global and Regional Changes in the Biosphere over the Past 300 Years*, edited by B.L. Turner, W.C. Clark, R.W. Kates, J.F. Richards, J.T. Matthews and W.B. Meyer, University of Cambridge Press, Cambridge, pp. 235–252.

Majors, J.E. 1989. Opportunities to protect instream flows and wetland uses of water in California, *Biological Report 89*, US Fish and Wildlife Service, Washington, DC.

Minckley, W.L. 1991. Native fishes of the Grand Canyon region: an obituary?, in *Colorado River and Ecology and Dam Management*, National Research Council, National Academy Press, Washington, DC, pp. 124–177.

Naiman, R.J., Magnuson, J.J., McKnight, D.M., Stanford, J.A. and Karr, J.R. 1995. Freshwater ecosystems and their management: a national initiative, *Science*, 270, 584–585.

National Research Council 1987. *River and Dam Management: A Review of the Bureau of Reclamation's Glen Canyon Environmental Studies*, National Academy Press, Washington, DC.

National Research Council 1992. *Restoration of Aquatic Ecosystems*, National Academy Press, Washington, DC.

National Research Council 1995. *Rediscovering Geography: New Relevance for the New Century*, National Academy Press, Washington, DC.

Nelson, J.E. and Pajak, P. 1990. Fish habitat restoration following dam removal on a warmwater river, in *The Restoration of Midwestern Stream Habitat*, Proceedings of a Symposium held at the 52nd Midwest Fish and Wildlife Conference, Rivers and Streams Technical Committee, 4–5 December 1990, Minneapolis, Minn., pp. 53–63.

Newbury, R. 1995. Rivers and the art of stream restoration, in *Natural and Anthropogenic Influences in Fluvial Geomorphology*, edited by J.E. Costa, P.J. Miller, K.W. Potter and P.R. Wilcock, American Geophysical Union, Washington, DC, pp. 137–150.

Office of the Federal Register 1983. *United States Organization Manual: 1983–1984*, US Government Printing Office, Washington, DC.

Palmer, T. 1993. *The Wild and Scenic Rivers of North America*, Island Press, Washington, DC, 338 pp.

Petts, G.E. 1979. Complex response of river channel morphology subsequent to reservoir construction, *Progress in Physical Geography*, 3, 329–362.

Petts, G.E. 1980. Morphological changes of river channels consequent upon headwater impoundment, *Journal of the Institute of Water Engineers and Scientists*, 34, 374–382.

Petts, G.E. 1984. *Impounded Rivers: Perspectives for Ecological Management*, Wiley, New York, 326 pp.

Pickup, G. 1980. Hydrologic and sediment modeling studies in the environmental impact assessment of a major tropical dam project, *Earth Surface Processes*, 5, 61–75.

Pickup, G. and Warner, R.F. 1984. Geomorphology of tropical rivers, II, channel adjustment to sediment load and discharge in the Fly and Lower Purari Rivers, Papua New Guinea, *Catena*, supplement 5, 13–46.

Pitlick, J. 1992. Stabilizing effects of riparian vegetation during an overbank flow, Trinity River, California, EOS, Supplement 73, 231.

Richards, K.S., Brunsden, D., Jones, D.K.C. and McCaig, M. 1987. Applied fluvial geomorphology: river engineering project appraisal in its geomorphologic context, in *River Channels: Environment and Process*, edited by K.S. Richards, Blackwell, London, pp. 348–382.

Schmidt, J.C. and Graf, J.B. 1990. Aggradation and degradation of alluvial sand deposits, 1965–1986, Colorado River, Grand Canyon National Park, Arizona, *US Geological Survey Professional Paper 1493*, 32 pp.

Simons, R.K. and Simons, D.B. 1994. An analysis of Platte River channel changes, in *The Variability of Large Alluvial Rivers*, edited by S.A. Schumm and B.R. Winkley, American Society of Civil Engineers, New York, pp. 341–361.

Thomas, R.L. 1972. The distribution of mercury in the sediment of Lake Ontario, *Canadian Journal of Earth Science*, **9**, 636–651.

Thoreau, H.D. 1849. *A Week on the Concord and Merrimack Rivers*, reprinted 1961, Signet Classics, New York, 339 pp.

Tinkler, K.J. 1985. *A Short History of Geomorphology*, Barnes and Noble, Totowa, NJ, 317 pp.

US Army Corps of Engineers 1993. *Central and Southern Florida Project, Environmental Restoration of the Kissimmee River, Florida*, Final Integrated Feasibility Report and Environmental Impact Statement, US Army Corps of Engineers, Washington, DC.

US Bureau of Reclamation 1981. *Central Valley Project: Historical Background and Economic Impacts*, US Department of Interior, Washington, DC.

US Bureau of Reclamation 1991–93. *Annual Reports*, US Bureau of Reclamation, US Department of Interior, Washington, DC.

US Bureau of Reclamation 1994. *Final Environmental Impact Statement, Operation of Glen Canyon Dam*, 2 vols, US Burea of Reclamation, Salt Lake City.

US General Accounting Office 1991. *Hydroelectric Dams: Costs and Alternatives for Restoring Fisheries in the Elwha River*, report to the Chairman, Subcommittee on Oversight and Investigations, Committee on Energy and Commerce, US Government Printing Office, Washington, DC.

US House of Representatives 1992. Joint hearings on H.R. 4844, Elwha River Ecosystem and Fisheries Restoration Act, 102nd Congress, 2nd Session, Government Printing Office, Washington, DC.

US House of Representatives 1993. H.R. 1321, Black Canyon National Conservation Act, 102nd Congress, 2nd Session, Government Printing Office, Washington, DC.

US Office of the Assistant Secretary of the Army for Civil Works 1992. Kissimmee River restoration study, Letter from Chief of Engineers, Submitting a Report Pursuant to Section 116(h) of the Water Resource Development Act of 1990, US Army Corps of Engineers, Washington, DC.

US Senate 1984. Fish and wildlife restoration in the Trinity River, Senate Report 98-647, to Accompany H.R. 1438, Calendar No. 1295, 98th Congress, 2nd Session, Government Printing Office, Washington, DC.

US Senate 1992. Report to Accompany S. 2527, Elwha River Ecosystem and Fisheries Restoration Act, 102nd Congress, 2nd Session, Government Printing Office, Washington, DC.

Van Urk, G. and Smit, H. 1989. The Lower Rhine; geomorphological changes, in *Historical Change of Large Alluvial Rivers: Western Europe*, edited by G.E. Petts, H. Moller and A.L. Roux, Wiley, Chichester, pp. 167–184.

Warner, R.F. 1988. Human impacts on river channels in New South Wales and Victoria, in *The Fluvial Geomorphology of Australia*, edited by R.F. Warner, Academic Press, Sydney, pp. 343–363.

Webb, R.H., Rink, G.R. and Pringle, P.T. 1987. Debris flows from tributaries of the Colorado River, Grand Canyon National Park, Arizona, *US Geological Survey Open-File Report 87-118*.

Williams, G.P. and Wolman, M.G. 1984. Downstream effects of dams on alluvial rivers, *US Geological Survey, Professional Paper 1286*.

Williams, T. 1993. Freeing the Kennebec River, *Audubon*, **95**, 36–42.

Zhou, Z. and Pan, X. 1994. Lower Yellow River, in *The Variability of Large Alluvial Rivers*, edited by S.A. Schumm and B.R. Winkley, American Society of Civil Engineers, New York, pp. 341–361.

INDEX